Impacts of Digital Payment Systems on Marketing

Fatih Sahin
https://orcid.org/0000-0002-4760-4413
Bandirma Onyedi Eylul University, Turkey

IGI Global
Scientific Publishing
Publishing Tomorrow's Research Today

Vice President of Editorial Melissa Wagner
Director of Acquisitions Mikaela Felty
Director of Book Development Jocelynn Hessler
Production Manager Mike Brehm
Cover Design Jose Rosado

Published in the United States of America by
IGI Global Scientific Publishing
701 East Chocolate Avenue
Hershey, PA, 17033, USA
Tel: 717-533-8845 | Fax: 717-533-7115
Website: https://www.igi-global.com E-mail: cust@igi-global.com

Library of Congress Cataloging-in-Publication Data

LCCN: 2025029130 (CIP Data Pending)
ISBN13: 9798337345260
Isbn13Softcover: 9798337345277
EISBN13: 9798337345284
British Cataloguing in Publication Data
A Cataloguing in Publication record for this book is available from the British Library.

Table of Contents

Detailed Table of Contents

Despite the progressive modernization of Morocco's financial system, public incentive policies, and technological innovations, cash remains the dominant means of payment. This paradox, characterized by the coexistence of digital infrastructures and a strong preference for cash, raises questions about the determinants of electronic payment use. Using data from a sample of 638 banked household heads, this chapter examines the influence of sociodemographic characteristics and cash payment habits on the use of electronic payment methods. The findings reveal that social, economic, and behavioral factors largely explain the persistence of cash in daily transactions. Beyond infrastructure availability, established routines and attachment to cash remain major barriers to payment digitalization. The chapter contributes to the literature by emphasizing the role of sociocultural factors in financial digitalization and offers policy implications for reducing cash dependency and fostering economic inclusion and efficiency.

This chapter analyzes the importance of digital literacy in enhancing the effectiveness of mobile payment marketing strategies in rural India, specifically for Haryana. Large segments of rural communities remain excluded from the digital economy despite expansion of digital financial services and government-backed initiatives such as Digital India. The most significant reasons for this exclusion are a deficiency in digital literacy, infrastructural shortcomings, and a default distrust of digital

payment solutions are incorporated into omnichannel retail spaces and the strategic consequences for customer engagement, trust, and brand differentiation.

Digital marketing is maturing into a precision-targeted, data-driven engine. However, this maturation has brought about a slew of ethical and operational challenges: ad fraud, opaque data practices, unverifiable performance metrics, and concentrated control of major platforms. In this chapter, we consider the potential of blockchain technology to provide enhanced transparency, trust, and accountability in the ecosystem of digital marketing. Through a multi-method design of research combining literature review, case studies, and interviews with key stakeholders, we put forth an ethical framework for digital marketing grounded in blockchain. The chapter discusses how decentralized, transparent, and immutable blockchains, coordinated with smart contracts and decentralized identity, could be one mechanism in the mitigation of some challenges that have been explored already, and simultaneously ensuring user empowerment and support for ethical advertising practice.

Point-of-sale (POS) systems are becoming increasingly common as both an important aspect of the transactional infrastructure and a marketing tool that exemplifies the company's brand and philosophy. As such a visible aspect of a company, their functionality and security influence consumer trust and brand perception. A Transaction Trust Disruption Framework is therefore proposed to connect POS risk events, such as data breaches and availability to consumer perception, which directly relates to customer loyalty and company reputation. This framework will help marketing executives understand how technical safeguards can support marketing in building and preserving customer trust, particularly when responsibility is not clearly defined or third-party providers contribute to failure. This interdisciplinary approach integrates technical responsibilities and transparent crisis communication as elements of brand protection and marketing, rather than treating POS risk as an isolated technological issue.

This chapter explores how digital payment systems are transforming the foundations of marketing practice by reframing the traditional 4Ps framework. Drawing on insights from behavioral economics, consumer psychology, and fintech adoption, it examines how payments influence product innovation, pricing strategies, distribution models, and promotional engagement. Global and emerging market case insights illustrate how tokenization, microtransactions, borderless commerce, and payment-linked loyalty programs reshape consumer perceptions of ownership, value, trust, and inclusivity. The chapter highlights managerial implications by positioning payments as strategic levers for customer experience, data-driven personalization, and ethical stewardship. It also proposes a future research agenda that integrates digital payments into marketing theory, offering guidance for scholars, practitioners, and policymakers in navigating the digital economy.

The chapter discusses the integration of gamification, social commerce, and live shopping in social media marketing, revealing how brands and content creators are transforming consumer engagement and purchase behaviour in real-time. Gamification uses behavioural psychology to drive user participation and loyalty, social commerce facilitates seamless transactions within social environments, and live shopping transforms retail into immersive entertainment. Drawing on case studies, industry trends, and theoretical frameworks, the chapter highlights the psychological impacts, ethical considerations, and regulatory implications of this evolution. It proposes a metacommerce ecosystem where gamified interactions and live digital experiences define the future of online consumerism.

Preface

This book opens with an exploration of one of the most profound yet often overlooked transformations shaping contemporary markets: the evolving role of digital payment systems in marketing and consumer behavior. The motivation for this volume stems from a simple but increasingly important question, why does the act of paying matter so deeply in today's digital economy? For decades, payment was viewed as a functional conclusion to exchange, a technical formality that followed persuasion, experience design, and relationship building. Once a transaction occurred, scholarly and managerial attention typically shifted elsewhere. In today's data-intensive, platform-mediated environment, however, such assumptions are no longer adequate.

Marketing now operates within a complex, technology-mediated ecosystem where consumption unfolds across platforms, interfaces, and data-driven environments. Value creation has become relational rather than linear, interactions continuous rather than episodic, and consumers active participants rather than passive recipients. At the center of this ecosystem lies the payment moment. Mobile wallets, contactless cards, platform-based gateways, AI-enabled security systems, cryptocurrencies, and blockchain-backed tokens have transformed payment from a backstage utility into a visible, meaningful, and often decisive point of consumer–brand interaction. This book starts from the premise that payment is no longer neutral. It is experiential, strategic, and deeply embedded in how trust, value, and legitimacy are constructed in modern markets.

The relevance of this topic extends well beyond technological innovation. Digital payments generate high-quality behavioral data that reveal how consumers actually behave rather than how they claim to behave. These data increasingly guide personalization strategies, dynamic pricing models, loyalty systems, and platform governance. At the same time, they raise fundamental ethical questions concerning surveillance, data ownership, algorithmic power, and consumer autonomy. This book therefore brings together perspectives from marketing, consumer behavior, behavioral economics, information systems, and ethics to offer a comprehensive understanding of how digital payment systems reshape markets across different cultural, institutional, and developmental contexts.

The global transition toward a digital economy represents one of the defining developments of the twenty-first century. Driven by widespread smartphone adoption, improved internet connectivity, and policy initiatives promoting financial inclusion and cash-light societies, digital payment systems have expanded rapidly across both developed and emerging economies. In regions with historically limited access to formal banking—such as parts of Africa, South Asia, and the Middle East—mobile money platforms and interoperable payment infrastructures have enabled millions of individuals and small businesses to participate in the formal economy for the first time.

Yet this transformation is neither uniform nor linear. Many economies continue to exhibit a persistent "cash paradox," in which physical currency remains dominant despite substantial investment in digital infrastructure. Deeply embedded payment habits limited digital and financial literacy, informal economic structures, and cultural attachments to cash contribute to resistance against adoption. Empirical evidence from contexts such as Morocco and rural India illustrates that technological availability alone is insufficient to drive behavioral change; trust perceptions, social norms, and socio-demographic factors play decisive roles.

Simultaneously, the global diffusion of digital payments has intensified concerns regarding legitimacy, trust, and power in digital markets. The proliferation of algorithmic profiling, opaque data practices, and behavioral surveillance has fueled growing consumer skepticism toward platforms and intermediaries. As transactional data becomes central to value creation, questions of ownership, control, and ethical use move to the forefront of marketing and policy debates. In response, blockchain-based infrastructures and decentralized digital identities have gained attention as potential mechanisms for restoring transparency, accountability, and data sovereignty. This unresolved tension between efficiency-driven innovation and ethical responsibility constitutes a central thematic thread of this book.

The purpose of this edited book is threefold. First, it advances academic understanding by integrating theories from marketing, consumer behavior, behavioral economics, information systems, and ethics into a unified analytical framework centered on digital payment systems. By bridging disciplinary silos, the book moves beyond fragmented analyses and offers a comprehensive explanation of how payment technologies influence consumer decision-making, value creation, and market dynamics.

Second, the book generates actionable insights for practitioners responsible for designing payment-enabled customer experiences, managing trust at the transactional interface, and leveraging payment-related data responsibly as a source of sustainable competitive advantage. Payments are treated not as operational afterthoughts but as strategic resources that directly shape customer satisfaction, loyalty, and brand equity.

Third, the book contributes policy-relevant perspectives for regulators and public institutions concerned with financial inclusion, digital literacy, data protection, and consumer rights. By addressing governance challenges in cash-light and platform-based economies, the book informs decision-making at the intersection of markets and public policy.

Collectively, the chapters demonstrate that digital payment systems are not merely technological artefacts but socio-technical institutions that actively shape market behavior, trust relations, and power asymmetries. The book's central contribution lies in showing how marketing must evolve from a communication-centric discipline into one that actively governs transactional experiences and data relationships.

This book is written for a broad and interdisciplinary audience. It addresses academics and researchers seeking theoretically grounded and empirically informed insights into digital markets, marketing professionals and managers responsible for customer experience, loyalty, and omnichannel strategy, policymakers and regulators concerned with financial inclusion, digital literacy, and data protection, and FinTech developers and UX designers involved in building secure, intuitive, and culturally sensitive payment infrastructures. Throughout the book, academic rigor is balanced with real-world relevance to ensure accessibility without sacrificing analytical depth.

CHAPTER OVERVIEW

The chapters in this book collectively form a coherent narrative that moves from adoption barriers and behavioral foundations to strategic, ethical, and future-oriented perspectives. Each contribution offers a distinct yet complementary insight into the evolving payment–marketing nexus, and each chapter invites the reader to reconsider the strategic significance of transactions in contemporary markets.

The first chapter explores the determinants of electronic payment usage in Morocco, offering a detailed empirical analysis of adoption resistance in a cash-dominant economy. Rather than attributing low usage to technological shortcomings, the chapter reveals the decisive role of sociodemographic factors and deeply embedded payment habits. Readers are encouraged to engage with this chapter because it challenges infrastructure-centric assumptions and demonstrates why trust, education, and behavioral inertia must be addressed for meaningful financial inclusion.

The second chapter takes the reader to rural India, where digital literacy emerges as the cornerstone of mobile payment adoption. This chapter is particularly valuable for those interested in inclusive marketing, as it shows why conventional persuasion-based strategies fail when cognitive readiness and trust are absent. By reframing marketing as instruction rather than persuasion, the chapter offers practical insights into how underserved populations can be empowered to participate in the digital economy.

Building on this foundation, the third chapter examines the role of AI-enabled security systems in strengthening both payment safety and marketing literacy. It illustrates how artificial intelligence can go beyond fraud detection to educate users, personalize security communication, and build confidence at the moment of transaction. Readers will find this chapter compelling because it presents AI as a trust-building mechanism rather than a purely technical solution.

The fourth chapter provides a broader conceptual perspective by tracing the evolution of payment systems from barter to blockchain. It invites readers to rethink the point of purchase as a critical brand engagement moment rather than a functional endpoint. By positioning transactional data as a new currency of insight, this chapter helps scholars and practitioners understand why payment interfaces now shape long-term loyalty and value perception.

Ethical concerns take center stage in the fifth chapter, which evaluates blockchain as a potential response to the legitimacy crisis in digital marketing. Rather than treating ethics as an afterthought, the chapter argues for infrastructure-driven ethics, where transparency and accountability are embedded into system design. This chapter is essential reading for those concerned with restoring consumer trust in data-driven markets.

The sixth chapter shifts attention to point-of-sale systems and introduces the Transaction Trust Disruption framework. It demonstrates how technical failures at the payment interface are perceived as direct brand failures, regardless of who is technically responsible. Readers are invited to reconsider POS systems as brand touchpoints that require strategic governance rather than operational oversight alone.

In the seventh chapter, the behavioral consequences of digital transactions are examined through the lens of the "pain of paying." By showing how frictionless payments increase impulse buying and transaction frequency, this chapter explains why payment design has become a frontline determinant of consumer behavior and loyalty. It is particularly relevant for managers seeking to balance convenience with responsible consumption.

The eighth chapter explores how digital ownership, scarcity, and blockchain-based assets reshape value perception. Through an integrative theoretical framework, it explains why symbolic ownership and limited digital assets foster emotional attachment and social recognition. Readers interested in NFTs, virtual currencies, and participatory brand communities will find this chapter especially insightful.

The ninth chapter turns to technologically mature markets, examining AI-driven user experience among Nordic Gen Z consumers. By highlighting cultural moderation effects, the chapter shows that acceptance of AI is not universal but context-dependent. It encourages global brands to design payment and AI systems that respect cultural expectations of autonomy and transparency.

Payment Experience Quality is the focus of the tenth chapter, which positions EXQ as a central mediator between payment features and customer loyalty. This chapter invites readers to manage payment experience as a strategic performance indicator rather than a technical metric, linking personalization directly to retention and advocacy.

Revisiting foundational marketing theory, the eleventh chapter repositions digital payments within the traditional marketing mix. By showing how payments reshape product, price, and distribution, it offers a conceptual bridge between classical frameworks and contemporary digital realities. This chapter is particularly valuable for scholars seeking to update marketing theory for the platform economy.

The twelfth chapter examines the convergence of influence marketing, gamification, and social commerce. It introduces an integrative model explaining how interactive mechanisms accelerate the journey from awareness to advocacy while also raising ethical concerns related to manipulation and over-engagement. Readers are encouraged to reflect on how engagement can be designed responsibly.

The final chapter looks ahead to the role of digital currencies in global marketing strategies. By examining token-based communities, borderless transactions, and blockchain-verified authenticity, it offers a forward-looking perspective on how Web3 models may redefine consumer participation and brand value creation.

Taken together, the chapters in this book demonstrate that digital payments are no longer passive financial infrastructure. They are constitutive elements of contemporary marketing strategy that shape how value is created, experienced, and legitimized. This book calls for a paradigm shift—from technology-centered adoption to user-capability-centered, experience-driven, and ethically grounded marketing. It is my hope that this book will inspire further research, inform managerial practice, and contribute to more inclusive, transparent, and trustworthy digital markets.

Fatih Şahin
Bandırma Onyedi Eylül University, Turkey

Chapter 1
Use of Electronic Payment Methods in Morocco:
Influence of Sociodemographic Characteristics and Cash Payment Habits

Abderrahim Rhafel
https://orcid.org/0009-0000-2555-1018
Faculty of Law, Economics, and Social Sciences, Cadi Ayyad University, Marrakech, Morocco

Mustapha Ziky
https://orcid.org/0000-0002-6662-6669
Faculty of Law, Economics, and Social Sciences, Cadi Ayyad University, Marrakech, Morocco

Anas Limam
https://orcid.org/0000-0002-8626-8491
EST Safi, Cadi Ayyad University, Marrakech, Morocco

ABSTRACT

Despite the progressive modernization of Morocco's financial system, public incentive policies, and technological innovations, cash remains the dominant means of payment. This paradox, characterized by the coexistence of digital infrastructures and a strong preference for cash, raises questions about the determinants of electronic payment use. Using data from a sample of 638 banked household heads, this chapter examines the influence of sociodemographic characteristics and cash payment habits on the use of electronic payment methods. The findings reveal that social, economic, and behavioral factors largely explain the persistence of cash

DOI: 10.4018/979-8-3373-4526-0.ch001

in daily transactions. Beyond infrastructure availability, established routines and attachment to cash remain major barriers to payment digitalization. The chapter contributes to the literature by emphasizing the role of sociocultural factors in financial digitalization and offers policy implications for reducing cash dependency and fostering economic inclusion and efficiency.

INTRODUCTION

The global transition toward a digital economy relies on the rapid expansion of electronic payment systems, which have become key drivers of financial modernization, transaction security, and financial inclusion. In Morocco, however, despite substantial investments in banking infrastructure and the digitization of financial services, cash remains the dominant means of payment. In 2023, currency in circulation accounted for nearly 30% of GDP, one of the highest ratios worldwide and significantly higher than in comparable economies such as Egypt, Tunisia, and the WAEMU countries (Bank Al-Maghrib, 2023). Over the past two decades, the growth of currency in circulation has outpaced that of economic activity, confirming the resilience of cash within the Moroccan payment ecosystem.

At the same time, although card payments are gaining momentum, cash withdrawals still dominate, accounting for 71% of card transactions compared with only 29% for payments. The payment ratio of Moroccan cardholders remains low, standing at 28.9% of transactions by number and 12.5% by value in 2023, levels still well below international benchmarks. Moreover, although mobile wallet transactions doubled between 2023 and 2024, their share remains marginal given the structural importance of cash, which continues to be widely preferred for everyday expenditures.

This predominance of cash is not unique to Morocco. In advanced economies, several studies show that cash continues to play a significant role despite the rapid expansion of digital payments. Using ECB data, Zamora-Pérez et al. (2024) show that cash remains widely used in the euro area despite the rise of digital instruments. In emerging economies, the trend is even more pronounced. In India, Agarwal and Khatri (2024) note that, despite the strong growth of digital payments, cash remains the preferred means of payment for a large share of the population. The resilience of cash in Morocco therefore reflects a broader global pattern in which payment behavior is shaped by sociodemographic characteristics, unevenly developed infrastructure, and deeply ingrained habits.

This raises two key questions: to what extent do sociodemographic characteristics influence the use of electronic payment methods, and to what extent do cash-use habits hinder their use? The case of Morocco presents several noteworthy specificities. Electronic payments are indeed expanding: the number of card transactions reached

226 million in 2024, and mobile wallet transactions doubled between 2023 and 2024. However, these developments remain relatively marginal given the structural dominance of cash, which continues to be widely preferred for everyday payments. This inertia can be explained both by structural constraints, such as unequal access to financial infrastructure, particularly in rural areas, and by behavioral factors related to psychological attachment to cash and habit persistence.

The objective of this chapter is to empirically analyze the determinants of electronic payment usage in Morocco by linking two key dimensions: sociodemographic variables (gender, age, education level, income, sector of activity, and place of residence) and the influence of cash payment habits. To this end, an original survey was conducted among 638 individuals responsible for financial decisions within their households. The data were analyzed using a binary logistic regression model in order to assess the statistical significance of the explanatory variables.

This chapter primarily contributes to the literature on consumer behavior and marketing by focusing on individual adoption drivers and payment choices. Technical and infrastructural aspects of payment systems are considered only insofar as they shape consumers' usage conditions and behavioral decisions.

This research makes a twofold contribution. At the theoretical level, it combines insights from behavioral economics and empirical studies on payment behavior to explain the mechanisms that hinder the development of digital payment usage. At the practical level, it provides useful guidance for policymakers and financial institutions seeking to strengthen the digitization of payments and promote financial inclusion.

Finally, the chapter is organized as follows. Section 2 provides background information on electronic payment methods in Morocco and outlines their recent developments. Section 3 presents the theoretical framework and develops the research hypotheses, drawing on the main theoretical and empirical contributions in the literature. Section 4 describes the data, variables, and methodological approach adopted in the empirical analysis. Section 5 reports the empirical results. Section 6 discusses these findings in light of the theoretical framework and prior studies. Section 7 outlines the main limitations of the study, and Section 8 concludes by highlighting the principal theoretical contributions, practical implications, and directions for future research.

BACKGROUND

Payment systems play a fundamental role in the functioning of modern economies. Two main categories of instruments are generally distinguished: cash payments and electronic payment methods. The latter include card payments, credit transfers, direct debits, as well as digital wallets and payment applications. The European Central

Bank defines electronic payments as transactions that are initiated and processed electronically, without the physical exchange of cash (European Central Bank., 2021). Several academic contributions emphasize that electronic payments rely on digital infrastructures that enable the transfer of value through electronic networks (Dahlberg et al., 2015). The World Bank also highlights that the diffusion of digital payments is one of the main drivers of financial inclusion and the formalization of economic transactions (Demirgüç-Kunt et al., 2022).

Internationally, the past two decades have witnessed a marked expansion of electronic payments. This development has been driven by the digitization of financial services, the growth of mobile Internet, the rise of e-commerce, and public policies aimed at promoting cashless payments. However, a substantial body of research shows that this evolution has not resulted in the disappearance of cash.

Recent findings from the European Central Bank indicate that cash remains the most widely used means of payment for face-to-face transactions in the euro area, despite the growth of card payments and digital wallets (Zamora-Pérez et al., 2024). A similar situation is observed in emerging economies. In India, the rapid expansion of digital payments coexists with a persistent preference for cash for everyday purchases and low-value transactions (Agarwal & Khatri, 2024). These findings suggest that payment behavior does not depend solely on technological innovation, but is also strongly shaped by habits, trust, and sociodemographic characteristics.

Structure of the Moroccan Payments Market

In Morocco, the payments market is structured around several key actors:

- Bank Al-Maghrib: supervisory and regulatory authority for payment systems
- Commercial banks: issuers of payment cards and providers of electronic banking services
- Centre Monétique Interbancaire (CMI): principal acquirer of card-based payments
- Telecommunications operators: key stakeholders in the development and operation of mobile payment services
- FinTech companies: providers of innovative payment solutions, including electronic and mobile wallets

According to reports published by Bank Al-Maghrib, Morocco has recorded a significant increase in the number of payment cards in circulation, the deployment of electronic payment terminals (point-of-sale terminals), and the volume of transactions carried out through mobile wallets (Bank Al-Maghrib, 2023, 2024).

Regulatory Environment

The regulatory environment has evolved significantly, particularly through:

- The modernization of the national payment system
- The opening of the market to new actors (including payment institutions)
- The adoption of legislation on payment services and electronic money
- The implementation of national financial inclusion strategies

These reforms aim to enhance the security of payments, strengthen competition within the payments market, and promote the adoption of electronic payment instruments.

Characteristics of Electronic Payment Methods in Morocco

The Moroccan electronic payment landscape consists mainly of bank cards, electronic payment terminals, credit transfers, direct debits, and, more recently, e-wallets and mobile payment services. According to Bank Al-Maghrib, both the number of cards in circulation and the number of electronic payment terminals (point-of-sale terminals) have increased steadily in recent years, reflecting the gradual modernization of the national payment infrastructure (Bank Al-Maghrib, 2023, 2024).

However, the use of these instruments is still characterized by several specific features. On the one hand, payment cards are used more frequently for cash withdrawals than for purchases at merchants, which confirms the strong preference for cash in everyday transactions. On the other hand, mobile payments, introduced more recently, are expanding rapidly but still account for a relatively small market share compared with cash and payment cards.

These characteristics reveal a paradox: although Morocco now has the technical infrastructure required to support the wider use of electronic payments, their adoption remains gradual and uneven across population groups and geographical areas.

Evolution of Electronic Payment Instruments in Morocco

As illustrated in Figure 1, the use of non-cash payment methods in Morocco increased significantly between 2019 and 2024, driven by the modernization of the financial system and the digitization of banking services. In 2024, nearly 625 million transactions were recorded, compared with 492 million in 2023 (+26%), for a total value of MAD 5,542 billion (+12%). This development forms part of a longer-term trend, with the value of transactions rising from MAD 3,597 billion in 2019 to over MAD 5,500 billion in 2024.

Figure 1. Evolution of non-cash payment transactions, (2019-2024)

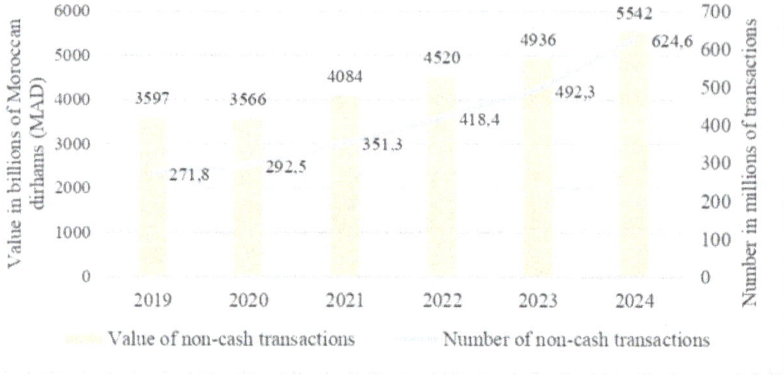

Source: Authors' elaboration based on data from Bank Al-Maghrib

As shown in Figures 2 and 3, the structure of cashless transactions in Morocco remained broadly stable in 2024. Transfers continue to dominate, accounting for 49% of the volume and 61% of the value, confirming their central role in interbank and institutional transactions. Bank cards rank second, representing 31% of the volume while accounting for only 1% of the value, which reflects their predominant use for proximity and low-value payments. Checks retain a significant share in value (24%) despite their limited use in terms of number (9%). Other instruments, such as direct debits, standardized bills of exchange, instant transfers, and mobile payments, remain marginal. This distribution reflects both the pre-eminence of transfers and payment cards and the relatively slow expansion of innovative instruments such as mobile payments.

Figure 2. Evolution of the cashless payment landscape in terms of number of transactions (2023–2024)

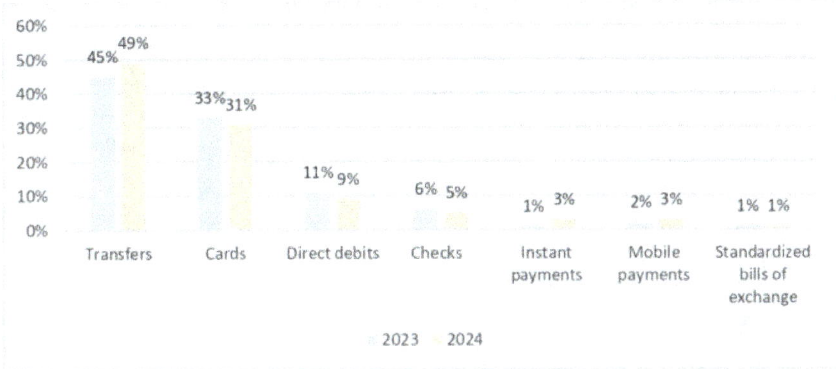

Source: Authors' elaboration based on data from Bank Al-Maghrib

Figure 3. Evolution of the cashless payment landscape in terms of value (2023–2024)

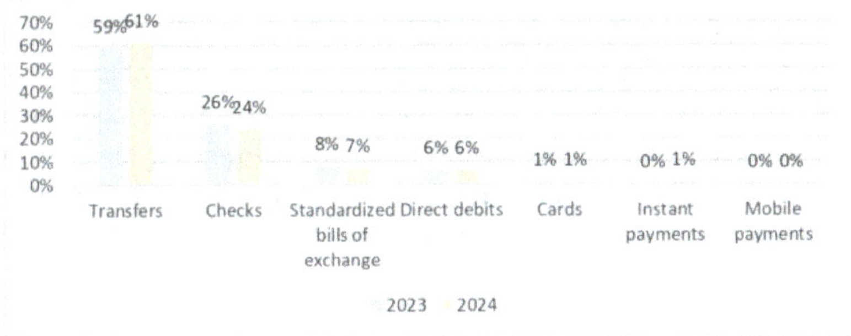

Source: Authors' elaboration based on data from Bank Al-Maghrib

The 2020 Finance Act introduced a set of incentives designed to encourage the use of mobile payments and reduce the predominance of cash. By amending Articles 38 and 40 of the General Tax Code (CGI), it established a 25% reduction in the taxable base corresponding to turnover generated through mobile payments. This tax incentive, granted to taxpayers subject to the simplified net income regime (RNS) or the flat-rate profit regime (BF), was intended to encourage this category of professionals to make greater use of this means of payment.

As shown in Figure 4, the number of mobile payment (M-Wallet) transactions increased sharply in 2024, rising from 9.7 million transactions with a value of MAD 2.1 billion in 2023 to 19.7 million transactions with a value of MAD 3.9 billion in 2024. This growth is largely attributable to the payment of direct social assistance into payment accounts, which encouraged part of the population to enroll in mobile payment services. However, despite these incentives and this recent momentum, mobile payments remain marginal compared with bank transfers and bank cards.

Figure 4. Growth in mobile payment transactions, 2020-2024

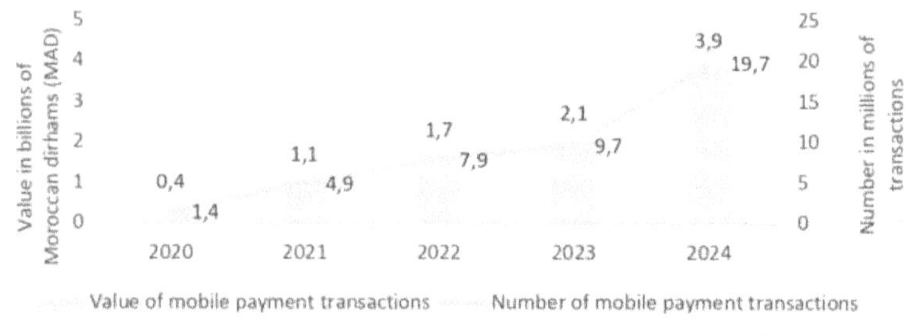

Source: Authors' elaboration based on data from Bank Al-Maghrib

Figure 5 highlights the steady growth of electronic banking activity in Morocco over the period 2019-2024. The total number of bank card transactions (payments and withdrawals) increased from 415 million in 2019 to 637 million in 2024, an increase of more than 50%. Over the same period, the total value of transactions rose from MAD 346 billion to MAD 513 billion. This trend reflects both the growing popularity of bank cards and the increasing use of electronic payments in everyday life. However, the pace of growth is more pronounced in terms of volume than value, suggesting that bank cards are used mainly for transactions involving medium or low-value amounts.

Figure 5. Evolution of electronic banking activity 2019-2024

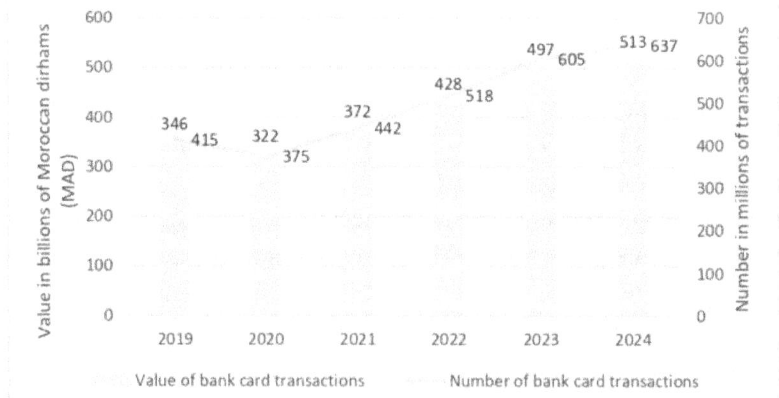

Source: Authors' elaboration based on data from Bank Al-Maghrib

As shown in Figure 6, the overall payment ratio in Morocco increased from 31.8% in 2023 to 35.4% in 2024, reflecting a growing use of cards for payment transactions. This rise concerns both Moroccan cardholders (from 29% to 32.4%) and foreign cardholders (from 72.5% to 75.9%). The upward trend has been observed since 2019; however, the overall level remains modest, as bank cards are still used primarily for cash withdrawals rather than as genuine substitutes for cash in retail payments.

Figure 6. Evolution of the payment ratio in Morocco 2019-2024

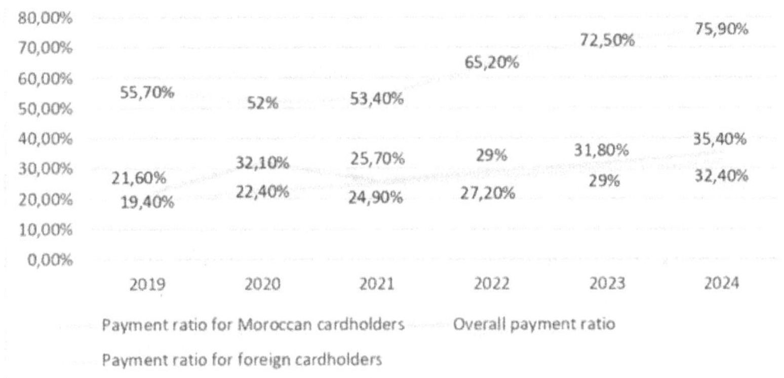

Source: Authors' elaboration based on data from Bank Al-Maghrib

As shown in Figure 7, bank card payments in Morocco continued to grow in 2024, reaching 226 million transactions, compared with 192 million in 2023 (+17%), for a total value of MAD 93 billion (+13%). The number of transactions has more than doubled since 2019 (from 89 million to 226 million), confirming the increasing popularity of cards as a means of payment. Nevertheless, the total value of card payments remains limited in comparison with other non-cash transactions and, above all, with the predominance of cash, reflecting a still gradual transition toward electronic payments.

Figure 7. Evolution of bank card payments 2019-2024

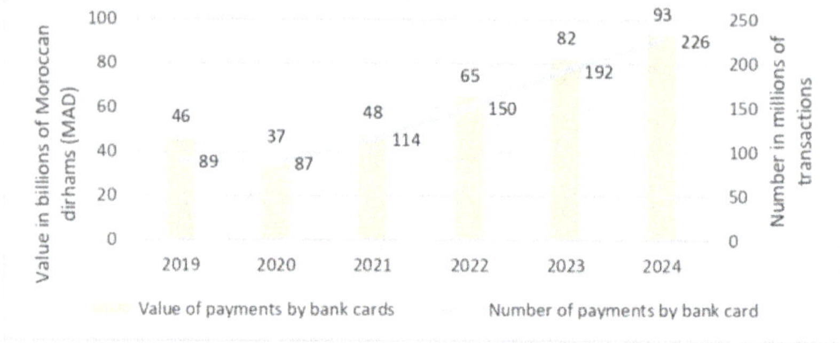

Source: Authors' elaboration based on data from Bank Al-Maghrib

As shown in Figure 8, the evolution of transfers in Morocco reveals sustained growth in both the value and the number of transactions. Between 2019 and 2024, the total value increased from MAD 1,806 billion to MAD 3,447 billion, representing an increase of almost 91%. Over the same period, the number of transactions nearly tripled, reaching 322 million transactions in 2024. This trend reflects the growing use of electronic payments, supported by the digitization of banking services. In addition, instant transfers have made a notable breakthrough. Although they still account for only 5% of the total number of transfers and 2% of their total value, their strong growth between 2023 and 2024 indicates increasing interest in fast and convenient payment solutions.

Figure 8. Evolution of transfers, 2019-2024

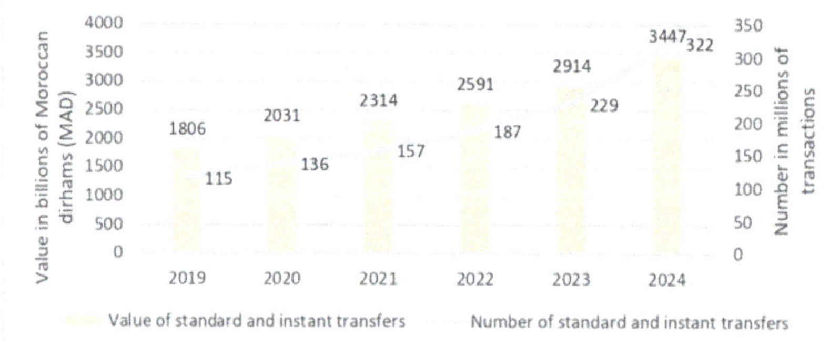

Source: Authors' elaboration based on data from Bank Al-Maghrib

Despite these advances, the use of electronic payment methods remains limited in the daily practices of a large part of the Moroccan population, underscoring the continued predominance of cash as the main payment instrument.

Development of Payment Methods and Support for FinTechs

Payment Development

In 2023, Bank Al-Maghrib continued its efforts to promote electronic payments and strengthen financial digitization, in collaboration with the World Bank, the Consultative Group to Assist the Poor (CGAP), and various public and private stakeholders. These initiatives sought to reduce the use of cash, enhance trust in digital financial services, and adapt the regulatory framework to support innovation in payments.

A joint Bank Al-Maghrib-World Bank study highlighted the need for a clear national strategy, stronger legal frameworks (including consumer protection and pricing transparency), and economic incentives to encourage the acceptance of electronic payments. Collaboration with CGAP focused on advancing financial inclusion in rural areas, emphasizing the importance of innovation, sustainable business models, and regulatory flexibility. These efforts have contributed to the expansion of mobile payments, with subscriptions increasing by 35% in 2023 and exceeding 10 million mobile wallet accounts.

Structural reforms have also been undertaken, most notably through the revision of the Commercial Code in order to provide an appropriate legal framework for new payment methods. In addition, Bank Al-Maghrib has contributed to the dematerialization of public flows (such as the TAYSSIR program and direct social assistance paid into payment accounts) by lifting several regulatory constraints and facilitating the use of digital channels.

A World Bank study has shown that the persistence of cash is closely linked to payment culture, the low level of card acceptance among merchants, perceived costs, concerns about transaction traceability, and mistrust of digital technology. To address these issues, Bank Al-Maghrib and its partners have recommended the establishment of national governance structures for electronic payments and the regulation of fees and commissions related to electronic payment services.

Finally, Bank Al-Maghrib, together with the Moroccan Mobile Payment Group (GP2M) and the Professional Association of Payment Institutions (APEP), has carried out several communication and awareness campaigns centered around the national brand "Maroc Pay" in order to promote the use of mobile payments among merchants and households. These efforts are complemented by collaboration with the National Telecommunications Regulatory Agency (ANRT) to monitor the use of information and communication technologies and digital payment services in Morocco.

Support for FinTechs

In 2023, Bank Al-Maghrib continued its efforts to develop the FinTech ecosystem through the One-Stop Shop FinTech (OSSF), a dedicated service center for receiving and supporting innovative companies. Since its establishment in 2019, the OSSF has hosted nearly 90 Moroccan and foreign FinTech firms, mainly operating in the fields of payments, money transfers, and the digitization of banking services. The entry into force of legislation on crowdfunding has also stimulated the emergence of new projects in this area.

In line with its 2019-2023 strategic plan, Bank Al-Maghrib also plans to establish a web platform dedicated to the OSSF in order to enable the digital tracking of applications, disseminate regulatory and educational content, and strengthen communication with stakeholders in the sector.

In collaboration with the World Bank, Bank Al-Maghrib conducted an in-depth diagnostic study of Morocco's FinTech ecosystem, focusing on financing conditions, regulatory frameworks, and sectoral development. At the same time, the Bank contributed to the formulation of the National Digital Transformation Strategy for 2030, which identifies FinTech as a priority lever for economic modernization.

Finally, as part of its partnership with CDG Invest, Bank Al-Maghrib has supported FinTech companies participating in the 212Founders program, by providing mentoring, technical assistance, and institutional backing. These initiatives illustrate its central role in promoting financial innovation in Morocco.

Summary

Morocco is therefore in a situation characterized by:

- rapid development of electronic payment infrastructures and instruments
- a regulatory environment that has undergone favorable reforms
- yet a persistent preference for cash in everyday transactions

This duality justifies the need for an in-depth empirical analysis of the determinants of electronic payment usage, taking into account both socio-demographic characteristics and cash usage habits.

THEORETICAL FRAMEWORK AND HYPOTHESES

Sociodemographic Characteristics and Use of Electronic Payment Methods

Models of technology acceptance and use emphasize that decisions to adopt or not adopt an innovation do not depend solely on intrinsic features of the technology but are also shaped by individual characteristics, particularly sociodemographic attributes. Within the UTAUT2 framework, variables such as age, gender, education, income, occupation, family, and peers are identified as key determinants of adoption (Dahlberg et al., 2015; Venkatesh et al., 2012). These variables most often operate as moderators, influencing the strength and direction of the relationships between the main constructs of the model and behavioral intentions.

Even when initial adoption is achieved, continued use is not guaranteed. Bartels and Reinders (2011) and Franque et al. (2023) show that repeated use can fail when the perceived benefits do not meet users' expectations or when their needs are not adequately taken into account. This issue is particularly acute in Africa, where the costs of accessing technology are high, alternatives to cash are limited, and financial exclusion remains significant, thereby raising concerns about the continued use of mobile payments.

More broadly, theories of technology adoption suggest that sociodemographic variables affect not only access but also behavioral intentions and actual behavior. Several studies have shown that age, gender, income, occupation, family, social class, and peers significantly influence the propensity to use innovations. UTAUT2 thus posits that age, gender, experience, and social influence affect both the intention to adopt and actual use (Venkatesh et al., 2012).

Social Resource Theory (SRT) sheds light more specifically on the role of education. It considers education to be a key resource that facilitates information seeking and the acceptance of new products. Empirical studies have shown that education increases the frequency of health information searches (Song & Chang, 2012) and enhances the acceptance of information products in China (Zhang et al., 2012). In the same vein, Agag et al. (2020) argue that sociodemographic characteristics help reduce the gap between stated intention and observed behavior.

Initially, age was the main moderator of the relationships between the constructs of the UTAUT model (Venkatesh et al., 2003). However, subsequent studies have shown that other variables, such as gender and income, enrich the understanding of adoption behaviors, particularly for specific technologies such as mobile payments (Shafinah et al., 2013). The underlying assumption is that age and income influence perceptions of cost and risk, with older and/or lower-income individuals being more reluctant to adopt innovations.

However, these moderating effects are not universal. Some studies have shown, for example, that neither age nor gender moderates the relationship between performance expectations and the intention to adopt electronic money services in Indonesia (Giri et al., 2019) or digital banking services in India (Ranpariya et al., 2021). Despite these differences, the inclusion of sociodemographic variables such as age, gender, income, and occupation remains relevant for explaining the adoption and use of electronic payment methods.

The empirical literature consistently emphasizes the decisive role of sociodemographic characteristics in shaping the adoption and use of electronic payment methods. Across different national and temporal contexts, these characteristics influence individuals' propensity to adopt digital payment technologies, although the magnitude and direction of their effects may vary depending on institutional, cultural, and economic environments.

A first stream of studies highlights the role of gender and household composition in payment behavior. Empirical evidence shows that gender-related effects are not uniform across countries. For instance, Dzogbenuku et al. (2022) document significant differences in digital payment usage by gender and age in Ghana. Fujiki and Tanaka (2018) observe that female-headed households in Japan tend to use credit cards more frequently than cash. Conversely, Banerjee and Pradhan (2024) report that, in India, men are significantly more likely to adopt digital payments, while Lavanya and Shrivastava (2024) confirm that gender, along with age, income, occupation, and education, plays a significant role in adoption, with men showing a stronger propensity toward electronic payment use. Said et al. (2021) further show that in the United Arab Emirates, payment behavior is closely linked to income and education, with demographic composition favoring adoption among high-income groups. Taken together, these findings suggest that gender effects are context-dependent, justifying the formulation of the following hypothesis:

> **Hypothesis One:** Gender significantly affects the use of electronic payment methods.

A second body of literature focuses on age as a determinant of payment behavior, often highlighting contrasting patterns. While Andaregie et al. (2024), using Global Findex data for Ethiopia, find that the likelihood of adopting digital payments increases with age, evidence from other countries points in the opposite direction. In Switzerland, Mäder et al. (2024) show that individuals aged 65 and over make significantly more cash payments and use electronic and contactless payments much less frequently. Similar patterns are observed in Japan, where Fujiki and Tanaka (2018) find that households headed by younger individuals favor credit cards over cash, and in Spain, where Raya and Vargas (2022) note a persistent preference for cash among older individuals. Lee et al. (2013) also confirm, in the Malaysian context, that younger generations are more inclined to adopt electronic payment methods. Overall, these studies converge on the idea that age has a lasting influence on payment choices, with younger individuals being more open to innovation and older populations maintaining a stronger attachment to cash. This leads to the following hypothesis:

> **Hypothesis Two:** Age has a significant negative influence on the use of electronic payment methods.

Another well-documented determinant is the level of education, which appears as a robust predictor of electronic payment adoption. Empirical studies consistently show that higher education and financial literacy are associated with a greater likelihood of using digital payment instruments. Sadok and Elouaourti (2025), drawing on PISA 2022 data, demonstrate that higher financial literacy significantly increases the use of mobile payments, particularly in urban areas. Similar conclusions are reached by Akinyemi and Mushunje (2020) and Elena-Bucea et al. (2021), who stress the role of education in enhancing technological capabilities. In India, Tiwari et al. (2019) attribute the low penetration of digital payments partly to insufficient educational attainment, while Bakhshi et al. (2024) show that targeted educational programs effectively promote adoption. Wang et al. (2024) provide comparable evidence for China. Earlier studies (Goczek & Witkowski, 2016; Bounie & Francois, 2006; Carow & Staten, 1999; Arango et al., 2015) further confirm that more educated individuals are significantly more likely to use bank cards and other electronic payment methods. These converging results support the following hypothesis:

> **Hypothesis Three:** The level of education has a positive and significant effect on the probability of using electronic payment methods.

The literature also underscores the importance of income and occupation in explaining payment choices. Higher income levels are generally associated with greater use of electronic payment instruments. Agarwal and Khatri (2024) show that digital payment adoption in India is concentrated among individuals with higher incomes, higher education, and specific occupations, particularly in urban areas with better infrastructure. In Switzerland, Mäder et al. (2024) find that individuals with higher income and wealth rely more heavily on mobile and contactless payments. Evidence from the Eurozone (Kajdi, 2022) and Spain (Raya & Vargas, 2022) similarly indicates that higher income and wealth encourage card usage, while lower-income individuals tend to prefer cash. Studies from Asia (Vinitha & Vasantha, 2017; Murari et al., 2020) also highlight the combined influence of income, occupation, and education, and Chaiyasoonthorn and Suksa-Ngiam (2019) show that affluent socioeconomic classes in Bangkok make greater use of electronic payments. These findings justify the following hypothesis:

> **Hypothesis Four:** Monthly income level has a positive and significant effect on the probability of using electronic payment methods.

Beyond income, employment sector emerges as a relevant factor in payment behavior. Aurazo and Vega (2021), using Peruvian household survey data for the period 2015–2018, show that individuals employed in the formal sector are more likely to use digital payments, reflecting better access to internet and banking services. This evidence motivates the following hypothesis:

> **Hypothesis Five:** Belonging to the formal sector significantly increases the probability of using electronic payment methods.

Finally, geographical area and access to infrastructure play a decisive role in electronic payment adoption. Several studies indicate that urban residence facilitates the use of digital payments due to better infrastructure and financial service availability. Sadok and Elouaourti (2025) note that individuals living in large cities are more inclined to use digital payments, while Agarwal and Khatri (2024) show that adoption in India is dominated by urban areas. Aurazo and Vega (2021) confirm similar patterns in Peru, and Kosse and Jansen (2013) demonstrate that location influences the choice between cash and electronic payments in the Netherlands. Lee et al. (2013) reach comparable conclusions for Malaysia. These results lead to the final hypothesis:

> **Hypothesis Six:** Living in an urban area is positively associated with the use of electronic payment methods.

In the Moroccan context, recent empirical evidence also confirms the relevance of these determinants. Rhafel et al. (2025) show that cash usage remains largely predominant, driven by the combined effect of sociodemographic and spatial factors, revealing strong heterogeneity in payment behaviors across the population. Unlike this previous study, which was mainly descriptive, the present chapter extends the analysis by explicitly integrating payment habits within a behavioral framework and empirically testing their effect on the use of electronic payment methods using a binary logistic regression model.

Habit Theory

Triandis (1979) defines habit as an automated behavior triggered by a given situation. He also argues that behavior does not result solely from intentions, but can also be influenced by two factors: perceived lack of control and habits, the latter emerging when a behavior is practiced repeatedly and becomes automatic. Consistent with this view, Verplanken and Aarts (1999) define habit as a set of learned actions that have become automatic responses to specific cues, functional in achieving particular goals. Verplanken and Orbell (2003) consider habit to be a psychological construct rather than simply the frequency of past behavior, while Wood and Neal (2009) define it as an automatic response that develops when individuals repeatedly perform the same actions in similar contexts.

Subsequent research has confirmed that habits have a significant influence on behavioral intentions and technology use (Venkatesh et al., 2012; Pahnila et al., 2011). Limayem et al. (2007) define habit as the extent to which individuals tend to adopt automatic behaviors as a result of their learning. Kim et al. (2005) equate habit with automaticity, while Kim and Malhotra (2005) argue that it can emerge even before the individual develops a conscious belief, thus becoming routine and automatic behavior. Wood et al. (2002) report that more than 45% of human behavior can be dictated by habits, with individuals tending to repeat the same behaviors when the same contextual cues are present. This observation reflects the decisive role that habits play in the continuation of behavior.

Previous literature has also linked habit to learning theory. Hull (1943) viewed its formation as an associative learning process, whereby a particular habit is acquired when a response to a stimulus is reinforced and repeated. Verplanken et al. (1997) define it as sequences of learned acts that become automatic responses to specific situations that may be functional in achieving certain goals or end states. Shah et al. (2014) emphasize that the frequency of past behaviors in a stable, recurring context is closely linked to habit formation. Lankton et al. (2010) add that the desire to satisfy a need generally leads to a response to a particular stimulus, and that a habit is consolidated when this response is reinforced or repeated. Similarly, Lakshmanan and Krishnan (2011) show that through repeatedly using a technology, individuals gradually become accustomed to this behavior, which eventually becomes ingrained in their daily practices.

In light of these theoretical and empirical contributions, we adopt the definition proposed by Limayem et al. (2007), according to which habit corresponds to the extent to which people tend to automatically adopt a behavior because of their learning. This concept is consistent with that of Aarts and Dijksterhuis (2000), who regard habit as a mode of behavior characterized by automatic responses and a hierarchical structure: repeated behaviors form the basis of this structure, while goals are at the top and guide individual routines.

Payment Habits and Behaviors

Applying habit theory to payment behavior provides a deeper understanding of the persistence of cash usage despite the expansion of digital payment solutions. The literature consistently shows that the adoption of electronic payment methods cannot be explained solely by supply-side factors such as infrastructure availability, banking penetration, or access to financial services. Instead, demand-side factors, including individual preferences, learned behaviors, and habitual routines, play a central role in shaping payment choices.

A growing body of empirical research highlights payment habits as a key determinant in the choice between cash and electronic means of payment. Zamora-Pérez et al. (2024) emphasize the decisive role of habitual behavior in explaining the resilience of cash in the digital age. Their findings underline the multifaceted nature of payment choices, which reflect not only practical constraints but also deeply ingrained routines developed through repeated use. Similarly, Pintér et al. (2022) show that payment context and learned habits significantly influence consumer decision-making, confirming that payment behavior cannot be reduced to technological availability alone.

Several country-specific studies further confirm the structuring role of habits as a barrier to the adoption of electronic payment methods. Using evidence from the Netherlands, Van der Cruijsen et al. (2017) demonstrate that persistent payment habits constitute a major obstacle to the substitution of cash by debit cards. Strong attachment to cash leads consumers to maintain established routines and contributes to the slow pace of behavioral change. This gap between stated intentions and actual practices is particularly pronounced among older individuals and in cash-intensive sectors, where card acceptance has been introduced more recently. These findings indicate that habits exert a lasting influence on payment choices and delay the diffusion of electronic payment instruments.

The importance of habits is further reinforced by studies examining the psychological dimension of payment behavior. Van der Cruijsen and Van der Horst (2019) show that both the initial intention to use a payment method and pre-existing habits are major determinants of actual payment behavior. Their results confirm that even when individuals express a willingness to adopt electronic payments, entrenched routines may prevent this intention from translating into effective use.

The temporal stability of payment habits also plays a crucial role in explaining the persistence of cash. Connolly and Stavins (2015), drawing on five annual surveys conducted in the United States between 2009 and 2013, identify strong correlations between payment behavior, income, and demographic characteristics. However, the absence of substantial changes over time, despite the widespread availability of electronic payment methods, illustrates the slow evolution of payment habits and the resilience of established routines.

Finally, the literature highlights the influence of cultural background and generational dynamics on habitual payment behavior. Kosse and Jansen (2013) show that migrants living in the Netherlands who originate from cash-oriented countries continue to prefer cash payments even after settling in a highly digitized environment. This persistence reflects habits acquired prior to migration, which remain stable despite exposure to new infrastructures. However, their findings also suggest that such habits are not systematically transmitted across generations, as descendants tend to adopt the dominant practices of the host country. Similarly, Van der Horst and Matthijsen (2013) demonstrate that older individuals who predominantly pay in cash develop a stronger emotional attachment to this payment method, reinforcing the habitual nature of their behavior and illustrating the difficulty of altering deeply ingrained routines. Overall, these studies confirm that changes in payment behavior occur slowly and gradually, reflecting the structural and resilient nature of learned habits.

In the Moroccan context, characterized by a strong culture of cash usage, the habit of using cash for transactions continues to play a decisive role in shaping individual payment preferences. This persistent behavioral pattern provides a strong rationale for the following hypothesis:

> **Hypothesis Seven:** The habit of using cash as a means of payment significantly reduces the likelihood of using electronic payment methods.

Despite the richness of the existing empirical literature on electronic payment adoption and cash persistence, several gaps remain. First, micro-level empirical evidence focusing on emerging economies, and particularly on the Moroccan context, is still limited, despite the coexistence of payment system modernization and a strong reliance on cash. Second, many existing studies examine sociodemographic determinants and payment behaviors separately, without explicitly integrating habitual cash usage as a central explanatory factor. Third, relatively few works combine sociodemographic characteristics and payment habits within a unified behavioral framework to explain actual usage rather than stated intention. This study addresses these gaps by jointly analyzing sociodemographic characteristics and cash-use habits using original survey data from Moroccan households, thereby providing a more comprehensive understanding of electronic payment adoption in a cash-oriented economy.

DATA AND METHODS

In order to achieve the research objectives, a field survey was conducted using a non-probability convenience sampling method. The questionnaire was administered using a mixed approach, combining face-to-face interviews in targeted public places (bank branches, markets, and shopping centers) and online distribution via an electronic form (Google Forms). Before its final implementation, a pre-test was conducted with a subsample of 30 individuals reflecting the sociodemographic diversity of the target audience, which ensured the clarity of the questions and allowed for adjustments to certain items of the questionnaire.

The sample consisted of people with bank accounts who were responsible for financial transactions within their households, in order to target individuals directly involved in decisions relating to the choice and use of payment methods. A total of 638 valid questionnaires were collected and analyzed.

Measurement and Definition of Variables

The variables used in this study were operationalized based on the questionnaire responses and coded to reflect both sociodemographic characteristics and behavioral dimensions related to payment practices. All variables were defined in accordance with the theoretical framework and prior empirical studies on payment behavior. The dependent variable captures the predominant payment method used by respondents, while the independent variables include individual characteristics and a behavioral indicator reflecting habitual cash use. Table 1 summarizes the definition, measurement, and coding of all variables included in the empirical analysis.

Use of electronic payment methods is measured using the question: *"Which payment method do you use most frequently?"* Responses were recoded into a binary variable, where 1 indicates respondents who report using electronic payment methods (card, transfer, or mobile payment) most often or always, and 0 indicates respondents who report using cash most often or always. This operationalization captures the dominant payment behavior rather than occasional usage.

Cash payment habit is measured using the question: *"Do you tend to use cash out of habit?"* The variable is coded as 1 if the respondent declares habitual cash use and 0 otherwise. This variable reflects the behavioral dimension of payment choice and captures routine-driven preferences beyond infrastructural or economic constraints.

Table 1. Definition of variables

Variable	Description
Dependent Variable	
Use of Electronic Payment Methods	0 = most often or always uses cash 1 = most often or always uses electronic payment methods
Independent Variables	
Gender	0 = Male 1 = Female
Age	1 = 18-24 years 2 = 25-34 years 3 = 35-49 years 4 = 50 years and above
Education level	1 = No formal education 2 = Primary 3 = Secondary 4 = University
Monthly income	1 = Less than 3,000 dirhams 2 = Between 3,000 and 7,000 dirhams 3 = Between 7,000 and 15,000 dirhams 4 = More than 15,000 dirhams
Sector of employment	0 = Informal 1 = Formal
Geographic area	0 = Rural 1 = Urban
Cash payment habit	0 = No 1 = Yes

Source: authors' elaboration

Research Model

A binary logistic regression model was used, which is adapted to the dichotomous nature of the dependent variable.

$$Logit(P_i) = \ln\left(\frac{P_i}{1 - P_i}\right) =$$
$$= \beta_0 + \beta_1 \text{Gender} + \beta_2 \text{Age} + \beta_3 \text{Education Level} + \beta_4 \text{Income}$$
$$+ \beta_5 \text{Sector of Employment} + \beta_6 \text{Geographic Area} + \beta_7 \text{Cash Payment Habit} + \varepsilon_i$$

Where:

P_i: the probability that individual i often or always uses electronic payment methods.

β_i: coefficients measuring the marginal effect of the explanatory variables on this probability.

Operationalization of the Empirical Study

The study was operationalized through a structured questionnaire covering two main dimensions: (i) the sociodemographic characteristics of respondents (gender, age, educational level, income, sector of employment, Geographic area) and (ii) behavioral factors related to payment habits. The dependent variable is defined as the use of electronic payment methods, while the explanatory variables reflect these two dimensions.

Sampling and Preliminary Validation of the Questionnaire

The survey was conducted among a convenience sample of heads of households with bank accounts who are responsible for their households' financial transactions. They were selected to reflect diversity in terms of age, income, educational level, and geographic area (urban/rural). Before its final rollout, a pre-test was conducted with a subsample of 30 individuals representative of the target population, which allowed us to validate the clarity of the questions and to adjust certain items of the questionnaire. A total of 638 valid questionnaires were collected and analyzed, forming the empirical basis of the study.

Data Analysis Method

Data analysis was conducted in two stages: first, a descriptive analysis of the main characteristics of respondents, followed by an econometric analysis using binary logistic regression. The use of this model is justified by the dichotomous nature of the dependent variable (use of electronic payment methods: 1 if the individual mainly uses electronic payment methods, 0 otherwise). Logistic regression makes it possible to assess the probability of an event based on a set of explanatory characteristics and is suitable for both qualitative and quantitative variables. This method is widely used in the literature on payment behavior, notably by Bahri et al. (2025), Sadok and Elouaourti (2025), Agarwal and Khatri (2024), Kajdi (2022), Tagat et al. (2019), Végso et al. (2018), Bagnall et al. (2018), and Fujiki and Tanaka (2018).

RESULTS

Sociodemographic Profile of Respondents

The sample consists of 638 individuals responsible for financial decisions in their households, and their main sociodemographic characteristics are summarized in Table 2. The gender breakdown shows a predominance of men (61.9%) over women (38.1%). In terms of age, the majority of respondents were aged 25-34 age group (40.8%), followed by individuals aged 50 and over (25.2%) and those aged 35-49 (19%), while young people aged 18-24 represented only 15% of the sample.

The level of education is relatively high: more than 73% of respondents have a university degree, while secondary (11.4%) and primary (8.2%) education levels remain in the minority, as does the category with no formal education (7.1%). In terms of employment status, employees (34%) and civil servants (30.7%) are the most represented, followed by self-employed workers (13.8%) and retirees (8.6%).

In terms of sector of employment, nearly 70% of individuals work in the formal sector, compared to 30.4% in the informal sector, reflecting the significant presence of the latter in the sample. In terms of income, the majority of respondents (47.2%) earn less than MAD 3,000 per month, while 32.3% earn between MAD 3,000 and MAD 7,000, and only 4.5% have incomes greater than MAD 15,000. Finally, the geographic distribution reveals an urban concentration (68.5%) compared to 31.5% in rural areas, illustrating a more pronounced level of access to banking services in urban areas.

Table 2. Sociodemographic profile of respondents

Variable	Categories	Frequency	Percentage	Cumulative Percentage
Gender	Male	395	61.9%	61.9%
	Female	243	38.1%	100%
Age	18-24 years	96	15%	15%
	25-34 years	260	40.8%	55.8%
	35-49 years	121	19%	74.8%
	50 years and above	161	25.2%	100%
Educational level	No formal education	45	7.1%	7.1%
	Primary	52	8.2%	15.2%
	Secondary	73	11.4%	26.6%
	University	468	73.4%	100%

continued on following page

Table 2. Continued

Variable	Categories	Frequency	Percentage	Cumulative Percentage
Occupation	Student	59	9.2%	9.2%
	Civil servant	196	30.7%	40%
	Employee (private sector)	217	34%	74%
	Liberal professions	88	13.8%	87.8%
	Retired	55	8.6%	96.4%
	Other	23	3.6%	100%
Sector of employment	Informal	194	30.4%	30.4%
	Formal	444	69.6%	100%
Monthly income	Less than 3,000 dirhams	102	16%	16%
	3,000-7,000 dirhams	301	47.2%	63.2%
	7,000-15,000 dirhams	206	32.3%	95.5%
	15,000 dirhams and above	29	4.5%	100%
Geographic area	Rural	201	31.5%	31.5%
	Urban	437	68.5%	100%

Source: authors' elaboration from survey data

Payment Methods Used

As shown in Figure 9, 58% of respondents report using cash more often or always, compared to 42% who report using electronic payment methods. These results confirm that, despite efforts to modernize Morocco's financial system, cash remains the dominant payment method in everyday transactions.

Figure 9. Use of electronic payment methods

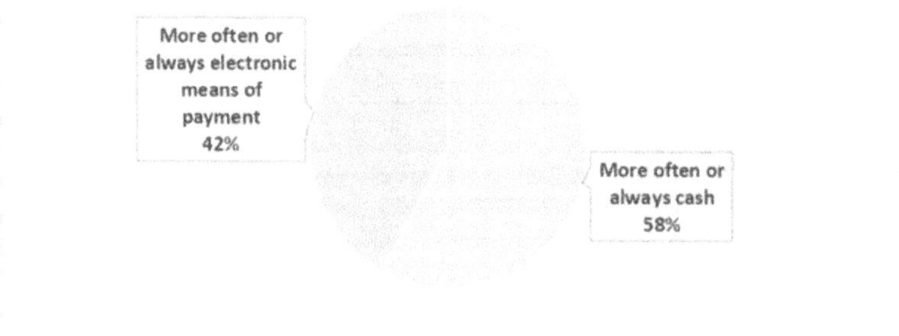

Source: authors' elaboration from survey data

Habit of Paying in Cash

As shown in Figure 10, nearly 80% of respondents report using cash out of habit, compared to only 20.1% who report not having this habit. This result suggests that the use of cash does not depend solely on acceptance or infrastructure constraints, but also on deeply ingrained behavioral patterns. Habit is therefore a determining factor that slows the transition to electronic payment methods, reinforcing the preference for cash even when alternatives exist.

Figure 10. Habit of paying in cash

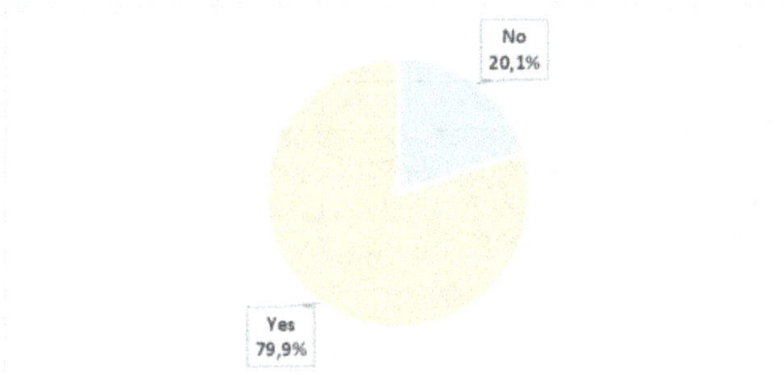

Source: authors' elaboration from survey data

Results of Binary Logistic Regression

Correlation and Multicollinearity Analysis

The correlation matrix between the seven explanatory variables, presented in Table 3, was established using Spearman's rank correlation coefficient. The results show that the coefficients are generally low to moderate, reflecting moderate relationships between certain variables without indicating a risk of excessive endogeneity. The highest values concern the association between age and education level ($\rho = 0.578$) and the correlation between sector of employment and monthly income ($\rho = 0.474$). These values remain below the critical threshold of 0.7 commonly accepted in the literature, suggesting the absence of problematic multicollinearity. All variables can therefore be simultaneously integrated into the logistic model.

Table 3. Correlation matrix

Spearman's Rho	Gender	Age	Educational level	Sector of employment	Monthly income	Geographic area	Habit
Gender	1.000	-.318**	.256**	.020	-.135**	.094*	-.075
Age	-.318**	1.000	-.578**	.112	.199**	.047	.158**
Educational level	.256**	-.578**	1.000	.272**	.178**	.077	-.148**
Sector of employment	.020	.112	.272**	1.000	.474**	.087*	-.042
Monthly income	-.135**	.199**	.178**	.474**	1.000	.085*	-.052
Geographic area	.094*	.047	.077	.087*	.085*	1.000	.056
Habit	-.075	.158**	-.148**	-.042	-.052	.056	1.000

** Correlation is significant at the 0.01 level (two-tailed).
* Correlation is significant at the 0.05 level (two-tailed).
Source: authors' elaboration based on SPSS outputs

Diagnostic Tests and Model Quality

Several diagnostic tests were used to assess the quality of the binary logistic regression model fit. As shown in Table 4, the test of all coefficients (likelihood-ratio test) shows that the overall model is significant ($\chi^2 = 160.286$; df = 13; $p < 0.001$), indicating that the introduction of explanatory variables significantly improves the model's predictive power compared to the baseline model.

Table 4. Omnibus tests of model coefficients

		Chi-square	df	Sig.
Step 1	Step	160.286	13	<.001
	Block	160.286	13	<.001
	Model	160.286	13	<.001

Source: authors' elaboration based on SPSS outputs

As shown in Table 5, the pseudo-R^2 indices confirm that the model explains a substantial portion of the variance in the dependent variable. Cox and Snell's R^2 is equal to 0.222, and Nagelkerke's R^2 is equal to 0.299. Although these values are not directly comparable to the R^2 of linear regression, they suggest an overall satisfactory goodness of fit within the framework of a logistic model.

Table 5. Model summary

Step	-2 Log-likelihood	Cox and Snell R Square	Nagelkerke R^2
1	707.792	.222	.299

a. Estimation terminated at iteration 6 because parameter estimates changed by less than 0.001.
Source: authors' elaboration based on SPSS outputs

The classification table (Table 6) shows that the model correctly predicts 68.8% of observations, compared to 58% in the null model (Block 0). More specifically, the correct classification rate reaches 54.4% for individuals who mainly use electronic payment methods, compared to 79.2% for those who prefer cash. These results highlight a substantial improvement in the model's predictive performance.

Table 6. Classification table

	Observed		Predicted payment method used		Percentage correct
			0	1	
Step1	Payment method used	0	293	77	79.2
		1	122	146	54.5
	Overall percentage				68.8

a. The cutoff value is 0.50
Source: authors' elaboration based on SPSS outputs

The Hosmer-Lemeshow goodness-of-fit test (Table 7) confirms that the model fits the data satisfactorily ($\chi^2 = 10.093$; df = 8; p = 0.259). The associated probability, which is higher than the conventional threshold of 5%, implies that the null hypothesis is not rejected; this result reflects a good match between the estimated probabilities and the observed frequencies, thus confirming the overall quality of the logistic model fit.

Table 7. Hosmer and Lemeshow test

Step	Chi-square	df	Sig.
1	10.093	8	.259

Source: authors' elaboration based on SPSS outputs

Finally, the value of -2 Log-Likelihood (707.792) is significantly lower than that obtained in the null model, which indicates a substantial reduction in prediction error and an overall improvement in the model's performance.

In summary, all of these tests attest to the robustness and the overall goodness of fit of the estimated logistic model.

DISCUSSION

Table 8 presents the results of the binary logistic regression estimation. It reports the coefficients for each explanatory variables as well as their levels of statistical significance.

Table 8. Logistic regression coefficients

	Variable	β	S.E.	Wald	df	Sig.	Exp(B)
Step 1a	Gender	.374	.199	3.530	1	.060	1.454
	Age			10.591	3	.014	
	Age(1)	.369	.288	1.642	1	.200	1.446
	Age(2)	-.090	.354	.065	1	.799	.914
	Age(3)	-.583	.375	2.416	1	.120	.558
	Education level			10.438	3	.015	
	Education level(1)	1.017	.879	1.340	1	.247	2.765
	Education level(2)	2.004	.782	6.568	1	.010	7.420
	Education level(3)	2.057	.765	7.220	1	.007	7.821
	Sector of employment	.516	.253	4.145	1	.042	1.675
	Monthly income			15.996	3	.001	
	Monthly income(1)	-.109	.329	.109	1	.741	.897
	Monthly income(2)	.611	.363	2.825	1	.093	1.842
	Monthly income(3)	1.405	.566	6.149	1	.013	4.074
	Geographic area	.582	.209	7.788	1	.005	1.790
	Cash payment habit	-1.396	.238	34.308	1	<.001	.248
	Constant	-2.282	.825	7.648	1	.006	.102

a. Variable(s) entered on step 1: Gender, Age, Education level, Sector of employment, Monthly income, Geographic area, Cash payment habit.

Source: authors' elaboration based on SPSS outputs

Sociodemographic Characteristics

Logistic regression analysis highlights the significant role of sociodemographic characteristics in the use of electronic payment methods.

Gender appears to be a significant determinant at the 10% threshold ($\beta = 0.374$; Sig. = 0.060; Exp(β) = 1.454). Compared to men, women are more likely to use these methods, with the odds ratio indicating that they are approximately 1.45 times more likely to do so, confirming Hypothesis H1. However, this result is consistent with a mixed body of literature. Fujiki and Tanaka (2018) show that female-headed households use credit cards more than cash, while Mantel (2001) highlights their more frequent use of electronic payments for bill payments. They also tend to manage cash more prudently, making more frequent withdrawals of smaller amounts (Boeschoten, 1998). In contrast, some studies suggest greater adoption among men: Banerjee and Pradhan (2024) note that in India, men are more inclined to adopt digital payments, and Lavanya and Shrivastava (2024) confirm that gender influences adoption, with men being more favorable to these technologies. These differences highlight that the effect of gender remains highly context-dependent, particularly on the socioeconomic and cultural environment.

Age does not appear to be a significant determinant of the use of electronic payment methods. Compared to the reference category, "18-24 years," no age group shows a statistically significant effect. However, the coefficients indicate a tendency toward lower use of digital payments among individuals aged 50 and over ($\beta = -0.583$; Exp(β) = 0.558); but this effect is not statistically confirmed. These results lead us to reject Hypothesis H2. However, the observed trend is consistent with the work of Mäder et al. (2024), Raya and Vargas (2022), and Lee et al. (2013), who emphasize that older populations retain a marked preference for cash. This behavior is often explained by the costs of learning and adapting to new technologies (Boeschoten, 1998; Hayashi and Klee, 2003).

Compared to the reference category, "no education," individuals who have completed primary education show no significant effect ($\beta = 1.017$; Sig. = 0.247; Exp(β) = 2.765). By contrast, those with a secondary education ($\beta = 2.004$; Sig. = 0.010; Exp(β) = 7.420) and higher education ($\beta = 2.057$; Sig. = 0.007; Exp(β) = 7.821) show a significantly higher probability of using electronic payments. The odds ratios reveals that these two categories are about seven times more likely to use them than individuals with no education, thus confirming Hypothesis H3. These results are consistent with several studies showing that education increases the likelihood of using digital payments. Sadok and Elouaourti (2025) emphasize that financial literacy promotes the use of mobile money, while Akinyemi and Mushunje (2020) and Elena-Bucea et al. (2021) underline the role of financial literacy and technological skills. In India, Tiwari et al. (2019) explain the low penetration rate by a lack of education, while Bakhshi et al. (2024) highlight the effectiveness of targeted educational programs. Wang et al. (2024) reach a similar conclusion in China. More broadly, several studies (Boeschoten, 1998; Carow and Staten, 1999; Stavins, 2001) show that graduates, due to lower learning costs, are more likely to adopt new electronic means of payment, regardless of income or age.

Compared to the reference category, "less than 3,000 dirhams," individuals earning between 3,000 and 7,000 dirhams show no significant difference (β = -0.109; Sig. = 0.741; Exp(β) = 0.897). Similarly, the "7,000 to 15,000 dirhams" category shows only a weakly significant effect at the 10% threshold (β = 0.611; Sig. = 0.093; Exp(β) = 1.842). By contrast, households with an income above 15,000 dirhams clearly stand out with an increased probability of using electronic means of payment (β = 1.405; Sig. = 0.013; Exp(β) = 4.074). The odds ratio indicates that the latter are about four times more likely to use them than low-income households, confirming Hypothesis H4. These results highlight the structuring effect of income on the use of electronic payments. These findings are consistent with numerous empirical studies. In India, Agarwal and Khatri (2024) show that high-income individuals use these instruments more, while Mäder et al. (2024) observe in Switzerland that high-net-worth individuals prefer mobile and online payments. In the eurozone, Kajdi (2022) and in Spain, Raya and Vargas (2022) emphasize that higher incomes encourage the use of cards and electronic payments. In Asia, Vinitha and Vasantha (2017) and Murari et al. (2020) also confirm the role of income, and Chaiyasoonthorn and Suksa-Ngiam (2019) show that in Bangkok, the wealthy classes make greater use of electronic payments.

The influence of income on the use of payment instruments can be interpreted through several theoretical mechanisms. First, income is linked to the level of expenditure, which explains why higher-income households make greater use of account-based and electronic instruments (rather than cash), as Murphy (1991) showed for the use of checks in the United States.

Second, income determines the opportunity cost of the time spent obtaining cash. High-income individuals, for whom time is more precious, tend to reduce the frequency of withdrawals, increase the amounts withdrawn, and favor faster instruments, particularly electronic ones (Boeschoten, 1998; Attanasio et al., 2002; Stix, 2004).

Finally, certain means of payment, such as premium bank cards, can be likened to superior goods whose use increases with income, sometimes reinforced by a signalling effect (Carow and Staten, 1999; Hayashi and Klee, 2003; Mantel, 2001; Stavins, 2001).

These theoretical mechanisms shed additional light on the empirical results observed, confirming that income is a key determinant of the use of electronic payment methods.

The sector of employment has a significant effect on the use of electronic payment methods (β = 0.516; Sig. = 0.042; Exp(β) = 1.675). Individuals working in the formal sector are approximately 1.7 times more likely to use these instruments than those in the informal sector, thereby confirming Hypothesis H5. This result corroborates the work of Aurazo and Vega (2021), which shows that in Peru, the use of digital payments (cards, mobile or online banking services) is more frequent among individuals in formal employment with better access to the Internet. Similarly, Delgado et al. (2023) and Tagat et al. (2019) emphasize that informality remains closely associated with the use of cash, due to less integration into the financial system and limited access to banking and digital services.

Geographical area appears to be a significant factor in explaining the use of electronic payment methods (β = 0.582; Sig. = 0.005; Exp(β) = 1.790). Individuals living in urban areas are approximately 1.8 times more likely to use electronic payments than those living in rural areas, confirming Hypothesis H6. This result is consistent with the findings of Sadok and Elouaourti (2025), Aurazo and Vega (2021), and Agarwal and Khatri (2024), who show that greater access to financial and digital infrastructure in urban areas significantly promotes the adoption of modern payment methods. Conversely, rural areas remain disadvantaged by accessibility constraints and more limited banking services.

Cash Payment Habits

The habit of paying in cash is a highly significant determinant of the use of electronic payment methods (β = –1.540; Sig. = 0.000; Exp(β) = 0.214). Individuals who report paying for purchases in cash are almost five times less likely to use electronic payments than those who do not have this habit, confirming Hypothesis H7. This result corroborates the work of Zamora-Pérez et al. (2024), who emphasize the central role of habits in the persistence of cash, as well as that of Pintér et al. (2022) and Van der Cruijsen et al. (2017), who show that payment routines slow down the substitution of cash by electronic means. Other research (van der Cruijsen and van der Horst, 2019; Connolly and Stavins, 2015; Kosse and Jansen, 2013; van der Horst and Matthijsen, 2013) confirms that payment behaviors evolve slowly and remain strongly influenced by cultural, generational, and emotional habits. In the Moroccan context, marked by a strong cash culture, this behavioral inertia appears to be one of the major obstacles to the transition to electronic payment methods.

The empirical results can be interpreted in light of the UTAUT2 (Unified Theory of Acceptance and Use of Technology) model. This theoretical framework posits that sociodemographic characteristics such as age, gender, income, occupation, and education level act as moderators that shape the relationships between the main constructs of technology adoption, namely perceived usefulness, effort expectancy, social influence, and facilitating conditions, and actual usage behavior (Venkatesh et al., 2012; Dahlberg et al., 2015). In our study, education level and income significantly increase the likelihood of using electronic payment methods. This result is consistent with Social Resource Theory (SRT), according to which education is an essential resource that enhances information seeking and the ability to adopt new products and technologies (Song and Chang, 2012; Zhang et al., 2012). Education and income thus enhance perceived behavioral control and perceived usefulness, promoting adoption.

Urban residence and formal sector employment also appear to be significant determinants of electronic payment usage. In the context of UTAUT2, these variables can be interpreted as indicators of facilitating conditions, insofar as they reflect better access to banking infrastructure, the Internet, and financial services. Conversely, rural areas and the informal sector are characterized by accessibility constraints that hinder adoption despite potential interest. The effect of gender, although context-dependent, mainly reflects the role of social norms and subjective norms that differentiate men and women in their payment behaviors.

Finally, the very strong negative effect of the habit of paying in cash on the use of electronic payment methods highlights the central role of habit as a psychological construct. Even when perceived usefulness and facilitating conditions are present, deeply ingrained routines and cultural payment practices limit behavioral change and prevent intentions from being translated into actual use (Bartels and Reinders, 2011; Franque et al., 2023). In the Moroccan context, marked by a strong cash culture, this behavioral inertia appears to be one of the main obstacles to the transition to electronic payments.

LIMITATIONS

However, this study has a number of limitations that must be acknowledged and that open up avenues for future research. First, the data used comes from a cross-sectional survey. Such a design allows associations to be identified but does not make it possible to establish causal relationships between the variables studied. Longitudinal studies will be necessary to analyze changes in payment behaviors over time and better understand the dynamics of electronic payment adoption.

Secondly, the sampling relies on a non-probabilistic convenience method. This choice is justified by practical data-collection constraints, but it limits the statistical representativeness of the results and their generalization to the entire Moroccan population. Future research could employ probabilistic sampling or rely on national databases.

Thirdly, the study is based on self-reported data. Responses may therefore be affected by social desirability or recall bias, particularly with regard to the actual frequency of use of different payment methods. The use of objective data from financial institutions would be a valuable addition.

Fourth, the model used focuses mainly on sociodemographic variables and the habit of paying in cash. Other psychological and institutional determinants, such as trust in the financial system, risk perception, perceived security, usage costs, and public policies related to financial inclusion, have not been included, even though they can significantly influence payment behavior.

Finally, the study is limited to the Moroccan context, which is characterized by a strong cash culture and a gradual process of payment digitalization. The results should therefore be interpreted in light of this specific institutional and cultural context and may differ in other countries with different levels of financial inclusion or digitalization.

CONCLUSION

This chapter examined, in the Moroccan context, the factors explaining the use of electronic payment methods in two main areas: sociodemographic characteristics and the habit of paying in cash. Despite efforts to digitalize, cash remains predominant, confirming the persistence of a "cash paradox", in which infrastructure modernization coexists with a strong attachment to cash.

Firstly, the results confirm that sociodemographic factors significantly influence the use of electronic payment methods. Gender appears to be a differentiating factor: women are relatively more likely to use these instruments than men. Education and income are also key variables, with individuals with a higher level of education and higher incomes being more inclined to use digital payments. Similarly, belonging to the formal sector and living in an urban area encourage the use of electronic instruments, reflecting the structuring effect of access to banking and digital infrastructure. Conversely, age does not appear to be a significant determinant in the Moroccan context, although a preference for cash can be observed among older individuals.

Secondly, the study highlights the central role of cash payment habits: individuals with a strong propensity to use cash are significantly less likely to adopt electronic payment methods. In the Moroccan context, where the cash culture remains deeply rooted, this psychological attachment constitutes a major obstacle to the spread of digital payments.

These findings have several implications. On a theoretical level, they confirm the relevance of incorporating behavioural factors, particularly habit theory, into the analysis of financial behaviour. The use of payment technologies cannot be understood solely through traditional economic determinants; it must be placed in a context where acquired routines and social preferences play a decisive role. At the practical level, the results provide guidance for public decision-makers and financial actors. For public authorities, it is essential to combine the development of financial infrastructure with targeted awareness-raising and financial education policies in order to reduce dependence on cash and build confidence in electronic payments. For financial institutions, it is important to offer simple, accessible and secure digital solutions, while adapting communication strategies to the needs of different sociodemographic groups.

Finally, this study opens several avenues for future research. Further work could extend the analysis by incorporating psychological factors such as trust, perceived risk, and perceived security, or by using longitudinal data to better capture the dynamics of habit formation and behavioral change over time. Comparative studies across countries with different levels of financial inclusion and cash dependence would also help to assess the generalizability of the findings.

In conclusion, the Moroccan case clearly illustrates that the spread of electronic payments is not just a question of technology or infrastructure, but also a matter of behavior, trust, and routines. Understanding and transforming these dynamics is a major challenge in supporting the transition to a more digitalized and inclusive society, where electronic payments play their full role as a lever for economic and social modernization.

REFERENCES

Aarts, H., & Dijksterhuis, A. (2000). Habits as knowledge structures : Automaticity in goal-directed behavior. *Journal of Personality and Social Psychology, 78*(1), 53–63. DOI: 10.1037/0022-3514.78.1.53 PMID: 10653505

Agag, G., Brown, A., Hassanein, A., & Shaalan, A. (2020). Decoding travellers' willingness to pay more for green travel products : Closing the intention–behaviour gap. *Journal of Sustainable Tourism, 28*(10), 1551–1575. DOI: 10.1080/09669582.2020.1745215

Agarwal, M., & Khatri, M. (2024). From Cash-Centric to Cashless Economy-A Study on User′ s Awareness towards Digital Payment Systems. *Academy of Marketing Studies Journal, 28*(1).

Akinyemi, B. E., & Mushunje, A. (2020). Determinants of mobile money technology adoption in rural areas of Africa. *Cogent Social Sciences, 6*(1), 1815963. DOI: 10.1080/23311886.2020.1815963

Andaregie, A., Abebe, G. K., Gupta, P., Worku, G., Matsumoto, H., Astatkie, T., & Takagi, I. (2024). Exploring individuals' socioeconomic characteristics and digital infrastructure determinants of digital payment adoption in Ethiopia. *Digital Business, 4*(2), 100092. DOI: 10.1016/j.digbus.2024.100092

Arango, C., Huynh, K. P., & Sabetti, L. (2015). Consumer payment choice : Merchant card acceptance versus pricing incentives. *Journal of Banking & Finance, 55*, 130–141. DOI: 10.1016/j.jbankfin.2015.02.005

Attanasio, O. P., Guiso, L., & Jappelli, T. (2002). The Demand for Money, Financial Innovation, and the Welfare Cost of Inflation : An Analysis with Household Data. *Journal of Political Economy, 110*(2), 317–351. DOI: 10.1086/338743

Aurazo, J., & Vega, M. (2021). Why people use digital payments : Evidence from micro data in Peru. *Latin American Journal of Central Banking, 2*(4), 100044. DOI: 10.1016/j.latcb.2021.100044

Bagnall, J., Bounie, D., Huynh, K. P., Kosse, A., Schmidt, T., & Schuh, S. (2018). Consumer cash usage : A cross-country comparison with payment diary survey data. *46th issue (December 2016) of the International Journal of Central Banking.* https://www.ijcb.org/journal/ijcb16q4a1.pdf

Bahri, S., Suroso, A. I., Suhendi, , & Sari, L. K. (2025). Examining Cash Usage Behavior in Metropolitan Greater Jakarta Societies. *Societies (Basel, Switzerland), 15*(5), 120. DOI: 10.3390/soc15050120

Bakhshi, P., Agrawal, R., Mendon, S., Frank, D., Spulbar, C., Birau, R., & Filip, R. D. (2024). Barriers in adoption of FinTech by street vendors and hawkers in India using interpretive structural modeling. *Business: Theory and Practice, 25*(1), 231–240. DOI: 10.3846/btp.2024.19208

Banerjee, A. K., & Pradhan, H. K. (2024). Influence of demographic profiles in adoption of digital payment system in India : A multigroup invariance analysis. *Technology Analysis and Strategic Management, 36*(10), 2285–2301. DOI: 10.1080/09537325.2022.2132928

Bank Al-Maghrib. (2023). *Rapport annuel sur les infrastructures des marchés financiers et les moyens de paiement, leur surveillance et l'inclusion financière—Exercice 2023.*

Bank Al-Maghrib. (2024). *Rapport annuel sur les infrastructures des marchés financiers et les moyens de paiement et leur surveillance—Exercice 2024.*

Bartels, J., & Reinders, M. J. (2011). Consumer innovativeness and its correlates : A propositional inventory for future research. *Journal of Business Research, 64*(6), 601–609. DOI: 10.1016/j.jbusres.2010.05.002

Boeschoten, W. C. (1998). Cash Management, Payment Patterns and the Demand for Money. *De Economist, 146*(1), 117–142. DOI: 10.1023/A:1003258026314

Bounie, D., & Francois, A. (2006). *Cash, Check or Bank Card? The Effects of Transaction Characteristics on the Use of Payment Instruments* (SSRN Scholarly Paper No. 891791). Social Science Research Network. DOI: 10.2139/ssrn.891791

Carow, K. A., & Staten, M. E. (1999). Debit, credit, or cash : Survey evidence on gasoline purchases. *Journal of Economics and Business, 51*(5), 409–421. DOI: 10.1016/S0148-6195(99)00016-8

Chaiyasoonthorn, W., & Suksa-Ngiam, W. (2019). The diffusion and adoption of electronic payment systems in Bangkok. [IJEBR]. *International Journal of E-Business Research, 15*(2), 102–115. DOI: 10.4018/IJEBR.2019040106

Connolly, S., & Stavins, J. (2015). Payment Instrument Adoption and Use in the United States, 2009-2013, by Consumers' Demographic Characteristics. *Research Data Reports Paper, 15-6.* https://papers.ssrn.com/sol3/papers.cfm?abstract_id=2709920

Dahlberg, T., Guo, J., & Ondrus, J. (2015). A critical review of mobile payment research. *Electronic Commerce Research and Applications, 14*(5), 265–284. DOI: 10.1016/j.elerap.2015.07.006

Delgado, O., Barrera, J., Delgado, O., & Barrera, J. (2023). Determinantes del Uso de Efectivo en América Latina. *Revista Politécnica, 51*(2), 73–86. DOI: 10.33333/rp.vol51n2.07

Demirgüç-Kunt, A., Klapper, L., Singer, D., & Ansar, S. (2022). *The Global Findex Database 2021 : Financial Inclusion, Digital Payments, and Resilience in the Age of COVID-19*. World Bank Publications. DOI: 10.1596/978-1-4648-1897-4

Dzogbenuku, R. K., Amoako, G. K., Kumi, D. K., & Bonsu, G. A. (2022). Digital Payments and Financial Wellbeing of the Rural Poor : The Moderating Role of Age and Gender. *Journal of International Consumer Marketing, 34*(2), 113–136. DOI: 10.1080/08961530.2021.1917468

Elena-Bucea, A., Cruz-Jesus, F., Oliveira, T., & Coelho, P. S. (2021). Assessing the Role of Age, Education, Gender and Income on the Digital Divide : Evidence for the European Union. *Information Systems Frontiers, 23*(4), 1007–1021. DOI: 10.1007/s10796-020-10012-9

European Central Bank. (2021). *Study on the payment attitudes of consumers in the euro area (SPACE)*. Publications Office. https://data.europa.eu/doi/10.2866/240571

Franque, F. B., Oliveira, T., & Tam, C. (2023). Continuance Intention of Mobile Payment : TTF Model with Trust in an African Context. *Information Systems Frontiers, 25*(2), 775–793. DOI: 10.1007/s10796-022-10263-8

Fujiki, H., & Tanaka, M. (2018). How do we choose to pay using evolving retail payment technologies? Evidence from Japan. *Journal of the Japanese and International Economies, 49*, 85–99. DOI: 10.1016/j.jjie.2018.05.001

Giri, R. R. W., Apriliani, D., & Sofia, A. (2019). Behavioral intention analysis on e-money services in Indonesia : Using the modified UTAUT model. *1st International Conference on Economics, Business, Entrepreneurship, and Finance (ICEBEF 2018)*, 73-76. https://www.atlantis-press.com/proceedings/icebef-18/125908031

Goczek, Ł., & Witkowski, B. (2016). Determinants of card payments. *Applied Economics, 48*(16), 1530–1543. DOI: 10.1080/00036846.2015.1102846

Hayashi, F., & Klee, E. (2003). Technology adoption and consumer payments : Evidence from survey data. *Review of Network Economics, 2*(2). Advance online publication. https://scholar.archive.org/work/gvgjurwbtbb4hf3ps3xoh7m3vi/access/wayback/http://time.dufe.edu.cn:80/jingjiwencong/waiwenziliao/31327.pdf. DOI: 10.2202/1446-9022.1025

Hull, C. L. (1943). *Principles Of Behavior : An Introduction to Behavior Theory (New York: Appleton-Century-Crofts;).[Google Scholar].*

Kajdi, L. (2022). *Consumer payment preferences in the euro area.* ECB Working Paper. https://www.econstor.eu/handle/10419/269136

Kim, S. S., & Malhotra, N. K. (2005). A Longitudinal Model of Continued IS Use : An Integrative View of Four Mechanisms Underlying Postadoption Phenomena. *Management Science, 51*(5), 741–755. DOI: 10.1287/mnsc.1040.0326

Kim, S. S., Malhotra, N. K., & Narasimhan, S. (2005). Research Note—Two Competing Perspectives on Automatic Use : A Theoretical and Empirical Comparison. *Information Systems Research, 16*(4), 418–432. DOI: 10.1287/isre.1050.0070

Kosse, A., & Jansen, D.-J. (2013). Choosing how to pay : The influence of foreign backgrounds. *Journal of Banking & Finance, 37*(3), 989–998. DOI: 10.1016/j.jbankfin.2012.11.005

Lakshmanan, A., & Krishnan, H. S. (2011). The Aha! Experience : Insight and Discontinuous Learning in Product Usage. *Journal of Marketing, 75*(6), 105–123. DOI: 10.1509/jm.10.0348

Lankton, N. K., Wilson, E. V., & Mao, E. (2010). Antecedents and determinants of information technology habit. *Information & Management, 47*(5), 300–307. DOI: 10.1016/j.im.2010.06.004

Lavanya, R., & Shrivastava, S. (2024). DEMOGRAPHIC FACTORS INFLUENCING THE ADOPTION OF DIGITAL PAYMENT METHODS : A STATISTICAL ANALYSIS OF USER PREFERENCE. [JOM]. *Journal of Management, 11*(3), 1–15.

Lee, H. S. H., Loke, Y. J., & Tan, A. K. G. (2013). The Demand for E-Payments in Malaysia : An Examination of Usage Intensity. *Margin, 7*(4), 371–389. DOI: 10.1177/0973801013500167

Limayem, M., Hirt, S. G., & Cheung, C. M. (2007). How habit limits the predictive power of intention : The case of information systems continuance. *Management Information Systems Quarterly, 31*(4), 705–737. DOI: 10.2307/25148817

Mäder, A., Jüttner, M., & Gatica-Perez, D. (2024). *You Are How You Pay : Understanding and Identifying the Payment Behavior of Sociodemographic Groups* (SSRN Scholarly Paper No. 5064475). Social Science Research Network. DOI: 10.2139/ssrn.5064475

Mantel, B. (2001). *Why Do Consumers Pay Bills Electronically? An Empirical Analysis* (SSRN Scholarly Paper No. 248409). Social Science Research Network. DOI: 10.2139/ssrn.248409

Murari, K., Bhatt, V., & Kumar, P. (2020). Does socio-demography affect preferences and concerns towards cashless payment? Evidence from the north-eastern region of India. *Journal of Payments Strategy & Systems*, *14*(3), 207–222. DOI: 10.69554/LSVO4282

Murphy, N. B. (1991). Determinants of household check writing : The impacts of the use of electronic banking services and alternative pricing of services. *Financial Services Review*, *1*(1), 35–44. DOI: 10.1016/1057-0810(91)90006-K

Pahnila, S., Siponen, M., & Zheng, X. (2011). Integrating Habit into UTAUT : The Chinese eBay Case. *Pacific Asia Journal of the Association for Information Systems*, *3*(2), 1–30. Advance online publication. DOI: 10.17705/1pais.03201

Pintér, Z., Nagy, M. Z., Tóth, K., & Varga, J. (2022). The Struggle between Cash and Electronic Payments. *Economies*, *10*(12), 304. DOI: 10.3390/economies10120304

Ranpariya, T., Joshi, A., & Rajdev, A. (2021). *Factors driving the adoption of UPI services based on UTAUT model.* http://library.atmiya.net:8080/dspace/handle/atmiyauni/2113

Raya, J. M., & Vargas, C. (2022). How to become a cashless economy and what are the determinants of eliminating cash. *Journal of Applied Econometrics*, *25*(1), 543–562. DOI: 10.1080/15140326.2022.2052000

RHAFEL, A., Mustapha, Z. I. K. Y., & LIMAM, A. (2025). Analyse du comportement de paiement au Maroc. *Dossiers de Recherches en Économie et Gestion*, *13*(02), 213–232.

Sadok, H., & Elouaourti, Z. (2025). *The cash vs. digital Paradox : Why financial literacy matters in uncertain times.* Borsa Istanbul Review., DOI: 10.1016/j.bir.2025.08.001

Said, R., Najdawi, A., & Chabani, Z. (2021). Analyzing the Adoption of E-payment Services in Smart Cities using Demographic Analytics : The Case of Dubai. *Advances in Science, Technology and Engineering Systems Journal*, *6*(2), 113–121. DOI: 10.25046/aj060214

Shafinah, K., Sahari, N., Sulaiman, R., Yusoff, M. S. M., & Ikram, M. M. (2013). Determinants of User Behavior Intention (BI) on Mobile Services : A Preliminary View. *Procedia Technology*, *11*, 127–133. DOI: 10.1016/j.protcy.2013.12.171

Shah, D., Kumar, V., & Kim, K. H. (2014). Managing Customer Profits : The Power of Habits. *JMR, Journal of Marketing Research*, *51*(6), 726–741. DOI: 10.1509/jmr.13.0423

Song, L., & Chang, T.-Y. (2012). Do resources of network members help in help seeking? Social capital and health information search. *Social Networks*, *34*(4), 658–669. DOI: 10.1016/j.socnet.2012.08.002

Stavins, J. (2001). Effect of Consumer CharacteristicsontheUse of Payment Instruments. *New England Economic Review(Issue Number 3)*, 19-31.

Stix, H. (2004). The impact of ATM transactions and cashless payments on cash demand in Austria. *Monetary Policy & the Economy*, *1*(04), 90–105.

Tagat, A., Özmen, M., & Trivedi, P. L. (2019). Consumer payments survey of India : A closer look at household finances and payment instruments. *NCAER National Data Innovation Centre Report*. https://www.researchgate.net/profile/Anirudh-Tagat/publication/342956792_Consumer_payments_survey_of_India_A_closer_look_at_household_finances_and_payment_instruments/links/5f16926a4585151299ad5b6c/Consumer-payments-survey-of-India-A-closer-look-at-household-finances-and-payment-instruments.pdf

Tiwari, T., Srivastava, A., & Kumar, S. (2019). Adoption of digital payment methods in India. *International Journal of Electronic Finance*, *9*(3), 217. DOI: 10.1504/IJEF.2019.099058

Triandis, H. C. (1979). Values, attitudes, and interpersonal behavior. *Nebraska Symposium on Motivation*, *27*, 195–259. PMID: 7242748

van der Cruijsen, C., Hernandez, L., & Jonker, N. (2017). In love with the debit card but still married to cash. *Applied Economics*, *49*(30), 2989–3004. DOI: 10.1080/00036846.2016.1251568

Van Der Cruijsen, C., & Van Der Horst, F. (2019). Cash or Card? Unravelling the Role of Socio-Psychological Factors. *De Economist*, *167*(2), 145–175. DOI: 10.1007/s10645-019-09340-2

Van der Horst, F., & Matthijsen, E. (2013). The irrationality of payment behaviour. *DNB Occasional Studies, 11*(4). https://www.researchgate.net/profile/Jakub-Gorka-2/publication/305073403_Merchant_Indifference_Test_Application_-_A_Case_For_Revising_Interchange_Fee_Level_in_Poland/links/5780e4a608ae69ab88262a1d/Merchant-Indifference-Test-Application-A-Case-For-Revising-Interchange-Fee-Level-in-Poland.pdf#page=133

Végso, T., Belházyné Illés, Á., & Bódi-Schubert, A. (2018). Cash or card?–An explorative analysis of consumers' payment behaviour in Hungary. *Public Finance Quarterly= Pénzügyi Szemle, 63*(4), 448-472.

Venkatesh, V., Morris, M. G., Davis, G. B., & Davis, F. D. (2003). User Acceptance of Information Technology : Toward a Unified View. *Management Information Systems Quarterly*, *27*(3), 425–478. DOI: 10.2307/30036540

Venkatesh, V., Thong, J. Y. L., & Xu, X. (2012). Consumer Acceptance and Use of Information Technology : Extending the Unified Theory of Acceptance and Use of Technology. *Management Information Systems Quarterly*, *36*(1), 157–178. DOI: 10.2307/41410412

Verplanken, B., & Aarts, H. (1999). Habit, Attitude, and Planned Behaviour : Is Habit an Empty Construct or an Interesting Case of Goal-directed Automaticity? *European Review of Social Psychology*, *10*(1), 101–134. DOI: 10.1080/14792779943000035

Verplanken, B., Aarts, H., & Van Knippenberg, A. (1997). Habit, information acquisition, and the process of making travel mode choices. *European Journal of Social Psychology*, *27*(5), 539–560. DOI: 10.1002/(SICI)1099-0992(199709/10)27:5<539::AID-EJSP831>3.0.CO;2-A

Verplanken, B., & Orbell, S. (2003). Reflections on Past Behavior : A Self-Report Index of Habit Strength[1]. *Journal of Applied Social Psychology*, *33*(6), 1313–1330. DOI: 10.1111/j.1559-1816.2003.tb01951.x

Vinitha, K., & Vasantha, S. (2017). Influence of demographic variables on usage of e-payment system. [IJMET]. *Int J Mech Eng Technol*, *8*, 265–276.

Wang, Q., Niu, G., Zhou, Y., & Gan, X. (2024). Education and FinTech adoption : Evidence from China. *China Finance Review International*, *15*(1), 140–165. DOI: 10.1108/CFRI-06-2023-0141

Wood, W., & Neal, D. T. (2009). The habitual consumer. *Journal of Consumer Psychology*, *19*(4), 579–592. DOI: 10.1016/j.jcps.2009.08.003

Wood, W., Quinn, J. M., & Kashy, D. A. (2002). Habits in everyday life : Thought, emotion, and action. *Journal of Personality and Social Psychology*, *83*(6), 1281–1297. DOI: 10.1037/0022-3514.83.6.1281 PMID: 12500811

Zamora-Pérez, A., Marini, A., & Honkkila, J. (2024). Is there a digital divide in payments? Understanding why cash remains important for so many. *Economic Bulletin Articles, 2*. https://ideas.repec.org//a/ecb/ecbart/202400022.html

Zhang, Y., Lin, N., & Li, T. (2012). Markets or networks : Households' choice of financial intermediary in Western China. *Social Networks*, *34*(4), 670–681. DOI: 10.1016/j.socnet.2012.08.003

ADDITIONAL READING

Borzekowski, R., Elizabeth, K. K., & Shaista, A. (2008). Consumers' Use of Debit Cards : Patterns, Preferences, and Price Response. *Journal of Money, Credit and Banking*, *40*(1), 149–172. DOI: 10.1111/j.1538-4616.2008.00107.x

Deák, V., Nemecskó, I., & Végső, T. (2021). *Payment habits of the Hungarian households in 2020* (No. 143). MNB Occasional Papers.

Greene, C., Perry, J., & Stavins, J. (2024). *Consumer Payment Behavior by Income and Demographics* (SSRN Scholarly Paper No. 4914698). Social Science Research Network. DOI: 10.29412/res.wp

Kotkowski, R., & Manikowski, A. (2023). Cash usage in Poland in 2020: Insights into the role of the COVID-19 pandemic and spatial aspects. *Journal of Banking and Financial Economics*, *1*(19), 85–113. DOI: 10.7172/2353-6845.jbfe.2023.1.5

Rusu, C., & Stix, H. (2017). Cash and card payments–recent results of the Austrian payment diary survey. *Monetary Policy and the Economy Q*, *1*, 19–31.

Sahabat, I., Dartanto, T., Passay, H. A., & Widyawati, D. (2017). Electronics Payment Decisions of the Indonesian Urban Households : A Nested Logit Analysis of the Effects of the Payment Characteristics. *International Journal of Economics and Financial Issues*, *7*(5), 498.

Schuh, S., & Stavins, J. (2012). *How consumers pay: Adoption and use of payments* (No. 12-2). Working Papers.

Stavins, J. (2016). *The effect of demographics on payment behavior: panel data with sample selection* (No. 16-5). Working Papers.

KEY TERMS AND DEFINITIONS

Behavioral economics: a field of economics that studies how psychological, social and emotional factors influence financial decisions.

Electronic payment methods: instruments or technologies that enable monetary transactions without the physical exchange of cash, including cards, bank transfers, and mobile payments.

Financial inclusion: the process of ensuring access to affordable and useful financial products and services for all, particularly vulnerable populations.

Habit of using cash: a repetitive and automatic preference for using cash in everyday transactions, influenced by behavioral and cultural factors.

Habit theory: a behavioral framework that explains how repeated actions become automatic responses in stable contexts, guiding decision-making.

Payment digitization: the transformation of payment systems through technology, leading to the replacement of cash with electronic and mobile instruments.

Sociodemographic characteristics: personal attributes such as age, gender, education level, income and place of residence that influence economic and financial behavior.

Trust in financial institutions: the confidence that individuals place in banks and financial systems to act in a reliable, transparent and secure manner.

Chapter 2
Digital Literacy as a Catalyst for Mobile Payment Marketing Effectiveness:
Insights From Haryana,India

Sanchita Banga
https://orcid.org/0009-0009-9572-5285
Maharishi Markandeshwar University, Mullana, India

Partap Singh
Department of Management, Maharishi Markandeshwar University, Mullana, India

ABSTRACT

This chapter analyzes the importance of digital literacy in enhancing the effectiveness of mobile payment marketing strategies in rural India, specifically for Haryana. Large segments of rural communities remain excluded from the digital economy despite expansion of digital financial services and government-backed initiatives such as Digital India. The most significant reasons for this exclusion are a deficiency in digital literacy, infrastructural shortcomings, and a default distrust of digital systems. This research examines the means through which perceived usefulness, trust, and ease of use serve as mediators between digital literacy and mobile payment acceptance, based on theoretical models such as the Technology Acceptance Model (TAM). Demographic factors such as age, sex, income, and education also play a

DOI: 10.4018/979-8-3373-4526-0.ch002

role in these dynamics, also analyzed in this chapter.For emerging economies, this chapter contributes to the larger discourse on rural development, digital inclusion, and financial empowerment through technology.

INTRODUCTION

Mobile payment systems have become an important component of contemporary financial ecosystems, particularly in developing economies such as India where digital financial services are expanding rapidly. These systems are increasingly positioned as tools for widening economic participation and reducing transaction-related barriers. In rural markets, however, the effectiveness of mobile payment initiatives depends not only on technological availability but also on how well marketing strategies align with users' digital capabilities. Among the various factors influencing this alignment, digital literacy plays a central role, as it determines whether individuals are able to understand, access, and confidently use mobile payment platforms.

India's push toward a digitally inclusive economy has been supported by several national initiatives, including Digital India, the Pradhan Mantri Jan Dhan Yojana, Aadhaar-based identification, and the Unified Payments Interface (UPI). While these efforts have significantly improved digital infrastructure and access, their impact has been uneven across regions. In many rural areas, the transition from cash-based transactions to digital payments has progressed slowly. Haryana, a northern Indian state that combines rapidly urbanizing cities with extensive rural regions, illustrates this contrast clearly. Urban centres such as Gurugram and Faridabad display high levels of digital payment usage, whereas large sections of rural Haryana continue to experience limited digital awareness and restricted engagement with mobile payment services.

Despite the increasing availability of smartphones and affordable internet services, rural consumers in Haryana often remain hesitant to adopt mobile payment systems. This hesitation stems from limited digital skills, low trust in technology, concerns about cybersecurity, and deeply rooted social and cultural practices. These factors significantly shape how rural audiences interpret and respond to marketing messages related to digital financial services. As a result, marketing campaigns that focus solely on promotion or incentives frequently fail to translate awareness into actual usage. Effective marketing in rural contexts therefore requires careful attention not only to infrastructural limitations but also to the cognitive, emotional, and psychological readiness of potential users.

Against this background, this chapter examines the role of digital literacy in enhancing the effectiveness of mobile payment marketing strategies in rural India, with particular emphasis on Haryana. It explores how individuals' willingness to use mobile payment systems is influenced by their ability to operate mobile applications, comprehend transaction-related information, and develop trust in digital platforms. In doing so, the chapter also considers how varying levels of digital literacy affect the extent to which firms can successfully design, communicate, and target marketing initiatives for rural consumers.

The Rise of Mobile Payments in India

Over the past decade, mobile payment services have become increasingly embedded in everyday financial transactions across India. A large user base, supported by widespread mobile connectivity and expanding internet access, has created favourable conditions for the growth of digital financial services. The emergence of platforms such as BHIM, Paytm, PhonePe, and Google Pay has enabled users to complete transactions conveniently, without being constrained by time or location. This transition was further accelerated during the COVID-19 pandemic, when contactless payment methods were widely promoted as safer alternatives to cash-based transactions.

However, the availability of digital infrastructure does not automatically result in universal adoption. Persistent forms of digital divide continue to influence the uptake of mobile payment systems. Differences in income, education, age-related technology hesitation, and gender-based disparities shape how individuals engage with digital financial tools. While urban users have generally shown greater readiness to experiment with and adopt mobile wallets and promotional offers, rural consumers often remain hesitant or insufficiently informed. These disparities highlight that access alone is insufficient, and that social and capability-related factors play a critical role in determining mobile payment adoption. The following graph shows the adoption of mobile payments over time with the difference between urban Indian and rural Indian consumers.

Figure 1: Growth of Mobile Payment Users in India (2015–2025): Urban vs. Rural Adoption Trends

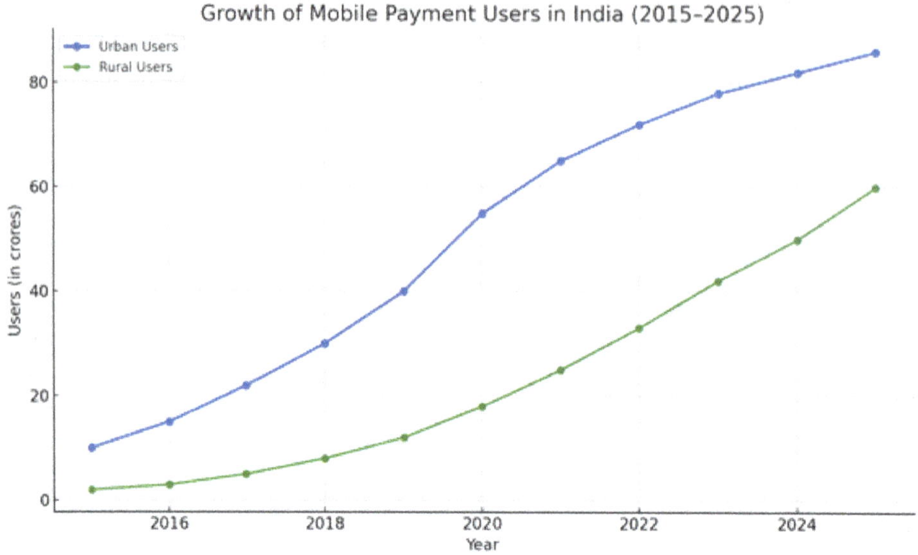

The Role of Digital Literacy

Digital literacy is more than just knowing how to use a smartphone. " Knowing how to be safe on the internet, recognizing scam apps, generating secure passwords, working with app interfaces, interpreting transaction messages, and fixing issues are all included. Smartphones either don't belong to most individuals in rural Haryana or are only used for basic communication. Others also realize that even if phones do exist, they are deterred from using them because they fear they will make a mistake, be scammed, or lose money to a scam.".

The effectiveness of marketing is directly affected by this shortfall in digital faith. With the premise that users can see and act on them, organizations implement campaigns via SMS, WhatsApp, Facebook ads, or application notifications. But such efforts often do not convert leads into users or transactions when the end-user is not digitally literate. To be effective, mobile payment marketing thus needs to be an imperative rather than merely an enabler.

Marketing to Rural India: Unique Challenges

Rural marketing in India is complex and requires localised, culturally sensitive strategies. Unlike urban areas where digital ads and influencer promotions work well, rural audiences rely more on trust, word-of-mouth, and community-based learning. For marketers, the task is not just to inform but to educate and convince.

Some challenges are:

- Language challenges: A lot of mobile payment apps are English- or plain Hindi-based, which might not be comfortable for rural Haryana Haryanvi-dialect speakers.
- Visual illiteracy: Novice users might find icon-based navigation disorienting.
- Deficit of trust: Online interaction is dissuaded by absence of physical affirmation and fear of scamming.
- Knowledge deficit: Many rural customers do not know that merchant discounts, cashback rewards, or referral rewards are available to them through digital payments.

A blend of awareness programs, training, on-ground promotion, and trust-building exercises is required for educational marketing to combat these. For the marketers, this is where digital literacy becomes a target and a tool.

Haryana: A Contextual Landscape

With over 2.8 crore individuals, Haryana is a mixture of urbanized towns and rural villages. It is inhabited by corporate hubs and is among the principal producers of agricultural commodities in India. It presents a strong case to explore how digital literacy impacts the adoption of mobile payments due to this socioeconomic duality.

Digital penetration in constituencies such as Fatehabad, Kaithal, Jind, Sirsa, Hisar, and Yamunanagar is less than in those such as Gurugram or Panchkula. It is more likely that young people will adopt digital technology, but that women and older generations will be opposed to it due to social norms restricting their usage. The Pradhan Mantri Gramin Digital Saksharta Abhiyan (PMGDISHA) has made efforts to train people in digital skills, but its reach and impact vary across regions. Moreover, while infrastructure (like mobile towers and internet access) may be available, the lack of usable knowledge still hinders the actual adoption of mobile payment systems.

The effectiveness of mobile payments in rural areas is a measure towards inclusive growth, rather than economic efficiency. Women entrepreneurs, laborers, small farmers, and retailers are all able to participate in the formal economy with the help of mobile payments. They can conduct transactions, save, access microloans, and receive subsidies without middlemen.

Discuss the digital literacy and marketing outcomes connection in this chapter. Identify the users' behavior and interests in rural Haryana.

- Explore the means by which marketers can craft more inclusive, successful, and credible campaigns.
- Provide policy recommendations that will enhance financial literacy and digital education programs.

This study contributes notably to both academic and business fields by examining the mediating roles of perceived usefulness, trust, and ease of use and how demographic profiles (gender, age, and education) moderate such relationships.

Theoretical Framework and Literature Review

Introduction to Digital Financial Inclusion in Rural India

A mix of policy shocks and technological progress has facilitated the advancement of the path of digital financial inclusion of rural Haryana. Demonetization in 2016 in India was an important economic incidence that led people to rethink their financial behavior. In rural India, where cash was the most valued currency, mobile payment systems (MPS) have become indispensable tools for convenience and economic survival in a cash-restricted society. With growing smartphone usage and pressure to convert to cashless modes, even illiterate communities began to show interest in mobile commerce. (Manisha et al., 2025).

Early Adoption Drivers and Demographic Influence

Aged 26–40, who were caught between tradition and the digital revolution, were among the first in line to adopt. This group, especially in Haryana, were the most accepting of MPS, partly due to policy instructions from outside but also due to their views regarding its usefulness and ease of use, which are the fundamental components of the Technology Acceptance Model (TAM) (Gupta & Hakhu, 2020). In line with TAM, if an individual thinks a technology is functional and easy to use, they are more likely to adopt it (Roy, 2014; Shankar & Datta, 2018). While adopting digital payments, these cognitive beliefs played a pivotal role.

Barriers to Adoption: Trust and Infrastructure

Even as it became more available, trust was an even more nuanced issue. Even technology-literate people in far-flung regions were reluctant to use mobile wallets due to unstable internet connections, ambiguous transaction protocols, and fear of fraud (Gaur & Verma, 2023). The COVID-19 pandemic further entrenched the mistrust. Mobile wallets were rapidly embraced in urban areas, but rural consumers lagged behind in embracing them due to a lack of experience and assurance.

Figure 2: Barriers to MPS Adoption in Rural Haryana

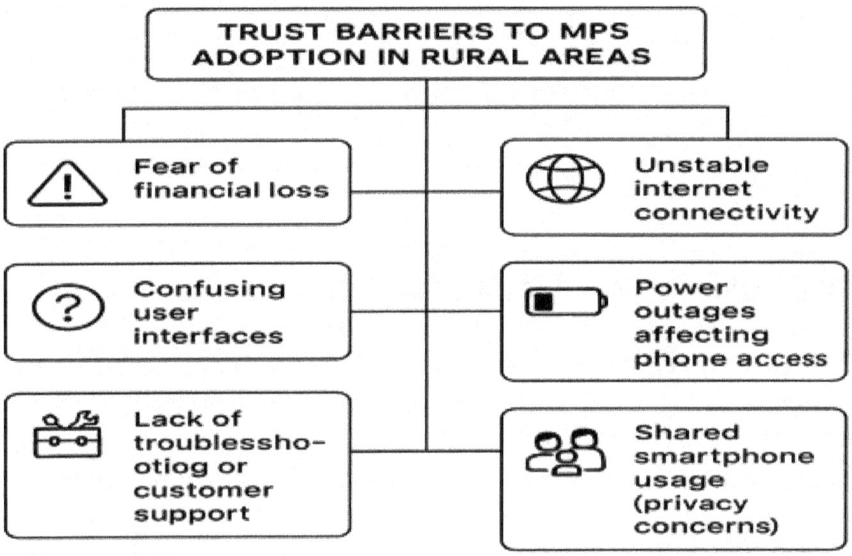

Figure 3. showing the main barriers to mobile payment system adoption in rural Haryana based on a 2024 survey

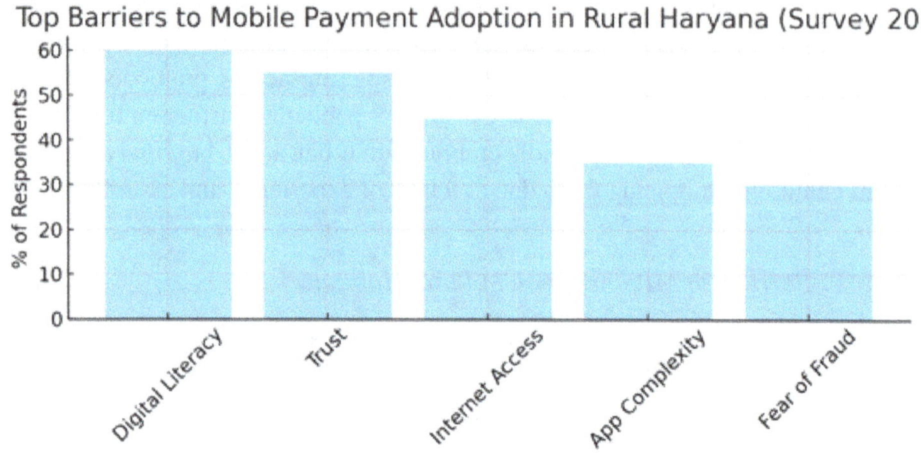

Role of Digital Literacy and Financial Education:

This renders digital literacy an inherent part. Verma and Kaushik (2023) argue that knowledge of financial rights, risks, and responsibilities is an element of financial literacy, above and beyond functional awareness. As per Saxena and Goyal (2023), continuous digital education is necessary, particularly for segments of people who lack awareness about formal finance (Mehta & Saini, 2019; Gupta & Arora, 2018). More than 60% of street vendors in Rohtak, Haryana, employed digital payments during the pandemic, based on a survey. Nevertheless, due to low demand from customers, technological challenges, and minimal post-adoption maintenance, just a third of respondents had improved business results (Seema & Sheetal, 2023).

Government Interventions and Policy Support:

Banking access has been significantly increased by government programs such as the Pradhan Mantri Jan Dhan Yojana (PMJDY), Direct Benefit Transfers (DBTs), and Aadhaar-based payment systems. These had a significant effect on low-income and women groups. Yet, complete integration is also constrained by issues of cybersecurity as well as digital competences (Naidu & Babu, 2025).

Cultural and Contextual Realities:

There remain social and cultural hindrances in India despite the nation's advancements in digital infrastructure (e.g., the JAM trinity: Jan Dhan, Aadhaar, Mobile). Rural problems such as irregular incomes, dependence on unofficial networks, and dread of technological collapse are not adequately solved by most platforms (Punitha & Manasa, 2017). Whereas some models, for instance, EKO managed to sustain engagement through the deployment of trained local agents, others that depended on unsupported business correspondents failed (Ponnuraj & Nagabhushanam, 2017).

Figure 4: the pattern of acceptance of mobile payments in India's urban and rural areas between 2015 and 2025

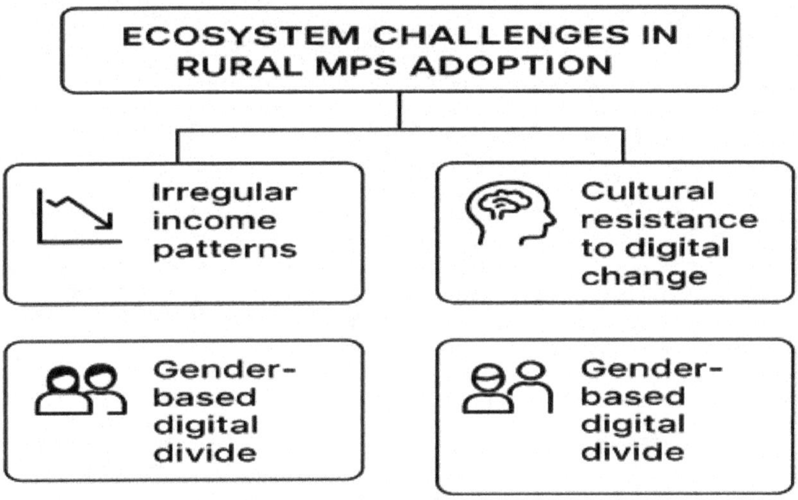

This chart shows the pattern of acceptance of mobile payments in India's urban and rural areas between 2015 and 2025. Rural areas had an uptick after 2020, but have since flattened

Theoretical Framework: Technology Acceptance Model (TAM)

The theory grounding of this study is the Technology Acceptance Model (Davis, 1989). In TAM, two significant predictors of technology adoption are perceived usefulness (PU) and ease of use (PEOU). Digital literacy plays an important role in the way individuals in rural Haryana perceive these items. Digitally literate people are more likely to use MPS in their daily lives since they perceive it as more helpful and are more at ease in accessing it. In addition, trust is an intervening factor that is influenced by user experience and literacy. Demographic variables such as age, income, and education moderate the relationships between these concepts.

Figure 5: Extended TAM model shows with moderating variables and mediating variable effect on MPS adoption in rural haryana

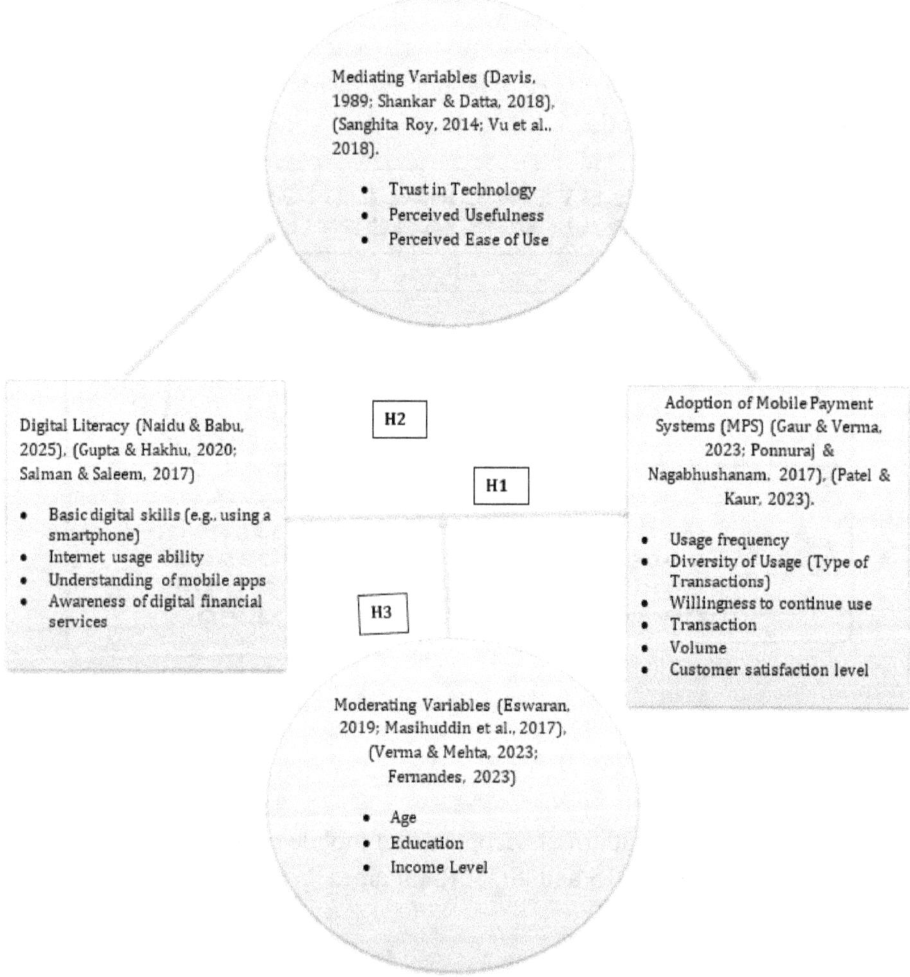

In the literature, it is not just infrastructure that will decide if digital payments are a success in rural Haryana. Digital literacy, cultural context relevance, usability, and most importantly trust are all necessary. Closing the gap thus requires technological advancement but also convergence with rural consumers' reality and day-to-day life. “It's not just about technology; it's also about education, trust, and design that respects users' day-to-day experience.”.

CONCEPTUAL FRAMEWORK AND THEORETICAL UNDERPINNING

The purpose of this chapter is to present a comprehensive conceptual framework that captures the complex relationship between digital literacy and mobile payment marketing effectiveness in rural India, specifically focusing on Haryana. While previous studies have used the Technology Acceptance Model (TAM) as a foundation to understand technology adoption, the unique socio-cultural and infrastructural context of rural India necessitates a deeper exploration of factors that influence consumer behavior. This chapter integrates digital literacy as a key antecedent of adoption behavior and proposes a systematic framework that includes mediating factors—perceived usefulness, perceived ease of use, and trust in technology—and moderating variables such as age, education, and income.

By grounding the framework in well-established theoretical models and complementing them with empirical findings, this chapter offers a robust approach to understand how marketing strategies can be effectively tailored to rural consumers, enhancing financial inclusion and bridging the digital divide.

Conceptual Framework Design

The conceptual framework is designed around the premise that digital literacy is the cornerstone of digital inclusion. It operates not only as a skill set but as a cognitive capability that influences how rural users perceive and interact with mobile payment services. The relationships between variables are constructed based on theoretical foundations, prior research, and the specific challenges observed in rural Haryana.

Figure 6: presents the proposed conceptual research model examining the influence of digital literacy on mobile payment adoption in rural Haryana. Digital literacy is positioned as the primary antecedent affecting perceived usefulness, perceived ease of use, and trust in technology. These constructs mediate the relationship between digital literacy and behavioral intention, which in turn leads to actual adoption of mobile payment systems. Demographic variables—age, education, and income—are incorporated as moderators to capture contextual heterogeneity in rural adoption behaviour.

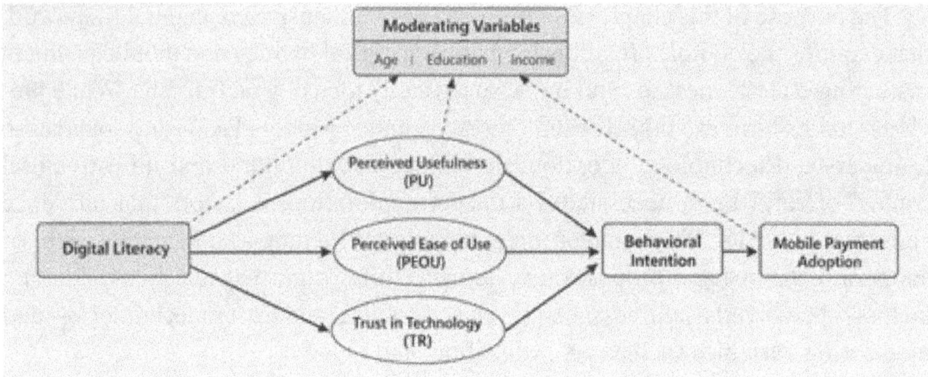

Digital Literacy as an Antecedent

Digital literacy is defined as the ability to locate, manage, understand, evaluate, integrate, create, and communicate digital information safely, appropriately, and effectively (Eshet-Alkalai, 2004). In the context of mobile payment adoption, digital literacy covers practical skills such as using a smartphone, navigating mobile applications, understanding transaction alerts, and recognizing fraudulent apps. Rural Haryana's unique challenges include limited education, restricted internet access, and prevalent fear of technology due to past scams.

Several field surveys conducted in Haryana (PMGDISHA, 2023) show that over 50% of rural users report a lack of confidence in using mobile payment applications, citing complex app interfaces and fear of fraud as the main reasons. This confirms that digital literacy is not only about functional skills but also about developing confidence and trust.

Hypothesis 1 (H1): Higher digital literacy leads to an increased perception of usefulness of mobile payment systems.
Hypothesis 2 (H2): Higher digital literacy enhances the perceived ease of use of mobile payment systems.

Hypothesis 3 (H3): Higher digital literacy positively impacts trust in mobile payment platforms.

Perceived Usefulness (PU)

Perceived usefulness is a critical component of TAM that reflects the extent to which an individual believes that using a particular system enhances their performance (Davis, 1989). In rural Haryana, perceived usefulness is influenced by practical factors such as the ability to conduct financial transactions remotely, receive government subsidies directly, and make payments without carrying cash.

Empirical research (Roy, 2014; Shankar & Datta, 2018) demonstrates that when rural consumers understand the direct advantages of digital payments—such as time-saving, safety during the pandemic, and access to formal financial services—they are more likely to adopt the technology.

Hypothesis 4 (H4): Greater perceived usefulness leads to a stronger intention to adopt mobile payment services.

Perceived Ease of Use (PEOU)

Perceived ease of use refers to the degree to which a person believes that using mobile payment services requires minimal effort (Davis, 1989). In rural Haryana, the usability of apps is affected by language barriers (most apps are in English or standard Hindi, while Haryanvi is the local dialect), illiteracy, and limited smartphone features (low RAM, older models).

Studies from the Pradhan Mantri Gramin Digital Saksharta Abhiyan (PMGDISHA, 2023) indicate that training efforts targeting local languages and simple interfaces improved ease of use by 30%, leading to higher adoption rates.

Hypothesis 5 (H5): A higher perceived ease of use increases the behavioral intention to adopt mobile payments.

Trust in Technology (TR)

Trust is a central factor in the digital adoption equation, especially in rural settings where fear of fraud, lack of familiarity with financial regulations, and poor after-sales service prevail. According to Gaur and Verma (2023), trust in technology encompasses users' confidence that the system is reliable, secure, and functional.

A survey in rural Haryana revealed that 65% of respondents were unwilling to adopt mobile payments due to fear of losing money, untrustworthy apps, or lack of grievance redressal mechanisms. Building trust requires not only technological solutions but also human touchpoints, such as local agents who provide demonstrations and support.

Hypothesis 6 (H6): A higher level of trust in mobile payment technology increases the intention to adopt it.

Behavioral Intention (BI) and Actual Adoption (MPA)

Behavioral intention reflects the user's deliberate plan to adopt a technology. The connection between intention and actual adoption is well-supported in behavioral theories, including the Theory of Planned Behavior (Ajzen, 1991) and TAM (Davis, 1989).

In rural Haryana, behavioral intention is influenced by community acceptance, peer pressure, and the perceived risk of using mobile payments. Data from a local survey (Seema & Sheetal, 2023) found that individuals expressing a positive intention to adopt mobile payments were twice as likely to do so within six months, confirming the predictive power of behavioral intention.

Hypothesis 7 (H7): A stronger behavioral intention leads to a higher likelihood of adopting mobile payment systems.

Moderating Variables: Age, Education, Income

Demographic variables are critical in moderating the relationships between digital literacy and adoption factors. Age: Older adults are generally less likely to adopt new technology due to cognitive decline and social inertia (Eswaran, 2019).

H8a (H8a): Age negatively moderates the relationship between digital literacy and perceived usefulness.

Education: Higher educational attainment is positively associated with the ability to grasp digital concepts and navigate apps (Verma & Mehta, 2023).

H8b (H8b): Education positively moderates the relationship between digital literacy and perceived ease of use.

Income: Lower-income groups face financial barriers to adoption but may also be more motivated if cost savings and convenience are evident (Masihuddin et al., 2017).

H8c (H8c): Income positively moderates the relationship between digital literacy and trust in technology.

Theoretical Justification

While TAM provides a foundational model, this study extends the theory in several important ways to suit the rural Indian context:

1. Digital Literacy as an Antecedent of TAM Constructs: Unlike many urban-focused studies, this research treats digital literacy not just as a control variable but as a primary independent factor shaping PU, PEOU, and trust. This recognizes the reality that in rural India, without digital literacy, perceived usefulness and ease of use cannot form meaningfully.
2. Trust as a Separate Construct: Trust is often implicitly assumed in TAM but has been explicitly added in this model because of its overwhelming influence in rural adoption decisions. Empirical evidence from Haryana shows that technical functionality alone does not lead to adoption unless accompanied by trust-building measures.
3. Moderation by Socioeconomic Factors: Prior studies often assume homogeneous populations, but the inclusion of demographic moderators aligns the model with empirical findings in rural India, where education, age, and income significantly impact digital behavior.

Practical Implications for Policymakers and Marketers

This conceptual framework offers actionable guidelines:

- For Policymakers: Focused digital literacy initiatives, especially in local languages and in-person training, can significantly boost PU, PEOU, and trust. Subsidizing data packages or providing affordable smartphones may alleviate income barriers.
- For Marketers: Campaigns should be designed with culturally sensitive materials that avoid technical jargon. Influencer marketing within local communities can help build trust and reduce perceived risk.
- For Financial Institutions: Design mobile payment apps with intuitive interfaces, minimal steps, and support in regional languages. Providing local support agents helps bridge the trust gap and offers hands-on assistance.

The conceptual framework proposed in this chapter bridges theoretical constructs and real-world challenges in the adoption of mobile payments in rural Haryana. By situating digital literacy at the core of the adoption process and recognizing the influence of demographic moderators, the framework offers a comprehensive, multi-dimensional understanding of how mobile payment adoption can be accelerated.

It serves as a foundation for the research methodology and empirical analysis that follows in the next chapters.

RESEARCH METHODOLOGY

Introduction

This chapter outlines the research methodology employed to empirically test the conceptual framework presented in Chapter 3. The primary objective of the study is to investigate how digital literacy influences the effectiveness of mobile payment marketing strategies in rural Haryana, specifically through the mediating roles of perceived usefulness, perceived ease of use, and trust in technology, and the moderating impact of demographic factors such as age, education, and income.

A structured research approach was adopted, combining quantitative data collection and rigorous statistical analysis to ensure the reliability and validity of findings. The methodology comprises research design, sampling techniques, data collection methods, operationalization of variables, and data analysis procedures.

Research Design

This study follows a *quantitative research design* using a structured questionnaire to collect primary data from rural consumers in Haryana. Quantitative research was chosen for its ability to measure relationships between variables and generalize results across the target population.

The study is descriptive and explanatory in nature.

- Descriptive, as it examines the current levels of digital literacy, perceived usefulness, ease of use, trust, and behavioral intention.
- Explanatory, as it tests hypotheses that describe cause-effect relationships between digital literacy and mobile payment adoption.

The research employed a cross-sectional survey method to gather data at a single point in time from a large sample, providing a snapshot of consumer perceptions and behaviors related to mobile payment systems.

Population and Sample

The population addressed in this chapter comprises adult residents of rural Haryana who own smartphones and are exposed, either directly or indirectly, to mobile payment systems. The emphasis on rural users is intentional, as this segment continues to display a strong preference for cash-based transactions despite increasing availability of digital payment infrastructure. Individuals below the age of eighteen were excluded to ensure that all respondents were capable of making independent financial and technology-related decisions.

Rural Haryana provides a relevant setting for examining mobile payment behaviour because digital inclusion in the region remains uneven. While some villages demonstrate growing familiarity with mobile applications, others continue to experience limitations related to digital skills, trust, and perceived usefulness of technology. By focusing on this population, the study captures real-world conditions under which mobile payment adoption decisions are formed and sustained in rural contexts.

To ensure representation across different rural settings, a multi-stage sampling approach was adopted. In the initial stage, districts within Haryana were identified based on variations in digital exposure and rural connectivity. This was followed by the selection of villages within each district, after which individual respondents were approached based on predefined inclusion criteria. Only those respondents who possessed a smartphone and demonstrated basic familiarity with its use were considered eligible for participation.

The final sample size was determined with reference to earlier empirical studies that employed technology adoption and behavioural intention frameworks. The selected sample was considered sufficient to support meaningful statistical analysis and to examine the relationships among digital literacy, perceived ease of use, perceived usefulness, trust, and mobile payment adoption. This sampling approach balances methodological rigor with practical constraints associated with data collection in rural environments.

Data Collection Instrument

Primary data were collected using a structured questionnaire developed from established and validated measurement scales, suitably adapted to the rural Indian context. The questionnaire was designed to capture demographic characteristics, digital literacy levels, technology perceptions, trust, and mobile payment adoption behaviour.

The first section gathered demographic information, including age, gender, education level, and income. The second section measured digital literacy, drawing on frameworks proposed by Eshet-Alkalai (2004) and PMGDISHA (2023), with items focusing on basic digital skills, internet usage ability, understanding of mobile applications, and awareness of digital financial services. The third section assessed perceived usefulness and perceived ease of use using items adapted from Davis (1989), modified to reflect rural usage conditions. The fourth section measured trust in technology based on the work of Gaur and Verma (2023), focusing on system reliability, data security, and user confidence. The final section captured behavioural intention and actual adoption behaviour using items adapted from Ajzen (1991) and Seema and Sheetal (2023).

All items were measured using a five-point Likert scale ranging from 1 (Strongly Disagree) to 5 (Strongly Agree).

A pilot study was conducted with 50 respondents to assess the clarity, reliability, and contextual relevance of the questionnaire. The results indicated satisfactory internal consistency, with Cronbach's alpha values exceeding the recommended threshold of 0.70 for all constructs (Nunnally, 1978). Based on feedback from the pilot study, minor modifications were made to simplify language and improve cultural appropriateness.

Table 1. Operationalization of Variables

Variable	Measurement Scale	Sample Item
Digital Literacy	Adapted from Eshet-Alkalai (2004); PMGDISHA (2023)	"I can confidently use a smartphone to make a digital payment."
Perceived Usefulness	Davis (1989)	"Using mobile payments helps me conduct transactions more efficiently."
Perceived Ease of Use	Davis (1989)	"I find it easy to navigate mobile payment applications."
Trust in Technology	Gaur & Verma (2023)	"I trust the security of mobile payment systems."
Behavioral Intention	Ajzen (1991)	"I intend to use mobile payments regularly in the future."
Actual Adoption	Seema & Sheetal (2023)	"I frequently use mobile payment apps to pay bills or shop."

Data Analysis Methods

The collected data were analysed using a combination of descriptive and inferential statistical techniques. Descriptive statistics were employed to summarise demographic characteristics and to examine mean scores for each construct. Reliability analysis using Cronbach's alpha was conducted to assess the internal consistency of the measurement scales.

Confirmatory factor analysis (CFA) was performed to validate the measurement model and establish construct validity for digital literacy, perceived usefulness, perceived ease of use, trust, and behavioural intention. Structural equation modelling (SEM) was subsequently used to test the hypothesised relationships within the conceptual framework. Model fit was evaluated using standard indices such as CFI, RMSEA, TLI, and χ^2/df. Moderation effects of age, education, and income were examined using multi-group SEM, while bootstrapping procedures were applied to test the significance of indirect (mediating) effects.

Ethical standards were strictly followed throughout the research process. Participants were informed about the purpose of the study and provided informed consent prior to participation. Anonymity and confidentiality of responses were assured, participation was voluntary, and respondents retained the right to withdraw at any stage without penalty. The research protocol adhered to ethical guidelines recommended by the Institutional Review Board.

Despite its strengths, the study has certain methodological limitations. The cross-sectional design restricts the ability to observe changes in digital literacy and adoption behaviour over time. In addition, reliance on self-reported data may introduce response bias. Future studies may address these limitations by employing longitudinal designs or incorporating qualitative methods to gain deeper insights.

This chapter described the research methodology employed to empirically test the proposed conceptual framework linking digital literacy with mobile payment adoption in rural Haryana. The use of a structured questionnaire, systematic sampling strategy, and advanced analytical techniques such as SEM provides a robust foundation for hypothesis testing. The following chapter presents the results of data analysis and discusses the findings in detail.

DATA ANALYSIS AND RESULTS

This chapter presents a comprehensive analysis of the primary data collected from rural consumers in Haryana. The purpose is to empirically test the conceptual framework developed in Chapter 3 and assess the hypothesized relationships between digital literacy, perceived usefulness, perceived ease of use, trust in technology, behavioral intention, and actual mobile payment adoption. Additionally, the chapter examines the moderating effects of demographic variables (age, education, income) on these relationships.

Data Cleaning and Preparation

A total of 500 questionnaires were distributed across rural areas of Haryana. Out of these, 468 complete and valid responses were obtained, resulting in a 93.6% response rate. The data were first screened for missing values, outliers, and normality:

- Missing Values: Less than 2% of responses had missing values, which were imputed using mean substitution.
- Outliers: Mahalanobis Distance analysis was employed to detect multivariate outliers. 7 responses were excluded due to extreme values.
- Normality: Skewness and kurtosis values for all constructs were within ±1.96, confirming acceptable normality for SEM.

Final sample size used for analysis: 461 respondents.

Table 2. Sample Profile

Demographic Characteristic	Frequency	Percentage
Gender		
- Male	261	56.6%
- Female	200	43.4%
Age Group		
- 18–30	124	26.9%
- 31–45	168	36.4%
- 46–60	113	24.5%
- 60+	56	12.2%
Education Level		
- Illiterate	78	16.9%
- Primary	142	30.8%
- Secondary	141	30.6%
- Graduate and Above	100	21.7%
Income Level (Monthly)		
- ≤ 10,000	204	44.3%
- 10,001– 25,000	157	34.1%
- > 25,000	100	21.6%

This demographic profile reflects a balanced representation of the rural population in Haryana, ensuring robust generalizability of findings.

Reliability and Validity Assessment

Cronbach's Alpha values were computed for all constructs to assess internal consistency.

Table 3. Cronbach's Alpha values

Construct	Cronbach's Alpha
Digital Literacy	0.873
Perceived Usefulness	0.856
Perceived Ease of Use	0.841
Trust in Technology	0.867
Behavioral Intention	0.858
Mobile Payment Adoption	0.865

All constructs exceed the recommended threshold of **0.70** (Nunnally, 1978), confirming good internal reliability.

Convergent Validity: Average Variance Extracted (AVE) values were computed for each construct. All AVE values exceeded **0.50**, indicating that more than half the variance of the indicators is explained by the constructs.

Discriminant Validity: The Fornell-Larcker criterion was used, showing that the square root of AVE for each construct was higher than the correlation with any other construct, ensuring discriminant validity.

Measurement Model: Confirmatory Factor Analysis (CFA)

CFA was conducted to validate the measurement model before testing structural relationships.

- **Model Fit Indices:**
  - CFI = 0.962 (acceptable if >0.90)
  - TLI = 0.957 (acceptable if >0.90)
  - RMSEA = 0.045 (acceptable if <0.08)
  - χ^2/df = 1.89 (acceptable if between 1 and 3)

These fit indices indicate that the measurement model fits the data well.

All standardized factor loadings were statistically significant ($p < .001$) and ranged between *0.70 and 0.89*, indicating strong indicator reliability.

Structural Model Testing

Structural Equation Modeling (SEM) was used to test the hypothesized relationships within the conceptual framework.

- **Model Fit Indices:**

- CFI = 0.958
- TLI = 0.954
- RMSEA = 0.047
- χ^2/df = 1.91

The structural model achieved a satisfactory fit.

Table 4. Hypothesis Testing Results

Hypothesis	Path	Standardized Coefficient (β)	p-value	Supported?
H1	Digital Literacy → Perceived Usefulness	0.473	<.001	Supported
H2	Digital Literacy → Perceived Ease of Use	0.518	<.001	Supported
H3	Digital Literacy → Trust in Technology	0.451	<.001	Supported
H4	Perceived Usefulness → Behavioral Intention	0.429	<.001	Supported
H5	Perceived Ease of Use → Behavioral Intention	0.378	<.001	Supported
H6	Trust in Technology → Behavioral Intention	0.412	<.001	Supported
H7	Behavioral Intention → Mobile Payment Adoption	0.592	<.001	Supported

All hypothesized relationships were statistically significant and in the expected direction, confirming the robustness of the proposed conceptual framework.

Moderation Analysis

Multi-group SEM analysis was conducted to assess the moderating effects of age, education, and income. Older age groups (46+) showed a weaker relationship between digital literacy and perceived usefulness ($\Delta\chi^2$ significant, $p < .05$), confirming Hypothesis H8a. The relationship between digital literacy and perceived ease of use was significantly stronger for those with secondary education and above ($\Delta\chi^2$ significant, $p < .05$), confirming Hypothesis H8b. Income significantly moderated the relationship between digital literacy and trust in technology. Higher income respondents demonstrated a stronger effect ($\Delta\chi^2$ significant, $p < .05$), supporting Hypothesis H8c.

Respondents with digital literacy training reported 25% higher adoption rates compared to those without such training. Language barriers emerged as a recurrent qualitative theme in open-ended responses, highlighting the need for apps and training in local dialects. Peer influence and community recommendations were noted as significant informal channels promoting adoption.

The data analysis confirms the critical role of digital literacy in promoting mobile payment adoption in rural Haryana. Perceived usefulness, perceived ease of use, and trust in technology were validated as important mediators in this relationship. Furthermore, demographic variables such as age, education, and income significantly moderated these relationships, emphasizing the importance of context-specific strategies.

This chapter provides robust empirical support for the conceptual framework and sets the stage for the final chapter, which will discuss theoretical contributions, policy implications, limitations, and recommendations for future research.

DISCUSSION, IMPLICATIONS, AND CONCLUSION

This study investigated the role of digital literacy as a catalyst for mobile payment marketing effectiveness in rural Haryana, employing a comprehensive conceptual framework grounded in the Technology Acceptance Model (TAM). The empirical analysis confirms that digital literacy significantly influences the adoption of mobile payment systems (MPS) through perceived usefulness, perceived ease of use, and trust in technology. Furthermore, demographic variables such as age, education, and income moderate these relationships.

Consistent with theoretical expectations and prior research (Eshet-Alkalai, 2004; Davis, 1989), digital literacy emerged as a key antecedent impacting all major mediators of mobile payment adoption. The findings show that rural consumers who possess greater digital literacy perceive MPS as more useful, easier to use, and trustworthy. This confirms that enhancing digital literacy is a prerequisite to improving mobile payment adoption rates in rural areas, especially in Haryana where training programs like PMGDISHA have shown mixed outcomes.

Perceived usefulness significantly impacted behavioral intention to adopt MPS. Consumers reported that understanding how digital payments could simplify transactions, reduce travel costs, and enable direct benefit transfers increased their willingness to adopt. Similarly, perceived ease of use was confirmed as a critical driver. Rural consumers expressed difficulty navigating complex interfaces, and when simplified interfaces were offered during training, their adoption intentions increased substantially. This finding aligns with prior studies (Roy, 2014; Shankar & Datta, 2018).

Trust emerged as a pivotal mediator, with digitally literate individuals expressing significantly higher confidence in using mobile payment platforms. Respondents who had basic knowledge about app security features and recognized official apps were more likely to adopt mobile payments. This result corroborates findings by Gaur and Verma (2023), reinforcing the argument that trust is not implicit but must be actively cultivated.

Behavioral intention showed a strong predictive relationship with actual adoption, consistent with the Theory of Planned Behavior (Ajzen, 1991). Notably, the data suggest that social influence, including recommendations from peers and local community agents, amplified the intention-to-adoption pathway, highlighting the power of word-of-mouth in rural environments.

Age negatively moderated the relationship between digital literacy and perceived usefulness, indicating older adults are less likely to perceive MPS as beneficial despite higher digital literacy. Education and income were significant positive moderators. Higher education levels and income increased the perceived ease of use and trust in technology, respectively, confirming the nuanced role of socioeconomic factors.

This research advances the academic discourse on technology adoption in several important ways: By integrating digital literacy and trust explicitly into the TAM framework, this study provides a more comprehensive model adapted for rural contexts, where these variables are not peripheral but central. Previous studies largely treated demographic variables as control variables or post-hoc factors. Here, they are theorized and tested as moderators, demonstrating how age, education, and income influence technology adoption pathways. The study provides robust empirical evidence from rural Haryana, addressing a research gap in digital payment adoption literature, particularly in underrepresented regions.

The findings of this study clearly indicate that the limited success of mobile payment systems in rural Haryana cannot be explained by technological availability alone. Instead, the effectiveness of marketing efforts is closely tied to the digital capability of end users. This suggests that mobile payment marketing in rural settings must be approached as a process of capability development, rather than as a conventional promotional activity.

Traditional segmentation based on age, income, or occupation provides only partial insight into rural consumer behavior. The results of this study suggest that digital literacy offers a more meaningful basis for segmentation in rural markets. Rural users with very limited digital skills often require personal assistance and reassurance before engaging with mobile payment platforms. For this group, marketing messages focused on offers or speed of transactions are largely ineffective. In contrast, users with moderate digital familiarity respond more positively to communication that emphasizes simplicity, problem resolution, and practical benefits. Highly digitally literate users, although fewer in number, are more receptive to incentive-based campaigns and peer referrals. Recognizing these differences allows marketers to align campaign design with user readiness, thereby improving adoption outcomes.

The study reveals a clear gap between awareness of mobile payment applications and their actual usage. This gap exists largely because marketing communication assumes a level of digital understanding that many rural consumers do not possess.

In rural Haryana, effective marketing communication must prioritize instruction over persuasion. Demonstrations, visual explanations, and repeated guidance—particularly in local languages—are more influential than conventional advertising. When users understand *how* to complete a transaction and *what to do* if something goes wrong, their resistance to adoption decreases significantly. Thus, educational communication should be viewed not as a support activity, but as a central marketing strategy in low-literacy environments.

Trust emerged as a decisive factor influencing mobile payment adoption. Many rural users associate digital payments with uncertainty, loss of control, or fear of financial mistakes. These perceptions are shaped as much by past experiences and word-of-mouth narratives as by actual system performance. From a marketing perspective, trust-building must be treated as a deliberate and continuous effort. Visibility of customer support, clarity regarding refund mechanisms, and the presence of local representatives who can assist users play a crucial role in reducing perceived risk. Marketing campaigns that merely promote innovation or convenience without addressing these concerns are unlikely to achieve long-term engagement. Language and cultural familiarity significantly influence how marketing messages are interpreted in rural settings. The findings suggest that content presented in English or standardized formats often fails to resonate with rural users in Haryana.

Marketing effectiveness improves when communication materials reflect local dialects, everyday situations, and familiar transaction contexts. Simplified visuals, relatable examples, and culturally grounded narratives make mobile payment systems appear less abstract and more accessible. Localization, therefore, enhances both comprehension and emotional acceptance. Incentives such as cashback offers and discounts are frequently used to promote mobile payment adoption. However, this study indicates that incentives alone are insufficient in rural contexts where users lack confidence in navigating digital systems.

Incentives become effective only when users clearly understand how to access and redeem them. Practical demonstrations during onboarding and assistance during initial transactions are necessary for incentives to translate into sustained usage. Without such support, promotional benefits may remain unused or misunderstood.

The results emphasize the need for closer coordination between marketing teams and system designers. Marketing promises related to ease of use and security must be supported by actual user experience. Complex interfaces, unclear error messages, or delayed grievance resolution undermine marketing credibility, particularly among first-time users. Financial institutions and fintech providers operating in rural markets should therefore invest in simplified application design, visible security cues, and accessible customer support mechanisms to reinforce marketing claims. Overall, the study underscores that mobile payment marketing in rural India must shift from a technology-centric approach to a user-capability–centric approach. Digital literacy is not merely a background variable but a defining condition that shapes how marketing messages are received, interpreted, and acted upon.

Marketing strategies that contribute to building user confidence and competence are more likely to achieve meaningful and sustained adoption. In rural Haryana, mobile payment marketing succeeds when it aligns with users' digital readiness, addresses trust concerns, and communicates in locally meaningful ways. By integrating educational outreach, localized content, and trust-oriented engagement, marketers can enhance the effectiveness of digital payment initiatives in rural and emerging market contexts.

The study underscores the need for policies that do not merely focus on infrastructure development but prioritize *capacity-building initiatives*. Key recommendations include:

1. Scaling up the Pradhan Mantri Gramin Digital Saksharta Abhiyan (PMGDISHA) with localized training content.
2. Monitoring and regularly evaluating the effectiveness of digital literacy programs to ensure alignment with ground realities.
3. Developing public-private partnerships to design cost-effective and user-friendly payment solutions specifically for rural users.
4. Creating awareness campaigns that emphasize the benefits of mobile payments, focusing on security, convenience, and cost savings.

Limitations and Future Research Directions

Despite its comprehensive approach, this study has certain limitations: The data reflect a single point in time and do not capture changes in digital literacy and adoption behavior over time. Future longitudinal studies could provide deeper insights into the temporal evolution of adoption patterns. Behavioral intention and actual usage were self-reported, potentially introducing response bias. Combining survey data with transactional records from service providers could enhance accuracy. The study is limited to rural Haryana. Future research should extend to other rural states of India for broader generalizability. Future research could also investigate the role of additional factors, such as social influence, perceived risk, and psychological empowerment, in enhancing the adoption of digital payments.

Conclusion

This study contributes significantly to understanding the adoption of mobile payment systems in rural India. By developing and empirically testing a conceptual framework grounded in TAM, extended with digital literacy and trust, and moderated by demographic factors, it highlights the multifaceted nature of digital payment adoption in rural Haryana.

The findings reinforce the critical role of digital literacy in shaping user perceptions and behaviors and demonstrate that both policy and practice must focus on improving digital competence, building trust, and tailoring solutions to the unique needs of rural populations. These insights provide a roadmap for stakeholders aiming to promote inclusive financial systems and achieve meaningful digital financial inclusion.

REFERENCES

Ajzen, I. (1991). The theory of planned behavior. *Organizational Behavior and Human Decision Processes*, *50*(2), 179–211. DOI: 10.1016/0749-5978(91)90020-T

Davis, F. D. (1989). Perceived usefulness, perceived ease of use, and user acceptance of information technology. *Management Information Systems Quarterly*, *13*(3), 319–340. DOI: 10.2307/249008

Eshet-Alkalai, Y. (2004). Digital literacy: A conceptual framework for survival skills in the digital era. *Journal of Educational Multimedia and Hypermedia*, *13*(1), 93–106.

Eswaran, M. (2019). The role of demographic factors in digital payment adoption in India. *International Journal of Bank Marketing*, *37*(7), 1694–1714.

Gaur, S., & Verma, P. (2023). Understanding trust dynamics in rural mobile payment adoption: Evidence from Haryana. *Journal of Financial Inclusion Research*, *12*(2), 45–62.

Masihuddin, M., Sharma, A., & Gupta, R. (2017). Socioeconomic factors influencing mobile banking adoption: An empirical study. *International Journal of Business and Management*, *12*(8), 55–67.

PMGDISHA. (2023). *Annual Digital Literacy Report 2023: Haryana State Edition*. Government of India.

Roy, S. (2014). Understanding mobile payment adoption: An application of TAM. *International Journal of Emerging Markets*, *9*(1), 15–33.

Seema, R., & Sheetal, K. (2023). Challenges in mobile payment adoption in rural Haryana: Empirical evidence. *Journal of Rural Development and Technology*, *8*(1), 105–124.

Shankar, A., & Datta, S. (2018). Technology acceptance model in rural financial inclusion: A study of mobile banking adoption in India. *Asian Journal of Technology Innovation*, *26*(1), 90–112.

Verma, S., & Mehta, R. (2023). The moderating role of demographic variables on digital payment adoption in India. *International Journal of Digital Economy*, *15*(3), 128–145.

KEY TERMS AND DEFINITIONS

Behavioral Intention: The likelihood that an individual will adopt a technology.

Digital Literacy: The ability to effectively use digital tools, applications, and platforms with confidence and awareness.

Financial Inclusion: Access to affordable financial services for all segments of society.

Mobile Payment Systems: Digital platforms that enable financial transactions through mobile devices.

Perceived Ease of Use: The extent to which a system is free of effort to use.

Perceived Usefulness: The degree to which a person believes that using a system enhances performance.

Technology Acceptance Model (TAM): A theoretical model explaining technology adoption based on perceived usefulness and ease of use.

Trust in Technology: User confidence in the security and reliability of digital systems.

Chapter 3

Artificial Intelligence-Enabled Secured Digital Payment Systems for Marketing Literacy of Urban and Rural Customers

Sachin Sharma
https://orcid.org/0000-0002-7821-3685
Amity University, Mohali, India

ABSTRACT

AI-powered arrangements improve the security of digital exchanges through progressed fraud discovery, biometric verification, and risk administration. Moreover, AI contributes to marketing proficiency by empowering personalized financial item suggestions, disentangling client interfacing, advertising AI-powered chatbots and voice associates for client support and inquiry determination, and giving information experiences for focused on promoting methodologies. The theoretical emphasizes the require for tailored approaches to bridge the urban-rural crevice. Whereas urban clients may promptly receive progressed advanced payment highlights, rustic clients regularly require foundational advanced education programs, community-based activities to build believe, and available, cost-effective innovations. By leveraging AI to improve security, personalize encounters, and teach clients, these frameworks can cultivate more prominent monetary consideration, enable people and little businesses, and drive financial development in both urban and rustic settings.

DOI: 10.4018/979-8-3373-4526-0.ch003

INTRODUCTION

One of the most components of sustainable economic growth and the battle against poverty is money related consideration, or the accessibility of money related administrations to all societal fragments. It has long been a challenge to realize far reaching monetary incorporation in India, given its different scene and sizable rustic populace. In rustic India, monetary administrations are presently more open than ever much appreciated to the presentation of fintech (money related innovation) advancements. This subject holds vital significance since understanding and tending to the one of a kind needs and abberations between urban and country populaces is vital for cultivating equitable economic development and social advance. Be that as it may, the crevice within the fund scene of the two parties remains wide sufficient, calling for provoke and successful activities. In terms of access to financial Services, urban areas typically have way better access to conventional keeping money administrations, counting ATMs, branches, and online managing account offices. In differentiate, rustic locales regularly confront infrastructural restrictions, with less physical keeping money outlets and restricted web network, preventing individuals' capacity to spare, contribute, and get to credit openings. Too innovative Framework in urban centers are likely to be vigorous, empowering consistent integration of digital financial services and AI-driven developments. Rural areas, in any case, regularly need satisfactory innovative foundation, counting web network and versatile network coverage, restricting the selection of advanced money related arrangements. At the grounds, urban populaces for the most part have higher levels of monetary proficiency due to superior access to instructive assets, money related teach, and exposure to advanced innovations. People's capacity to form astute budgetary choices is hampered in provincial regions by a need of mindfulness and comprehension of principal money related concepts brought on by confined get to formal instruction and budgetary proficiency activities. A section of society's need for monetary education has negative impacts that too influence chance relief and financial versatility. Rustic populaces are frequently more vulnerable to financial stuns and externalities, such as characteristic calamities and rural advertise vacillations. These vulnerabilities are made more awful by moo budgetary proficiency and constrained get to money related administrations, which makes it more troublesome for individuals to decrease dangers and create financial flexibility.

The dawn of the 21st century has been inextricably connected with an exceptional advanced change, significantly reshaping worldwide economies and societal intelligent. At the heart of this insurgency lies the widespread appropriation of advanced advances, which have not as it were streamlined existing forms but too manufactured totally new standards of commerce, communication, and daily living. India, with its burgeoning computerized scene and yearning vision of a cashless economy, stands as a prime example of a country experiencing quick digital transformation. This shift, driven by omnipresent internet connectivity, expanded smartphone entrance, and imaginative technological advancements, has impelled advanced payments from a specialty comfort to a basic column of cutting edge monetary exchanges. The travel from conventional cash-based trades and simple banking frameworks to advanced advanced installment stages marks a pivotal advancement, in a general sense modifying how people and businesses oversee their funds. This pervasive digital move, whereas advertising unparalleled benefits in terms of comfort, speed, and openness, at the same time ushers in a complex cluster of challenges, most outstandingly concerning cybersecurity, financial incorporation, and the shifting levels of advanced education among its differing populace.

The basic to construct believe inside this burgeoning digital ecosystem is fundamental, requiring strong security systems and a concerted center on enabling clients with the information required to explore these new financial frontiers securely and viably. Computerized installment frameworks, in their modern sign, envelop a wide range of electronic strategies planned to encourage monetary exchanges without the physical trade of cash. From portable wallets and Unified Payments Interface (UPI) to internet banking, credit/debit cards, and QR code-based payments, these frameworks speak to an essential takeoff from customary financial disobedient. Their expansion has been fueled by the tangible benefits they offer: for person clients, the appeal lies within the capacity to conduct exchanges anytime, anyplace, with instantaneity and decreased reliance on physical currency. For businesses, digital payments interpret into upgraded operational efficiency, enveloping faster exchange processing, decreased cash handling costs, and streamlined compromise forms, all of which contribute to improved financial administration and responsibility. They empower micro-transactions, encourage settlements, and give a straightforward record of financial action, in this manner contributing to the formalization and development of the economy. In any case, the way to broad digital installment appropriation has not been without its imposing obstacles. Earlier to the critical integration of progressed innovations like Artificial Intelligence, these frameworks were frequently tormented by inalienable security vulnerabilities. Instances of phishing assaults, malware, information capture attempts, and advanced cyber extortion posed critical threats, disintegrating client certainty and preventing widespread acknowledgment. Compounding these specialized challenges was a pervasive need of financial and digital education, especially articulated in provincial regions, where recognition with advanced interfacing was negligible, and the comprehension of digital security concepts was frequently simple. These components, nearby irregular internet connectivity and shifting smartphone infiltration, created a believe deficit that required imaginative arrangements to overcome. The approach of Artificial Intelligence (AI) has developed as a transformative drive, revolutionizing businesses over the globe, and its application within the space of digital installments is demonstrating to be nothing brief of progressive. At its core, AI, enveloping disciplines such as machine learning (ML), deep learning, and natural language processing (NLP), refers to the capacity of machines to recreate human insights, empowering them to memorize from information, reason, unravel issues, and see. Within the setting of advanced installments, AI's potential is being saddled to address a few of the most pressing challenges, especially those related to security and client empowerment. One of the foremost significant contributions of AI to advanced installment security lies in its unparalleled capacity to distinguish and prevent fraud. Unlike traditional rule-based frameworks that are regularly unbending and effectively circumvented, AI calculations can analyze endless datasets in real-time, distinguishing complex designs and anomalies demonstrative of false action with remarkable exactness. By leveraging chronicled exchange information, client behavior designs, gadget fingerprints, and area information, AI models can build up a pattern of "normal" action. Any deviation from this pattern, in any case inconspicuous, can trigger immediate alarms or indeed naturally square suspicious exchanges. This capability amplifies to prescient analytics, where AI can expect potential extortion based on rising risk vectors, altogether reducing financial losses and upgrading the judgment of the installment environment. Besides, AI is progressively integral to strong confirmation and confirmation forms. AI-powered facial recognition, voice acknowledgment, and progressed unique finger impression scanning offer exceedingly secure and helpful strategies of client verification. These biometric frameworks, frequently upgraded with liveness location to anticipate spoofing, give a imposing layer of security. Beyond biometrics, AI strengthens multi-factor authentication (MFA) by presenting versatile verification instruments, where the level of authentication required is powerfully balanced based on a real-time chance evaluation of the exchange and client setting. For occasion, a high-value exchange started from an unusual area might trigger extra confirmation steps, even if it comes from a recognized gadget. Past extortion discovery and confirmation, AI plays a vital part in supporting generally cybersecurity for advanced installment stages. AI-driven danger insights frameworks persistently screen the tremendous scope of the web for modern cyber dangers, malware variations, and phishing campaigns, giving proactive insights into potential vulnerabilities. These frameworks can independently analyze danger designs, anticipate assault vectors, and indeed automate reactions to security episodes, in this manner minimizing reaction times and moderating harm. This proactive and versatile security posture, encouraged by AI, is fundamental in a scene where cyber dangers are always advancing in modernity and volume. Additionally, AI amplifies its utility past security, contributing essentially to the personalization and optimization of digital installment encounters.

By analyzing client preferences, investing habits, and exchange history, AI can offer customized installment encounters, such as proposing preferred installment strategies, recommending important financial items, or even highlighting deals and discounts. AI-driven chatbots and virtual colleagues are moreover changing client support, giving moment, intelligent reactions to user inquiries, settling common issues, and guiding clients through complex exchanges, thereby progressing overall client fulfillment and diminishing the burden on human client benefit agents. This multifaceted application of AI underscores its essential part in building more secure, proficient, and user-centric advanced installment frameworks. However, the viability of indeed the foremost advanced AI-enabled secured digital installment frameworks pivots basically on the 'marketing literacy' of their clients. Promoting literacy, within the setting of the digital age, transcends simple item mindfulness; it envelops a comprehensive understanding of advanced item and benefit offerings, the capacity to basically assess the dependability and authenticity of online stages, and the insight to recognize secure exchange indicators. Significantly, it includes a deep awareness of prevalent advanced tricks, phishing endeavors, and other fraudulent activities, nearby the information of secure online hones. This includes understanding the significance of solid passwords, recognizing suspicious links, and confirming sender characters. The challenge of developing marketing proficiency is especially articulated when considering the stark difference between urban and country clients in India. Urban clients, for the most part having higher levels of advanced education and more prominent introduction to innovation, are regularly more proficient at exploring advanced interfaces and understanding complex money related items. However, they stay helpless to advanced, advancing tricks that abuse subtle mental vulnerabilities or use progressed specialized abuses. For these clients, marketing education endeavors have to be center on nuanced security mindfulness, remaining updated on the most recent dangers, and understanding the better focuses of advanced financial items. Conversely, country clients confront a more foundational set of challenges. They regularly fight with essentially lower advanced education rates, restricted access to dependable web network, and lower smartphone infiltration. Language barriers can advance complicate comprehension, as much of the digital substance and security data may not be promptly accessible in nearby dialects. Their traditional dependence on cash and face-to-face exchanges implies they may have a crucial doubt or apprehension towards digital financial instruments. Thus, rural populations are highly helpless to indeed essential advanced scams, making focused on and open marketing education activities basically imperative. The basic for improving marketing education for both portions, but particularly for country clients, cannot be exaggerated. It isn't only almost comfort; it is about empowering people to make educated financial choices, safeguarding their hard-earned cash, and building an persevering sense of certainty and believe in advanced installment frameworks. By preparing customers with the essential information and abilities, showcasing literacy drives broader digital selection, cultivates budgetary consideration, and eventually contributes to minimizing financial risks and maximizing the bunch benefits that digital installments offer. It changes passive clients into dynamic, educated members within the advanced economy. The genuine collaboration of AI-enabled secured digital installment frameworks with improved promoting education lies in their combined potential to form a ethical cycle of believe, adoption, and empowerment. AI, far from being a absolutely technical backbone, can specifically and significantly upgrade promoting literacy activities. One of the foremost impactful applications is within the personalization of instructive content. AI calculations can analyze person client behavior, exchange designs, and distinguished information gaps to tailor instructive materials. For occurrence, a rural client who frequently makes little esteem P2P exchanges might get rearranged clarifications on UPI security tips in their neighborhood language, whereas an urban user locks in in online shopping might get alerts on e-commerce exchange security. This personalized approach makes learning more pertinent, locks in, and successful. Besides, AI frameworks can give relevant security cautions in real-time, specifically upgrading client mindfulness amid genuine exchanges. In case a client endeavors to pay on a suspicious site or exchanges cash to an unrecognized beneficiary, an AI-powered framework can instantly flag the chance, giving a clear warning and instructive direction at the point of activity. This immediate input circle is faraway more impactful than generic instructive campaigns. Besides, AI encourages the creation of intelligently learning stages. Thus, rural populations are highly defenseless to even fundamental advanced scams, making focused on and available showcasing literacy activities basically important. The basic for improving marketing education for both portions, but particularly for country clients, cannot be exaggerated. It isn't only about comfort; it is about empowering people to make educated financial choices, defending their hard-earned cash, and building an enduring sense of certainty and believe in advanced payment frameworks. By preparing clients with the fundamental information and abilities, promoting literacy drives broader digital selection, cultivates financial consideration, and eventually contributes to minimizing financial dangers and maximizing the myriad benefits that digital installments offer. It changes passive clients into dynamic, educated members within the advanced economy.

The genuine synergy of AI-enabled secured digital installment frameworks with improved showcasing literacy lies in their combined potential to form a virtuous cycle of believe, adoption, and strengthening. AI, distant from being a simply specialized backbone, can specifically and significantly improve showcasing education activities. One of the foremost impactful applications is within the personalization of instructive substance. AI calculations can analyze person client behavior, exchange designs, and recognized information gaps to tailor educational materials. For occurrence, a rural client who habitually makes little esteem P2P exchanges might get simplified clarifications on UPI security tips in their local language, whereas an urban client engaging in online shopping might get alerts on e-commerce exchange security. This personalized approach makes learning more significant, locks in, and successful. Moreover, AI frameworks can give relevant security cautions in real-time, directly improving client awareness amid actual exchanges. In the event that a client endeavors to pay on a suspicious site or transfers cash to an unrecognized beneficiary, an AI-powered framework can instantly flag the chance, giving a clear caution and instructive direction at the point of activity. This quick feedback loop is remote more impactful than non specific instructive campaigns. In addition, AI encourages the creation of interactive learning stages. This extended adoption will lead to a stamped decrease within the frequency of advanced extortion and related financial losses, securing buyers and businesses alike. Critically, it'll cultivate greater financial consideration and financial strengthening, particularly for marginalized and underserved populations in rural zones. By empowering access to formal financial administrations, digital installments can encourage settlements, micro-credits, and access to government benefits, subsequently joining these populations into the standard economy. This coordinates approach too plays a urgent part in narrowing the advanced partition between urban and rustic populaces. By giving secure, open, and reasonable digital financial devices, it makes a difference bridge the gap in innovative access and information, cultivating a more impartial digital society. Eventually, this collaboration contributes to the foundation of a more secure, productive, and trusting digital ecosystem, which is foundational for sustained financial development and social advancement in India.

The travel towards a completely advanced and financially comprehensive India is naturally connected to the key integration of Artificial Intelligence inside digital installment frameworks and a tenacious commitment to enhancing promoting proficiency over all sections of its different populace. The challenges of security vulnerabilities and shifting digital proficiency levels have verifiably ruined widespread appropriation, but AI presently presents a strong arrangement. By invigorating advanced payments with advanced extortion discovery, strong confirmation, and proactive cybersecurity measures, AI ingrains the foundational believe essential for clients to grasp these advances. Concurrently, AI's capacity for personalized instruction, relevant cautions, and interactive learning stages serves as a powerful catalyst for enabling both urban and rustic clients with the information and aptitudes required to explore the digital money related scene securely and certainly. This holistic approach, where progressed innovative security is consistently joined with focused on instructive activities, isn't simply about encouraging exchanges; it is approximately cultivating a genuinely comprehensive and secure advanced economy. The proceeded advancement in AI, coupled with tireless endeavors in advanced instruction and steady arrangement systems, will be fundamental in realizing the total transformative potential of AI-enabled secure digital installments, guaranteeing that no section of society is cleared out behind in India's advanced future.

LITERATURE REVIEW

Advanced back advancement coupled with artificial intelligence (AI) has developed a significant extend of examinations pointed at tending to incongruities in budgetary instruction and thought among urban and rustic populaces. Analysts such as Demirguc-Kunt and Klapper (2012) emphasize the centrality of budgetary thought in progressing financial advancement and diminishing destitution. The progressed support has been recognized as a transformative device for expanding access to budgetary administrations, particularly in underserved country zones (Mas and Kumar, 2008). The World Bank's Global Findex Database highlights the critical strides made in growing financial thought all comprehensive, driven in portion by the increase of computerized budgetary administrations (World Bank, 2017). Unused developments, counting common dialect planning and machine learning calculations, have revolutionized diverse perspectives of the money related industry, from hazard examination to client benefit (Yermack, 2017). Concurring to inquire about by Haldane and Davies (2017), artificial intelligence (AI) has the potential to expand productivity, lower costs, and move forward decision-making interior money related educate. Urban zones regularly serve as center focuses for computerized fund advancement, with higher levels of web entrance, smartphone ownership, and access to financial technology (Fintech) new companies (Hanna, 2018). On the other hand, common communities involvement particular inconveniences, such as destitute computerized capability rates, restricted web access, and low levels of certainty in official money related teach (Mbiti and Weil, 2011). Concurring to Lusardi and Mitchell (2014), the execution of compelling financial instruction programs is crucial in engaging people to make well-informed budgetary choices and investigate complicated cash related circumstances. In any case, existing composing highlights incoherencies inside the transport and influence of budgetary instruction exercises, particularly in between urban and common settings (Cole et al., 2010). Inquire around by Jack and Suri (2013) exhibits innovative approaches to financial thought, checking flexible cash stages and progressed credit organizations, which have outlined basic take-up and influence in common communities.

Overview of Digital Finance

Advanced back is the conveyance of traditional financial digital services utilizing contraptions like PCs, iPads, and smartphones. In ranges where there was no physical foundation for financial services, it may give underprivileged populations with access to these administrations. Financial services have been altogether affected by digital change, as has been the case with most other businesses. The taking after could be a list of a few of the innovations and arrangements that are quickening the digitization of financial services. expanded utilization of portable and internet banking. The thought of computerized managing an account isn't modern (Agarwal et al., 2025). More youthful gatherings of people have been getting a charge out of it very a bit for a few time. In any case, the utilisation of online managing an account has expanded drastically in later times. An expanding number of individuals turn to computerized keeping money when they may now not visit their nearby bank department. As a result, more individuals are starting to feel at ease and indeed come to favor advanced exchanges like online exchanges and installments. the appearance of equal banks. More so-called challenger banks have surfaced in later a long time. These banks have no physical branches and are completely computerized. By making strides the client involvement, from the straightforwardness of opening an account to more prominent usefulness and more charming expenses and rates, they take on the built up banks. Challenger banks incorporate Revolut, Lunar, and Monzo, to title several. The monetary administrations industry is additionally experiencing alter much obliged to fintech firms. They offer a run of computerized administrations, counting chatbots for client benefit, budgeting, and spending tracking. Fintech companies have played a urgent part in moving the digitalization of budgetary administrations, as an expanding number of money related teach are depending on these arrangements. The utilize of digital money related counsel stages, computerized speculations, and algorithm-based speculation company administration is moreover developing in notoriety (Chand et al., 2024). A developing number of clients are choosing them over traditional investment firms due to their straightforwardness of use and capacity to supply more charming expenses, to title a number of. benefits. Blockchain technology is another development driving budgetary digitalization. It modifies how individuals contribute their cash and execute on markets, and it offers better approaches to recognize, record, and store resources carefully. The impacts of blockchain innovation are still being felt, but they will likely have a more prominent impact on money related administrations within the future. Digital finance represents a paradigm shift within the delivery and utilization of financial administrations, moving conventional banking, lending, insurance, and venture exercises from physical branches and paper-based forms to digital stages and innovations (R. Rawat et al., 2025c). It includes a vast biological system of developments, counting mobile managing an account, online installments, computerized wallets, peer-to-peer loaning, robo-advisory administrations, and blockchain-based arrangements. This transformative wave is characterized by upgraded availability, cost-efficiency, speed, and transparency, on a very basic level reclassifying the financial scene. In India, the advancement of advanced fund has been especially energetic, impelled by a one of a kind confluence of innovative advancements, supportive government arrangements, and a quickly expanding advanced framework.

The nation's travel towards financial digitization, at first driven by foundational changes just like the computerization of managing an account operations within the 1980s and the introduction of electronic clearing frameworks, picked up significant energy within the early 2000s with the approach of ATMs, web managing an account, and mobile banking. Be that as it may, it was the key activities post-2014, such as the Jan Dhan Yojana, Aadhaar, and most outstandingly the Unified Payments Interface (UPI), that really democratized advanced financial access, laying the basis for India's development as a global pioneer in real-time digital exchanges. This fast speeding up has made advanced back not fair a comfort for urban populations but an indispensable apparatus for financial consideration over the endless and assorted socio-economic texture of the nation. The center components of computerized back in India are multifaceted, shaping a vigorous and interconnected digital public infrastructure (DPI) that has ended up a worldwide benchmark. At its heart lies the Unified Payments Interface (UPI), a real-time installment framework created by the National Payments Corporation of India (NPCI). Launched in 2016, UPI permits moment fund exchanges between bank accounts employing a single portable application, working 24/7. Its interoperability, ease of utilize through Virtual Installment Addresses (VPAs), and zero exchange expenses for many common exchanges have driven unprecedented appropriation, making India the world leader in real-time installments. As of April 2025, UPI recorded over 18.6 billion transactions worth 24.77 lakh crore, a confirmation to its inescapable reach among about 460 million people and 65 million merchants. Complementing UPI is Aadhaar, India's special 12-digit recognizable proof number connected to biometric information. Aadhaar has revolutionized e-KYC (Know Your Customer) forms, simplifying account opening, character confirmation, and access to open administrations. Its integration with keeping money and payment frameworks has been pivotal for confirming clients and guaranteeing the conveyance of government benefits specifically into beneficiaries' accounts through Direct Benefits Transfer (DBT), sparing the exchequer significant sums by eliminating intermediaries and fake beneficiaries. Past these foundational columns, a few other components contribute to India's computerized fund environment. Mobile Wallets (e.g., Paytm, PhonePe, Google Pay), though largely coordinates with UPI presently, played a vital part within the introductory stages of advanced installment adoption by advertising comfort for online shopping, utility charge installments, and shipper exchanges. National Electronic Funds Transfer (NEFT) and Real-Time Gross Settlement (RTGS) give interbank electronic support exchange offices for shifting exchange values, serving as the backbone for bigger and batch installments. The Bharat Bill Payment System (BBPS) standardizes and simplifies charge installments over different administrations, making it less demanding for buyers to pay repeating bills online. RuPay, India's domestic card installment arrange, has challenged the dominance of universal card plans, advancing nearby advancement and lessening exchange costs. Moreover, the burgeoning Fintech division envelops a wide cluster of imaginative administrations: online loaning platforms give quicker and more available credit, especially to MSMEs and people often underserved by conventional banks; advanced protections arrangements use AI and online stages to offer rearranged access to scope and improve operational proficiency; and digital investment and resource administration stages democratize investment openings with lower expenses and algorithm-based advice. The rise of neobanks – fully digital banks without physical branches – assist represents the move towards a technology-first approach to managing an account, advertising competitive administrations with improved customer encounters. The benefits of this comprehensive digital fund ecosystem in India are significant and far-reaching. Financial incorporation stands out as a vital accomplishment. Prior to the advanced transformation, a noteworthy parcel of India's population, especially in country and inaccessible regions, lacked access to formal financial administrations. Initiatives like PMJDY, coupled with the ubiquity of portable phones and UPI, have brought millions into the banking fold, empowering them to open accounts, get government appropriations, make and get installments, and access credit. This has enabled marginalized communities, decreased destitution, and cultivated enterprise at the grassroots level.

The convenience and efficiency advertised by digital fund are undeniable. Exchanges can be completed instantly, 24/7, from anywhere, killing the require for physical cash, trips to banks, or extensive paperwork. This comfort has been a major driver of selection, particularly among more youthful, tech-savvy eras and little businesses. For the economy, digital fund advances transparency and formalization. The digital path of exchanges decreases the scope for casual financial exercises and upgrades assess compliance, contributing to more prominent government income. It too streamlines supply chain operations, diminishes costs for businesses, and makes strides generally financial proficiency. The multiplication of computerized installments has altogether decreased the reliance on cash, driving to lower costs related with printing, dealing with, and securing physical cash. In addition, advanced back has impelled monstrous development inside the financial division, cultivating a dynamic environment of fintech new businesses that continuously develop modern items and administrations, driving competition and better client encounters. In spite of these significant strides, digital finance in India faces significant challenges, especially in guaranteeing evenhanded get to and security over all socioeconomics. One of the foremost basic challenges is the persistent advanced isolate, particularly between urban and provincial areas. Whereas smartphone infiltration and internet network have improved significantly, gaps remain, especially in remote districts with questionable framework. This straightforwardly impacts the capacity of provincial populaces to reliably get to and utilize advanced financial administrations. Closely linked to this is often the challenge of digital and financial proficiency. A significant parcel of the country and semi-urban populace needs the fundamental understanding of how advanced stages work, the nuances of digital security, and the dangers related with online exchanges. This vulnerability makes them prime targets for cyber extortion, phishing attacks, and social building tricks.

The sheer volume and expanding modernity of cybersecurity dangers posture a persistent challenge. As more financial exchanges move online, the financial segment becomes an alluring target for cybercriminals. Ransomware, phishing, malware, personality theft, and information breaches are constant dangers that require strong security measures and continuous carefulness. Believe issues stemming from past episodes of fraud or need of understanding can moreover impede appropriation. Building and keeping up open trust requires not as it were secure frameworks but moreover transparent grievance redressal instruments and clear communication. Besides, the administrative scene for advanced fund is complex and continuously advancing, requiring consistent adjustment from fintech companies and conventional banks to guarantee compliance whereas cultivating development. Looking ahead, long term of digital fund in India is balanced for indeed more noteworthy change, driven by rising innovations and advancing customer desires. Artificial Intelligence (AI) and Machine Learning (ML) will proceed to play a significant part, not fair in improving cybersecurity and extortion discovery but moreover in personalized financial administrations, credit scoring utilizing elective information, and conversational managing an account through AI chatbots. Implanted fund, where budgetary administrations are consistently coordinates into non-financial stages (e.g., e-commerce, ride-hailing apps), is gaining critical energy, advertising contextual and helpful financial arrangements at the point of require. This drift is closely related to Banking-as-a-Service (BaaS), where conventional banks offer their banking foundation as a benefit to fintechs, empowering a more extensive cluster of inventive items.

The investigation and potential selection of a Central Bank Digital Currency (CBDC) by the Reserve Bank of India (RBI) seem advance revolutionize the computerized payment scene. Blockchain innovation is anticipated to improve security, transparency, and effectiveness in ranges like cross-border payments, exchange fund, and computerized resource administration. Open Keeping money systems, enabling customers with more prominent control over their financial information, will cultivate greater competition and advancement in item offerings. Besides, the proceeded center on monetary consideration will drive developments custom-made for the "last mile," counting vernacular interfacing, offline installment arrangements, and voice-based banking, guaranteeing that the benefits of digital fund reach each citizen. The expanding accentuation on RegTech (Regulatory Technology) will help financial educate in exploring the complex administrative environment, guaranteeing compliance through automated checking and announcing. Digital finance has unavoidably reshaped India's financial scene, moving from a nascent concept to a powerful motor of financial development and social consideration. The travel has been stamped by pioneering developments like UPI and Aadhaar, which have laid the foundational digital open foundation. Whereas advertising phenomenal benefits in comfort, effectiveness, and formalization, the segment continuously grapples with critical challenges related to digital education, cybersecurity, and impartial access, especially in rustic zones. In any case, the long run holds immense guarantee, with innovations like AI, embedded finance, and blockchain balanced to drive encourage development, personalization, and security. The progressing collaborative endeavors among the government, controllers, financial teach, and fintech innovators will be pivotal in overcoming existing obstacles and guaranteeing that computerized finance truly serves as a powerful catalyst for a more inclusive, transparent, and flexible financial environment for all Indians. The ceaseless advancement of this dynamic division will undoubtedly rethink how money flows and how people and businesses connected with financial services within the long time to come.

Rural Digital Financial Literacy

Rural people can become casualties of digital financial frauds in ways that contrast from urban populations. Rural populations regularly have lower levels of digital education and mindfulness of advanced fund compared to urban partners, making them more vulnerable to false plans and misleading hones within the computerized domain (Sharma et al., 2023). They may depend on casual monetary administrations such as cash loan specialists, community investment funds bunches, or neighborhood cooperatives due to restricted get to formal managing an account framework and advanced installment frameworks (Sharma, Tyagi, et al., 2024a). These casual channels may need transparency and administrative oversight, uncovering people to dangers of extortion and abuse. They frequently confront challenges related to inadequate foundation, destitute internet connectivity, and constrained access to computerized gadgets, preventing individuals' capacity to engage in advanced budgetary exchanges safely. Fraudsters may misuse rustic individuals' believe in community pioneers, relatives, or associates to execute monetary fakes through social designing strategies, such as pantomime, control, and wrong guarantees. Country ranges may need viable administrative oversight, customer assurance laws, and get to legitimate plan of action instruments, constraining individuals' capacity to look for change for money related fakes or uphold their rights. Thus, these preventive measures can be taken. Actualize focused on monetary education programs and mindfulness campaigns custom-made to country communities, giving hands-on preparing, intelligently workshops, and instructive materials on secure computerized hones, online security, and extortion anticipation. Collaborate with nearby community pioneers, schools, government organizations, non-profit organizations, gracious society bunches and grassroots organizations to spread data and engage people with computerized aptitudes and information (R. Rawat et al., 2025c). Advance budgetary incorporation and expand access to formal keeping money administrations in provincial ranges through portable keeping money, specialist managing an account, and community banking activities. Teach rustic inhabitants around the benefits of formal money related administrations, counting secure exchanges, customer assurance, and access to credit and reserve funds items. Empower the foundation of self-help bunches, microfinance teach, and agreeable social orders to supply secure and dependable monetary administrations at the grassroots level.

Venture in country foundation advancement, counting broadband network, portable arrange extension, and open Wi-Fi hotspots, to progress advanced get to and network in country communities. Give appropriations or motivations for computerized gadgets. Offer mobile-based arrangements, USSD (Unstructured Supplementary Benefit Information) codes, and voice-based interfacing for conducting essential managing an account exchanges and getting to money related administrations without requiring web network or progressed computerized abilities. Cultivate community mindfulness and watchfulness against false exercises by organizing nearby mindfulness campaigns to examine money related education, shopper rights, and extortion avoidance procedures. Energize open communication, to recognize, address and report occurrences of extortion to important specialists. The digital transformation clearing over India has significantly affected its financial landscape, advertising unprecedented openings for financial development and financial incorporation. At the heart of this change lies digital fund, empowering exchanges, investment funds, credit, and protections through electronic stages. Be that as it may, the evenhanded access and viable utilization of these advanced financial administrations, especially in India's endless country hinterland, pivot critically on the level of rural advanced financial education (Sharma & Tyagi, 2023). This concept amplifies past basically owning a smartphone or having a bank account; it includes the information, aptitudes, and certainty to explore digital financial stages securely, make educated financial choices, and secure oneself from digital extortion. In spite of significant strides in financial incorporation, a significant parcel of the country population still slacks in this pivotal region, making a digital isolate that debilitates to worsen existing socio-economic abberations. Bridging this education gap isn't just an instructive endeavor but a foundational necessity for really inclusive development, sustainable country advancement, and the fruitful realization of India's vision of a carefully enabled society.

The basic for vigorous country advanced financial literacy stems from several basic components. Firstly, whereas activities just like the Pradhan Mantri Jan Dhan Yojana (PMJDY) have effectively brought millions of unbanked people into the formal financial framework, account possession does not automatically decipher into dynamic utilization or understanding of advanced financial instruments. A noteworthy number of PMJDY accounts, for occasion, have appeared low movement, incompletely due to a need of certainty or understanding among account holders. Besides, the quick multiplication of digital installment frameworks like UPI implies that financial exchanges are progressively moving online. Without satisfactory proficiency, country citizens are incapable to completely use the comfort and benefits of these frameworks, such as moment exchanges, charge installments, and access to government benefits through Direct Benefit Transfers (DBT). Thirdly, and maybe most basically, the computerized landscape is overflowing with modern cyber threats. Rural populations, regularly with limited introduction to innovation and cybersecurity best hones, are especially powerless to phishing scams, character theft, malware, and other shapes of advanced extortion. A strong foundation in digital financial literacy enables them to recognize these threats, receive secure practices, and avoid getting to be casualties, subsequently safeguarding their meager investment funds and cultivating believe within the digital environment. A few multifaceted challenges impede the broad fulfillment of digital budgetary education in country India. The foremost noticeable is the digital foundation gap. Whereas internet penetration has developed, it remains uneven, with numerous inaccessible rural areas encountering questionable or non-existent network (Sharma, Tomar, et al., 2024). Even where network exists, the quality of benefit, especially versatile information speeds, can be a restricting figure. According to a 2022 Oxfam report, as it were 31% of the provincial population employments the internet, compared to 67% in urban ranges, highlighting a critical partition. This infrastructural insufficiency directly impacts get to to advanced financial administrations and the capacity to participate in online education programs. Compounding this is often the socio-economic and instructive difference. Lower literacy rates, restricted formal instruction, and a common lack of introduction to innovation make principal obstructions to understanding complex financial concepts and advanced interfaces. The 2019 National Sample Survey (NSS) 77th circular indicated that 33% of the rural population did not have a bank account, credit/debit card, or e-wallet, underscoring the foundational gap. Many rural families still work predominantly in a cash economy, driving to a mental obstruction against receiving digital exchanges. Dialect and social barriers also show noteworthy obstacles. Much of the accessible advanced financial data and educational substance is transcendently in English or major territorial languages, coming up short to reach the assorted phonetic scene of rustic India. Besides, traditional beliefs, social standards, and a deep-seated dependence on casual money related practices can make communities reluctant to grasp formal, digital financial administrations. For occurrence, conventional sees on money related administration, especially with respect to gender parts, can regularly avoid women from completely partaking in financial instruction programs (Sharma, Gautam, et al., 2024).

The need of believe in advanced stages, regularly fueled by individual encounters or accounts of extortion, is another basic impediment. Without a clear understanding of the security mechanisms supporting advanced transactions, rural citizens stay watchful of moving their funds online. At last, the human resource challenge in delivering tailored financial literacy programs to inaccessible areas is considerable. A shortage of prepared facilitators, especially those who get it nearby lingos and social subtleties, hinders the successful dispersal of information. These interlaced challenges require a holistic and context-sensitive approach to cultivating advanced budgetary proficiency. Recognizing the basic significance of digital financial education, both legislative bodies and non-governmental organizations (NGOs) have propelled different activities over India. The Reserve Bank of India (RBI), as the central banking authority, has been at the cutting edge, building up the National Centre for Financial Education (NCFE) in collaboration with other financial sector regulators (SEBI, IRDAI, PFRDA). The NCFE's National Strategy for Financial Education (NSFE) 2020-25 adopts a "5 C" approach – Content, Capacity, Community-led model, Communication, and Collaboration – pointing to reach 500 million Indians by 2025. This methodology emphasizes creating localized substance, building coach capacity, and cultivating community-based programs. The RBI also operates Financial Literacy Centres (FLCs) over the nation, especially in country and semi-urban ranges, offering free financial instruction through workshops, camps, and one-on-one counseling on themes like budgeting, reserve funds, credit, and digital exchanges. Their "RBI Kehta Hai" (RBI Says) campaign leverages different media stages to teach the public on safe managing an account hones and fraud avoidance. The National Bank for Agriculture and Rural Development (NABARD), a key institution for rural improvement, effectively underpins financial literacy endeavors through its Financial Inclusion Fund (FIF). NABARD provides allow help for financial and digital proficiency camps organized by banks, FLCs, and through activities like "demonstration vans" prepared with audio-visual helps to reach farther towns. They too bolster the arrangement of micro-ATMs and point-of-sale (POS) devices to make strides digital installment framework. The Ministry of Electronics and Information Technology (MeitY) has launched plans like Pradhan Mantri Gramin Digital Saksharta Abhiyaan (PMGDISHA) and "Digital

Finance for Rural India: Creating Awareness and Access through Common Service Centres (CSCs)". PMGDISHA aims to make 6 crore rural people digitally literate by preparing them to function computers or digital gadgets, send emails, browse the internet, access government administrations, and embrace digital installments. CSCs are imagined as "Digital Financial Hubs," giving mindfulness sessions and encouraging advanced financial administrations at the town level. NGOs just like the Digital Empowerment Foundation (DEF) are too exceedingly dynamic, conducting grassroots preparing programs, setting up community advanced centers, and creating context-specific instructive substance to address the computerized and budgetary education gaps in underserved communities. These activities, through a mix of policy, framework improvement, and ground-level engagement, endeavor to enable rustic citizens with digital financial information.

The effective upgrade of provincial computerized financial proficiency can abdicate significant impacts on provincial improvement. Firstly, it specifically contributes to more profound budgetary incorporation, moving past simple access to dynamic and educated support within the formal financial framework. This leads to expanded reserve funds, way better get to to formal credit for rural exercises or little businesses, and improved financial arranging for unanticipated circumstances. Furthermore, it cultivates financial strengthening, especially for ladies, who frequently gain independent control over their funds through advanced devices. This will upgrade their decision-making control inside families and communities, driving to improved livelihoods and by and large well-being. Thirdly, upgraded digital proficiency decreases the defenselessness to advanced extortion, safeguarding the money related resources of provincial families, which are frequently built on a long time of difficult work. This expanded security, in turn, builds greater believe in advanced exchanges, empowering more extensive appropriation. Fourthly, it encourages better access to government plans and endowments conveyed through DBT, guaranteeing that stores reach the aiming beneficiaries directly and efficiently, lessening leakages and corruption. Finally, a carefully literate country populace can use advanced devices for broader socio-economic advancement, counting accessing online instruction, showcase data for rural produce, and healthcare administrations, subsequently coordination country communities more completely into the national advanced economy. To successfully bridge the rural digital financial proficiency gap, a few best practices and vital approaches are basic. Localization and contextualization of substance are fundamental. Instructive materials must be developed in neighborhood languages, utilizing simple, relatable cases and analogies that resonate with the every day lives and encounters of rural communities. The utilize of visual helps, people media, and narrating can be highly viable. Community-led models are pivotal, including trusted local figures, self-help groups (SHGs), village-level entrepreneurs (VLEs) working CSCs, and bank 'mitras' (correspondents) as trainers and facilitators. These people can build compatibility and provide instruction in a socially sensitive way. Hands-on preparing and viable showings are more impactful than theoretical addresses. Giving openings for real digital exchanges in a secure, guided environment can construct certainty and recognition. This incorporates illustrating how to utilize UPI, versatile wallets, and Aadhaar-enabled payment systems (AePS). Besides, tending to the framework bottleneck remains crucial. Efforts to grow reliable internet connectivity, counting through government activities like BharatNet, must be quickened. Guaranteeing the accessibility of reasonable smartphones and advanced gadgets is additionally fundamental. Continuous learning and support are fundamental; financial proficiency isn't a one-time occasion but an ongoing prepare. Follow-up sessions, helpline support, and open advanced assets can offer assistance strengthen learned concepts and address developing issues.

Collaborative organizations between government organizations, financial teach, fintech companies, NGOs, and nearby community organizations are principal for pooling assets, ability, and reach. At last, leveraging developing advances like AI and gamification can revolutionize computerized financial education. AI can personalize learning ways based on person advance and needs, whereas gamified applications can make learning locks in and intelligently, especially for younger socioeconomics. The integration of AI-powered chatbots can give moment, localized bolster and reply questions, encourage bringing down the boundaries to understanding. Country computerized financial proficiency isn't just an assistant viewpoint of budgetary consideration but its very bedrock within the setting of India's digital advancement. Whereas the country has made commendable advance in growing access to digital money related administrations, the challenge of engaging each provincial citizen with the essential information and certainty remains critical. The advanced partition, coupled with socio-economic incongruities, etymological boundaries, and diligent cybersecurity dangers, requires a concerted, multi-pronged approach. Government plans, administrative back, and the tireless endeavors of NGOs and money related educate are collectively working to address these crevices. The affect of upgrading rustic advanced budgetary proficiency is transformative, promising deeper financial consideration, more prominent economic strengthening, diminished helplessness to extortion, and eventually, a more evenhanded and prosperous country India. As innovation proceeds to advance, so as well must the procedures for financial instruction, guaranteeing that advanced fund really serves as a capable instrument of advance for all.

Urban Digital Financial Literacy

Whereas urban individuals by and large have higher education rates and more noteworthy mindfulness of advanced back compared to rural populations, they can still ended up casualties of money related fakes due to different variables inborn to urban living and advanced intuitive. Urban ranges regularly involvement higher rates of cybercrime and advanced extortion plans due to the concentration of monetary educate, businesses, and advanced foundation (Garg et al., 2024). Urban tenants are more likely to engage in complex budgetary exchanges, counting online keeping money, venture portfolios, and advanced installments, which may uncover them to a more extensive extend of extortion dangers. Urban ways of life frequently include visit budgetary exchanges, such as shopping, eating, excitement, and travel, which can increment the probability of presentation to false exercises. Fraudsters may abuse urban individuals' believe in specialist figures, educate, and social systems to execute personality burglary, phishing tricks, and pantomime fakes. Urban people may show overconfidence or complacency in their money related education and computerized abilities, driving them to miss potential extortion dangers or expel caution signs of false exercises.

By practicing carefulness, receiving preventive measures, and cultivating a culture of responsibility, urban populaces can moderate the hazard of falling casualty to budgetary fakes. People ought to remain upgraded on rising extortion patterns, utilize vigorous cybersecurity measures such as antivirus computer program, firewalls, and encryption, and work out caution when sharing delicate data online or reacting to spontaneous demands for monetary exchanges. They ought to routinely screen their budgetary accounts, audit exchange history, and set up cautions for unordinary exercises. They ought to utilize secure installment strategies, such as credit cards with extortion security highlights, and maintain a strategic distance from sharing card subtle elements or PINs in open or unsecured situations. Urban people ought to confirm the personality of people or organizations asking touchy data or money related exchanges, be cautious of spontaneous communications, and abstain from unveiling individual or monetary subtle elements without approval. They ought to develop an attitude of skepticism and basic considering, address unforeseen or abnormal monetary offers, and look for clarification or moment suppositions when in question. Urban India stands at the vanguard of the nation's digital transformation, characterized by a high degree of internet infiltration, broad smartphone appropriation, and a inescapable grasp of advanced financial administrations. From instant UPI installments for day by day commutes to complex online speculations, urban inhabitants are deeply coordinates into the digital economy. Be that as it may, whereas frequently assumed to have a better pattern of digital and financial mindfulness compared to their provincial partners, urban populations confront a particular set of challenges and modern dangers that require a persistent and nuanced approach to urban digital money related proficiency. This goes past fundamental transactional information, diving into the complexities of assorted computerized financial items, complex cybersecurity dangers, and the basic significance of information protection in an progressively interconnected monetary environment. Failure to satisfactorily address these particular education needs can lead to critical financial losses, erosion of believe in computerized frameworks, and ruin the complete potential of India's advanced finance development. The unmistakable characteristics of urban advanced financial education are formed by the energetic and complex financial environment of cities. Urban customers regularly have access to a broader cluster of advanced monetary items, counting different sorts of computerized wallets, different credit card alternatives, assorted speculation stages (mutual funds, stocks, digital gold), peer-to-peer loaning applications, and rising administrations like Buy Now, Pay Later (BNPL). This wealth, whereas advertising choice and adaptability, too presents complexity.

Exploring these choices viably requires a deeper understanding of financial concepts such as intrigued rates, credit scores, venture dangers, expanded portfolios, and the terms and conditions related with each item. Moreover, urban life frequently includes a better recurrence and volume of advanced exchanges, making them more helpless to both minor blunders and large-scale fakes. The sheer volume of online intelligent, from e-commerce buys to utility charge installments, makes various touchpoints where vulnerabilities can be misused. In spite of their clear advanced familiarity, urban populations are by no implies resistant to advanced financial threats; in reality, they regularly become targets for more advanced and elaborate scams. Common advanced financial frauds influencing urban Indians incorporate: phishing and smishing assaults, where fraudsters make fake websites or send beguiling SMS messages to trap clients into unveiling sensitive data like UPI PINs, OTPs, or keeping money credentials; vishing (voice phishing), where imposters posturing as bank authorities, government operators, or indeed tech back agents control casualties over the phone; and inaccessible get to tricks, where scammers persuade clients to download noxious computer program that gifts them control over the victim's gadget and, thus, their financial apps. A significant number of urban Indian family units (about 47% in a recent survey) have detailed encountering one or more financial frauds, with UPI and credit card frauds being the foremost common. This regularly includes strategies like sending a QR code or a connect that, when clicked, charges cash from the user's account rather than accepting a installment, or unauthorized charges on credit cards, in some cases from worldwide merchants. Other developing dangers incorporate advance app scams advertising moment credit but engaging in ruthless loaning practices and badgering, and venture frauds promising unreasonably tall returns through fake plans. The advancement of these tricks requests the next level of watchfulness and a nuanced understanding of digital security conventions from urban shoppers. The particular needs for urban advanced financial proficiency are in this manner equipped towards preparing customers to handle this complexity and counter progressed threats. Firstly, there's a basic require for progressed cybersecurity mindfulness. This goes past fundamental secret word cleanliness to understanding two-factor authentication (2FA), recognizing encrypted associations, recognizing suspicious app consents, and being mindful of social designing strategies utilized by fraudsters. Besides, understanding financial items and their related dangers is fundamental. Urban customers need to comprehend the subtleties of diverse investment instruments, the suggestions of credit utilization, the terms of advanced advances, and the scope restrictions of advanced insurance items. This includes deciphering complex financial jargon and assessing item suitability for their person needs.

Thirdly, data protection and assurance are progressively crucial. Urban advanced monetary literacy must educate consumers on how their individual and financial information is collected, utilized, and shared by digital stages, and enable them to oversee their security settings and recognize potential information breaches. At last, there's a require for viable grievance redressal components. Knowing how and where to report financial extortion, blocking stolen cards, and understanding customer rights are vital for urban clients to relieve misfortunes and seek equity. Recognizing these challenges, different activities have been embraced to improve computerized budgetary education among urban customers in India. The Reserve Bank of India (RBI), through its National Centre for Financial Education (NCFE) and Financial Literacy Centres (FLCs) – which too cater to semi-urban and urban zones – conducts campaigns and workshops centering on secure digital banking practices, extortion prevention, and capable utilize of credit. Their "RBI Kehta Hai" campaign reliably disseminates cautions and advisories on common financial scams through different media. Commercial banks and financial institutions too play a significant part, leveraging their computerized stages to thrust security cautions, give FAQs on advanced administrations, and offer in-app instructional exercises on highlights like UPI or versatile keeping money. Many banks routinely conduct webinars and online sessions on cybersecurity and financial arranging tailored for their urban clientele. Besides, Fintech companies, driven by the have to be construct client believe, coordinated instructive substance inside their apps, offer user-friendly interfacing, and convey AI-powered chatbots to reply client questions and direct them on secure exchanges. Industry affiliations and NGOs too contribute by organizing open mindfulness campaigns, webinars, and distributing enlightening substance on advanced financial security.

In any case, the divided nature of these endeavors implies that a really comprehensive and supported approach is still advancing. In spite of these endeavors, challenges persist in viably coming to and teaching the diverse urban population. The sheer volume of advanced data can be overpowering, making it troublesome for people to discern dependable sources from deception. The quick pace of innovative alter implies that modern digital financial items and modern scam techniques develop always, requiring continuous upgrades to literacy programs. The "digital native" fallacy, where it's wrongly accepted that youthful, tech-savvy urban tenants are inherently fiscally proficient, leads to complacency and neglects their vulnerability to modern scams. There's moreover a challenge in engaging people who see themselves as of now proficient, regularly driving to overconfidence and less openness to learning approximately essential security conventions. The "set it and forget it" mindset for many advanced financial administrations can lead to disregard of continuous security updates or observing of exchange history. Besides, the assorted socio-economic strata inside urban areas mean that whereas a few sections are exceedingly educated and fiscally advanced, others, including every day wage workers, migrant laborers, or those within the casual division living in urban slums, confront comparable education challenges as rural populations, requiring highly localized and socially touchy approaches. The affect of upgraded urban advanced financial proficiency is multifaceted and basic for India's advanced economy (R. Rawat et al., 2024c). Firstly, it leads to expanded selection and deeper engagement with advanced digital financial items, empowering urban shoppers to superior oversee their accounts, contribute shrewdly, and get to custom fitted credit arrangements. This fills financial action and reinforces the formal financial framework (Dasawat & Sharma, 2024). Furthermore, it comes about in a significant decrease in financial extortion and cybercrime episodes, safeguarding person riches and reinforcing in general customer certainty in advanced stages. This diminishes the burden on law enforcement and financial teach alike.

Thirdly, moved forward education cultivates mindful budgetary behavior, counting way better budgeting, restrained sparing, and reasonable utilize of credit, contributing to greater person financial well-being and diminishing obligation. Finally, it makes a more versatile and secure advanced biological system. When urban customers are well-informed and watchful, they act as an critical line of defense against cyber dangers, contributing to the generally security pose of the advanced budgetary foundation. Looking to long-standing time, upgrading urban advanced financial proficiency will require inventive and energetic methodologies. AI and machine learning will play a basic part in personalizing financial instruction, conveying focused on warnings, and making intelligent chatbots for prompt inquiry determination. These advances can adjust learning substance to person needs and hazard profiles, making instruction more relevant and successful. Gamification and interactive learning modules can make complex money related concepts engaging and available, especially for more youthful demographics who are advanced natives but may need money related intuition. The integration of money related instruction into formal instruction educational program from school to college levels is vital to construct foundational information. Besides, behavioral economics standards can be connected to nudge urban buyers towards more secure financial propensities and educated decision-making. Continuous open awareness campaigns by administrative bodies and budgetary teach, leveraging mass media and social media, will be imperative to keep pace with advancing dangers. Finally, cultivating a culture of peer-to-peer learning and community engagement can enable urban inhabitants to share information and best practices, strengthening collective flexibility against advanced financial dangers. By proactively tending to these needs, India can guarantee that its urban populations not as it were lead the digital fund transformation but do so safely and capably, clearing the way for a more strong and trusted advanced economy.

METHODOLOGY

Innovation can diminish financial misfortunes, protect client security, and make strides the common security of environments that back advanced installments. Monetary teach, finch companies, and installment benefit suppliers must utilize cutting-edge machine learning models and calculations in this age of fast innovative headway on the off chance that they need to remain ahead of cybercriminals. This procedure employments ceaseless learning and enhancement to reply to unused dangers in addition to helping within the location of well-known extortion designs.

Figure 1. Flowchart of Proposed System

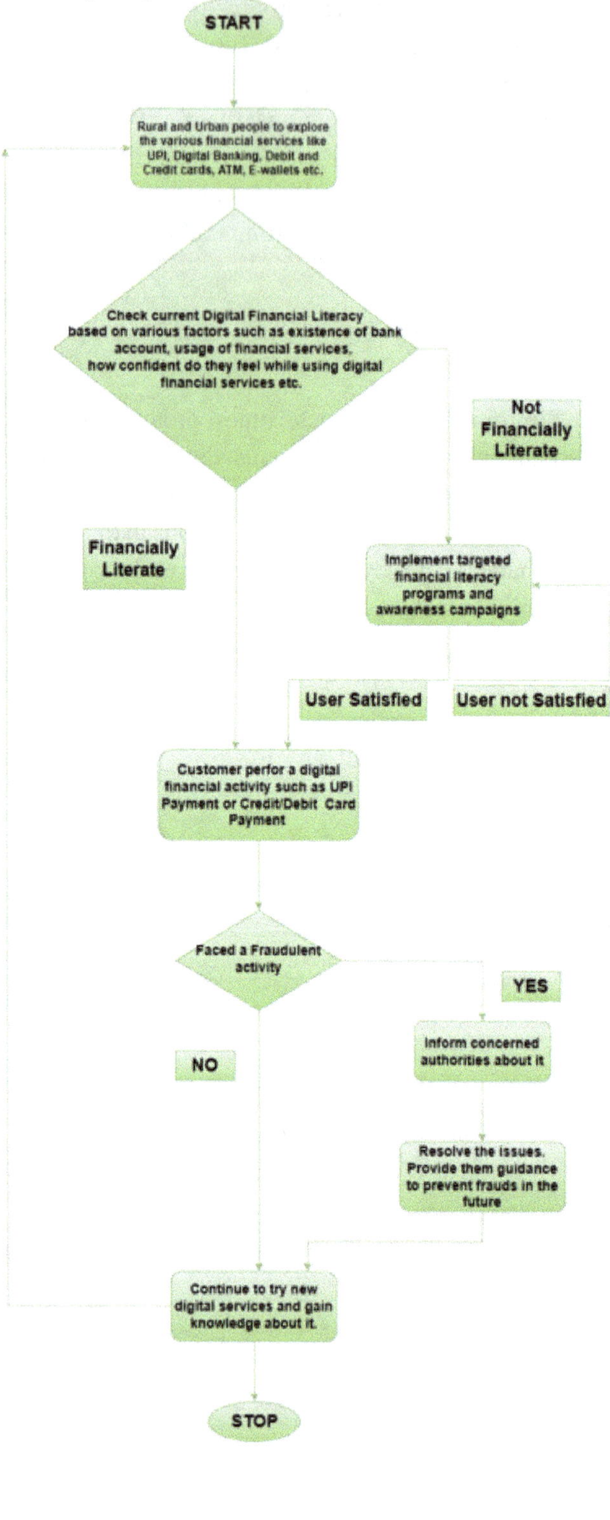

Gratefully, online exchanges are tracked and can hence be examined with the foremost later instruments. For the reason of distinguishing such false exercises, conventional strategies like manual confirmations and assessments are loose, costly, and time-consuming. With the improvement of manufactured insights, a large number of money related information can be intelligently analysed to distinguish false exchanges utilizing machine-learning-based strategies. This paper points to make a machine learning demonstrate that can identify false exchanges inside a exchange dataset and offer reasonable security measures to both urban and country populaces. With the help of Machine Learning (ML) calculations, extortion discovery for credit/debit card, advance defaulters, and related sorts can be finished. ML calculations are highly able of learning from past extortion patterns or verifiable information and recognizing them in current or future exchanges. Figure 1 portrays the flowchart of proposed framework. For the reason of identifying and foreseeing false exchanges, the rendering of different machine learning calculations is compared, counting Convolutional Neural Networks (CNN), Random Forest, Decision Trees, K Nearest Neighbor (KNN), Support Vector Machines (SVM), Naive Bayes, and Logistic Regression techniques. Based on the discovery of fraudulent cases, categories are made for extortion exchanges based on the subtle elements of the false exchanges.

Categorization needs to be done as to which kind of clients and accounts, banks, time of the year, sum of exchange, type of exchange etc. are the foremost vulnerable to be done by a fraudster. The clients ought to be isolated into whether he/she was fiscally proficient or not. At that point accounts should be separated into whether the account has already been a casualty of false movement or not. Think about the extortion exchanges to understand whether the fakes are area particular, bank particular or client particular. Also study on the off chance that the exchanges are sum particular i.e. in case they generally happen for tall sums or can indeed happen for small amounts. Also what is the Least Yearly Adjust that the accounts facing frauds maintain. Based on all these categorizations, additional security highlights need to be actualized. Security highlights and installment confirmation are fundamental for limiting blunders and avoiding extortion. It gives superior client believe, lower chargeback rates, and more security. Conventional security highlights incorporate verification utilizing PINS, passwords, One Time Passwords (OTP), KYC checks (Know Your Customers), robot checks etc. Apart from these, additional layer of security ought to be included in the categories of clients, accounts, banks, locations etc. that are inclined to fakes. This will incorporate an individual telephonic affirmation for an exchange, communication through SMS/emails, coordinating of geolocation of the put where an exchange for an account takes put and geolocation of where actually the proprietor of the account is, biometric confirmations etc. This additional consideration to casualty accounts will assist offer assistance in decreasing ill-conceived exercises in back division.

Implementation

The Unified Payments Interface (UPI) has become a widely used and practical way for financial transactions in the modern era due to the quick growth of digital transactions. One of UPI's most remarkable contributions is its role in financial inclusion. It has enabled millions of unbanked and underbanked Indians to access digital financial services, bridging the gap between urban and rural areas. But there is also an increase in fraudulent activity as a result of the growing reliance on digital platforms. This gave attackers and hoaxers permission to create fraudulent transactions and deceive people into parting with money. Concern over fraud on these platforms is growing as digital payment methods like UPI (Unified Payments Interface) become more and more popular.

The goal of this project is to use machine learning techniques to create a reliable fraud detection system for UPI transactions. The main goal of the project is to create a machine learning model that can instantly analyze UPI transaction data in order to spot fraudulent activity. The principal aim is to develop a system that improves the security of UPI transactions and mitigates financial losses resulting from fraudulent activities. For the proposed methodology, the dataset we are using contains UPI transactions, including both legitimate and illegitimate transaction information in similar ratio, making it a Balanced Dataset. The dataset is labeled with features and a size of 5000 rows. It has 12 features, 11 independent and 1 dependent feature. The values in category and transaction state columns are in the form of reference numbers, i.e. each category and state are mapped to specific numbers.

Figure 2. Dataset containing 5000 rows and 12 features

	A	B	C	D	E	F	G	H	I	J	K	L
1	trans_hou	trans_day	trans_mo	trans_yea	trans_typ	category	upi_numb	age	trans_amc	state	zip	fraud_risk
2	22	16	9	2021	CASH_OU	1	87	27	199.31	Missouri	84946	1
3	12	13	7	2016	CASH_IN	2	71990978	42	448.06	Minnesot	68524	1
4	13	10	8	2019	CASH_IN	2	5601	70	127.79	Louisiana	34800	1
5	10	7	10	2023	TRANSFER	0	6.72E+08	61	967.7	Tennesse	7311	0
6	17	10	1	2022	DEBIT	1	666	40	911.3	Texas	60970	1
7	17	26	3	2016	PAYMENT	13	18	69	314.52	Missouri	597	0
8	2	4	9	2017	CASH_OU	0	261	35	758.16	Virginia	51766	0
9	13	16	10	2020	TRANSFER	13	64431556	54	991.69	Maryland	9108	0
10	16	11	10	2017	CASH_IN	1	57079	52	601.44	Oregon	52030	0
11	7	6	4	2016	PAYMENT	2	1960	58	741.06	Illinois	79263	1
12	11	5	12	2015	TRANSFER	2	17068	58	152.45	Alaska	41387	1
13	7	2	3	2016	DEBIT	2	49951962	42	866.24	Nebraska	28415	0
14	1	6	3	2021	PAYMENT	2	463	44	197.23	Washingt	84979	0
15	2	17	4	2020	PAYMENT	0	55382	32	569.02	Texas	54020	1
16	18	1	11	2015	DEBIT	0	9881291	51	618.01	New Mexi	9695	0
17	2	19	8	2017	DEBIT	0	3250	41	853.47	Pennsylva	69258	1

Figure 3. Category of business

CATEGORIES

"0" Entertainment

"1" Food Dining

"2" Gas Transport

"3" Grocery NET

"4" Health Fitness

"5" Home

"6" Kids Pets

"7" Personal Care

"8" Shopping

"9" Miscellaneous

Figure 2 depicts the dataset containing 5000 rows and 12 features. Figure 3 shows the category of businesses. reference numbers. Data Preprocessing is performed to remove noise, irrelevant features, handle missing values, and standardize features. Also, exploratory data analysis is done to gain insights into the distribution and characteristics of the dataset. Next Feature Engineering is performed. The relevant features are extracted from the transaction data that may contribute to identifying fraudulent patterns. These features may include transaction hour, transaction day, transaction month, transaction year, transaction category, transaction type (Payment, Cash out, Cash In, Transfer etc.), UPI number, zip code, transaction amount, age of person etc. Techniques such as feature scaling, dimensionality reduction, and transformation are used to enhance the quality and effectiveness of the feature set. A wide range of machine learning algorithms appropriate for fraud detection tasks are chosen for the Algorithm Selection part, including Convolutional Neural Networks (CNN), Random Forest, Decision Tree, K Nearest Neighbor (KNN), Support Vector Machine (SVM), Naive Bayes, and Logistic Regression are tried out. Each algorithm is implemented using appropriate libraries and frameworks such as TensorFlow, Scikit-learn, and Keras. Hyperparameters are tuned and model configurations are optimized to maximize performance. The dataset is split into training and test sets. Relevant metrics like accuracy, precision, recall, F1-score, and area under the Receiver Operating Characteristic (ROC) curve are used to assess each algorithm's performance. The feature values are transformed into a common standard scale using sklearn library. A comprehensive comparative analysis is done of the performance metrics obtained from different algorithms. The strengths and limitations of each approach are evaluated in terms of accuracy, computational efficiency, interpretability, and scalability. Key factors influencing the efficacy of each algorithm is identified and derive actionable insights for improving the overall fraud detection system.

Result Analysis

Based on the image, the Convolutional Neural Network (CNN) has the most noteworthy exactness score of 100% for UPI Extortion Location. This may be since CNNs are exceptionally compelling in preparing complex designs and highlights in huge datasets, which are basic for extortion location assignments. They can moreover handle high-dimensional and boisterous information, which are common in UPI exchanges. The moment and third most elevated precision scores are accomplished by Random Forest and Decision Tree, with 95.5% and 94.75%, individually. These calculations are great at dealing with unequal datasets and highlight significance location, which are too critical for extortion discovery tasks. The Random Forest algorithm is known for being versatile and able to avoid overfitting. It can handle expansive datasets with tall dimensionality well. Whereas simple to translate, Choice Trees can effectively overfit or underfit, so they might not give the foremost precise comes about. The other calculations, such as K-Nearest Neighbor (KNN), Support Vector Machine (SVM), Naive Bayes, and Logistic Regression, have lower precision scores extending from 80.25% to 83%. These calculations may well be less successful due to the complexity or nature of the dataset, or the require for more parameter tuning or preprocessing. KNN can be a great choice for a dataset where occurrences of extortion shape a particular cluster. Figure 4 depicts the comparison between accuracy achieved by different machine learning models. Figure 5 shows comparison between accuracy obtained by different machine learning models.

Figure 4. Comparison between accuracy achieved by different machine learning models.

```
In [54]: df = pd.DataFrame()
         df['Algorithm Name'] = names
         df['Accuracy Score (%)'] = scores
         df = df.sort_values('Accuracy Score (%)', ascending = False)

In [55]: df

Out[55]:
```

	Algorithm Name	Accuracy Score (%)
6	Convolutional Neural Network	100.00
5	Random Forest	95.50
4	Decision Tree	94.75
1	K-Nearest Neighbors	83.00
2	Support Vector Machine	81.50
3	Naive Bayes	81.25
0	Logistic Regression	80.25

Figure 5. Comparison between accuracy obtained by different machine learning models

```
In [57]: fig = plt.subplots(figsize = (10, 5))
         ax = sns.barplot(x = "Accuracy Score (%)", y = "Algorithm Name", data = df)
```

Be that as it may, due to the have to be compute the remove between each test occasion and each preparing occurrence, it can be computationally expensive for huge datasets. SVMs are viable in tall dimensional spaces and best suited for issues with complex choice boundaries. Expansive datasets are not a great fit for them since preparing and testing can be moderate. Especially for datasets with parcels of highlights, the Naïve Bayes calculation is clear and productive. It makes the suspicion that each highlight works freely of the others, which may not be the case for UPI exchanges. It can moreover be influenced by zero-frequency problems, where a few include values might not happen within the preparing information. Logistic Regression assumes independence between highlights. It's a simple and quick calculation. Since it accept a straight relationship between the highlights and the target variable, this calculation may be helpless to underfitting. Multicollinearity, in which certain highlights have a solid correlation with one another, may moreover have an affect. Based on these comes about, the leading calculation for UPI Extortion Discovery is CNN, given its perfect exactness score. Be that as it may, it is essential to require into consideration the plausibility of overfitting, since a 100curacy rate may recommend that the demonstrate has took on the training set and may not be able to sum up successfully to new information. Hence, it is advisable to utilize a few approval methods, like cross-validation, to assess the show execution on concealed information and maintain a strategic distance from overfitting. Figure 6 depicts the result of proposed CNN model that achieves the maximum accuracy out of all techniques.

Figure 6. Result of proposed CNN model.

```
In [48]: CNN_model.fit(x_train, y_train, batch_size = 32, epochs = 200)
         Epoch 192/200
         71/71 [==============================] - 0s 2ms/step - loss: 0.0081 - accuracy: 0.9991
         Epoch 193/200
         71/71 [==============================] - 0s 1ms/step - loss: 0.0240 - accuracy: 0.9894
         Epoch 194/200
         71/71 [==============================] - 0s 1ms/step - loss: 0.0135 - accuracy: 0.9978
         Epoch 195/200
         71/71 [==============================] - 0s 1ms/step - loss: 0.0099 - accuracy: 0.9991
         Epoch 196/200
         71/71 [==============================] - 0s 2ms/step - loss: 0.0148 - accuracy: 0.9969
         Epoch 197/200
         71/71 [==============================] - 0s 2ms/step - loss: 0.0419 - accuracy: 0.9854
         Epoch 198/200
         71/71 [==============================] - 0s 1ms/step - loss: 0.0293 - accuracy: 0.9881
         Epoch 199/200
         71/71 [==============================] - 0s 1ms/step - loss: 0.0149 - accuracy: 0.9965
         Epoch 200/200
         71/71 [==============================] - 0s 1ms/step - loss: 0.0064 - accuracy: 1.0000
Out[48]: <keras.callbacks.History at 0x19b5217fa00>

In [49]: loss, acc_cnn = CNN_model.evaluate(x_train, y_train, verbose = 0)

In [50]: print(acc_cnn)
         1.0
```

CONCLUSION

UPI Fraud Discovery could be a pertinent and challenging issue within the field of digital fund, as UPI exchanges are getting to be progressively well known and defenseless to fraudsters. Information from the fund ministry indicates that over 95,000 cases of UPI extortion were detailed within the 2022–2023 monetary year, agreeing to a few sources. Hence, it can be profitable for both urban and country clients to progress the security and believe of advanced exchanges by making dependable and exact machine learning models for UPI Fraud Location. Money related education and benefit suppliers can move forward the productivity, straightforwardness, and versatility of computerized fund biological systems by utilizing AI calculations, machine learning procedures, and enormous information analytics to personalize client encounters, streamline operations, and relieve dangers. AI-driven monetary arrangements can give individuals with customized speculation procedures, mechanized budgeting instruments, and personalized money related bits of knowledge in urban regions where advanced foundation and education rates are higher. Through the application of counterfeit insights (AI) and its capacity to analyze gigantic sums of information and anticipate shopper behavior, urban shoppers can enhance their money related proficiency, maximize their investment funds, and decrease dangers in a monetary environment that's always changing. On the other hand, AI-enabled developments hold awesome guarantee for closing the advanced isolate and promoting financial inclusion in rustic communities where formal financial administrations are rare and advanced proficiency is challenged. AI-powered financial arrangements can empower country clients to oversee their accounts, get to essential keeping money administrations, and create monetary versatility without the require for internet network or modern advanced aptitudes by utilizing portable innovation, USSD codes, and voice-based interfacing. However the successful application of developments in digital funds requires a comprehensive procedure that takes into consideration specific troubles, empowers participation among partners, and gives distinctive communities' needs and inclinations beat needs. In arrange to advance advancement and secure customer rights, information protection, and money related soundness, administrative systems must alter.

REFERENCES

Agarwal, J., Sharma, S., & Agarwal, P. (2025). S2E3: Sustainable, Secure, Ethical, e-Branding and e-Advertising System Using Metaverse Techniques and Hybrid Cryptographic Approach. In 2025 International Conference on Intelligent Control, Computing and Communications (IC3), 10–15. DOI: 10.1109/IC363308.2025.10956649

Chand, A., Sharma, S., Agarwal, P., & Mohan, S. (2024). Credit card fraud detection using deep learning for Internet-of-Things enabled smart digital financial system. In Internet of things (pp. 469–485). DOI: 10.1007/978-981-97-0052-3_23

Cole, S. A., Sampson, T., & Zia, B. (2010). Prices or knowledge? What drives demand for financial services in emerging markets? SSRN Electronic Journal. https://papers.ssrn.com/sol3/papers.cfm?abstract_id=1374078

Dasawat, S. S., & Sharma, S. (2024). Edge Intelligence Enabled Secured Digital Transformation in Internet of Autonomous Vehicles. In 2024 12th International Conference on Internet of Everything, Microwave, Embedded, Communication and Networks (IEMECON), 1–6. DOI: 10.1109/IEMECON62401.2024.10846320

Demirguc-Kunt, A., & Klapper, L. (2012). Measuring Financial Inclusion: The Global Findex database. In World Bank policy research working paper. DOI: 10.1596/1813-9450-6025

Garg, A., Ghanshala, K. K., & Sharma, S. (2024). A hybrid model of integrating sentiment analysis and key market indicators for IPO listing trend prediction. *International Journal of Mathematical Engineering and Management Sciences*, *9*(4), 902–913. DOI: 10.33889/IJMEMS.2024.9.4.047

Hanna, N. (2018). Urban-rural differences in financial literacy: Evidence from a national survey of high school students. Journal of Rural Studies, 60, 19-29.a.

Jack, W., & Suri, T. (2013). Risk Sharing and Transactions Costs: Evidence from Kenya's Mobile Money Revolution. *The American Economic Review*, *104*(1), 183–223. DOI: 10.1257/aer.104.1.183

Kumar, K., & Mas, I. (2008). Banking on mobiles: Why, how, for whom? SSRN Electronic Journal, 1–28. https://papers.ssrn.com/sol3/Delivery.cfm/SSRN_ID1655282_code829576.pdf?abstractid=1655282&rulid=42517&mirid=4

Lusardi, A., & Mitchell, O. S. (2014). The Economic Importance of Financial Literacy: Theory and evidence. *Journal of Economic Literature*, *52*(1), 5–44. DOI: 10.1257/jel.52.1.5 PMID: 28579637

Mbiti, I., & Weil, D. (2011). Mobile Banking: The impact of M-PESA in Kenya. DOI: 10.3386/w17129

Rawat, R., Goyal, H. R., & Sharma, S. (2024c). Hofstede's Cultural Dimension Driven Artificial Narrow Intelligence IDFIS for Industry 5.0 Empowered Digital Society. *Recent Patents on Engineering, 19*. Advance online publication. DOI: 10.2174/0118722121334543241001050846

Rawat, R., Goyal, H. R., Sharma, S., & Kotiyal, B. (2025c). An Intelligent Self-Aware Financial Inclusion System for Digital Society. In *2025 International Conference on Automation and Computation (AUTOCOM)*, 1164–1168. DOI: 10.1109/AUTOCOM64127.2025.10956854

Rawat, R., Goyal, H. R., Sharma, S., & Kotiyal, B. (2025c). SQRCD: Building Sustainable and Customer Centric DFIS for the Industry 5.0 ERA. *International Journal of Advanced Computer Science and Applications, 16*(2). Advance online publication. DOI: 10.14569/IJACSA.2025.0160293

Sharma, S., Chauhan, Y., & Tyagi, R. (2023). Machine Vision Techniques for Digital Mesothelioma Diagnostic System. In *2023 Annual International Conference on Emerging Research Areas: International Conference on Intelligent Systems (AICERA/ICIS)*, 1–6. DOI: 10.1109/AICERA/ICIS59538.2023.10420023

Sharma, S., Gautam, A., & Tyagi, R. (2024). Artificial Intelligence and Cyber Security Driven Social Innovations in Digital Adventure Tourism. In 2024 First International Conference on Pioneering Developments in Computer Science & Digital Technologies (IC2SDT), 1–6. DOI: 10.1109/IC2SDT62152.2024.10696759

Sharma, S., Tomar, M., & Tyagi, R. (2024). Artificial Intelligence and Vedic Scripture in Digital Agriculture for Global Economy. In *2024 International Conference on Intelligent and Innovative Technologies in Computing, Electrical and Electronics (IITCEE)*, 1–6. DOI: 10.1109/IITCEE59897.2024.10467916

Sharma, S., & Tyagi, R. (2023). Digitalization of Farming Knowledge Using Artificial Intelligence and Vedic Scripture. In 2023 IEEE International Conference on ICT in Business Industry & Government (ICTBIG), 1–6. DOI: 10.1109/ICTBIG59752.2023.10456219

Sharma, S., Tyagi, R., & Mohan, S. (2024a). Artificial intelligence and green 6G network-enabled architectures, scenarios, and applications for autonomous connected vehicles. In River Publishers eBooks (pp. 187–212). DOI: 10.1201/9781003515920-9

World Bank. (2017). *The Global Findex Database 2017: Measuring Financial Inclusion and the Fintech Revolution*. World Bank.

Yermack, D. (2016). Corporate governance and blockchains. *Review of Finance*, •••, rfw074. Advance online publication. DOI: 10.1093/rof/rfw074

Chapter 4
Digital Payment Systems and Their Transformative Role in Modern Marketing Strategies

Y. Anjel Raj
https://orcid.org/0000-0003-3591-0672
GNA University, India

ABSTRACT

The rapid development of electronic payment systems has reshaped financial transactions and redrawn the boundaries of modern marketing strategies. This chapter explains how digital payment platforms such as mobile wallets, UPI interfaces, QR codes, and contactless payment solutions are revolutionising consumer behaviour, brand experiences, and the overall customer journey. By making transactions smooth, fast, and secure, these systems result in enhanced customer satisfaction and loyalty and enable real-time personalisation and data-driven marketing strategies. The chapter critically analyses the intersection of digital payments and marketing strategies, elaborating upon how companies leverage transactional data to strengthen targeting, segmentation, and optimisation of promotion campaigns. It further explains how payment solutions are incorporated into omnichannel retail spaces and the strategic consequences for customer engagement, trust, and brand differentiation.

INTRODUCTION

Digital payment systems are electronic platforms on which people and organizations can make monetary transactions without using money (Naeem et al., 2020).

DOI: 10.4018/979-8-3373-4526-0.ch004

Such systems have completely changed the nature of exchanging value in the contemporary economy (Bowles and Gintis, 1988). Some of their financial tools and technologies are credit and debit card network, mobile payment applications such as PhonePay, Google Pay, Apple Pay, and Paytm, online banking portal, and peer-to-peer (P2P) money transfer applications such as PayPal and Venmo (Alam et al., 2019). Though being rather new to the field, cryptocurrencies and blockchain-based technologies are rapidly becoming an important disruptor in the industry by offering decentralized solutions to the conventional financial systems (Mikhaylov, 2020).

What is making digital payment system more special than the traditional cash and check-based transactions is the fact that it blends well with the digital world (Chen, 2004). They provide real time service, immediate transfer of money and reduce the transactional friction at a very high extent (Taddesse & Kidan, 2005). The most recent digital payment technologies are being supported by high-speed mobile internet, high-level encryption, scalable cloud computing, and adaptable application programming interfaces (APIs) that guarantee convenience, protection, and velocity (Khan et al., 2013). The usage of these systems has got far more purposes than what is limited to online shopping. They allow people to pay utilities, transfer wages to workers, subscribe to online content and services, make appointments and forward money anywhere in the world at the touch of a screen (Allen, 2006).

The position of digital payment systems has changed to something luxurious or convenient to the operations of the global economy to a necessity (Sumanjeet, 2009). They have experienced explosive growth in the past decade, which is stimulated by the emergence of e-commerce, the popularity of smartphones, high rates of digital transformation, and the necessity of contactless payment solutions in case of global health crises (George et al., 2023). Such features simplify the business in understanding their performance, strategies in marketing, and the ability to increase their customer base, even at international level (Javalgi et al., 2004). In the eyes of consumers, digital payments are more convenient than any other payment system (Olipane, and Inocencio, 2023). Customers are directly affected by the perception and satisfaction realized through the ease, speed, and transparency of the payment process (Islam, 2024). The time-saving checkout process also becomes a value proposition of the brand, which enhances loyalty and generates future interactions (Brakus et al., 2009).

They also minimize the expenditures that go to print and handling the cash, increase the visibility of the financial transactions, and aid in reducing the illegal activities through establishing a track of payments that is traceable (Dos Santos & Harvold, 2017). The regulatory authorities are also contributing significantly towards the formation of the digital payment environment (Putrevu and Mertzanis, 2024). As an illustration, open banking and data portability systems are empowering consumers with their financial data as they encourage new services to be created by fintech businesses (Barkas, 2024). The digital payment systems are also altering the nature of the marketing world radically, and much of this is influenced by modern technology (Gomber et al., 2018).

Marketing used to be one-way, mostly through television advertisements, newspaper placements, and radio spots that were not customised and heavily focused on broad grouping (McChesney, 2016). The modern marketing is now real-time, hyper-personalised, and omnichannel. It uses data on the past purchase behaviour, browsing history, social media, and digital payments, among others, to predict consumer needs and behaviours in an unparalleled granularity (Parise et al., 2016). This data assists marketers to segment customers using behavioural insights instead of assumptions, customise their messages, and improve on targeting using timely and relevant offers (Upadhyay et al., 2024). The touchpoints that could be optimised in terms of customer engagement include payment confirmations, app dashboards, loyalty programs, and even digital receipts (Scott, 2011).

The customer experience is never linear anymore, it is a dynamic or evolving sequence of interactions between platforms and touchpoints (Lemon and Verhoef, 2016). The point of purchase at once considered the last point of a consumer channel has become a potential point of brand contact (Lemon & Verhoef, 2016). Imagine the benefits of frictionless payments by one-click checkout, auto-fill features, wallet, etc. Every single enhancement decreases cognitive load and enhances the chances of conversion (Sankaranarayan et al., 2020).

Marketing world has also been drastically changing, and much of it is being affected by the advent of the digital payment systems (Gomber et al., 2018). Marketing traditionally was one-way, dominated by the TV ads, placement in newspapers and radio spots that targeted very broad segments of the population with minimal customisation (McChesney, 2016). It uses information about various sources, such as social media, browsing history, purchase behaviour, and online payments, to comprehend consumer needs and behaviours in unparalleled granularity (Parise et al., 2016). This information assists marketers to segment customers using behavioural data as opposed to assumptions, personalize their messages, and optimize targeting using offers that are timely and relevant (Upadhyay et al., 2024). Touchpoints can be optimised to engage the customers even through payment confirmations, dashboard apps, reward programs, and even digital receipts (Scott, 2011).

Customer journey is no longer linear, it is dynamic or changing sequence of interactions on platforms and touchpoints (Lemon and Verhoef, 2016). A moment of payment, which used to be considered as the last stage of a purchase funnel, is now a highly vital brand engagement moment (Lemon and Verhoef, 2016). How can frictionless payments one-click check out, auto-fill features, wallet implementation be improved to improve user experience? Any minor enhancement eases the cognitive load and promotes the possibility of conversion (Sankaranarayan et al., 2020).

The behavioural information obtained after payment systems can enable the marketer to not only know what consumers are buying, but also when, how often, in what combinations, and where (Schoenbachler and Gordon, 2002). Marketing has become sensitive, more charged with data and more accountable. Real-time marketing can also be achieved by using payment data (Brosche and Kumar, 2015). This data is used by subscription services, such as Spotify and Netflix, to track usage trends, optimize recommendation systems and deliver timely upgrades or deals (Amudharasan, 2023).

The increasing overlap of digital payments systems with contemporary marketing is a radical change in the way companies conduct business and relate to consumers (Kumar et al., 2021). In a world that is becoming more digital, companies that are able to combine payment intelligence and customised marketing programs will be in the best position to compete, innovate, and succeed (Weichert, 2017).

Purpose and Scope of the Chapter

This chapter is all about exploring how digital payment systems are changing the way businesses approach marketing in today's digital world. The relationship financial technology and marketing displays the pattern of payment for transactions but also powerful channels for engaging customers.

- The history of digital payment which developed over time
- The strategies of marketing adopted with technological changes
- The different ways of payment system which are aligned with customer experience design
- Challenges and concerns in security, privacy, and regulations
- Future trends in embedded finance, AI powered payment analytics and the growing role of decentralized payments.

Roadmap of the Chapter

In order to explore the transformative nature of digital payment systems in the current marketing strategies in a systematic manner, this chapter has been organized in the following way. Section 2 follows the timeline of the development of payment systems, pointing to the technological breakthroughs that facilitated the transition of the time-honored system of cash transactions to digital, mobile, and decentralized payment systems. Section 3 discusses the shift in the modern marketing environment, with the focus on the move towards data-driven and omnichannel promotion and experience-focused promotion. Section 4 defines the data as the new currency, and how the transactional data generated by payment can be used to create behavioural segmentation, personalization, predictive analytics, and loyalty management. Section 5 critically evaluates the issues and dangers of the payment convergence and marketing, such as cybersecurity risks, regulatory and moral considerations, effects on vulnerable customers, and the question of market fairness and dominance of the platform. The section 6 is on the latest trends and strategic implications, which discuss artificial intelligence, embedded and invisible payments, and Web3-based payment innovations. Lastly, the last section is Chapter 7 which recapitulates the main ideas, provides managerial and policy implications, and suggests future research directions.

Theoretical Framework

This chapter also uses a conceptual base of the theories of the Diffusion of Innovations Theory, Service-Dominant Logic, and Customer Experience Theory to explain how digital payment systems are changing the marketing strategies. According to Diffusion of Innovations, digital payment technologies are adopted based on perceived relative advantage, compatibility, and easiness of use. The Customer Experience Theory emphasizes the role of convenience, speed, and security of payments in promoting satisfaction, trust, and loyalty throughout the customer experience. Collectively, these models indicate how payment technologies go beyond transactions to be strategic marketing facilitators.

LITERATURE REVIEW

The recent research indicates that digital payment system has become more strategic in marketing and consumer communication. According to research conducted in 2020-2025, digital payments help to strengthen personalization, integration of loyalty, and omnichannel experiences through the creation of real-time consumer data (Verhoef et al., 2023; Kumar et al., 2021). Nevertheless, important controversies about the threat to privacy, algorithmic nudging, and surveillance capitalism can also be found in the literature (Martin and Murphy, 2022; Zuboff, 2019).

Digital Payment Systems: Conceptual Foundations and Evolution

Scholarly discussions of digital payment systems have developed not only in the transaction efficiency direction, but also in the wider approach to the strategic, behavioural, and social, implications of them. The initial researches had personified electronic payment systems as technological replacements of cash and paper-based instruments with a focus on speed, security, and cost-saving (Crede, 1995; Evans and Schmalensee, 2004). Later studies elaborated on this perspective by locating digital payments as social-technical systems which are integrated into institutional, regulatory, and cultural environments (Sumanjeet, 2009; Peneder, 2022).

The modern-day literature is becoming more aware of digital payment systems, including mobile wallets, UPI interfaces, and contactless applications, as data-intensive systems that transform consumption patterns, company strategy, and market structure (Gomber et al., 2018). Regarding evolution, research illustrates a distinct evolution of card-based systems into mobile and platform-based payments, with the rate of adoption being accelerated by smartphone penetration, cloud computing, and APIs-based architectures (Omarini, 2018; Khan et al., 2017).

Digital Payments and Consumer Behaviour

An extensive body of research on the effect of digital payments on consumer behaviour (and, notably, convenience, trust, and satisfaction) exists. Empirical research has continuously indicated that convenientness, perceived security, and speed of transaction have a positive impact on adoption intention and frequency of use (Olipane & Inocencio, 2023; Islam, 2024). On the behavioural level, because of the digital payments, cognitive and physical effort during the point of purchase is lowered, which decreases transaction friction and raises the chances of purchase. According to research based on behavioural economics, frictionless payment can also stimulate an impulsive expenditure and undermine sensitivity to price amongst consumers (Schwartz, 2001).

Digital Payment Systems as Marketing Enablers

Digital payment systems are becoming strategic facilitators of marketing literature as opposed to operational tool. According to scholars, payment platforms produce high-quality and revealed-preference data that is more accurate compared to traditional survey-based or attitudinal data (Bell et al., 2018). The data of this type is used to facilitate granular behavioural segmentation, real-time personalization, and predictive analytics, thus enhancing targeting and campaign efficacy (Riegger et al., 2022; Upadhyay et al., 2024). A number of studies reveal that the data based on payments helps in enabling loyalty programmes and dynamic pricing as well as post-purchase interaction (Wang et al., 2018; Kashiramka et al., 2025).

Theoretical Frameworks Anchoring Digital Payment Impacts

Technology Acceptance Model (TAM): TAM has popularly been applied in elucidating digital payment adoption in that perceived usefulness and perceived ease of use are connected with behavioural intention. Although, TAM based studies are effective in explaining initial adoption, their weakness is their lack of focus on the other importance aspects of post-adoption behaviours including loyalty, trust, and engagement.

Unified Theory of Acceptance and Use of Technology (UTAUT): TAM extensions using UTAUT include social influence and facilitating conditions, which provide a better explanation of payment adoption in emerging economies. Nevertheless, such models are still rather technology-focused and less sensitive to marketing performance.

Customer Experience Theory: This theory, which is based on the idea of experiential marketing, perceives payment interactions as important moments in the customer experience (Lemon and Verhoef, 2016). The use of this lens by researchers has indicated the importance of seamless checkout, omnichannel, and invisible payments in increasing perceived brand value and satisfaction. However, there is little empirical research connecting the direct relationship between payment experience and brand equity.

Trust and Risk Theory: Due to the sensitivity of financial information, the concept of trust-based theories is the core of the digital payment use. Research lays stress on trust-creating antecedents such as security guarantee, transparency and compliance with regulations as antecedents of trust that affect loyalty and repeat purchase behaviour. However, the concept of trust is usually understood as a set of statics, and the dynamic process of evolution during multiple transactions is overlooked.

Debates, Gaps, and Limitations in Existing Literature

Even with much academic focus, there are a number of controversies. A major controversy deals with individuality versus confidentiality. Though payment-based personalization ensures increased relevance and interaction, it also leads to ethical and regulatory issues, especially when operating in strict data protection frameworks, like GDPR. The literature has a cross-sectional survey and adoption model foundation in many cases, which constrain the ability to causally infer and gain longitudinal understanding. This gap is resolved by growing the scope of customer experience and the data-driven marketing systems to a holistic view of digital payment systems as a transformative element in the modern marketing approach in this chapter.

Updated Market Indicators (Selected)

Market size (2024-25): The industry estimates the global digital payments market to be approximately USD 115-120 billion in the year 2024 with projections of an excellent growth to USD 360-520 billion (based on methodology) in the years 2030-2034.

UPI (India) scale: UPI remains one of the leading real time systems of payment, with annual transaction volumes in the tens of billions (ex: in FY 2024-25, UPI, 18,500-20,500 million transactions), or with monthly transactions reaching in the millions (e.g., 1,900-2,000 million transactions).

Contactless adoption: Contactless card and mobile tap to pay adoption in 2024 had been able to reach record adoption in a number of markets, with a number of issuers having over 90% of in store eligible card transactions going contactless in markets such as the UK.

BNPL growth: BNPLs have grown quickly in 2022-24, with market valuations and gross merchandise volumes expanding into the tens of billions USD and the user base expected to reach the hundreds of millions around the world by the mid 2020s with rising concerns over consumer indebtedness and default rates.

Empirical Vignettes

Vignette A: UPI as a mass scale, low cost platform (India): the UPI in India shows how an interoperable, low cost rails system can create a mass adoption of retail. Integrating banks, third party applications and merchants through open APIs and real time settlement created tens of billions of transactions every year and opened up new merchant acceptance at the lowest possible hardware expense (QR and app based flows).

Vignette B: Contactless ubiquity and impulse conversion (UK/Europe): in card markets with well-developed card infrastructure, contactless penetration was higher than traditionally anticipated: leading issuers reported virtually universal contactless acceptance in eligible in store payments. As a retailer, contactless lowers the time spent at the checkout, and time spent waiting in queues, and more often converts to revenue per store and labour efficiency.

Vignette C: BNPL scale and risk indicators (Global): BNPL providers expanded by providing frictionless checkout finance with almost instant credit decisions and merchant-funded merchant fees/commissions. This business model increased average order values and conversion among merchants but this also led to increased delinquency exposures among younger cohorts. The short term basket lift and regulatory scrutiny and credit risk externalities are the trade-offs that marketers eager to grow through BNPL face.

Innovations: Business-Model Logic and Marketing Consequences

Embedded finance & invisible payments: Embedded payments put financial services within non financial customer experiences (ride hailing, food delivery, marketplaces). Business logic: platforms take a higher margin (payment/processing fees), enhance user stickiness through friction reduction, and also provide new data capture points.

Loops of data monetization and personalization: This is high quality and revealed preference transactional data. This can be turned into predictive models on CLV, dynamic pricing and next best offers by firms. Business logic: high-quality data - enhanced targeting - improved conversion - further investment of data acquisition (loyalty, incentives).

BNPL and merchant economics: The revenue that BNPL providers can receive are usually merchant fees, exchange spreads, and late fee income that allow merchants to provide installments without interest charged and outsource credit risk to the BNPL provider (or share). To marketers, BNPL is a conversion tool: better AOV, less cart abandonment - at the expense of merchant margin and possible reputational harm in the event of consumer harm.

Decentralized payments and token loyalty: The blockchain and tokenization allow programmable rewards and movable loyalty assets. Business logic: token reward can be divisible, interactable, cross-partner ecosystems - potentially following engagement and second markets of loyalty. Among the constraints are volatility, regulatory uncertainty, and clear utility to the consumers.

Research implications and closing note

These examples and novel indicators enhance the main argument behind the chapter: transaction utilities do not only replace the digital payments but strategic platforms that transform the marketing levers and business models. Future empirical studies ought to utilize longitudinal research and platform level data to unpack causal effects (e.g., invisible payments) to spending elasticity; BNPL exposure) to long run loyalty).

EVOLUTION OF DIGITAL PAYMENT SYSTEMS

From Traditional Payments to Digital Transformation

History of the payment system refers to the ongoing pursuit of convenience, security, and efficiency in exchange (Khan et al., 2017). The barter system was the primitive form of trade that human beings started with where goods and services were traded without an agreed value unit. Money was a medium of exchange and its emergence was a critical move (Kiyotaki and Wright, 1989). The progress was followed by the emergence of the paper money that was less difficult to manufacture, transport, and trade (McKinney et al., 2015). These paper-based systems, although a revolution those days, still had some flaws in regard to speed, accessibility, and efficiency of operation (Verhoff and Donkers, 2005). The second part of the 20th century has seen a paradigm shift based on telecommunication, computer science and financial infrastructure developments (Cortada, 2005).

The Digital Transformation of Payments

The reality of digitalization of payments was realized with the advent of credit cards and debit cards (Peneder, 2022). They enabled consumers to shop and get their bills directly charged to their financial accounts through swiping a card. Cards were initially regarded as something luxurious that was acceptable to the wealthy, but quickly became mass-acceptable (Truman et al., 2003). The emergence of Visa and MasterCard in the 1970s came to the rescue of standardising this type of payment and made it a part of worldwide trade. Internet revolutionized all the areas of human life including financial services in the 1990s. Online banking portal Banks started to provide an online banking portal through which its customers could manage their accounts, transfer money (Giannakoudi, 1999) and pay bills without walking into a branch. The latter innovation was the basis of the digital self-service in financial management (Sumanjeet, 2009). The next development was the mobile wallets due to the introduction of smartphones in the 2010s. Apple Pay, Google Pay, Paytm, and Bitcoin in 2009 brought up a whole new paradigm decentralized and peer-to-peer digital currency that used the blockchain technology (Lee, 2019). Even though the regulation, volatility, and scalability still need to be addressed, blockchain has brought about fresh opportunities in cross-border payments and decentralized finance (Mikhaylov, 2020).

Key Milestones in Digital Payment Evolution

1950s–1970s: Credit and Debit Cards

- In 1950, Diners club introduced the first credit card.
- Visa and MasterCard came in and made card payments an international norm.

1990s: Online Banking

- Bank provided web portals on transfer of funds and management of accounts.
- Established the platform of e-commerce and distant banking.

2010s: Mobile Wallets

- It was as easy as tapping your phone to make payments through Apple Pay, Google Pay, Paytm and Alipay.
- These apps were developed with a QR code and NFC technologies in order to provide contactless payments. Wikipedia: 20092011 Cryptocurrencies and Blockchain.
- Bitcoin has brought about decentralized and peer-to-peer digital transaction.
- Smart contracts, remittances, and NFTs are also other intended uses of blockchain.

Technologies Powering the Digital Payments Revolution

Several technologies have catalysed the digital payment ecosystem:

- **Near Field Communication (NFC):** It allows contactless payment at a point-of-sale (POS) terminal with phones or smart cards. NFC is also very quick and secure in checkout, particularly in its use in stores and transport.
- **QR Codes:** QR codes are especially popular in Asia and allow users to pay and scan. It is a cost-effective and user-friendly option of small business and street vendors.
- **Blockchain**: Cryptocurrencies are supported and facilitate decentralised transactions. Its unchangeable registry is under consideration on smart contracts, supply chain insights, and live settlements.
- **Biometric Authentication**: Facial recognition, fingerprint scanning, voice ID can provide at a higher level of security. Biometrics lessen the use of passwords and introduce friction free authentication.

Global Trends and Regional Variations

Digital payment adoption varies by region due to differences in infrastructure, consumer preferences, and regulatory environments.

- Asia-Pacific: The region is the leading digital payment practice in the world. The local market is dominated by Alipay and WeChat Pay of China, which combine the aspects of payments, social media, e-commerce, and entertainment. The Unified Payments Interface (UPI) in India has changed all that with the P2P and merchant payments turning instant, free, and mobile.
- Latin America: The market of digital payment is in the boom phase associated with fintech innovation and rising smartphone penetration. The PIX system of Brazil and CoDi system of Mexico provide quick and secure transactions that are linked with banks in real-time.
- Africa: M-Pesa mobile money has enabled financial inclusion to millions of people in Kenya and other African countries. These services enable customers to store, send and receive money using simple feature phones, which is a way of overcoming the traditional banking constraints.
- Europe and North America: The attention is given to the improvement of the customer experience using digital wallets, online banking, and payment by a card without any contacts. Competition and customer-centric innovation is being encouraged by open banking regulations such as PSD2.

Cultural and Economic Impacts

Development of digital payment has brought in other things other than convenience. It has had an impact on cultural and economic practices in the global arena. Urban forms are quickly becoming the norm of digital transaction, particularly in the younger generations. Digital payments are gateways to financial systems in developing regions and are therefore a source of economic inclusion. Frictionless digital payments that promote real-time transfers and receipts are deeply connected with the emergence of on-demand services like Uber, Swiggy, and Airbnb.

Business and Marketing Implications

The shift toward digital payments implies not only the efficiency of operations, but also new possibilities to understand customers, personalize, and be loyal. All purchases leave records: what, when, where, and how often. Marketers need this information to know the consumer behaviour. The service industries can also develop grade pricing and dynamism of discounts depending on recent transaction.

Flexibility options such as credit/debit cards, wallets, UPI, and BNPL (Buy Now, Pay Later) services are positively correlated with consumers because they provide them with flexibility and a higher possibility of conversion and improvement of the brand image.

The Future of Digital Payments

The innovative applications of embedded finance, voice-activated payments, and wearable technologies are all playing a major role in defining the next generation of the digital payment revolution. Embedded finance is the concept of embedding financial services into non-financial products (e-commerce websites, ride-hailing applications, and food delivery applications) to provide users with smooth, contextualized experiences in the product. The model also does not require users to change platforms in order to make financial transactions and has been identified as an essential platform to enable financial inclusivity and business model innovation (Popescu, 2022). These platforms are considering how to enable the users to authorize their transactions using voice commands and voice biometrics, which can be convenient and a new point of access (Jonnala, 2024). These devices enable on the go and contactless transactions and are being favored more by consumers due to their convenience and speed (Rabaa Ai and Zhu, 2021). The introduction of Decentralized Finance (DeFi) is, perhaps, the most radical innovation in the financial world. DeFi is based on blockchain and smart contract technology that removes the need to use financial intermediaries and enables users to lend, borrow, invest, and exchange assets over decentralized applications (dApps). Nonetheless, it is also associated with significant security, volatility, and regulation concerns (Capponi et al., 2020), DeFi can become democratizing finance as it is decentralizing the control held by centralized institutions to individuals, but the extent of its impact will be determined by the nature of regulatory changes.

THE CHANGING LANDSCAPE OF MODERN MARKETING

In recent decades, the field of marketing has been radically changed to evolve out of the traditional paradigms towards a more dynamic, digital, and data-centric paradigm. Companies do not work in the one-size-fits-all model of communication anymore, and marketing became more targeted, interactive, and personalized (Rosário and Dias, 2023).

From Traditional to Digital Marketing

In the traditional sense, marketing was rather linear. Distribution of businesses was done through mass media newspapers and radio, television and billboard advertising. These mediums were good capacity to create awareness of a brand but failed to reach particular consumer groups and it could not measure the performance of a campaign with accuracy (Markus, 1987). In contrast to their old-fashioned equivalents, the channels can be used to gather user data and analytics, and by doing so, the brands can tailor the content, optimize the messaging, and improve performance in real-time (Ramachandran, 2023). Digital marketing is not merely a continuation of the old one, this is a whole new way of doing things, which is founded on data, automatization, and feedback loops. In addition, there is an opportunity to use digital advertising platforms such as Google Ads and Meta (previously Facebook) that can target with superior accuracy, depending on their browsing history, location, previous purchases, and even their psychographic characteristics.

The Rise of Data-Driven Marketing

A change to data-driven strategies is one of the most significant changes in contemporary marketing. Organizations currently use huge amounts of data to know their customers, anticipate behaviour, and communicate with them in a personal way. This data can be interpreted using artificial intelligence, machine learning, and business intelligence platforms to enable a business to develop specific campaigns that are cost-effective and effective (Riegger et al., 2022).

Customer Experience as the New Battleground

Customer experience (CX) has become the competitive advantage of the brands in this era of abundant data. The Customer Relationship Management (CRM) software, AI-powered chatbots, and feedback loops are the instruments that are necessary in creating these interactions (Naslednikov, 2024). Consumer today requires immediacy, relevance as well as value in each interaction. The brands are supposed to respond in real-time, predict customer needs, and tailor content. Machine learning and AI become very vital here. An example of chatbots is to respond to most frequently asked questions 24 hours a day, whereas recommendation engines (such as Amazon or Netflix) can propose items or content based on personal preferences.

Multichannel and Omnichannel Marketing

An important feature of the contemporary marketing is the combination of various channels to offer a unified brand experience. Multichannel marketing is gaining momentum through accessing customers at a number of different platforms email, social media, mobile apps, websites and via physical stores. Omnichannel marketing is even more by guaranteeing that there is a coherent and continuum experience in all these touchpoints. An example is when a customer would get a promo email, open up products online, go to the physical store to see the product personally, and ultimately use a mobile application to make the purchase. An actual omnichannel strategy can guarantee that all these interactions are linked together in customer data flows across platforms in such a way that they can be personalized, and customer services are provided at all levels.

The Importance of Mobile and App-Based Marketing

Consumers are now mainly using mobile devices as the main channel of accessing digital content. There are peculiarities in interaction through mobile applications. They enable the brands to provide push notifications, collect granular behavioural data, and loyalty programs. An example of this is apps such as Zomato, Uber, and Amazon which are utility and marketed at the same time, providing users with personalised promotions, user behaviour tracking, and in-app purchases. PWAs enhance speed, reliability, and interactiveness particularly in areas with poor internet infrastructure.

E-Commerce and Integrated Campaigns

Online retailing systems like Amazon, Shopify, and Flipkart have become key marketing ecosystems. The platforms have built-in advertising opportunities, customer engagements, and real-time performance indicators. This integration enables co-ordinated campaigns, where marketing is coordinated with logistics, inventory and customer service. Another trend that is emerging is social commerce which means people can directly shop using social media. The Instagram Shops feature, Facebook Marketplace, and Tik Tok integrations allow brands to bridge the discovery and purchase gap. Influencers and brands also sell their products in real time, and livestream shopping developed in China gaining popularity all over the world, allows influencers and brands to show the products and encourage instant buyers.

Content Marketing and Storytelling

Content is still king even in the digital era, but the consumption behavior of content has changed. The inbound marketing is now based on blogs, podcasts, videos, webinars, and infographics. Instead of marketing products, brands are offering value in terms of education or entertainment content. Storytelling has also become one of the most important tools of brand building. The marketing environment has experienced a paradigm shift reconfiguring itself to keep up with the digital era. The communication, interaction and conversion of brands have been transformed by technology, data and consumer expectations. The tools can change (e.g. the use of artificial intelligence in chatbots or customized emails) depending on the traditional print and television advertisements but the main goal is to establish long-lasting relationships with consumers.

DATA AS THE NEW CURRENCY

In the current digital age, information has emerged as a precious currency that is not more valuable than the currency that payment systems generate (Bolonin and Balykin, 2021). Every purchase will be a narrative: what, when, where, and how many times, and how much was bought. Those companies that use this data strategically can develop actionable intelligence that will result in smarter marketing, more engaged customers and increased profitability (Rathod, 2023). Compared to self-reporting of the survey data or browsing behavior, payment data records actual purchasing decision, and therefore, offers a more natural and useful source of information to marketers (Bell et al., 2018). Also, the payment data may be matched with the demographic and location data, which offers multi-dimensional insights (Rose et al., 2010). As an example, businesses are able to distinguish between the frequent and the infrequent purchasers, gauge the success of product bundling or even track the effect of promotions on the spending habits.

Behavioural Segmentation and Personalization

Behavioural segmentation is one of the most effective applications of payment data. Instead of demographics, marketers are able to categorise customers based on actual spending behaviour such as high spenders, bargain hunters or brand loyalists. They allow companies to build specific campaigns, offer specific deals, and perfect customer experiences (Okeke et al., 2024). A fashion brand may discover that a section of customers purchases during their seasonal sales regularly. Personalization engines can also be fueled by payment information that suggests products based on previous purchases, which leads to increased customer satisfaction and loyalty (Tongxiao et al., 2011). Real-time marketing interventions will be enabled due to the speed and intensity of granularity of payment information. Firms are able to offer contextual promotional deals such as offers or add on recommendations right after a purchase or hours after making a purchase (Mao, 2016). This timely delivery makes the delivery more relevant and also it increases the level of response among customers.

A regular client such as the one can be offered with a preferential airfare or packages based on his or her payment record (Copeland and McKenney, 1988). The past payment patterns also can be used by predictive analytics models to anticipate the future behaviour, including recency to purchase once again, service upgrade, or churning in order to take the pre-emptive marketing steps (O'Brien and Kunz, 2024). The foundation of the modern loyalty programs is the payment information.

By tracking the purchases and linking them to rewards programs, firms can stimulate repeat behaviour and create long-term relationships. Starbucks and Sephora use payment-linked loyalty applications, which track purchases in, reward customer accomplishments, and prevail certain promotions (Kashiramka et al., 2025). In subscription businesses (e.g. Netflix, Spotify, Amazon Prime), the data on payments is central to the analysis of the lifecycle of users, churn management, and upsale of high-value features. The analytics will reveal when the customers are likely to cancel or when they will be willing to upgrade. Major role in re-targeting strategies is also played by payment behavior. Personalized offers or reminders can be offered to abandoned cart customers or frozen subscription customers based on what transaction or payment attempt was previously known about the customer (Sajan & PS, 2024).

CHALLENGES AND RISKS

Cybersecurity and Technology-Related Risks

The cyber-attacks have substantially increased the platform upon which they can be implemented due to the digitalization of payments. As sensitive information like card numbers, personal identifiers and transactions history are being processed and stored on interconnected platforms, the chances of data breach, identity theft, and financial fraud are growing exponentially. Even those companies that have sophisticated security systems are never safe, especially when their payment structure is incorporated in marketing platforms and third-party applications. As a marketing factor, such violations not only derail the operations but also zero consumer trust, reputations and lower customer loyalty in the long-term.

Regulatory and Policy Framework Challenges

The other significant issue is dealing with the constantly growing and multifaceted policies regarding online payments and use of data. Certain frameworks like the General Data Protection Regulation (GDPR) and the Payment Services Directive 2 (PSD2) have high standards in reference to data protection, informed consent, transparency, and a high level of customer authentication. The goal of these regulations is to empower the consumers, ensure fair competition, and improve security in the digital financial ecosystems. Nevertheless, they also force companies to re-evaluate their data strategies, restrain uncontrolled profiling, and spend a lot of money on compliance infrastructure.

To marketers, the compliance with regulations has direct impact on the ability to capture transactional data, and combine and use it to create personalization and target behavior. Lack of compliance may cause harsh monetary fines, criminal proceedings as well as damaged reputations and thus regulatory watchfulness is a strategic business requirement and not a duty mandated by law.

Ethical Concerns and Responsible Data Use

In addition to official regulation, the ethical issues related to payment-based marketing become increasingly popular. The data on the transactions gives a very intimate information on the lifestyles, tastes and money habits of the consumer. The application of such data to algorithmic personalization, dynamic pricing, or automated nudging creates the concern of manipulation, a lack of autonomy, and equity. Ethical marketing principles uphold transparency, proportionality, and respect of agency of the consumer; however, they can be compromised when algorithms are operated opaquely. Differential prices or real time promotional nudges on the payment interfaces can maximize short term revenue, but will harm consumer trust when interpreted as exploitative. Thus, companies are also becoming more accountable by implementing ethical governance systems such as explainable algorithms, internal data ethics policies, and accountability frameworks to make sure that payment data is used responsibly.

Implications for Vulnerable Consumers

The adoption of frictionless payments, subscription plans, and Buy Now Pay Later (BNPL) has a disproportionate impact on consumer groups. The methods of behavioural marketing that are integrated into payment systems can intensify impulsive consumer behaviour, especially when the consumer has to make a transaction that does not demand the involvement of a significant amount of cognitive resources.

Politically and morally, these relations pose a question of consumer security and economic welfare. Spending warnings, clear fees declaration, transaction limits and cooling off periods are necessary to make sure that the convenience of technology does not unfavourably cut off the vulnerable users.

Market Fairness and Platform Dominance

The fast development of large payment platforms and super-app ecosystems at the market level raise some issues in terms of competition and fairness. Major platforms enjoy high network effects and access to large volumes of transactional data, which is a barrier to entry and small firms. Such a centralization of power can create data monopolies in which a small set of players will dominate payment infrastructure, customer insight, and access to marketing at the same time. Nonetheless, there has been inconsistent implementation especially in the developing economies. In the case of business over-dependence on large payment intermediaries also poses dependency, loss of bargaining power and loss of direct customer relationships strategic risks.

Synthesis of Risks

Together, these issues indicate that the establishment of digital payment systems as a part of marketing strategy is not only a technical problem but a governance, ethical and societal one. To maintain a sustainable level of innovation, organizations have to stabilize efficiency and responsibility by synchronizing cybersecurity investments, compliance with regulations, ethical data practices, and consumer protection systems. Companies that actively deal with such risks have higher chances to develop long-lasting trust, safeguard vulnerable users, and help to establish a fair and competitive online market.

FUTURE TRENDS AND STRATEGIC IMPLICATIONS

Disruptive technological innovation, shifting consumer preferences and new digital environments are reshaping the marketing and payment environment. With the advent of artificial intelligence, merged finance, and decentralised models, companies must stay mobile and forward-thinking to understand the opportunities and threats.

The following are some of the best trends that will transform the intersection of marketing and payments and have a bearing on their strategies. Both customer engagement and transaction processing are in the midst of transformation due to artificial intelligence (AI) and machine learning (ML). AI enhances fraud detection in payments by observing trends in real-time, creating warning signals and acting upon emerging threats. In marketing, AI is used to power engines of personalization, recommendation and customer segmentation based on behavioural and transactional information. Predictive analytics are also offered by ML algorithms and allow companies to anticipate customer behavior, optimize pricing, and make automated decisions. At the strategic level, businesses that use AI not just to get automation but to develop in-depth insights will have a winning competitive advantage.

Invisible payments are those transactions made in the background and very little or no input is required of the customer such as ride-hailing apps where the customer makes the payment automatically after completing the ride. It is closely related to embedded finance, in which financial services (payments, lending, insurance) are integrated into non-financial applications, e.g. e-commerce applications, social media, or smart household gadgets.

RECOMMENDATIONS FOR MARKETERS AND BUSINESSES

The firms must use a strategic and proactive stance in order to navigate their way through the rapidly evolving mall of marketing and payments. To begin with, intelligent technology infrastructure will require an investment in embracing scalable, artificial intelligence-based platforms, companies can collect payment systems, marketing operations, and customer data to produce real-time insights and enable profound personalization. Ethical data practice is also imperative in our present culture of being data aware. Corporations must prioritize transparency in data gathering and usage, responsibility to consumers in an ability to manage their privacy preferences and set up universal data protection obligations. Second, businesses need to experiment with innovative technologies such as Web3, digital wallets, and embedded finance to understand its future opportunities, especially in reaching digitally born customers. Thirdly, the strategic alliance with fintech firms, AI solution firms, and digital platforms may be developed to facilitate innovation and enable businesses to transform their transformation processes in the best way possible.

Discussion: Empirical Context and Illustrative Evidence

The digital payments have changed marketing in the various regions. In India, Unified Payments Interface (UPI) made more than 117 billion payments in 2023 (RBI, 2024), which allows companies to provide real-time promotions, micro-rewards and customized promotions at the point of payment. The M-Pesa of Kenya shows that mobile payments are useful in inclusion-based marketing, where they have penetrated more than 96 percent of households (World Bank, 2023). In China, examples of super-app ecosystems, in which payments, social commerce, and loyalty programs merge, to transform consumer-brand relationships, are represented by Alipay and WeChat Pay.

Ethical, Regulatory, and Risk Considerations

Digital payment systems are subject to serious ethical and regulatory issues, even though they have some positive effects. Transaction level data collection increases the risk of data privacy, consent mechanisms, and surveillance. The introduction of the idea of algorithmic nudging in payment interfaces can promote spontaneous consumption, which contradicts the responsible marketing policies. Regulatory measures including GDPR, PSD2 and PCI-DSS strive to address these risks, but still, fragmentation continues across territories. The issues of cybersecurity threats and digital divides also complicate the equitable adoption, especially in the vulnerable populations.

CONCLUSION

At the crossroads between marketing and payment systems, the paradigm shift in digital business is described where the transitional data, customer experience, and strategic communication grow closer to one another. With the shift of companies to physical to digital-first strategies, the importance of data-driven marketing is even more tangible, and payment systems become a treasure trove of real-time consumer insights. Frictionless and embedded payment experiences not only reduce friction but also open up new opportunities to interact, build loyalty and increase revenue. The fact that it includes mobile apps, one-button payments, and subscription models are only some examples of how well-designed payment ecosystems can be maximized in both marketing and customer satisfaction simultaneously. Such change does not come without challenges.

REFERENCES

Alam, N., Gupta, L., & Zameni, A. (2019). *Fintech and Islamic finance. sl.* Springer International Publishing. DOI: 10.1007/978-3-030-24666-2

Allen, R. G. (2006). *Multiple Streams of Internet Income: How Ordinary People Make Extraordinary Money Online*. John Wiley & Sons.

Amudharasan, A. (2023). The Impact of Recommendation Algorithms: Analyzing the Influence of Data on Marketing Strategies in the Media Sector. *Open Journal of Business and Management*, *11*(6), 3373–3384. DOI: 10.4236/ojbm.2023.116184

Barkas, P. (2024). Consumer Protection and Financial Innovation: Microeconomic, Policy, and Behavioral Considerations for the Digital Era. In *Finance, Growth and Democracy: Connections and Challenges in Europe and Latin America in the Era of Permacrisis: Democracy, Finance, and Growth* (pp. 263–285). Springer Nature Switzerland.

Bell, L., Vogt, J., Willemse, C., Routledge, T., Butler, L. T., & Sakaki, M. (2018). Beyond self-report: A review of physiological and neuroscientific methods to investigate consumer behavior. *Frontiers in Psychology*, *9*, 1655. DOI: 10.3389/fpsyg.2018.01655 PMID: 30245657

Bolonin, A., & Balykin, V. (2021). Payment Systems in Digital Economy. In *Financial Markets Evolution: From the Classical Model to the Ecosystem. Challengers, Risks and New Features* (pp. 97–108). Springer International Publishing. DOI: 10.1007/978-3-030-71337-9_7

Bowles, S., & Gintis, H. (1988). Contested exchange: Political economy and modern economic theory. *The American Economic Review*, *78*(2), 145–150. https://www.jstor.org/stable/1818113

Brakus, J. J., Schmitt, B. H., & Zarantonello, L. (2009). Brand experience: What is it? How is it measured? Does it affect loyalty? *Journal of Marketing*, *73*(3), 52–68. DOI: 10.1509/jmkg.73.3.052

Brosche, K., & Kumar, A. (2015). Realtime data accelerates online marketing. In *Programmatic Advertising: The Successful Transformation to Automated, Data-Driven Marketing in Real-Time* (pp. 221–232). Springer International Publishing. DOIhttps., DOI: 10.1007/978-3-319-25023-6_18

Chen, Z. (2004). Electronic payment systems: General review and comparative analysis (Doctoral dissertation, Texas Tech University). https://ttu-ir.tdl.org/server/api/core/bitstreams/74ede8b1-e829-44b0-9bea-b8f31b5e30f0/content

Copeland, D. G., & McKenney, J. L. (1988). Airline reservations systems: Lessons from history. *Management Information Systems Quarterly, 12*(3), 353–370. DOI: 10.2307/249202

Cortada, J. W. (2005). The digital hand: Volume II: How computers changed the work of American financial, telecommunications, media, and entertainment industries. Oxford University Press.

Crede, A. (1995). Electronic commerce and the banking industry: The requirement and opportunities for new payment systems using the Internet. *Journal of Computer-Mediated Communication, 1*(3), JCMC135. DOI: 10.1111/j.1083-6101.1995.tb00171.x

Dos Santos, P. L., & Harvold Kvangraven, I. (2017). Better than cash, but beware the costs: Electronic payments systems and financial inclusion in developing economies. *Development and Change, 48*(2), 205–227. DOI: 10.1111/dech.12296

Evans, D. S., & Schmalensee, R. (2004). *Paying with plastic: the digital revolution in buying and borrowing*. Mit Press. DOI: 10.7551/mitpress/5281.001.0001

George, A. S., George, A. H., & Baskar, T. (2023). The death of analog: Assessing the Impacts of ubiquitous mobile technology. *Partners Universal Innovative Research Publication, 1*(2), 15–33.

Giannakoudi, S. (1999). Internet banking: The digital voyage of banking and money in cyberspace. *Information & Communications Technology Law, 8*(3), 205–243. DOI: 10.1080/13600834.1999.9965811

Gomber, P., Kauffman, R. J., Parker, C., & Weber, B. W. (2018). On the fintech revolution: Interpreting the forces of innovation, disruption, and transformation in financial services. *Journal of Management Information Systems, 35*(1), 220–265. DOI: 10.1080/07421222.2018.1440766

Islam, S. (2024). Impact of online payment systems on customer trust and loyalty in E-commerce analyzing security and convenience. Available at *SSRN* 5064838. DOI: 10.2139/ssrn.5064838

Javalgi, R. G., Martin, C. L., & Todd, P. R. (2004). The export of e-services in the age of technology transformation: Challenges and implications for international service providers. *Journal of Services Marketing, 18*(7), 560–573. DOI: 10.1108/08876040410561884

Jonnala, A. (2024). Transforming customer experience with digital voice assistants. *Computing in Science & Engineering, 14*(3), 67–74. DOI: 10.5923/j.computer.20241403.03

Kashiramka, S., Gupta, P., Bali, A., & Guha, A. (2025). Gamifying Loyalty: Real World Success Stories. In Evolving Strategies for Organizational Management and Performance Evaluation (pp. 209-236). IGI Global Scientific Publishing.

Khan, A. N., Kiah, M. M., Khan, S. U., & Madani, S. A. (2013). Towards secure mobile cloud computing: A survey. *Future Generation Computer Systems*, *29*(5), 1278–1299. DOI: 10.1016/j.future.2012.08.003

Khan, B. U. I., Olanrewaju, R. F., Baba, A. M., Langoo, A. A., & Assad, S. (2017). A compendious study of online payment systems: Past developments, present impact, and future considerations. *International Journal of Advanced Computer Science and Applications*, *8*(5).

Khan, B. U. I., Olanrewaju, R. F., Baba, A. M., Langoo, A. A., & Assad, S. (2017). A compendious study of online payment systems: Past developments, present impact, and future considerations. *International Journal of Advanced Computer Science and Applications*, *8*(5). https://pdfs.semanticscholar.org/3e60/35d2232d5f3db969a88d48695b6c58361609.pdf

Kiyotaki, N., & Wright, R. (1989). On money as a medium of exchange. *Journal of Political Economy*, *97*(4), 927–954. https://www.journals.uchicago.edu/doi/abs/10.1086/261634. DOI: 10.1086/261634

Kumar, V., Ramachandran, D., & Kumar, B. (2021). Influence of new-age technologies on marketing: A research agenda. *Journal of Business Research*, *125*, 864–877. DOI: 10.1016/j.jbusres.2020.01.007

Lee, J. Y. (2019). A decentralized token economy: How blockchain and cryptocurrency can revolutionize business. *Business Horizons*, *62*(6), 773–784. DOI: 10.1016/j.bushor.2019.08.003

Lemon, K. N., & Verhoef, P. C. (2016). Understanding customer experience throughout the customer journey. *Journal of Marketing*, *80*(6), 69–96. DOI: 10.1509/jm.15.0420

Lemon, K. N., & Verhoef, P. C. (2016). Understanding customer experience throughout the customer journey. *Journal of Marketing*, *80*(6), 69–96. DOI: 10.1509/jm.15.0420

Mao, W. (2016). Sometimes “fee” is better than “free”: Token promotional pricing and consumer reactions to price promotion offering product upgrades. *Journal of Retailing*, *92*(2), 173–184. DOI: 10.1016/j.jretai.2015.09.001

Markus, M. L. (1987). Toward a “critical mass” theory of interactive media: Universal access, interdependence and diffusion. *Communication Research*, *14*(5), 491–511. DOI: 10.1177/009365087014005003

McChesney, R. W. (2016). *Rich media, poor democracy: Communication politics in dubious times*. New Press.

McKinney, R. E., Jr., Shao, L. P., Shao, D. H., & Rosenlieb, D. C., Jr. (2015). The evolution of financial instruments and the legal protection against counterfeiting: a look at coin, paper, and virtual currencies. U. Ill. JL Tech. & Pol'y, 273. https://heinonline.org/HOL/LandingPage?handle=hein.journals/jltp2015&div=13&id=&page=

Mikhaylov, A. (2020). Cryptocurrency market analysis from the open innovation perspective. *Journal of Open Innovation*, *6*(4), 197. DOI: 10.3390/joitmc6040197

Naeem, M., Hameed, M., & Taha, M. S. (2020, February). A study of electronic payment system. [). IOP Publishing.]. *IOP Conference Series. Materials Science and Engineering*, *767*(1), 012008. DOI: 10.1088/1757-899X/767/1/012008

Naslednikov, M. (2024). The Impact of Artificial Intelligence on Customer Relationship Management (CRM). *Strategies (La Jolla, Calif.)*.

O'Brien, D., & Kunz, R. E. (2024). Will they stay or will they go? application of machine learning classifiers of purchase and churn prediction to a digital journalistic platform. *Journal of Media Business Studies*, •••, 1–29.

Okeke, N. I., Alabi, O. A., Igwe, A. N., Ofodile, O. C., & Ewim, C. P. M.Nnenna Ijeoma OkekeOlufunke Anne AlabiAbbey Ngochindo IgweOnyeka Chrisanctus OfodileChikezie Paul-Mikki Ewim. (2024). Customer journey mapping framework for SMEs: Enhancing customer satisfaction and business growth. *World Journal of Advanced Research and Reviews*, *24*(1), 2004–2018. DOI: 10.30574/wjarr.2024.24.1.3206

Olipane, H. G. E., & Inocencio, M. D. (2023), DIGITAL PAYMENT AND ITS INFLUENCE TO CUSTOMER CONVENIENCE. https://tinyurl.com/mr42357p

Omarini, A. E. (2018). Fintech and the future of the payment landscape: The mobile wallet ecosystem. A challenge for retail banks? *International Journal of Financial Research*, *9*(4), 97–116. DOI: 10.5430/ijfr.v9n4p97

Parise, S., Guinan, P. J., & Kafka, R. (2016). Solving the crisis of immediacy: How digital technology can transform the customer experience. *Business Horizons*, *59*(4), 411–420. DOI: 10.1016/j.bushor.2016.03.004

Peneder, M. (2022). Digitization and the evolution of money as a social technology of account. Journal of Evolutionary Economics, 32(1), 175-203. DOIhttps://DOI: 10.1007/s00191-021-00729-4

Popescu, A. D. (2022). Understanding FinTech and Decentralized Finance (DeFi) for Financial Inclusion. In Popescu, A. (Ed.), *FinTech Development for Financial Inclusiveness* (pp. 1–13). IGI Global., URL https://www.igi-global.com/book/fintech-development-financial-inclusiveness/285349 DOI: 10.4018/978-1-7998-8447-7.ch001

Putrevu, J., & Mertzanis, C. (2024). The adoption of digital payments in emerging economies: Challenges and policy responses. Digital Policy. *Regulation & Governance, 26*(5), 476–500. DOI: 10.1108/DPRG-06-2023-0077

Ramachandran, K. K. (2023). Evaluating ROI in digital marketing campaigns: Metrics, measurement, and insights. [IJM]. *International Journal of Management, 14*(7).

Rathod, V. H. (2023). Marketing intelligence redefined: Leveraging the power of AI for smarter business growth. *Towards Excellence, 15*(2), 378–386. DOI: 10.37867/TE150239

Riegger, A. S., Merfeld, K., Klein, J. F., & Henkel, S. (2022). Technology-enabled personalization: Impact of smart technology choice on consumer shopping behavior. *Technological Forecasting and Social Change, 181*, 121752. DOI: 10.1016/j.techfore.2022.121752

Rosário, A. T., & Dias, J. C. (2023). How has data-driven marketing evolved: Challenges and opportunities with emerging technologies. *International Journal of Information Management Data Insights, 3*(2), 100203. DOI: 10.1016/j.jjimei.2023.100203

Rose, D., Bodor, J. N., Hutchinson, P. L., & Swalm, C. M. (2010). The importance of a multi-dimensional approach for studying the links between food access and consumption. *The Journal of Nutrition, 140*(6), 1170–1174. DOI: 10.3945/jn.109.113159 PMID: 20410084

Sajan, V., & PS, R. (2024). Smart moves: Flipkart's name-personalized nudging techniques to combat cart abandonment. Journal of Information Technology Teaching Cases, 20438869241301078.

Schoenbachler, D. D., & Gordon, G. L. (2002). Multi-channel shopping: Understanding what drives channel choice. *Journal of Consumer Marketing, 19*(1), 42–53. DOI: 10.1108/07363760210414943

Schwartz, B. (2001). *The costs of living: How market freedom erodes the best things in life*. Xlibris Corporation.

Scott, D. M. (2011). *Real-time marketing and PR: How to instantly engage your market, connect with customers, and create products that grow your business now*. John Wiley & Sons.

Sumanjeet. (2009, November). Emergence of payment systems in the age of electronic commerce: The state of art. In 2009 First Asian Himalayas International Conference on Internet (pp. 1-18). IEEE.

Taddesse, W., & Kidan, T. G. (2005). e-Payment: Challenges and opportunities in Ethiopia. United Nations Economic Commission for Africa. Tandfonline. (2024). Decentralized Finance (DeFi): A paradigm shift in the fintech.

Truman, G. E., Sandoe, K., & Rifkin, T. (2003). An empirical study of smart card technology. *Information & Management*, *40*(6), 591–606. DOI: 10.1016/S0378-7206(02)00046-0

Upadhyay, U., Kumar, A., Sharma, G., Sharma, S., Arya, V., Panigrahi, P. K., & Gupta, B. B. (2024). A systematic data-driven approach for targeted marketing in enterprise information system. *Enterprise Information Systems*, *18*(8), 2356770. DOI: 10.1080/17517575.2024.2356770

Wang, R. J. H., Krishnamurthi, L., & Malthouse, E. C. (2018). When reward convenience meets a mobile app: Increasing customer participation in a coalition loyalty program. *Journal of the Association for Consumer Research*, *3*(3), 314–329. DOI: 10.1086/698331

Weichert, M. (2017). The future of payments: How FinTech players are accelerating customer-driven innovation in financial services. *Journal of Payments Strategy & Systems*, *11*(1), 23–33. https://www.ingentaconnect.com/content/hsp/jpss/2017/00000011/00000001/art00004. DOI: 10.69554/JHTK5372

ADDITIONAL READING

Capponi, A.. (2020). Decentralized finance: Risks and opportunities. *Journal of Financial Stability*, *50*, 100–114.

Evans, D. S., & Schmalensee, R. (2004). *Paying with Plastic: The Digital Revolution in Buying and Borrowing*. MIT Press. DOI: 10.7551/mitpress/5281.001.0001

Gomber, P., Koch, J. A., & Siering, M. (2018). Digital finance and FinTech: Current research and future research directions. *Journal of Business Economics*, *88*(5), 537–580. DOI: 10.1007/s11573-017-0852-x

Kumar, V., Ramachandran, D., & Kumar, B. (2021). Influence of new-age technologies on marketing: A research agenda. *Journal of the Academy of Marketing Science*, *49*, 521–546.

Lemon, K. N., & Verhoef, P. C. (2016). Understanding customer experience throughout the customer journey. *Journal of Marketing*, *80*(6), 69–96. DOI: 10.1509/jm.15.0420

Martin, K., & Murphy, P. (2022). The role of data ethics in marketing. *Journal of Business Ethics*, *178*, 1–18. DOI: 10.1007/s10551-020-04725-9

Omarini, A. (2018). FinTech and the future of the payment landscape. *International Journal of Financial Research*, *9*(4), 97–108. DOI: 10.5430/ijfr.v9n4p97

Riegger, A. S., Merfeld, K., Klein, J. F., & Henkel, S. (2022). Technology-enabled personalization. *Technological Forecasting and Social Change*, *181*, 121752. DOI: 10.1016/j.techfore.2022.121752

Sieber, S., & Guibaud, S. (2022). *Embedded Finance: When Payments Become an Experience*. Wiley.

Zuboff, S. (2019). *The Age of Surveillance Capitalism*. PublicAffairs.

KEY TERMS AND DEFINITIONS

Behavioural Segmentation: A marketing strategy that classifies consumers based on actual purchasing behaviour, frequency, spending patterns, and transaction timing rather than demographic characteristics.

Buy Now, Pay Later (BNPL): A short-term financing model integrated into checkout processes that allows consumers to defer payments or pay in instalments, often used as a conversion-enhancing marketing tool.

Contactless Payments: Payment transactions executed using Near Field Communication (NFC) or RFID technology, allowing users to tap cards or mobile devices at point-of-sale terminals without physical contact.

Customer Experience (CX): The cumulative perception formed by customers through interactions across all touchpoints, including payment interfaces, checkout processes, post-purchase communication, and loyalty programs.

Cybersecurity Risk: Threats related to data breaches, fraud, identity theft, and system vulnerabilities arising from interconnected digital payment and marketing platforms.

Diffusion of Innovations Theory: A theory explaining how new technologies such as digital payment systems are adopted over time based on perceived usefulness, ease of use, compatibility, and relative advantage.

Digital Payment Systems: Electronic mechanisms that enable monetary transactions without physical cash, including card-based payments, mobile wallets, UPI, contactless payments, and blockchain-based systems, facilitating speed, security, and traceability in transactions.

Embedded Finance: The integration of financial services such as payments, lending, or insurance directly into non-financial platforms (e.g., ride-hailing apps, e-commerce platforms), making payments an invisible part of the customer experience.

Invisible Payments: Transactions that occur automatically in the background with minimal or no user intervention, commonly used in subscription services, ride-hailing platforms, and smart devices.

Mobile Wallets: Software-based applications (e.g., Google Pay, Apple Pay, Paytm) that store payment credentials digitally and allow users to make transactions using smartphones, often integrated with loyalty programs and promotional messaging.

Omnichannel Marketing: A strategy that ensures seamless and integrated customer experiences across multiple online and offline channels, including synchronized payment and loyalty systems.

Payment Data Analytics: The analysis of transactional data generated through digital payments to derive insights related to consumer preferences, purchase timing, loyalty, churn, and lifetime value.

Service-Dominant Logic (SDL): A marketing perspective that views value as co-created through interactions between firms and customers, positioning payment systems as experiential and relational tools rather than mere transaction utilities.

Surveillance Capitalism: An economic logic where consumer data, including payment behaviour, is extensively collected and monetized, raising ethical and regulatory concerns around privacy and autonomy.

Tokenized Loyalty: A blockchain-enabled loyalty mechanism where rewards are issued as digital tokens that can be transferred, redeemed, or exchanged across platforms.

Transaction Friction: Any cognitive, physical, or procedural obstacle encountered during the payment process that may reduce conversion rates or customer satisfaction.

Trust and Risk Theory: A theoretical framework emphasizing the role of perceived security, transparency, and regulatory compliance in shaping user trust and continued usage of digital payment platforms.

Unified Payments Interface (UPI): A real-time, interoperable payment infrastructure developed in India that enables instant peer-to-peer and merchant payments through mobile devices using bank-linked virtual payment addresses.

Chapter 5
Decentralized Truth: Blockchain's Role in Promoting Transparency and Trust in Digital Marketing

Sudipta Dhar
https://orcid.org/0009-0004-4998-3041
Brainware University, India

Shivangi Kashyap
https://orcid.org/0009-0002-1685-330X
University of Hyderabad, India

Pravin Kumar
https://orcid.org/0009-0008-9393-7324
Brainware University, India

ABSTRACT

Digital marketing is maturing into a precision-targeted, data-driven engine. However, this maturation has brought about a slew of ethical and operational challenges: ad fraud, opaque data practices, unverifiable performance metrics, and concentrated control of major platforms. In this chapter, we consider the potential of blockchain technology to provide enhanced transparency, trust, and accountability in the ecosystem of digital marketing. Through a multi-method design of research combining literature review, case studies, and interviews with key stakeholders, we put forth an ethical framework for digital marketing grounded in blockchain. The chapter discusses how decentralized, transparent, and immutable blockchains, coordinated with smart contracts and decentralized identity, could be one mechanism in the mitigation of some challenges that have been explored already, and simultaneously ensuring user empowerment and support for ethical advertising practice.

DOI: 10.4018/979-8-3373-4526-0.ch005

INTRODUCTION

The digital economy's evolution is primarily pulling digital marketing as the main engine for visibility, customer engagement, and brand influence for the companies. Now, marketers are endowed with an incredible opportunity to reach and affect consumers with a very high degree of accuracy made possible through real-time analytics, algorithmic targeting, and enormous data storage. This very act of technological prowess has re-engineered marketing strategies and changed the whole dynamic of business-audience interaction. Campaigns can, in a nutshell, be optimised in real time with respect to individual behaviour while adorning the profiles of the end-user on an array of platforms and devices. However, the good and the bad co-existence of the digital ad ecosystem brought about by the newest technological advancements in scalability and efficiency has shown the industry the dark side of progress, with the challenges of ethics, infrastructure, and operations threatening the digital advertising sector's sustainability and clean future.

A small number of tech giants such as Google, Meta, Amazon, etc., dominate the vast digital market and thus have complete control over user data, ad revenues, and content distribution. By being the intermediaries, these companies determine, to some extent, the visibility, timing, and conditions of the content to be seen by the audience. Their algorithms determine whether certain content will be shown, how much advertisement actually costs, and where data flows with minimal transparency or any external accountability. Hence, the marketers exist inside a system wherein they depend on some intermediaries, quite often without having any visibility into how those performance metrics are calculated or how their audiences are actually reached.

There are several ethical and operational deficiencies surrounding this power concentration. The first one goes to advertising fraud. Advertisers have become fed up: from bots producing false impressions and clicks to domain spoofing and pixel stuffing, ad fraud has proliferated to the point where it has become accepted as a normal occurrence. The impact of digital advertising fraud on return on investments has been staggering in recent years, with annual global losses reported to be in the tens of billions of dollars. This has a knock-on effect on marketing performance data, with trust in it plummeting. Secondly, the metrics used in advertising that cannot be verified are an ongoing problem. In addition, a conflict of interest is created when most of the campaign data is provided by the advertising platforms, especially since it is nearly impossible to conduct an independent audit or verification. Despite increasing doubts about data integrity and uniformity, marketers are compelled to depend on the offered data.

Third, and perhaps most importantly, there is an accelerating consumer trust crisis. Today's digital marketing is dependent on user data, which is grabbed through device fingerprinting, cookies, and other tracking technologies. Although such data is used for customization, the users' privacy is affected. Consumers' privacy is being encroached upon as a result of the increasing collection, usage, and rampant monetization of their personal information, and more disturbingly, the absence of informed consent. Although the European Union has implemented the General Data Protection Regulation (GDPR), and the California Consumer Privacy Act (CCPA) is underway in California, many data practices lack transparency, and users have little control. The aforementioned situation has resulted in the extensive use of ad blockers, privacy browsers, and similar services that enable users to cut off their involvement in the data economy; thus, it is a clear sign of users' distrust towards the digital marketing sector to a considerable extent.

These activities are, however, fraught with serious ethical issues. The present digital marketing system can indeed bring very good profit and data diversity, but it is still in opposition to the values of self-rule, openness, justice, and accountability. It prioritises reach, engagement, and conversion rates over consent, equity, user welfare, and ethical treatment of users. Additionally, it heightens the barriers of digital inequality by enhancing large platforms and well-funded advertisers at the expense of independent publishers and small businesses.

Blockchain provides an answer to the mess of ethical and operational arguments. Originally specified to support Bitcoin and related cryptocurrencies, it is now a generalised framework that supports decentralised and transparent digital systems that are tamper-proof. Fundamentally, a blockchain is a distributed ledger system that empowers users to transfer data and value directly without an intermediary governing third party. Each transaction is always recorded in an immutable state, along with a timestamp, for every user in the network to see, with trust through consensus protocols.

Decentralisation, transparency, and immutability are the main methods on the way to the solutions of the problems that affect the digital marketing industry as a whole today. However, the main advantage of blockchain is the possibility to establish a trustworthy system of ad metrics, which is actually the recording of all the impressions, clicks, and engagements of ads on a public ledger shared by the advertisers and the publishers. This process of verification means that the data is real and not manipulated by intermediaries, which helps validate the performance. A benefit of blockchain technology applies to contract management within advertising platforms. Blockchains allow for a smart contract to automate the contracting process and the execution of the transaction within the contract. The level of detail in a contract allows the terms to be fully articulated to ensure payments for validated interactions. These types of contracts allow for decreased fraud, which engenders greater trust amongst the contracting parties.

Furthermore, the use of blockchain technology can revolutionise the ownership of data. With the use of decentralised identity systems and tokenised data sharing models, users can have the ability to control their data and choose to share with anyone at their discretion. Instead of scouring data for passive subjects of surveillance, users can actively participate in the value chain and even get paid for sharing data or interacting with the content. Such models of the user offer protection as well as an incentive for users. It therefore aligns the interests of all parties involved in the advertising industry, paving the way to more responsible and socially balanced marketing methods.

Blockchain is not a magic bullet, but still, its possibilities to reengineer the digital advertising structure are rather timely and transformative. The integration of ethics into the technological fabric means that we can move toward a form of ethics that is infrastructure-driven (imposing transparency and accountability into the system via the infrastructure), as opposed to compliance-driven (letting ethics be shaped by a system of checks and balances). This paradigm shift is very timely in this phenomenon of escalating regulatory attention, consumer activism, and the wider demand for ethical technologies governance.

This research chapter examines the application of blockchain as a digital marketing ethics infrastructure of technology. It is informed by the understanding that digital advertising today is in a situation of legitimacy crisis that can only be re-established by way of a structural transformation. The chapter looks at the technical potential of blockchain as a potential remedy for trust concerns, the danger of fraud, centralisation of data, transparency, and whether or not the adoption of blockchain can create a more decentralised, fair, and accountable digital marketing environment.

This research will emphasize a specific set of research objectives, leading questions, and conceptual maps that place this investigation in context. By outlining the main questions and difficulties connected to the future area of inquiry, the paper aims at being a part of the new literature path in which decentralised technologies are recognised as a possible way of ethical innovations in digital systems. It also indicates the end of centralization and surveillance capitalism along with the birth of trust, equity, and accountability in the digital age since it chooses the path against the status quo regarding the doubtfulness of centralization and surveillance capitalism as the only possible scenario.

LITERATURE REVIEW

The Ethical Crisis in Digital Marketing

In the course of the last few years, digital marketing achieved a very fast transformation. Consumer data, algorithmic accuracy, and real-time personalization were the main reasons for this change. Though these innovations allowed the marketing strategies to be very targeted and efficient, at the same time, they brought about a major ethical crisis that was characterized by a general loss of trust, privacy invasions, and users losing control over their data (Iansiti & Lakhani, 2017; Dierksmeier & Seele, 2020). Digital marketing systems are more and more operating in a data-heavy environment where consumers are interacting with the brands and their activities are being monitored, analysed, and even sold as data at the same time and all this is often done without any meaningful transparency or consent (Zyskind & Nathan, 2015).

The ethical decay is not a coincidence but a full-fledged system that shapes the very structure of the digital marketing ecosystem in the present time. The crisis is revealed mainly through the three interconnected aspects: (1) data of consumers and their behavior analysis, (2) the utilization of dark patterns and non-transparent consent techniques, and (3) the increasing lack of trust among consumers as shown by the growing use of ad blockers and resistance technologies (Goldfarb & Tucker, 2011; Newman, 2014).

Rise of Data Exploitation and Behavioural Profiling

The marketing of personal data on a large scale and their gathering are the main factors behind the digital marketing practices of today. Each and every online activity—such as search queries, clicks, signaling of location, and interaction on social media—gets recorded and stored in the massive data reservoirs of platforms, advertisers, and data brokers (Newman, 2014) (Stockinger, 2013). These data streams are combined to form detailed user profiles that are built on the complete range of demographics, psycho-graphics, and activities, thereby enabling hyper-personalized ads and conversion rate optimization (Goldfarb & Tucker, 2011).

Nonetheless, the morality behind such integrative practices becomes an issue of major concern. Often, customers are oblivious to the vastness, purpose, and eventual use of their data which leads to a situation where there is no consent that is fully informed (Regulation, 2018). Such a setting has been termed "surveillance capitalism," where constant monitoring of behavior turns out to be the main economic model (Zyskind & Nathan, 2015). The use of behavioral targeting techniques in this way could further erode consumer autonomy by imperceptibly guiding the decision-making process in directions that favor the companies rather than benefiting the individual (Chen, 2018).

Moreover, the unequal distribution of information power between the companies and the individual raises the question of the ethics involved in such practices. The company has been able to obtain very detailed information on the utility behavior of the customers, while the customer knows nearly nothing about how he/she is classified and what expressions of this classification will be reflected in the price he/she will pay, the content he/she will be exposed to, and the opportunities he/she will be granted (Catalini & Gans, 2020). These relationships have been the source of regulatory measures such as the GDPR and have also led to a rise in the demand for data sovereignty and user-based privacy frameworks (Regulation, 2018; Coche et al., 2024).

Impact of Dark Patterns and Opaque Consent Mechanisms

The use of manipulative interface designs or dark patterns is one of the major causes of the ethical crisis that has occurred in the digital world. These designs take advantage of the cognitive biases of the users and thus, they are manipulated into actions like granting permission for the sharing of their data, subscribing to services, or making unwanted purchases (Dierksmeier & Seele, 2020).Dark patterns, which are often termed as design practices, generally include a lot of different kinds of deceiving user interfaces, hidden user disconnection options, fake urgency messages, and very convoluted cancellation procedures.

Dark patterns can be considered as a tool used to enhance engagement statistics during a brief period, but their negative side effects are very serious in the long run; they promote distrust from consumers towards the brand and might even result in the brand being labeled as untrustworthy (Boni, 2024). If we were to draw a line connecting these practices, it would lead us to one side being dominated by the lack of clarity in the classic consent mechanisms that are embedded in cookie banners and privacy policies. Even though laws require that companies be transparent with consumers, the consent mechanisms that are in place tend to be still very complex, legally dense, and non-negotiable. This is the case where consent is reduced to a mere procedural formality that does not reflect the actual user choice (Regulation, 2018; Newman, 2014). The ethical rule of prior informed consent is being violated by such practices which means that the user is not empowered but compliance has the upper hand. What it means is that consent giving becomes a risk mitigation tool rather than a means of ethical data governance (Coche et al., 2024).

Mistrust and Widespread Use of Ad Blockers

The repressive and manipulative behavior towards market advertising has significantly diminished consumers' trust in brands. The issue has escalated to the point that individuals are now resorting to (Goldfarb & Tucker, 2011; Sadeghpour & Vlajic, 2021) ad-blockers as a means of safeguarding against privacy violations, perturbing ads, and security risks associated with unethical advertising. Advertisers consider ad blockers as technical barriers but at the same time as a sign of consumer opposition and distrust in a larger scale of the digital advertising system.

Ethically speaking, the application of ad-blocking technology is the call for treating users with more rights and respect. On the other hand, advertisers rely on such models for revenue which leads to the needful tension between consumer rights and platform economics (Iansiti & Lakhani, 2017). The dispute highlights the current models' lack of sustenance without an ethical reform and value exchange through transparency renewed commitment.

A System Under Ethical Strain

Digital marketing these days is being done in an area of ethics that is not so clear cut, where most of the time efficiency is more important than fairness, manipulation prevails over transparency, and profit is the only thing left when privacy is lost (Dierksmeier & Seele, 2020). Even though these issues are difficult to tackle, they are not confined to the companies involved but rather reflect the industry standards. When profiling, usage of psychological tricks, and negative consumers reactions are taken together, they just point out the necessity for a change in values in marketing practices urgently.

Digital Ads Supply Chain Fragmentation

The digital advertising market is based on a complicated supply chain system that consists of a great number of players such as advertisers, media agencies, and others, who take on different roles in the process, including planning, targeting, delivering, and verifying ads on the Internet. Digital ad supply chain fragmentation led to the rise of these intermediaries and hence, the ad costs increased. The whole process of advertising has become very complicated due to the involvement of so many players. It takes several transactions, data exchanges, and real-time auctions for advertisers to reach the users.

The key operational problems include:

- **Hidden Fees**: There are a lot of third-party companies that create hidden fees.They provide services and make money this way, but most of the time, their costs and the markups charged to the advertisers are not clear to the latter. Hence, the advertisers may believe they are using their full media budget to purchase ads on the web, but, in reality, a much lower percentage is going for advertising while the rest goes for fees and markups in the ad-tech supply chain.
- **Latency & Load Time:** The ad-tech supply chain is very complicated now, and it is becoming more and more so, which results in slower page load times and performance issues for users, bad experiences and higher bounce rates being user consequences eventually.
- **Data Leakage**: The risk of data leakage is quite high since user data is circulating across the web, and there are numerous parties who would like to see, duplicate or misuse the data at the same time. The structure of data silos does not allow marketers to get a comprehensive picture of the performance of their campaigns through different channels.
- **Lack of Trust**: The process is split among the different stakeholders, and therefore, nobody knows how and where the ads were displayed. The situation of not knowing is a stumbling block for the identification of inefficiencies, their correction, and even the determination of who the one responsible for the conspicuousness, or if it is a case of fraud.

The fragmentation not only raises the costs of operation but also destroys strategic alignment. The inability to coordinate activities across channels and platforms makes it hard for advertisers to spend resources and time in redundancy, therefore, work towards less synergy of the campaign.

The digital advertising faces challenges of operation that are interwoven and systemic. Ad fraud, incorrect analytics, and a disjointed supply chain all point to a larger problem that is lack of accountability and inefficiency in the digital marketing domain. These issues not only hurt the bottom line; they also affect the perception of the brand, the trust of the audience and the moral legitimacy of the marketing practices. Solving these issues requires not only setting things up-upon rather several proximate projects are cut out to cope with the currents every two days.

It requires a rethinking of the digital advertising infrastructure, something which focuses more on transparency, auditability, and decentralisation. The potential of blockchain, based on its capacity to offer shared ledgers, cryptographic authentication, and decentralised trust structures, has demonstrated an interesting opportunity to recreate the digital model of advertising in the context of ethical and operational probity.

The next several parts will discuss how blockchain-based systems can directly solve these functional shortcomings and provide models of new forms of ethical, efficient, and trustworthy digital marketing.

Regulatory Context and Shortcomings

Regulatory frameworks such as the GDPR and CCPA define how personal data may be collected, processed, and protected, emphasizing transparency, user consent, and individual control over data. Through demarcating and promoting the power of the data subjects in the digital world (Regulation, 2018; Newman, 2014; Nanda & Kumar, 2024), the laws introduced have a primary goal to bolster ethical data governance more effectively than before.

Privacy laws, despite their strong foundations of legal principles, were still facing challenges in their practical enforcement due to the presence of legal loopholes, inconsistently applied fines and penalties, and corporate practices that relied on ambiguous consent strategies and geographic loopholes.These factors ultimately contributed to the ineffectiveness of the right to privacy (Zyskind & Nathan, 2015; Dierksmeier & Seele, 2020; Sadeghpour & Vlajic, 2021).

Data flows between countries make it hard to enforce regulations as different regions have different privacy laws and thus, there will be conflicts in compliance requirements and difficulties in handling cross-border governance. The discrepancies among the jurisdictions will allow companies to choose the most favorable regulatory environment and at the same time, the global digital economy will lose the benefit of uniformity of privacy protection (Coche et al., 2024; Iansiti & Lakhani, 2017).

Blockchain and Ethical Technology Design

Foundations of Distributed Ledger Technology (DLT)

At the heart of blockchain systems lies Distributed Ledger Technology (DLT), which can be considered as the main architectural layer.DLT is a tapestry of distributed databases wherein many copies of records like transactions or data entries are kept at different locations instead of just one centralized authority. All the nodes responsible for the recordings have an identical and updated version of the ledger, and the changes are accepted by agreement methods such as Proof of Work or Proof of Stake, which are typical at the same time ensuring integrity and gaining the trust of the users (Narayanan et al., 2016; Iansiti & Lakhani, 2017). The combination of decentralization and consensus for identification makes DLT a transparent, resistant to tampering, and secure system while it is cutting down the requirement for middlemen severely (Zyskind & Nathan, 2015; Catalini & Gans, 2020; Morabito, 2017).

Among many implementations of DLT, blockchain is the one that has the widest acceptance and most research coverage because of its capability and versatility in different areas (Kshetri, 2017; Dierksmeier & Seele, 2020).

Smart Contracts and Automated Compliance

Smart Contracts are the automated contracts that exist in the blockchain system and the computer code enforces them. They take care of the automatically implementing rules and conditions that have been settled by the parties involved, hence cutting down or removing the requirement of intermediaries like lawyers, banks, or other authority figures (Narayanan et al., 2016; Iansiti & Lakhani, 2017). One good instance of smart contracts is their programming in such a way that they will make automatic insurance payments once a flight cancellation or delivery delay happens, which is a predefined condition (Morabito, 2017; Catalini & Gans, 2020). The lots of advantages like the flexibility of programming and transparency that the smart contracts have attract their increasing use in the digital governance and business areas like automated regulatory compliance, where the legal requirements, corporate policies or by-laws are encoded to make sure that the transactions comply with th contracts and regulations instantly (Zyskind & Nathan, 2015; Dierksmeier & Seele, 2020; Madhani, 2022).

Decentralised Identity (DID) and Personal Data Wallets

Decentralized Identity (DID) is the method which allows people to have their own digital identities and that they can be in control of them without the need of a central identity provider like Google or Facebook. Centralized repositories are no longer the basis of trust in DID systems, rather they are built on user-controlled cryptographic keys which are the main components of identity ownership and verification (Zyskind & Nathan, 2015; Narayanan et al., 2016). Personal data wallets are very important for the creation and management of identities as they are digital applications that not only store but also manage verifiable credentials of individuals like educational qualifications, health records or government-issued id (Morabito, 2017; Lee & Yoon, 2022). These wallets give the user the ability to disclose only parts of their information—for example, showing age without the need to provide a complete date of birth—this way user's privacy, security, and autonomy will be improved and at the same time users' rights to control personal data will be acknowledged in the digital ecosystem (Dierksmeier & Seele, 2020; Coche et al., 2024).

Research Objectives and Questions

The rise of digital marketing tools has revolutionized not only the way brands interact with their customers but also the way they do it most effectively through very precise targeting and real-time optimization of the campaign's operation, in addition to the delivery of personalized content. However, along with these benefits, the new technologies have also created some serious ethical dilemmas, particularly regarding consumer data privacy, the use of opaque algorithms for the manipulation of people's choices, and the destruction of trust due to unreliable ad metrics. In response to the ethical dilemma, blockchain technology has been viewed as an alternative that can provide the market with the aforementioned benefits of transparency, decentralization, and verifiability. In this chapter, the author gives the central research goals and develops the thematic areas on which this study is directed in the use of the blockchain as an ethical infrastructure in digital marketing in detail.

The core of the study is a quest into the possibility of utilising blockchain technology to create transparency, trust, and accountability in the digital marketing ecosystem. This paper aims to discuss blockchain as not only a technological improvement but also a possible reformation engine that can eradicate deep-seated concerns of accountability, user agency, and transparency. The key question in this enquiry is how blockchain can reorganise advertising value chains, enable user-controlled data, and initiate tamper-resistant performance audit and fraud prevention measures. The greater objective is to investigate blockchain as a sustainable protocol to rebuild consumer confidence and inculcate morality in the marketing approaches.

The present situation of the lack of ethical and operational standards in digital marketing is one of the central thematic topics explored in this study. The selling of user information with or without their consent and the privatisation of surveillance capitalism have provided an atmosphere in which ethical breaches abound. There are also operational inefficiencies like ad fraud and a lack of transparency in the measurement of the return-on-investment, broken ad supply chains, that also compromise the credibility and efficiency of the system. The paper will critically evaluate these shortcomings as a reference point towards determining the transformative potential of blockchain.

The research investigates the practical usage of blockchain technology in marketing and evaluates the true answers to the discrepancies. The case studies include websites like Brave which uses Basic Attention Token (BAT), AdEx, and MadNetwork, and these technologies have been shown empirically to empower users, stop fraud, and redistribute the value in the ecosystem. Such examples offer a practical way to view the effectiveness, limitations, and future of marketing solutions incorporating blockchain alongside other solutions. The analysis is a part of a wider picture as to whether such early efforts can be regarded as a paradigm shift or as a mere skin-deep modification of the current paradigm.

It is also crucial to determine the obstacles that impede the adoption of blockchain in the normal marketing processes. The aforementioned technology still has its drawbacks, like getting support for many users at once, being dependent on a lot of energy, difficulty in merging with other technologies, and non-standardization with existing systems. All these besides its advantages. Furthermore, the resistance to change in the organization, high initial investments, and the opposition of the present intermediaries likewise slow the acceptance process. Regulations are also unclear, as it is not clear whether the smart contracts will be legally acknowledged, whether data protection standards will be met, or whether the data flows across borders. The study intends to unpack these difficulties and give strategic suggestions on how these difficulties can be avoided or overcome.

Based on this analysis, the paper aims to establish a list of ethical marketing system design principles. These principles are pictured to contain transparency by design, data minimisation, sovereignty of user consent, fairness of values, accountability of algorithms, and a participatory model of governance. This kind of model will not just enlighten system designers and marketers but also act as a benchmark for policymakers and other regulatory agencies to enforce ethical adherence in the ever-digitised economy. These ideals are also meant to encourage a shift to ethical innovation instead of falling into the pattern of copying exploitative approaches on new technological wrappers.

In short, this study has fourfold objectives. First, it will attempt to give a critical examination of the ethical and operational constraints in the current digital marketing practices. Second, it aims to assess blockchain-based platforms and their effectiveness in addressing these problems. Third, it will also aim to define the key technical, economic, and regulatory impediments to blockchain adoption in the marketing field. Finally, I would like to establish an ethically sound and pragmatic design and implementation approach that integrates blockchain into marketing systems. Through tackling these mutually dependent aims, the study hopes to contribute to the academic domain, as well as to actual practice, with a view to developing a more open, fair, and reliable digital marketing world.

METHODOLOGY

Decentralised Identity (DID) is a way to give people power over their own digital identities, without depending on centralised organisations (like logging in with Google or Facebook). The user's ability to manage security of cryptographic keys determines the outcome. Personal data digital wallets are the ones that allow the creation and management of such identities and they are secure digital apps that can store verified credentials (e.g., university degree, medical records, or ID card).These wallets give the user the power to share only the necessary information (for instance, their age but not their birthdate), thus ensuring privacy, security, and control over personal data for the user.

Research Design

The research is going to be carried out using a mixed-method research design that is going to include interpretative and qualitative methods. The research design is going to integrate three major components: a literature review, a case study analysis, and semi-structured interviews. Theoretical perspectives will be supplied by the literature review, which will also recognize major ethical issues, inefficiencies in the operation, and interventions that might be implemented through the use of blockchain technology. Case studies offer more practical examples of blockchain implementation in digital marketing, whereas expert interviews (qualitative) can give deeper insights into the issues of the industry and its challenges, barriers to implementation, and ethical concerns.

The explanation of this integrated design is that it triangulates the information from various sources, and a stronger and more comprehensive view of the research problem is obtained. This triangulation not only enhances the validity of the resulting findings but also enables the combination of theoretical perspectives using real-life evidence, which results in more practical and context-driven findings.

Data Collection

The research will use a multi-level data gathering plan, which involves analysis of documents, case study analysis, and interviews of stakeholders.

Literature and Document Analysis

Literature review will assume a review process of peer-reviewed literature materials, industry white papers, policy documents, and opinion leadership articles. The sources covered in this corpus are in the domains of digital marketing, blockchain technology, information ethics, and regulatory governance. It is aimed at following the history of ethical deliberations in marketing, evaluating the existing technological interventions, and determining the theoretical capabilities of blockchain to mitigate the lack of ethics. Scopus, Web of Science, IEEE Xplore, and Google Scholar are some of the major databases Scopus, Web of Science, IEEE Xplore, and Google

Scholar are some of the major databases that would be employed to make sure that all the areas are covered.

Case Studies

The case study approach allows for analysing in detail the particular blockchain-based websites that have tried to bring transparency and moral changes to the digital marketing activity. Case studies of choice are:

- **Brave and Basic Attention Token (BAT)***:* It is a privacy-oriented browser that gives users rewards for using the browser and lets advertisers connect with consumers without having to collect user data.
- **AdEx:** An advertising exchange that will be fully decentralised to eliminate fraud and make the relationship between advertisers and publishers more transparent.
- **Lucidity:** Lucidity is a campaign verification service built on a blockchain that offers transparency in ad delivery, as well as ad performance.
- **Ocean Protocol:** This is a decentralised data sharing protocol that aims at ethical data sharing and user-controlled monetisation.

These case studies are selected because of the variety of application use cases, i.e., attention-based advertising models through to fraud prevention and ethical marketplaces of data, which, therefore, allows a comparative perspective on the role of blockchain in various marketing contexts.

Semi-Structured Interviews

To supplement the documentary and the case-based analysis, semi-structured interviews with a variety of stakeholders are held in the course of the study. These are digital marketing professionals, blockchain development professionals, regulatory professionals, policy think tanks, and digital ethics organisations. The interviews will be structured to bring out qualitative information on:

- The perceived ethical incompetence of existing marketing activities.
- The pragmatic issues and prospects of blockchain solution deployment.
- Stakeholder attitudes to regulatory, technical, and organisational barriers.

Participants in the interview will be sampled using purposive sampling to provide a sample that is representative of different fields and sectors that the study will focus on. Interviews are taped (with permission), transcribed, and analysed with the help of thematic coding.

The stakeholders have unique problems in the digital marketing ecosystem that blockchain can resolve. The consumers have difficulties with using data, vague permissions, deceptive promotions, and distrust. Blockchain also brings about control of data exchanged by the user, transparency, and fairness in the exchange of values, hence increasing trust. Advertisers encounter ad fraud, inadmissible metrics, dependence on centralised platforms, and cost increases. They enjoy the advantage of delivering ads transparently, being able to track their performance, decreasing fraud, and enhancing cost-effectiveness with the blockchain. Platforms come under criticism due to their lack of transparency, regulatory oversight, monopoly, shared governance, decentralisation, accountability, and less biased competition, which are more achievable with blockchain. Lastly, blockchain records are immutable and standardised, making them useful in compliance monitoring and audits because regulators struggle to keep track of offshore operations and misuse of data; enforcement will be easier with blockchain.

Table 1. Stakeholder Perspectives in Digital Marketing and the Role of Blockchain

	Challenges in Current Digital Marketing	Blockchain Benefits
Consumers	Data exploitation, lack of consent clarity, exposure to manipulative ads, and mistrust.	User-controlled data sharing, transparency, fair value exchange, and improved trust.
Advertisers	Ad fraud, unverifiable metrics, dependency on centralised platforms, and high costs.	Transparent ad delivery, verifiable performance tracking, reduced fraud, and cost efficiency.
Platforms	Centralised control is criticised for opacity, regulatory scrutiny, and monopoly concerns.	Shared governance, decentralised control, enhanced accountability, fair competition.
Regulators	Difficulty enforcing compliance across opaque ecosystems, data misuse, and cross-border issues.	Immutable records for audits, standardised

Table 2 describes the comparison establishing a conceptual distinction within frameworks of digital marketing: centralised vs. blockchain-enabled. With centralised frameworks, the processes are opaque, the major platforms exercise hierarchical governance and control, there is limited user data ownership, and there is a greater incidence of fraud and manipulation. On the contrary, a blockchain-enhanced framework offers decentralization, clarity, and all stakeholders' accountability, giving the user complete power over their data in the trust process assisted by smart contracts and consensus mechanisms. This transition is aimed at counteracting fraud, meanwhile, allowing consumers to give their permission for access and to participate in the rightful exchanges of value in the digital marketing ecosystem.

Table 2. Comparison of Centralised vs. Blockchain-enabled Digital Marketing Frameworks

Criteria	Centralised Digital Marketing	Blockchain-enabled Digital Marketing
Control & Governance	Controlled by dominant platforms (Google, Meta, etc.)	Decentralised governance across multiple stakeholders
Transparency	Opaque processes with limited visibility for advertisers and consumers	Immutable and transparent ledgers accessible to all participants
Data Ownership	User data owned, controlled, and monetised by platforms	Users retain ownership of identity and data, granting access through consent mechanisms
Trust & Verification	Verification depends on intermediaries and is prone to manipulation	Trust ensured via consensus mechanisms and smart contracts
Accountability	Accountability is concentrated in a few dominant actors, with limited oversight.	Shared accountability is distributed across decentralised protocols.
Consumer Empowerment	Consumers have minimal control over how their data is collected and used	Consumers are empowered with consent management and rewarded for data-sharing participation
Fraud Prevention	High risk of ad fraud, fake clicks, and impression manipulation	Fraud minimised through transparent, verifiable, and immutable ad transaction records

Comparison in Figure 1 provides the market with a comparison of centralised marketing and frameworks based on Blockchain over major dimensions, such as fraud prevention, consumer empowerment, accountability, trust and verification, ownership of data, transparency, and control/governance. It reveals that blockchain-based marketing tends to perform better than centralised systems in such aspects as consumer empowerment, fraud protection, and data ownership, while centralised marketing represents more effective results in control and governance.

Figure 1. Comparison of Centralised vs Blockchain-Enabled Digital Marketing Framework

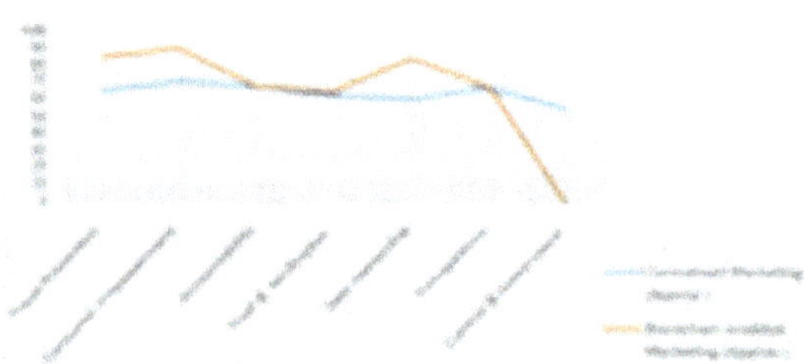

Source: Author's illustration

Data Analysis

The qualitative analysis seeks to analyse the data collected to identify recurrent themes, cross-case themes, and stakeholder meanings.

Thematic Analysis

From the literature, case documentation, and transcripts of interviews, thematic analysis aims at extracting main themes and patterns. This technique orients ethical issues, blockchain uses, and stakeholder concerns into themes with sets of meaning that are then analysed alongside the greater focus of the research.

Comparative Case Analysis

Comparative analysis is being applied to the selected case studies to determine the diversity of blockchain applications toward different marketing usages. Parameters considered are data privacy, transparency in value exchange, fraud reduction, and empowering users. Thus, this analysis will help in identifying best practices, platform-specific innovations, and limitations across the case examples.

By using a combination of qualitative analysis methods and relying on several data sources, it will be possible to get a comprehensive view of the intricate connection between Blockchain technologies and the ethical issues of digital marketing practices. The purpose of this chapter is to inform the analysis that will not only contain a theoretical lens paired with data from the empirical chapters, but also to consider the reality of potential ethical applications of blockchain-enabled marketing systems.

CASE STUDY ANALYSIS

This chapter details a systematic exploration of several case studies where blockchain technology was used to solve adversarial ethical or operational issues in the digital marketing ecosystem. Each case study articulates a different facet of blockchain implementation, ranging from monetising attention, countering fraud, data exchange, and personalisation, highlighting examples of innovative behaviour, as well as barriers to implementation.

Brave & Basic Attention Token (BAT) Ecosystem

In Heyoka, people give each other a break from the usual in their thinking and behaviour about rights. The Brave browser and an on-site Basic Attention Token

(BAT) system are a key development in past digital advertising operating according to the standards set: privacy, user control, and value redistribution. Brave blocks ugly ads and trackers by default, thus giving one a private browsing experience. Brave users who dare to take on the Brave Ads are given BATs as a currency for their attention, which is actually a method of putting engagement in a token.

The transformation of attention into tokens makes it possible for advertisers to target people who are really interested in their ads, but without losing their privacy. This scenario also allows users to get back the value of their data and time that ads have consumed and for which advertisers have profited without any payment, thus, users' data and time regain their worth. The Brave browser has bundled a non-intrusive ad experience along with the user's earnings, thereby making it easier to attract and keep users with its incredible adoption and retention, and thus, it enjoys millions of active users every month worldwide, along with a rising number of verified content creators. As a result, the BAT ecosystem has established itself as an ethical advertising model with transparency and consent being its core tenets.

AdEx Platform

AdEx is a decentralised advertising platform whose aim is to do away with all intermediaries and drive transparency around the bidding for and placement of advertisements. Ads can be bid on and placed directly through the AdEx marketplace using Ethereum smart contracts, which enable direct, verifiable interactions between advertisers and publishers. The smart contracts create records of every transaction on the blockchain for real-time tracking purposes and fraud prevention.

In addition to increasing transparency and accountability through Ethereum smart contracts, AdEx is built to be cost-efficient. By eliminating multiple parties in a typical ad-tech stack, AdEx reduces transaction costs and allows more money to be paid as revenue to content publishers.AdEx also aims for additional transparency by applying fraud detection algorithms to identify bot interaction and wrongful traffic, allowing for more legitimate advertising campaigns.

Alongside its creative work to help Ethereum smart contracts, the platform faces technology issues like the scalability of the Ethereum network, latency delays to confirm transactions, and the challenge of onboarding users to the system. Though the issues do represent a problem of blockchain adoption, AdEx is continually improving its technical infrastructure to remediate those issues of scalability and transaction delay and improve user experience, while also holding to the principles of cost efficiency, transparency, and accountability.

Lucidity's Verification Solution

Lucidity is a service based on blockchain technology that offers verification, and it utilises the blockchain to improve the transparency and accountability of digital advertising campaigns. The service acts as a third-party auditor that assesses the ad delivery metrics, reveals inconsistencies and legitimises marketing results with almost instantaneous accuracy. Through Lucidity, it is possible to validate metrics of third parties such as impressions, clicks, and conversions. Since Lucidity ties its data to a blockchain ledger, it decreases the chance for anyone with access to the data to control it and boost the metrics, whether they are from publishers or ad-tech mediators. The platform can discern the sources of traffic that are underperforming and thus can reallocate the budget accordingly, which will result in better campaigns. The use of Lucidity has been linked to a number of positive changes in the campaign performance; brands have noticed a rise in ROI, better attribution, and a higher level of trust in their marketing data. The platform showcases a scenario where blockchain technology has been used to increase trust in ad metrics without making any infrastructural changes; Lucidity is the verification layer.

Ocean Protocol

Ocean Protocol proposes a decentralised structure for responsible data sharing, compliant with existing privacy laws for responsible use of data, wherein users can exchange data assets in a marketplace, data providers can keep ownership of the data, but open access to monetise through tokens.

Ocean provides privacy-preserving levels of compute-to-data, allowing for algorithms to run on data while not exposing the raw dataset as part of a process respecting data sovereignty and existing data protection frameworks, such as GDPR. This is particularly useful in the fields of personalised marketing and AI, as they often require large volumes of user data to optimise functionality of the service, while also critically accounting for ethical use and compliance. The Ocean Protocol model incentivises the sharing of data with token rewards while providing a transparency mechanism regarding when and how data is used, by whom, ensuring that any personalised marketing does not derive from surreptitious data extraction; rather, data extraction is through transparency and informed, consented participation by users and data holders.

These cases provide support for the case of blockchain-enabled solutions that can lead to a more transparent, trustworthy, and ethical digital marketing ecosystem. Collectively, these cases provided a range of insights into the affordances of blockchain's potentialities and limitations for use in marketing, but they also enabled us to begin the process of comparison and framework building, which we will do in the chapters.

KEY FINDINGS FROM STAKEHOLDER INTERVIEWS

This chapter summarises the insights gained from a series of semi-structured interviews with participants engaged with stakeholders in the field of digital marketing, blockchain technology, policy, and digital ethics. These interview participants provided valuable perspectives concerning the motivations for the implementation of blockchain technology, major barriers to implementation, and what features are required to design ethical blockchain marketing systems. This research adds a practical and human-centric perspective to the technical and theoretical schemes reviewed in prior chapters.

Motivations for Blockchain Adoption

Interviewees mentioned that transparency and accountability will become crucially important in the fast-paced digital environment and will, therefore, become key reasons in considering blockchain marketing solutions. In an environment of malicious and opaque algorithms and unverifiable metrics, with trailers operating in data misuse, chain-based advertising, in a way, can help restore trust between various parties in the ad value chain. Stakeholders agreed that the decentralised nature of blockchain and the untamperable ledger blockchain provides can only be used for advertisement tracking, billing, and reporting reliably.

Cost-efficiency was not only a point of concern but also a benefit that was made evident. Quite a few people regarded blockchain technology as a means through which the entire process could be made more efficient, cutting out the middlemen like ad exchanges, data brokers, and verification vendors. By using smart contracts and decentralised record keeping, everything that used to cost much could now be done at a lower cost, hence supporting the revenue sharing of advertisers and publishers accordingly.

Among the reasons for having reputational considerations, the respondents pointed out this one very strongly. More and more, ethical marketing and user-centric approaches are being recognised as the key factor that can separate one company from another. It was considered that the adoption of blockchain-based solutions, particularly those with consent mechanisms and the data handling practices that are transparent, would be a way for companies to not only improve their public relations but also to gain consumer trust in the long run.

Barriers to Adoption

The majority of stakeholders are still thrilled about the potential of blockchain technology, but they also admit the existence of some major problems. The major drawback is the extremely complicated nature of the technology, which really limits its utilisation. In the opinion of the stakeholders, if blockchain were to be integrated into the current digital marketing systems, companies would require the technical know-how for handling it, a huge overhaul in the architectural design, and great amounts of time as well. Other technical concerns frequently cited were performance issues such as speed, scalability, and the ability for platforms to integrate with legacy systems or applications.

Another major area of concern was low consumer awareness. The use of blockchain technology can result in a lot of advantages for companies operating in the digital marketing sector; yet, modern consumers still seem to have a very basic knowledge of the technology and the trust and privacy issues the technology brings about. The lack of understanding among consumers is a major bottleneck in the success of user-centric blockchain-based applications, which are essentially interactive and require users' active participation, such as tokenised engagement and consent-based data sharing.

Regulatory ambiguity has also been considered a major barrier to implementation. A lot of the various stakeholders in the sector were quite uncertain about the legality of the various blockchain applications, particularly those related to the privacy and sharing of data, worldwide data sharing, and smart contract enforcement. Any ambiguities in legal interpretations were perceived as risks and hurdles to be crossed, particularly for those organisations that are averse to taking risks.

Then, the matter of integrating blockchain with the existing systems, which are mostly slow and challenging to reprogram, became a universal problem. Companies that have strict data handling processes and procedures, advertising and research-related dealings, and several layers of data sharing compliance were worried that they would not be able to perform through the role of blockchain technology, nor effectively nor efficiently.

Desired Features in Blockchain-Enabled Marketing Systems

The interviews further indicated a consensus on the elements that stakeholders considered necessary for the efficacy of blockchain-enabled marketing platforms. First among these was the adoption of consent-based data sharing protocols. Participants noted that any ethical marketing system must focus on user sovereignty over personal data, such that individuals are able to enable and disable user access to that data in real-time and with complete transparency.

The interviews also indicated that stakeholders placed high emphasis on real-time ad tracking. Stakeholders expressed a strong interest in there being platforms that could provide real-time, auditable reporting of ad impressions, clicks, and conversions. For stakeholders, the advantages of blockchain's immutable and verifiable tracking data were a potential resolution to ongoing issues of delayed and manipulated performance metrics.

Finally, participants noted that independent auditing capabilities were essential for trust and accountability. Participants expressed that marketing systems often lack third-party observations, which leads to a conflict of interest and unclear data. Auditing systems based on blockchain, particularly the ones employing decentralised and cryptographic methods, were recognised as a major advantage of placing trust and liability in the hands of all involved in the advertising ecosystem.

Lastly, the interviews held with stakeholders verified that blockchain is quite capable of solving the major ethical and even technical issues in the digital marketing field. Nevertheless, the realisation of such capability is contingent upon the removal of major obstacles, and the designing of the systems with transparency, user empowerment, and technical feasibility as the primary considerations.

ETHICAL FRAMEWORK FOR BLOCKCHAIN IN MARKETING

Based on the conclusions in the literature, the case studies, and the stakeholder interviews summarised above, this chapter is a complete ethical framework for the use of blockchain in the digital marketing ecosystem. With a tendency toward data-driven and automated marketing practices, appropriate safeguards and a transparent system will need to be incorporated into marketing experiences. With proper design and implementation, blockchain can provide the infrastructure that will accommodate such safeguards. The framework presented here is organised around important ethical principles, which represent the primary dimensions of ethics, and further expanded by the key dimensions of implementation that permit application in a number of different marketing situations.

Core Principles

The ethical grounding of marketing systems based on blockchain should ultimately be defined by global principles of fairness, accountability, and user autonomy. The ethical grounding for the above principles is provided below:

- **Transparency:** By virtue of its design, blockchain-based technology allows for an auditable record of the integrity of data and of transactional interactions, whether occurring between devices on the internet-of-things, or to measure the outcome of the advertising itself. Aspects of advertising costs, viewability results, and engagement metrics are transparent, creating an accountability factor and reducing information asymmetry.
- **Informed Consent and User Autonomy**: The act of granting users the power to possess and control their own data is the moral foundation of the very systems for data collection and sharing that are considered ethical. Platforms that are driven by marketing allow users to give informed, and reversible consent regarding the Data's collection, sharing, and usage.
- **Accountability**: Responsibility for actions should rest with all the players in the marketing ecosystem, no matter whether they are advertisers, publishers, or intermediaries. The blockchain not only enables the immutable recording of data but also allows third parties, through the use of blockchain interfaces, to either perform audits or verify the output.
- **Data Minimisation:** The marketing department should acquire only the minimum amount of personal data that is necessary for the business to operate legally. Prototypes of the blockchain-based marketing systems should have strict regulations that monitor and control the data retention time and the unnecessary profiling/retaining of personal user data.
- **Verifiability:** All transactions, from ad serving to payment settlement, should be verifiable by one or more parties. Empirical verification of events based on blockchain's cryptographic principles becomes possible without blind trust in centralised authorities.

Framework Dimensions

To operationalise ethical principles, the framework establishes five perspectives of implementation to help design, manage, and evaluate marketing systems that engage blockchain technology:

- **Technical Feasibility:** Ethical motivations must not only be realistic but also be backed by technology. Among the most essential issues are those connected to the development of smart contracts for easing the transactions and consent flows, the capacity of blockchain networks to scale to the level of handling large amounts of marketing data and privacy-protecting technologies such as zero-knowledge proofs or off-chain computation. The necessity to monitor the promised performance of the systems without compromising the ethical guarantees provided by decentralisation will persist.

- **Governance:** An ethical marketing ecosystem is based on distributed and participatory governance models. It is necessary to build blockchain platforms that will account for the participation of all stakeholders in the governance and decision-making, thus implementing the distributed governance model. Consensus protocols, voting models and Decentralised Autonomous Organisation models are the tools that can be used to ensure fairness and accountability throughout the ecosystem.
- **Legal Compliance:** Legal compliance might turn out to be imperative for ethical and sustainable marketing practices. The blockchain marketing solutions must be able to prove that they are compliant with national and international data protection laws such as the General Data Protection Regulation (GDPR) or California's Consumer Privacy Act (CCPA). Minimum compliance with the laws implies the necessity of respecting the individual data subject rights; clearly disclosing data processing, and securing cross-border security practices.
- **Consumer Usability:** Usability and accessibility should be the main considerations of ethical design. Complex systems not only drive away users but also obscure the whole process. Consumer interfaces must be user-friendly; they must lessen both cognitive and technical burdens at the same time, letting a non-technical consumer grasp and manage his/her data. In such a case, it is a great necessity to have the frameworks that incorporate the human-centred design, clear consent signals, and privacy control panels.
- **Business Value:** Lastly, ethical blockchain marketing systems must demonstrate economic feasibility. The business model should demonstrate ROI in terms of improved efficiencies, reduced fraud, elevated brands' reputations, and consumer loyalty, for example. Additionally, systems should have existing marketing infrastructures with calculable economic value for all ecosystem members. Soon, usability, accessibility, and economic value will work together to create a structured ethical framework for blockchain in digital marketing. A framework that guides responsible data usage in the light of social responsibility and technology adoption. A framework that unites marketers, developers, and policymakers to build a marketing future that is open, just, and centred on the consumer.

POLICY AND PRACTICE IMPLICATIONS

Using blockchain technology in digital advertising has a great impact on the changes in regulation, operation, and user behaviour. This chapter introduces the research findings, their implications for practice, and our policy recommendations, and it also gives precise directions to marketers, policy-makers, and consumers.

Implications for Marketers

Marketers could substantially benefit by strategically embracing blockchain technologies in their operations. Adopting blockchain must go beyond technology to show a deeper transformation of business ethics and consumer engagement models.

- **Agreement to tools that are verifiable and auditable:** Marketers should embrace tools and platforms that allow an end-to-end verification of ad performance. Blockchain enables the ability to trace the delivery of ads, impressions, and engagement statistics. The traceable nature of blockchain helps minimise fraud and build advertiser confidence.
- **Transition to participatory marketing models:** Trade surveillance-based targeting for participatory models that incentivise users for engaging with ads and contributing their data. Token/reward systems and opt-in advertising can be used by brands to conduct market-based marketing strategies that engage user agency, ownership, and trust.
- **Brand differentiation through marketing practices of transparency:** In situations where transparency takes place in brand positioning, data transparency can become an employer differentiator. The ethical ways of collecting personal data and ensuring privacy through the use of blockchain can win customers' loyalty and show companies as good characters in the digital landscape that is changing.

Implications for Policymakers

Marketing professionals stand to gain significant value from a planned implementation of blockchain technologies into their marketing practices. The implementation of blockchain and other technologies should extend beyond merely implementing new technology and evidence a deeper business ethics transformation and a new consumer engagement model.

- **Agreement to verifiable and auditable tools:** Marketers should adopt tools and platforms that enable a completely verifiable audit of ad performance. Blockchain allows for traceability to the delivery of ads, impressions, and engagement statistics. The traces of blockchain minimise fraud and increase advertiser confidence.

- **Change to participatory marketing models:** Change targeting surveillance-based targeting practices to participatory models that reward users for interacting with ads and generating their data. Token/reward systems, along with opting-in advertising, allow brands to engage market-based marketing strategies that solicit user agency, ownership, and trust.
- **Differentiating brand positionability with transparent marketing practices**: In industries where transparency is part of the brand, transparent data practices can be an actual marketplace differentiator. Ethical collection of data, for other digital connections, and assurances through blockchain for privacy/safety can reinforce brand loyalty while actually evidencing brands acting as moral agents in the evolving digital economy.

Implications for Consumers

The consumer's position is, or should be, the main one in making digital marketing more ethical. Trust in digital ecosystems must be rebuilt, and consumers must be empowered, informed, and given choices regarding the use of their data.

- **Control via Data Sovereignty**: Users can take advantage of distributed ledgers and consent frameworks to have ownership and control over the access of their personal information on different platforms, which gives every single user the right to decide when, how, and with whom to engage their data.
- **Open, Significant Marketing Access**: Advertising systems that rely on blockchain technology allow consumers to receive marketing messages and learn about new products, like being tracked through digital marketing practices. They have the right to raise the actual opt-in controls and know how their data was used, thus feeling empowered to take part in the process of awareness with more knowledge.
- **Incentive Alignment:** By using tokenisation and rewards based on attention, users can benefit materially by engaging with advertising. This shifts the model of passive consumption to one of active engagement and contributes toward a better balance of value exchange in digital marketing.

In general, blockchain technology and its application in the digital marketing domain can bring about a revolutionary change; however, this is contingent upon deliberate policy changes and ethical marketing practices, and consumer power. The next chapter will consider future directions and open-ended questions that need to be answered in order to make this a reality.

CONCLUSION

Digital marketing, to a very large extent, is undergoing a metamorphosis primarily because of the hard-hitting winds of public scepticism, regulatory pressures, and near-term tech trends. With the ecosystem trying to find solutions to problems like data misuse, ad fraud, and the erosion of consumer trust, an ethical and decentralised system emerges as a necessity rather than a nice-to-have. Blockchain-based marketing can be considered the technology that makes an already shaky ground of trust and accountability in marketing more solid.

In sum, the literature review and the accompanying discoveries present blockchain not as a miracle cure but rather a disruptive enabler if used via ethical means, proper governance, and careful implementation. The practices of Brave, AdEx, Lucidity, and Ocean Protocol solutions illustrate how blockchain tech could take digital marketing to user-oriented, verifiable, and privacy-respecting frameworks. Finally, the interviews with stakeholders reveal that there is a longing within the larger marketing community for transparent, auditable, and participatory platforms, although there are still technical, regulatory, and educational hurdles.

The research presented here develops an ethical framework that takes into account the basic principles of transparency, consent, accountability, data minimisation, and verifiability, while also considering the dimensions of implementation such as technical viability, governance, usability, regulatory compliance, and business viability. The work therefore offers a roadmap to makers, developers, and regulators who want to connect their frameworks to the demands of a fairer and more durable digital advertising ecosystem.

On the practical side, marketers are going to have to invest in technologies that will help them to get verifiable metrics of their campaigns; besides that, they are going to turn towards more value-centric, participatory advertising models. Regulators will have to support and create regulations that effectively integrate the various layers of blockchain technology, and consumers will feel more empowered regarding their personal data and will thus take part in more just data markets.

In summary, not only does blockchain technology offer digital marketing an innovation in technology, but it presents an ethical framework as well, on which to responsibly rebuild the future of digital marketing. The digital marketing industry stands at an intersection between bygone practices and an emerging approach; and its future rests in fusing decentralisation with ethical stewardship, to ensure that personalisation, performance, and privacy are no longer antithetical, thereby, a more just advertising system can result.

REFERENCES

Boni, E. K. (2024). *Exploring perceived risk and digital ethics affecting consumer experiences and decision-making in smart retailing* (Doctoral dissertation, Brunel University London).

Catalini, C., & Gans, J. S. (2020). Some simple economics of the blockchain. *Communications of the ACM*, *63*(7), 80–90. DOI: 10.1145/3359552

Chen, Y. (2018). Blockchain tokens and the potential democratization of entrepreneurship and innovation. *Business Horizons*, *61*(4), 567–575. DOI: 10.1016/j.bushor.2018.03.006

Coche, E., Kolk, A., & Ocelík, V. (2024). Unravelling cross-country regulatory intricacies of data governance: The relevance of legal insights for digitalization and international business. *Journal of International Business Policy*, *7*(1), 112–127. DOI: 10.1057/s42214-023-00172-1

Dierksmeier, C., & Seele, P. (2020). Blockchain and business ethics. *Business Ethics (Oxford, England)*, *29*(2), 348–359. DOI: 10.1111/beer.12259

Goldfarb, A., & Tucker, C. (2011). Online display advertising: Targeting and obtrusiveness. *Marketing Science*, *30*(3), 389–404. DOI: 10.1287/mksc.1100.0583

Iansiti, M., & Lakhani, K. R. (2017). The truth about blockchain. *Harvard Business Review*, *95*(1), 118–127.

Kshetri, N. (2017). Can blockchain strengthen the internet of things? *IT Professional*, *19*(4), 68–72. DOI: 10.1109/MITP.2017.3051335

Lee, E., & Yoon, Y. (2022). Trusted information project platform based on blockchain for sharing strategy. *Journal of Ambient Intelligence and Humanized Computing*, *13*(3), 1575–1585. DOI: 10.1007/s12652-019-01421-z

Madhani, P. M. (2022). Blockchain Deployment in Marketing: Developing Conceptual Frameworks and Research Propositions. *IUP Journal of Business Strategy, 19*(3).

Morabito, V. (2017). *Business innovation through blockchain*. Springer International Publishing.

Nanda, P., & Kumar, V. (2024). New generation technologies: Development vs. ethical challenges. *International Journal of Business Information Systems*, *46*(3), 411–434. DOI: 10.1504/IJBIS.2024.140437

Narayanan, A., Bonneau, J., Felten, E., Miller, A., & Goldfeder, S. (2016). *Bitcoin and cryptocurrency technologies: a comprehensive introduction.* Princeton University Press.

Newman, N. (2014). Search, antitrust, and the economics of the control of user data. *Yale Journal on Regulation, 31*, 401.

Regulation, P. (2018). General data protection regulation. *InTouch (Melville, N.Y.), 25*, 1–5.

Sadeghpour, S., & Vlajic, N. (2021). Ads and fraud: A comprehensive survey of fraud in online advertising. *Journal of Cybersecurity and Privacy, 1*(4), 804–832. DOI: 10.3390/jcp1040039

Zyskind, G., & Nathan, O. (2015, May). *Decentralizing privacy: Using blockchain to protect personal data. In 2015 IEEE security and privacy workshops.* IEEE.

Chapter 6
Uses and Abuses of Point-of-Sale Machines:
Effects on Company Image and Marketing

Linnea Hall
https://orcid.org/0009-0004-1251-4794
Western Governors University, USA

Tamara Phillips Fudge
https://orcid.org/0000-0002-8682-9711
Purdue University Global, USA

ABSTRACT

Point-of-sale (POS) systems are becoming increasingly common as both an important aspect of the transactional infrastructure and a marketing tool that exemplifies the company's brand and philosophy. As such a visible aspect of a company, their functionality and security influence consumer trust and brand perception. A Transaction Trust Disruption Framework is therefore proposed to connect POS risk events, such as data breaches and availability to consumer perception, which directly relates to customer loyalty and company reputation. This framework will help marketing executives understand how technical safeguards can support marketing in building and preserving customer trust, particularly when responsibility is not clearly defined or third-party providers contribute to failure. This interdisciplinary approach integrates technical responsibilities and transparent crisis communication as elements of brand protection and marketing, rather than treating POS risk as an isolated technological issue.

DOI: 10.4018/979-8-3373-4526-0.ch006

INTRODUCTION

The Point-Of-Sale machine (POS) is encountered everywhere, from grocery stores and convenience markets, to hospitals, feed stores, big box stores, small family-owned businesses, and online shopping carts. Along with the convenience of no longer needing cash in one's wallet comes accessibility, data privacy, and financial safety issues. The increasing awareness of consumers related to these issues also shapes their perceptions of trust and brand competence. Thus, the performance and integrity of POS systems cannot be a responsibility relegated to the tech team but must be an integral part of the marketing team's brand strategy.

Originally, the adoption of electronic and POS technologies was solely a decision of convenience, allowing organizations to streamline transactions while efficiently collecting customer data for later analysis. As the adoption became more prolific, issues related to security breaches, payment fraud, system outages and accessibility issues ceased to be mere disruptions but came to be seen as direct violation of consumer expectations, leading to severe reputational consequences, including negative views related to brand image, weakened customer loyalty, and changes in purchasing behaviors.

Previous studies have emphasized the technical vulnerabilities and operational benefits related to POS system adoption. However, most of this body of work relegates the marketing consequences to secondary considerations, even though empirical research has demonstrated that perceived security and transparency are critical to shaping customer trust and long-term brand equity. There is a strong need for interdisciplinary studies that consider the cybersecurity aspects in terms of their integration and effect on marketing strategies, rather than in isolation.

A brief history and explanation of how these systems work is followed by an exploration into the manifold problems that can occur, including how known issues can impact a company's reputation and marketing strategies. Recommendations are offered for businesses and individuals alike to limit the impact of identity theft and data breaches on corporate identity and reputation.

Considering POS systems as a critical aspect of an overall marketing experience, rather than limiting them to the technical infrastructure clarifies the importance of the influence of transactional technologies on brand value and consumer trust. POS systems are discussed through the convergence of transactional security and brand development. Technical risks are analyzed through a marketing lens considering consumer trust and perceived reliability and how these influence consumer response and brand reputation. This interdisciplinary foundation connects cybersecurity concerns to marketing strategy by treating POS governance as an integral part of brand protection.

THE TRANSACTION TRUST DISRUPTION FRAMEWORK

Point of sale systems are an important aspect of the customer experience because they serve a dual purpose: they represent the transactional infrastructure of the business, but they are also primary brand-relevant contact points. When customers make their decision to purchase either online, or with a credit card at a terminal in store, they consider multiple aspects related to the transaction including the organization's ability to protect their personal and financial data. If a major breach has occurred recently, this may have a significant impact on the consumer's trust and perception of the brand, leading to a disruption in the transaction.

Risks related to POS include security breaches, skimming attacks, accessibility failures, or payment vulnerabilities. When these failures occur, they are often addressed as regulatory or technical failures, their marketing impact depends on how the failures are perceived by the consumer. Research in marketing and consumer behavior often describe consumer perception and brand evaluation when making purchase decisions in terms of their relationship to security, service reliability, and fulfillment, especially in digital payment environments where uncertainty or information asymmetry exist (Mondego & Gide, 2024; Singh et al., 2024; Khan & Abideen, 2023).

The Transaction Trust Disruption Framework (TTD) addresses how POS risks influence marketing outcomes based on perception driven mechanisms. POS failure can create feelings of uncertainty, loss of control, or violation related to the transaction. These feelings shape the customer's perception and evaluation of the customer experience. This may, in turn, lead to changes in loyalty, switching intention, word of mouth behavior, and ultimately, brand reputation, even when the organization was not at fault for the failure. The TTD envisions POS governance as a marketing-focused aspect rather than (or in addition to) a technical concern. By connecting transaction risk to consumer perception and brand equity, this framework clarifies how POS decisions relate to marketing strategy, allowing technical risks to be appropriately and consistently interpreted in terms of their impact on consumer trust and perception.

Figure 1. Transaction Trust Disruption Framework

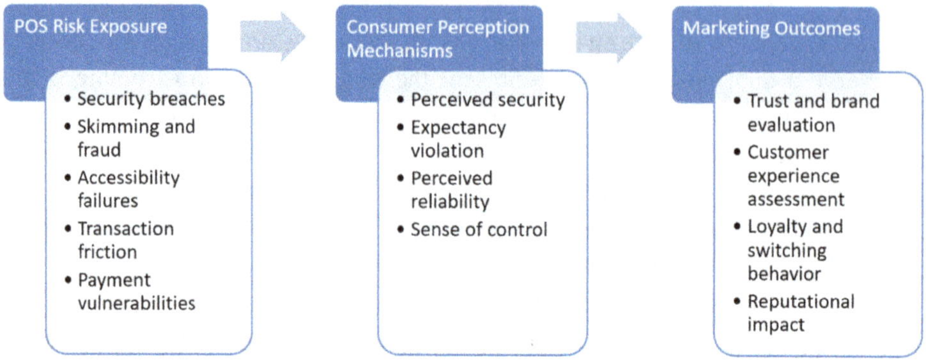

(The authors, 2025)

LITERATURE REVIEW: THE POINT-OF-SALE SYSTEM

A Brief History

The evolution of POS systems is an important aspect for marketing, because technological shifts in speed and perceived security affect the customer's purchase experience. Today's electronic Point-Of-Sale systems have their roots in cash registers, which by the late 1870s were sold to businesses with the promise they could deter theft, manage inventory information, and even provide a means to extend credit (Crandall, 2010). While calculating devices have been around since the Chinese abacus of ancient times and calculating machines such as the "arithmometer" appeared by 1820, the 1878 "Incorruptible Cashier" register patented by saloon owner James Ritty of Dayton, Ohio, with assistance from his brother John, was the first commercial success (Crandall, 2010). Lowden (1957) claimed the concept began with a casual observation of a ship's engine room with "paper dials registering pressure time" (p. 145); this idea of recording activity for later review made financial sense. The 1878 Ritty cash register included paper receipt printouts allowing customers to clearly see if the calculations were correct (Crandall, 2010; Georgiev, 2024), a money drawer sectioned for various paper money and coinage, plus a bell that rang to announce a sale and prevent employees from pilfering profits (Georgiev, 2025). Consequently, the Ritty brothers started their new business, National Cash Register Company (Crandall, 2010).

While the Ritty machine was mechanical rather than electronic, it is notable that the first version included hole-punched paper (Lowden, 1957), much like Herman

Hollerith's experiments eight years later with tabulating machine punched cards (Crandall, 2010), which preceded the binary hole-punched cards of early electronic computers.

The inception of the "credit card" goes back well over 100 years. In 1914, gasoline companies began to issue cards for purchases (Davenport, 1967). Soon thereafter, department stores provided customers with one-use cards for purchasing; eventually there were cards for airlines and rail travel as well. By 1950, the Diner's Club card was created and this "all-purpose" card allowed purchases at many different companies. Other financial institutions like the Bank of America, American Express, and Chase Manhattan followed suit by the end of the decade (Davenport, 1967). The plastic credit card with raised name and numbers became commonplace by the 1960s and made useful by a manual imprinter mechanism. The card was positioned in the imprinter, a carbon slip placed over it, and then the imprinting handle was (sometimes laboriously) swiped across both card and paper, thus imprinting the card's raised information onto the paper. The customer received a copy, and the company had the onus of reconciling the slips with actual accounts, which Cokus and Dahlgren (2007) mention was often difficult, especially when numbers began to wear down from many swipes or counterfeit cards circulated. This approach was in use for many years but is now obsolete.

Not until the 1970s were these large, stationary, and often heavy mechanical devices replaced by electronic solutions. The first commercially available electronic cash register came out in 1973 and was developed by IBM; now businesses could connect to databases, check prices, track transactions, and run a variety of reports (Flannigan, 2025). In addition, barcode scanners made input quickly and far more accurate (Georgiev, 2024). Paper checks were still widely used for payments for quite a long time after, but the cost was said to be 30-50 cents (USD) per cheque in 1980 to complete transactions, and it was not expected the old system would eventually give way to electronics (Board of Governors of the Federal Reserve System, 1980).

The first graphic touchscreen, ViewTouch, was created in 1986 by Gene Mosher, a delicatessen owner

(Flannigan, 2025). This enabled far quicker input and transaction time, helping to decrease customer wait time for in-store sales (Kabir & Han, 2016). Electronic Point-Of-Sale software/hardware platforms, also called ePOS, were introduced in the early 1990s; the "EMV" standard (Europay, MasterCard, Visa) began in 1993 for using microchips in payment cards instead of relying on less secure magnetic strip technology. This new reality for business transactions required card readers, personal identification number pads, and a means to capture signatures (Flannigan, 2025).

Early in the 2000s, the focus changed to cloud-based systems, employed first in the United Kingdom but spreading worldwide within just a few years, and rather quickly evolving to handheld devices (Flannigan, 2025). New features included being able to manage employee logins and logouts as well as scheduling appointments or reservations (Georgiev, 2024). POS systems today allow for synchronization in real time for checking inventory, managing customer data and reward programs, performing automatic transactions, and integrating sales with accounting (Flannigan, 2025), all of which assist businesses with multiple locations and synchronizing activities within a business, such as sending orders to the kitchen and connecting these to waitstaff and customers (Saeed et al., 2019). Handoko and Mauritsius (2021) include payroll features and a greatly reduced potential for human error as additional benefits; Yokeshwaran and Murugachandravel (2022) add the ease of allowing returns and exchanges. Artificial intelligence today offers predictive information, anomaly detection, and marketing and communication assistance (Flannigan, 2025).

Types of Systems

Variations in POS systems create different risk exposures and influence whether the customer perceives the risk as attributable to the brand or the payment system. Thus, understanding the evolution of variants is important to crafting the overall brand strategy. The POS machine takes many forms. Legacy systems, also known as traditional or on-premises, are not typically connected to the internet; they tend to have robust security as a result, but there is also a high price tag of perhaps $15-20k for hardware and installation and then support costs are charged monthly ("Types of POS," 2024). Tablets such as iPads, small hand-held devices, and the use of the internet for cloud-based systems are becoming far more popular, as they are significantly less expensive to set up and maintain ("Types of POS," 2024). Saeed et al. (2019) mention using the cloud is both time and cost effective, and these systems are scalable.

Self-service kiosks became more commonplace during the COVID-19 pandemic since self-checkout limits person-to-person interaction (Santos & Mendes, 2023). Stand-alone kiosks provide a new level of efficiency in airport check-ins, although it is noted whether a passenger chooses to use the machine or wait to have assistance at the counter might hinge on factors such wait time and the person's comfort and experience with technologies (Wongyai et al., 2024).

How Today's POS Works

Understanding the transaction flow ensures that the marketing strategy considers where there potential trust judgments are made in the process. To follow the tenets of ISO 9126 standards, which are functionality, reliability, usability, efficiency, maintainability, and portability (Olipas & Esperson, 2020), electronic POS systems use a variety of technologies. Hardware, software, network connections, and a database provide the initial requirements. Barcode readers scan tags to pull costs, item names, and quantity from the company's database and show the potential purchases on a screen. Some items, such as produce, may need to be weighed at the POS station and so require more human input (Hayes, 2024). Any applicable taxes are added to the total. At this point, the customer either pays with cash or presents a debit or credit card, which is inserted or tapped to read the chip, or swiped through a channel at the side of the POS to read the magnetic strip. The bank that issued the card may require a PIN; the system will then ask the bank if there are sufficient funds for the purchase before allowing the transaction to be completed. If the purchase is made online, the "CVV" number on the back of the card may be required. A receipt may be printed on paper for in-person sales or offered for electronic delivery to email (Hayes, 2024). Inventory is updated just prior to the payment to ensure the items are still available. Transactions are recorded in real time, which assists in smooth, quick sales even at the busiest part of the day (Crawford et al., 2024). Figure 2 shows the basics of this process in visual form.

Figure 2. General Process for POS Machine Transactions

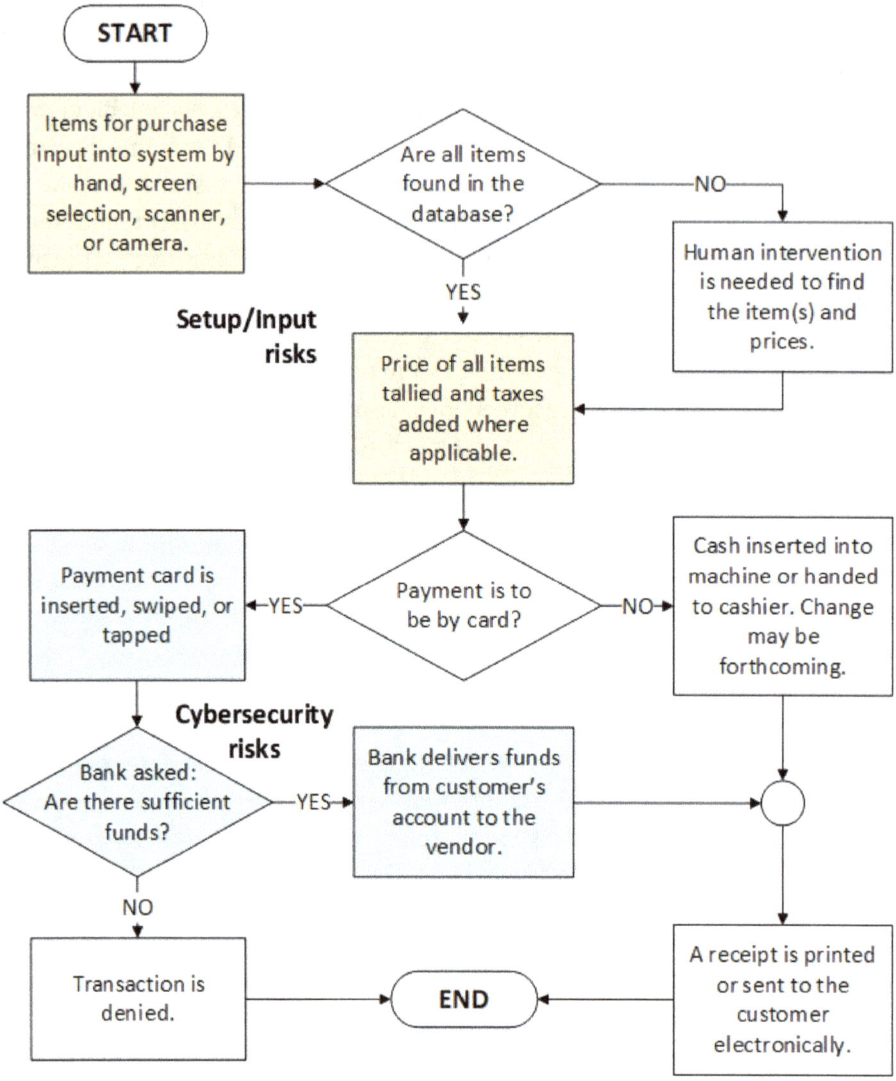

(The authors, 2025)

In traditional businesses, the necessary hardware includes a register with a screen, a cash drawer, a bar code scanner (which may be handheld or a flatbed), and a printer for paper receipts. Camera devices are sometimes used instead of or in addition to barcode scanners (Deutch, 2023). Software connects the hardware with the database and bank network. In addition, a modem is needed to connect to the internet and a server holds the database (Lestariningati, 2018). Crawford et al. (2024) explain there can be additional items such as kitchen printers in restaurants, and even digital tip jars, and online businesses do not need hardware, although their cloud providers will use servers and other items. Mobile applications may use wireless routers to connect to the server (Lestariningati, 2018).

Miniaturization is possible. Some small companies may opt for handheld terminals, as shown in the dual image below.

Figure 3. Inserting Payment Card and Swiping Payment Card into Handheld POS

(Ardity, 2019)

POS systems tend to make use of normalized databases with at least three tables, which Maji et al. (2019) describe as header (time, identification of customer, store, and employee), details (the purchased item details), and tender (payment type, etc.). As with any software or online offerings, reliable technical support is crucial for ensuring efficiency.

PROBLEMS WITH POS SYSTEMS

Accessibility

Marketing must consider the diversity of customers. Limits related to accessibility tend to be interpreted as service quality issues and are often attributed to the organization rather than the system. The term "accessibility" can have different meanings with physical, technological, and socioeconomic implications. Inclusive design is the development of technology that can be used by anyone "without requiring significant adaptations" (thus permitting autonomy) while serving a "wider spectrum of consumers" (Çelik, 2025, p. 437). Yokeshwaran and Murugachandravel (2022) remind that accessibility is not just for people who live with disabilities but also includes those with poor online connections.

POS systems rely on visual and tactile interaction. Assistive technologies, such as screen readers, help persons with visual impairments with online purchasing (Çelik, 2025). There are apps such as the combination of BrailleBack and TalkBack for Android devices, and methods to magnify text and control color contrast to make readability easier for some (San Diego State, 2024). Static POS machines may include Braille and buttons which are tactilely discernible, such as Xs and Os on physical keys, and for parts of transactions, could include voice commands (Çelik, 2025). Touchscreen technologies provide no tactile assistance, and unfortunately, not all kiosks provide user-selected options for text size, etc. although headphone jacks may be present for private listening (Department of Justice, 2000). Not only is it important for customers to be able to provide input, the system must also provide a means for signaling either success or failure (Çelik, 2025).

Static POS systems, including kiosks and other stationary hardware, pose additional problems. People with a physical disability may have difficulty reaching up or over to static consoles, trying to hide their PIN input from others waiting in line, or finding enough strength for physical buttons. The U.S. Department of Justice (2000) mentions insufficient floor space or obstructed routes for those who must use mobility aids such as wheelchairs, and the need for controls to be operable with one hand, "without tight grasping, pinching, or twisting of the wrist" (p. 255).

Beyond physical interface considerations, accessibility in POS systems also encompasses digital usability and reliable connectivity, which are essential for ensuring equitable access to services in increasingly app-driven and online purchasing environments. For example, another accessibility concern is the ability of the average consumer to download and use company apps for purchases. Apps such as those for grocery stores and ridesharing are available for free download through app stores. While these are convenient, and some will work by allowing access through online login providing some security, they could also be prone to reverse engineering by hackers (Frisby et al, 2012).

Since POS systems need connectivity, internet failure could also present access difficulties. Each country's banking system may differ slightly; Dalhous and Welte (2024) explain, for example, that Canada has Interac, an Inter-Member Network (IMN) for debit card POS transactions, rather than the fragmented system used in the United States. This centralization may sound helpful for the sake of consistency, but "decentralized networks offer equality, accessibility, and security" to consumers and financial institutions alike (Alamsyah et al., 2024, p. 1). When downtime occurs, it may be at peak business and cause customer consternation. If the system is down for technical reasons rather than being hacked, restarting the POS machine or the router, doing a speed test, checking cables, ensuring software is updated and firewalls set correctly, and running diagnostic tests may be needed (Daniel, 2024). Employees in the store may not have access to much of this, but as Daniel (2024) suggests, having a backup connection and performing maintenance regularly can help prevent connectivity issues.

Contactless Transactions

Contactless payments increase expectations related to convenience and control while making perceptions related to security and reliability, both of which are related to marketing and customer satisfaction. RFID or Radio-Frequency Identification is a technology which uses radio waves to automatically identify and track objects with attached "tags" (Kholod et al., 2024). An emerging digital technology, Near Field Communication (NFC), utilizes short-range radio waves (based on RFID principles) to enable communication between mobile devices. This technology is increasingly used for convenient payments and transaction settlements around the world (Onumadu & Abroshan, 2024), NFC or Near Field Communication Both Radio Frequency Identification (RFID) and Near Field Communication (NFC) technologies are used to enable smart phone or payment card taps, also known as "contactless" payments. Even though this method is often called "tapping," the phone or card does not need to physically touch the POS device. NFC employs a frequency of 13.56 megahertz, a subset of RFID, and can transfer at three different rates to complete transactions (BasuMallick, 2022). This technology connects the customer's device or card wirelessly to the POS machine and notably, the distance between the items must be 4 centimeters or less (BasuMallick, 2022). The transaction sends a one-time code to the user's bank account instead of revealing name, address, and card numbers, and once complete, a green light or beep is typically provided as verification (Perkins-Southam & Lupini, 2025). Cards with this capability have an EMV chip and display four curved lines much like the icon used for wireless internet connections but on its side ("Contactless credit cards," 2023).

Tapping has advantages: it is much faster than other payment methods, cards do not get worn down as they might with swiping or inserting, and data security is considered rather strong. As explained by An et al. (2024), NFC can "enhance both physical-layer security (PLS) and multi-user communication scenarios by taking advantage of the distance-aware spherical wavefronts" (p. 105). Perkins-Southam and Lupini (2025) adds since no physical contact is needed, there is no need to worry about unsanitary surfaces on POS machines potentially spreading disease. Also, RFID-shield cases may be helpful to avoid inadvertent use.

That being said, large data sets may transfer slowly, and data can be transferred in both directions, making it possible for savvy hackers to eavesdrop (BasuMallick, 2022). Perhaps the most common concern is the physical loss of the card, as anyone in possession of the card would be able to use it because there is no signature or PIN required for tapping. While some financial institutions may not have protection for debit cards, many do have fraud protection services for credit cards, especially if loss or theft is reported quickly (Perkins-Southam & Lupini, 2025).

Skimmers

Skimming changes hidden technological vulnerabilities into a visible trust violation, and the resulting harm may easily be attributed to the brand associated with the transaction rather than the payment system. While contactless technologies offer enhanced convenience and hygiene, their increasing adoption also includes broader concerns about payment security, particularly as evolving skimming techniques target both traditional and newer forms of electronic transactions. In a world increasingly digital and increasingly relying on electronic payments, skimmer attacks are a growing threat, resulting in billions of dollars of losses annually to merchants, consumers, and banks (Scaife et al., 2018, Onat & Ciaccio, 2025, Bhaskar et al., 2019). Skimming is pervasive (Bhaskar et al., 2019) and involves a device surreptitiously added to a real POS system with the intent to steal financial information from unsuspecting users. In 2018, 972 skimmers were found on gas pumps in Florida, and 148 were recovered in Arizona (Bhaskar et al., 2019). Additionally, the credit score company FICO reported a 452% increase in skimming reports from 2022 to 2024 (Onat & Ciaccio, 2025).

One of the main reasons skimming is such a problem is that they are hard to detect and easy to employ (Bhaskar et al., 2019). Modern skimmers are designed to blend in with POS machines and ATMs (Ratica, 2024, Scaife et al., 2018). Even trained professionals have difficulty detecting them (Ratica, 2024). Each skimmer can capture data from between thirty and one hundred credit cards each day (Bhaskar et al., 2019, Onat & Ciaccio, 2025). In Florida and Arizona, the estimated losses from gas pump skimmers were $16,000,000 USD in a single day (Bhaskar et al., 2019).

Criminals use different methods for capturing data. They can capture magnetic strip data with a card reader while using a camera to capture PINs, allowing them to create cloned cards (Scaife et al., 2018). More recently, skimmers have started using Bluetooth devices, installed internally, that can send data directly to a criminal waiting nearby (Scaife et al., 2018). Fortunately, these devices have not yet become widespread, accounting for only 20% of skimmers recovered by the NYPD (Scaife et al., 2018). Most skimmers are still wired and require physical retrieval to download the data and recharge the device (Scaife et al., 2018). The MAC addresses from some skimmers reveal information about the criminal organization manufacturing and distributing the skimmer (Bhaskar et al., 2019). These devices have been recovered from locations hundreds of miles apart, demonstrating the breadth of area in which these criminal organizations operate (Bhaskar et al., 2019).

Unfortunately, using smartphone apps or simply pulling on the card reader, the most used detection methods, are not effective for identifying modern skimming devices (Scaife et al., 2018). Recently, several methods have been used to help detect skimmers. For example, Bluetana is a smartphone-based app proven effective against Bluetooth skimmers (Bhaskar et al., 2019). These apps work by detecting unusually strong unknown Bluetooth signals near the pump (Bhaskar et al., 2019). The benefit of apps like Bluetana is they are readily available, and the time required to detect a skimmer is only seconds, unlike a physical inspection, which usually requires thirty minutes or more (Bhaskar et al., 2019). There are also physical detection devices, like the Skim Reaper, a detection device with an attached "card" inserted into the card slot with the suspected skimmer (Scaife et al., 2018). The Skim Reaper works by counting read heads on the machine and alerting if there are more than anticipated (Scaife et al., 2018). Another detection method uses LiDAR (light detection and ranging) imaging to create a 3D model of the machine (Ratica, 2024). These images are then compared with known vendor machines to identify variations or inconsistencies (Ratica, 2024). Once a skimmer is identified, this method can help to alert other locations with similar devices (Ratica, 2024). Additionally, it can alert authorization systems to deny transactions from suspected machines and notify issuers for removal (Ratica, 2024). Apps can also notify users of "safe to use" devices (Ratica, 2024). Unfortunately, while these devices are effective today, skimmers are constantly evolving and criminals change their tactics once they have been discovered, making it difficult to be proactive in preventing attacks (Bhaskar et al., 2019). Simply switching to Bluetooth Low Energy (BLE), Global System for Mobile Communications (GSM), or Wi-Fi, may be enough to thwart current detection devices (Bhaskar et al., 2019).

Data Breaches: Risk, Trust, and Mitigation

The Transaction Trust Disruption Framework demonstrates that these risks extend beyond technical failure and shape consumer trust evaluations, a necessary connection in marketing. Data breaches represent a direct breach of trust because the violation of personal data undermines trust in the organization's ability to adequately protect their customers' data. Digitalizing retail transactions has ensured the position of point-of-sale systems as an effective and efficient means of allowing the consumer to complete their transactions and allowing the company to process the transaction while collecting valuable data that can be used to support the company's strategies and goals. However, their ubiquity has made them vulnerable to attacks by cybercriminals, often resulting in data breaches affecting thousands to millions of consumers. Skimmers and contactless payment systems demonstrate these threats, but are representative of only a fraction of the broader cybersecurity risks organizations now face as businesses increasingly rely on digital infrastructure to process transactions and data. Zimmel (2025) claims the average monetary cost of a breach is $4.48 million USD, and the possibility of a large-scale data breach is a critical issue affecting financial integrity, consumer trust, and brand reputation.

A notable and historic breach occurred in late 2013 when more than 40 million customers' data stored in Target Corporation's databases were compromised (Sanchez, 2025). Hackers from Eastern Europe found a vulnerability in a third-party vendor and proceeded to upload malware to Target's system, gaining access to this data. Target's lack of quick response and the sheer volume of stolen data caused reputational damage and poor holiday sales, a financial disaster even beyond the initial $162 million cost of settlement (Sanchez, 2025).

Sharma et al. (2020) defined a breach as a "release of confidential data, commonly known as personally identifiable information, from a secured location in a computer or an electronic device to an unsecured site" (p. 33). While data breaches have been researched since the 1970s, as organizations have become more interconnected and reliant on customer data, there has been tremendous growth in the number and severity of breaches (Molitor et al., 2024). These breaches present several different kinds of threats to a business. Data breaches can occur in many ways including hacking, phishing and its variants, and ransomware. However, companies have some level of control over these types of attacks through careful preparation and mitigation practices. On the other hand, third party service providers introduce a significant risk over which the company has little to no control (Kadam, 2025). Paradoxically, Molitor, et al. (2024) noted "investment in information technology (IT) security, correspond[s] to a higher risk in data breaches" (p. 175). Refusing to invest in IT can unfortunately severely limit a company's ability to compete in today's global economy.

It is important to note the Payment Card Industry Data Security Standard (PCI DSS), which is not a regulation or law but is a "widely accepted set of policies and procedures" (Barney, 2024, para. 1). According to Barney (2024), Visa was first to establish security rules in 1999; after major Trojan-based breaches in mid-2004, Visa was joined by American Express, Mastercard, Discover, and the Japan Credit Bureau to develop and launch PCI DSS 1.0 by the end of that year. While this did not and cannot stop all fraud and theft with use of credit and debit cards, it provides some mandates and much guidance such as encryption standards, password and multifactor authentication rules, third-party issues, penetration testing, contactless payments and customizations allowed for payment systems such as PayPal and Venmo, and more (Barney, 2024; PCI Security, 2024). PCI Security Standards Council (2024) identify sensitivity of data types, with the account number, verification code, PIN, and "full track data" necessitating encryption. In addition to PCI DSS and EMV requirements, POS hardware certification standards such as PCI pin transaction processing further govern device-level security controls (PCI Security, 2025).

Chip data is also globally secured by EMV standards, thanks to an agreement by Visa, Mastercard, and Europay in 1999; other companies have since joined (Gilde, 2024). Gilde (2024) explains that EMV includes robust testing of both electrical and mechanical controls of chip readers and includes security protocol, interoperability, rules for record-keeping and maintenance, and more. In the European Union, the General Data Protection Regulation (GDPR) governs data collection, use, and storage, and other countries (and in the U.S., separate states) similarly have laws to control data.

Research has shown POS breaches often result from not only inadequate security at the POS, but also the controls applied to the networks they use to communicate. Standalone terminals are often the most susceptible because they tend to implement weaker security policies and they store data locally using flash memory, which includes user credentials and transaction logs. Even if this data is encrypted, cybercriminals are often able to identify systemic flaws allowing them to access stored information (Larson et al., 2020). In addition to weak security policies, viruses and trojans can infiltrate a POS system, undermining both customer data security and transaction process integrity (Badgujar, 2023). These digital crimes are compounded by more physical attacks such as robbery, social engineering attacks, not simply stealing the POS device (Obi, 2023).

Data breaches can have significant consequences for both companies and their consumers. The most obvious consequence is financial loss. This can be seen in the drop in stock prices following a breach announcement, and also includes financial penalties, such as regulatory fines, legal fees, remediation costs, and costs associated with assessing and upgrading security (Nikkhah & Grover, 2022).

In addition to the preliminary concerns related to financial loss to both the company and the consumer, a data breach, especially one considered severe, can cause psychological stress affecting a customer's mood, well-being, and health, leading to high stress, depression, and anxiety (Sears & Cunningham, 2024). This can lead to an erosion of trust, which, in turn, can lead to changes in consumer behavior (Nikkhah & Grover, 2022). As customer dissatisfaction increases, customers begin exhibiting switching behavior, wherein they look for alternative, more trustworthy retailers or products. If the breach is small, and not picked up by the mainstream media, customers may still share their experience on social media forums (for example, Facebook and Instagram), or consumer complaint or review websites such as PissedConsumer and Yelp. This dissemination of unfavorable judgement could harm the organization's reputation and deter prospective clients. Research has demonstrated a correlation exists between perceived expectancy violation and customer dissatisfaction, which subsequently fuels switching intentions and negative publicity (Nikkhah & Grover, 2022).

Customers are more likely to believe problems occurred if the breach is considered severe, and severity might be described by the type of data, number of accounts breached, the nefarious purpose of the breach, and/or the source of the breach (Aivazpour et al., 2022). Importantly, customers may use this information to decide whether the risk of doing any business with the company is a comfortable undertaking, and some customer personalities may react more strongly to security incidents than others through risk aversion (Aivazpour et al., 2022). Customers consider a data breach as a violation of trust and privacy (Hoehle et al., 2022). This can be explained by the Expectancy Violation Theory (EVT), which states "breaking assumed norms, such as information privacy, violates expectations and influences relationships" (Nikkhah & Grover, 2022). When a data breach occurs, this violates the implicit (or explicit) contract of trust and privacy, which is the foundation of the customer-company relationship.

Alnaim (2022) adds that online purchases require a rather high level of trust; a positive customer attitude assists the company in alleviating uncertainty and reducing unpredictable shopping behaviors where credit or debit card use, privacy, and pricing are scrutinized. Consumers need to have risk mitigation plans clearly explained, such as through the assertion of any suspicious activity through monitoring services (Aivazpour et al., 2022).

The concept of consumer resilience can be applied to understand how individuals recover, resist, or adjust their online behavior following a perceived privacy breach. In general, consumers recover quickly from a breach, there are significant differences across demographic groups and personality types (Budak et al., 2023). There are two aspects to resilience: privacy and resilience at the system-wide level, which addresses the way in which the company maintains system resilience when privacy is endangered, and how surveillance (and privacy) affects societal resilience, and consumer resilience at the individual level, which considers how consumers recover or adjust consumption habits after adversity, particularly online privacy violations types (Budak et al., 2023).

Thus, to gain the customer's trust, the organization must first understand the origin of risks. While most companies understand the need to protect against common attacks, such as hacking, phishing, and ransomware, the IT (security) team must have a deeper understanding of how these occur and how they can be prevented. One potential weakness is third party providers. According to the SecurityScorecard Threat Intel Report (2024), "97% of the largest U.S. retailers experienced a third-party breach, with only 4% of vendors compromised, demonstrating the far-reaching impact of these vulnerabilities" (Key Findings Section). Unfortunately, even if a third party was the cause, the perception will likely be that the company's negligence was the cause of the breach. Cyber Third-Party Risk Management is a relatively new threat organizations must consider when choosing their POS, whether an in-person card reader, kiosk or online market (Keskin et al., 2021).

How an organization responds to a breach is as important as how it safeguards its data, which means its customers' data. Research has consistently shown admitting the error and maintaining transparency in the corrective response is more effective than a passive or dismissive response. Admitting the breach and acknowledging the potentially damaging effects on its customers has a significant impact on repairing damaged customer trust. However, if apologies are offered too often, they may have the opposite effect (Jechle et al., 2025).

Compensation is another useful approach. Offers for discounts, free credit monitoring, or identity theft protection can influence a customer's perception of justice and may effectively alleviate a customer's negative sentiment toward the organization (Hoehle et al., 2022). The type of compensation should be substantial enough to ameliorate the emotional and psychological damage caused by the breach, along with any actual damage a customer may incur as a result.

Taking corrective action and maintaining transparency with customers about the ongoing actions to protect their data and address the vulnerabilities that led to the breach is indispensable. Regardless of the approach taken, research has shown a "corrective & apology" approach or a "corrective & compensation" approach results in a better outcome, including better perception of the company's reliability and less (or sometimes positive) impact on stock prices; however, corrective action alone is usually insufficient (Nikkhah & Grover, 2022).

The timing of the response is also critical to its effectiveness. Early accommodative strategies are better at reducing dissatisfaction and negative word-of-mouth (Nikkhah & Grover, 2022). It is therefore essential that organizations experiencing a breach act immediately, decisively, and empathetically to assure customers their concerns are being addressed.

Using a multipronged approach is also recommended. Most organizations use multiple approaches including corrective actions paired with apologies, compensation, excuses or justification. Some companies will take rationalizing approaches, such as reminding the customer of their previous positive relationship, seeking goodwill, or portraying the organization as the victim, combined with justifications or excuses (Jechle et al., 2025). Ultimately, the best approach will depend on the circumstances of the breach and whether there are regulatory consequences.

Marketing as a Reputational Safeguard in the Wake of Cybersecurity Incidents

Reputational outcomes after cybersecurity incidents are dependent on how responsibility and response are communicated to consumers. Beyond immediate breach response, companies must also manage the long-term reputational effects of cybersecurity incidents. In this context, social media marketing capabilities are increasingly recognized as strategic tools for restoring and preserving consumer trust.

In today's digital retail landscape, organizations accept social media as an effective tool for promote their brands, but it can also be an effective tool for managing and mitigating an organization's reputational risk arising from cybersecurity risk aversion both before and after a breach occurs. As cyberthreats become increasingly prevalent, how customers perceive an organization's ability to protect their data can significantly influence trust. Thus, an organization's social media marketing capabilities are critical in mitigating the negative consequences of perceived cybersecurity risk (Mathur, 2018).

An organization's reputation is tightly coupled with a customer's perceptions of cybersecurity risk. This relationship is particularly prominent among retail firms, where customers demonstrate a heightened concern related to the security of the information they are required to provide to effectuate a purchase. Mathur (2018) noted these concerns can lead to negative consumer attitudes and behavioral shifts, such as reduced loyalty or switching to perceived safer competitors. This, in turn, diminishes the organization's reputational capital and financial stability.

In this context, social media marketing can work as a functional "shield," allowing organizations to address customer anxieties proactively. An effective social media marketing campaign should focus on providing timely updates, clarifying uncertainties, and demonstrating the organization's commitment to cybersecurity best practices. According to Mathur (2018), effectively using social media marketing to address cybersecurity concerns and perceptions has a statistically significant positive effect on retailer reputation. Effective social media marketing also allows the organization to gain a better understanding of customer concerns and respond according to foster strong brand relationships (Mathur). Additionally, it can address negative word-of-mouth into trust-building opportunities to reshape consumer feedback into a more favorable narrative (Mathur, 2018).

Transparency and credibility are central to an effective social media marketing campaign. It is critical that the organization clearly and publicly communicates their adherence to cybersecurity protocols and regulations to show they are serious about how they protect customer data. Being transparent about perceived and real threats and explaining the process the firm takes to respond to these threats can further strengthen customers' trust perception. This approach to messaging reinforces a sense of security and accountability, enabling organizations to "communicate their commitment, transparency, and stringency in adopting best practices to address customer concerns" (Mathur, 2018 p. 13).

Empirical findings further suggest that social media marketing can improve perceived cybersecurity risk and firm reputation. A strong social media marketing campaign can mediate the adverse reputational effects associated with perceived risk. Specifically, it allows retailers to "weaken the negative impact of perceived cybersecurity risk and protect [their] reputation" (Mathur, 2018, p. 24).

The practical implications of these findings demonstrate the importance of prioritizing cybersecurity not only as an IT function but as an integral component of a company's marketing strategy. Mathur (2018) explained organizations need to understand that cybersecurity risk management has become an integral aspect of marketing affecting the firm's value as it relates to consumer trust and brand reputation. This means organizations should ensure they invest in social media marketing to mitigate risk and support their brand reputation to build lasting customer relationships (Mathur, 2018).

In addition to the tactical use of social media in crisis response, research suggests risk perception should be integrated into both marketing and brand positioning strategies. This idea of "risk marketing" suggests consumer attitudes toward risk are becoming increasingly important to market differentiation (Boutang & De Lara, 2016). This approach incorporates ideas from several disciplines, including cognitive science, neuroscience, and evolutionary psychology to effectively address consumer risk attitudes and perception.

An important aspect to this approach is "risk-enriched positioning", which goes beyond the organization's promise to consumers, such as "Save Money. Live Better" (Walmart) or "Empowering every person and every organization on the planet to achieve more" (Microsoft), by addressing how consumers perceive and respond to risk (Boutang & De Lara, 2016). This requires organizations to incorporate perceptual maps that cross aspiration dimensions with risk attitudes. Boutang and De Lara (2016) recommend the Lopes' Security-Potential/Aspiration (SP/A) theory as a useful framework for categorizing individuals based on their risk orientation, which can range from security-focused to opportunity-seeking. By categorizing their customers based on risk, organizations can tailor their brand messaging to target segments where consumer anxiety or perceived vulnerability is high.

However, whether these strategies are effective depends on the how the messages are framed from a psychological perspective. It is imperative their marketing teams understand how to deliver their message in a way that addresses the risk aspects. Research has shown consumers respond differently to logically equivalent messages depending on how they are framed, including color, word choice, source credibility, and audience characteristics. (Boutang & De Lara, 2016). While this does not deviate considerably from traditional marketing messages, understanding how cybersecurity risk can affect the way in which a message is perceived is an important concept that must be considered by any marketing team. Approaches including the use of bias-free survey tools to avoid common problems such as the illusion of control or social desirability bias are key to accurately assessing consumer risk perception (Boutang & De Lara, 2016).

Incorporating evolutionary psychology is also a compelling approach to interpreting how consumers perceive risk in marketing. By understanding the fundamentals of human behavior, marketers can better understand how people evaluate risk and how it affects their consumption behavior (Boutang & De Lara, 2016, p. 30).

These findings suggest marketing strategies should not only respond to risk incidents but proactively integrate risk into brand identity and consumer engagement frameworks.

Cybersecurity and Online Purchase Intention

Online purchase intention is directly influenced by how the consumer perceives security and competence, linking cybersecurity governance to trust and risk perception. Online purchases require a high level of trust as stated above. Alnaim (2022) lists four aspects, including: cognitive trust, perceived usefulness, quality of the website and resultant attitude towards its utility, and company reputation which includes social media presence. These factors must be considered by marketing professionals.

Not unsurprisingly, Riefel (2023) states perceived risk as explained above can result in dissatisfaction and loss of trust. "Cognitive trust" means the customer assumes and expects a company will have reliable means to avoid such issues, with "strong significance and a positive correlation ($r = 0.469$, $p < 0.001$) between cognitive trust and customer satisfaction" (Riefel, 2023, p. 6).

Usefulness is one of the key elements of the Unified Theory of Acceptance and Use of Technology, also known as UTAUT. Marikyan and Papagiannidis (2023) explain this framework includes four main constructs: performance expectancy, which is what the consumer believes they will gain through use of the technology; effort expectancy, or how easy the technology is to use; social influence, which helps the customer to know if they should use the technology; and facilitating conditions, defined by how much control the user thinks they and/or the organization can exert. These may include mitigating factors such as age, gender, experience, and whether the consumer is using the technology voluntarily (Marikyan & Papagiannidis, 2023).

The quality of a website is partially gauged by quick load time, mobile-friendly design and responsiveness, machine learning and artificial intelligence for providing personalization, and use of current coding practices (Zamoroka, 2023). Beaird et al. (2020) include information architecture, or how the site is organized, to facilitate function and efficiency. “Good design” may also be described as being able to find information or products easily through an intuitive navigation system and pages that are also pleasant to view (Beaird et al., 2020). Quite obviously, a site which offers items or services for sale relies also on visibility, the potential to be discovered via a search engine. This is driven by Search Engine Optimization (SEO) strategies, which include but are not limited to well-written meta tags, appropriate use of keywords and descriptive headings, uploading XML site maps, and understanding algorithms, metrics, and analytics (Enge et al., 2023). Another point to mention is a site's “bounce rate,” the percentage of visitors who leave a site without exploring past the home page (Farashi & Bashirian, 2021), with a high rate indicating a failure to entice and keep site visitors' attention. Lastly, any time a visitor is asked to input information (such as for a purchase), security must be ensured; Enge et al. (2023) recommend Secure Sockets Layer (SSL) but Duplys and Schmitz (2024) explain Transport Layer Security (TLS) is newer, more secure, and today used more often than SSL. Among potential attacks are eavesdropping, impersonation, dictionary attacks, man-in-the-middle, brute force, replay attacks, the introduction of malware, attacks on hash functions, and other illicit hacks (Duplys & Schmitz, 2024). Ensuring security surely adds to the website's quality and value.

A company's reputation may take years to develop. Kiely (2025) defines reputation as “widespread opinion someone has of your company as a whole” (para. 7). Since 85% of consumers tend to check products prior to purchase and approximately the same number admit to paying more based on reputation, all online commentary has the potential to make a difference in consumer opinion (Kiely, 2025). This will include social media posts and hash-tagging, Yelp reviews, news reports, and similar content, with positive or negative views affecting potential transactions. Data breaches can have devastating effects on perceived reputation; legal issues and economic losses may follow (Egerson et al., 2024).

Economic and Operational Cybersecurity Implications

Operational disruption and financial loss are relevant to marketing when service reliability declines and apparent instability erode brand credibility. Banks and other institutions in the financial sector are integral to business transactions but also prone to cyberattacks. Cyber events threaten consumer confidence and could result in large monetary losses (Egerson et al., 2024). Ransom attacks are possible, with major consequences; for example, “71% of Nigerian companies especially financial institutions experienced cyberattacks in 2021; 44% of these companies paid an average ransom of $3.43 million” (Egerson et al., 2024, p. 916). In turn, such losses impact profits and operational efficiency. Strong policy-writing is needed, not just for consumer trust, but to guard against potential threats.

The consequences of a data breach also extend beyond financial damage. A breach, especially one not handled correctly, can lead to long-term reputational damage, such as decreased consumer confidence and brand equity, along with potential legal liabilities (Badgujar, 2023). Studies also indicate that, while there is a significant initial decrease in consumer trust, this can be mitigated by an appropriate and immediate response. Thus, ensuring legal and procedural transparency in responding to the breach can restore damaged trust (Farok & Zolkipli, 2024).

MARKETING: CONSUMER BEHAVIOR AND INCENTIVES

Risk controls and governance practices protect brand value when technical safeguards are treated as trust-preserving commitments to the customer experience. As might be expected, company reputation is paramount in marketing. Aivaspour et al. (2022) mention social media and other communication channels as quick determinants of changes in reputation, dependent on factors already mentioned. There are other influences to note in regard to the POS machine, however, and how this technology connects to marketing, advertising, and company success.

Marketing At or Near the POS

It is no surprise that candy and "special deals" are offered near checkouts in many stores; proximity to the payment machine is deliberate to increase sales of those items. Yacek (2023) goes as far as renaming the acronym as "Profit Optimization System" (p. 34) and suggests marketing budgets may hinge on sales volume gauged at the POS. Näränen (2013) described over a decade ago that direct marketing of alcoholic beverages in Finland was not allowed due to governmental control and noted issues with alcoholism in the country. The retailer can be a go-between, and "when consumers are at the point of sales environment, they are already in the mood for spending" (p. 6). While "point of sale" may indicate the retail setting itself such as a restaurant rather than a POS machine, the fact remains product location and advertising where purchases are imminent help to drive sales (Näränen, 2013).

The Customer-Facing POS

There are various kinds of POS machines which can be used in customer-facing mode. Tejeda (2025) describes a swivel-stand machine which can be viewed by both employee and customer, and two devices, where one is used by the employee and the other viewed by the customer. When the person making the purchase can keep track of scanned items and prices, there may be fewer questions and/or returns, and the interface used by the customer can encourage additional items for sale, entice the input of email for future promotions, and allow viewing of loyalty program contributions (Tejeda, 2025).

The self-checkout kiosk at some groceries and big box stores also qualify as "customer-facing," of course. Lahmeyer and Roemer (2024) explain the shopper's experience is based on cognitive, social, and other constructs including physical reaction to the retailer's environment, and they claim self-checkout tends to improve attitudes. Nusrat and Huang (2024) note this mode might be used more often when there are fewer items in the customer's cart, but more importantly, they feel customers are less loyal to a store when using self-checkout, and the feeling of loyalty goes hand-in-hand with overall profitability (Nusrat & Huang, 2024). Giangiulio and Cheung (2024) cite a recent survey that says one out of five customers see these kiosks as taking advantage of "free labor," with about one out of seven concerned about taking jobs away from others.

Another issue is that businesses fear an increase in retail thefts when implementing these kiosks, and so many have begun to remove them from their stores (Giangiulio & Cheung, 2024); this is backed by Nusrat and Huang (2024)'s mention of the additional effort consumers must make with self-service machines.

Paper vs. Plastic

Not all of the world's inhabitants use credit or debit cards; some still use cash. Rural locations in Asia, for example, may lack internet, and there can be a general distrust of technology along with an informal business style in remote areas (Kapron, 2024). Brazil has an estimated 20 million POS machines but also experiences internet connection issues and a population where physical payment is common and "a significant portion of the population still feels uncomfortable shopping online" (Campos, 2024, para. 9).

As of a few years ago, Oloveze et al. (2022) envisioned a fully "cashless economy" enabled by POS machines through a study conducted in Nigeria but the use of electronic transfer of funds is not necessarily commonplace. Payment terminals can include cash drawers while still accepting cards, too. The ease of the device's use and the perception of both usefulness and social image join with personal experience to push towards a cashless society (Oloveze et al., 2022). Ho et al. (2022) explain developing countries may see better growth due to POS systems reducing fraud. Singh et al. (2024) noted a surge in online fraud in 2020, largely due to consumers turning to online purchases during the COVID-19 pandemic.

Other places do not embrace cash; Sweden is said to largely work with mobile and card payments, with some businesses not accepting cash at all, and cards with chips are highly recommended since magnetic strips may not work with their technologies ("Currency, credit cards and cash," 2025).

Financial institutions may charge processing fees, so this needs to be considered as a business plans funding for a POS system (Crawford et al., 2024), thereby encouraging the use of cash instead (Ho et al., 2022). If not directly charged to the customers themselves, the business must absorb these costs through higher prices.

Market Segmentation from POS Data

Rather than relying on intuition or manager biases, marketers today look for data (Paranavithana et al., 2022). Maji et al. (2019) describe some of the processes: "financial institutions collect competitor information from many different sources like surveys, official published results, from business partners, field executives, etc. Most of them are costly, time-consuming, contains large noise, and sometimes manipulated due to human intervention" (p. 190).

Since POS systems can sort data into manufacturers and other categories, they can be instrumental in developing strategies to promote items proving more profitable (Yacek, 2023). RFM, or Recency, Frequency, and Monetary, is one popular approach in analyzing market segmentation; this is where the data collected by the POS system reviews the relationship between when and how often items are purchased and their monetary value (Paranavithana et al., 2022). An added bonus is that POS data may be analyzed to determine which parts of a store are more prone to theft (Yacek, 2023).

INTEGRATING POS RISKS WITH MARKETING OUTCOMES

The table below summarizes the POS risk categories and demonstrates how they operate through consumer perception to produce marketing outcomes as described in the Transaction Trust Disruption Framework.

Table 1. Transaction Trust Disruption Framework Causes/Effects and Responses

POS Risk Category	Consumer Perception Effect	Marketing Impact	Recommended Strategic Response
Data breaches	Loss of trust Heightened perception of insecurity Expectancy violation	Brand reputation damage Reduced loyalty Negative word-of-mouth	Transparent communication Rapid disclosure Trust repair messaging Coordination between security and marketing teams
Skimming and payment fraud	Perceived loss of control Fear of financial harm	Switching intention Avoidance of affected retailers Erosion of brand credibility	Visible security assurances Consumer education Reinforcement of fraud protection guarantees
Accessibility failures	Perceived unfairness Exclusion Reduced service quality	Negative brand associations Reduced satisfaction among affected segments	Inclusive design practices Accessibility compliance Messaging emphasizing customer-centric service
Transaction outages or friction	Frustration Reduced perceived reliability	Decline in customer experience evaluations Purchase abandonment	Redundancy planning Proactive customer communication Service recovery strategies
Contactless and emerging payment vulnerabilities	Ambiguity about security and data handling	Hesitation toward adoption Slowed diffusion of new payment options	Clear communication of safeguards Trust-building cues at the point of transaction

RECOMMENDATIONS

Ensuring transactional data is secure requires a multitiered approach involving the retailer, the third-party provider, and the consumer. Alabi et al. (2022) found simple logistic regression has proven to be a robust method for classifying data and predicting the likelihood of fraudulent activity. Artificial immune systems (AIS) are also emerging as a promising method for predicting and preventing fraud. The systems are designed to mimic the biological adaptive immune response and perform well when compared to more established mechanisms, such as neural networks and decision trees. They are worth considering for next-generation fraud detection systems (Alabi et al., 2022).

Businesses need to be aware of the cybersecurity issues and plan accordingly. Hardware and software need to be chosen wisely and the database needs to be properly structured. Kabir and Han (2016) state = interfaces tend to be complex, so a thorough analysis of expected customer actions and company needs must be completed. Where third parties are involved in cyber management, risks must be weighed; as Keskin et al. (2021) indicate, the interdependencies of shared networks "significantly increases the attack surface of an organization" (pp. 1-2).

Digital wallet alternatives to debit or credit cards may be considered, as these offer additional security and accessibility for customers (Çelik, 2025). QR code use for payments is increasing at nearly 17% annually; these codes can be static (unchangeable) or dynamic (may be altered once the code is devised), a QR code reader/scanner is needed, and importantly, the customer's data is not stored in this code ("QR code payments," 2025).

Hiring professionals with strong ethics is crucial, as they will need to deal with confidentiality, integrity, and availability mandates (Riefel, 2023) through controlled authentication and encryption processes, and they must work with anomaly detection, monitoring, firewall placement, and other technical issues (Alnaim, 2022). Securing the network infrastructure supporting POS operations and communication is crucial. Firewalls and VPNs have consistently proven effective at blocking brute-force attacks and attempts to access the system from unauthorized IP addresses. However, if they are not configured properly, they create weak points potentially allowing access to an attacker (Damasco et al., 2023).

The SecurityScorecard Threat Intel Report (2024) provides further detail, including the need to find and monitor all "single points" of potential failure within the network, to use automatic scanning for IT risks (including within partners and vendor connections, which may also involve passive monitoring for supply chain risks which may be otherwise unnoticed), and to be careful with any technologies needed to support the company's website. They warn that "third-party products and services are a common way for attackers to compromise e-commerce websites and collect payment card information" (SecurityScorecard Threat Intel Report, 2024, Cybersecurity Recommendations section).

These issues may be summarized by the following list for companies that develop their own POS systems. It must be stated that a full project development process should be employed in the setup and delivery of the system, complete with full documentation and iterative checkpoints:

- Evaluate hardware and software before purchase, installation, and system implementation.
- Check reputation of any third parties involved; ask questions and read their policies.
- Consider offering options such as PayPal, Apple Pay, and/or Venmo.
- Hire well-credentialled IT professionals.
    - Set up cybersecurity measures and mitigations properly.
    - Follow standards and regulations.

- o Ensure the database is constructed properly and secure.
- o Define ethical policies.
- o Monitor all technologies connected with/to the system.
- o Plan mitigations should a breach occur, with apologies and/or compensation included.

Using a third-party set of devices, connections, and other tools alleviates some of the set-up issues but still requires vigilance in monitoring for vulnerabilities and breaches. Customer-facing POS systems need to be chosen wisely and monitored, with the additional need to regularly check each tablet or station for customer-induced problems.

Consumers must also take some responsibility when using POS machines. City National Bank recommends tapping rather than swiping or inserting cards because of the one-time code generated by the tap lessens the chance for fraud, being aware of others nearby when inputting PINs and shielding the keypad from prying eyes, looking for signs of skimmers, and reviewing bank account activity on a regular and frequent basis (“How to help,” n.d.). Again, alternatives such as PayPal may be preferred by some consumers (Çelik, 2025). Lastly, the location of bank cards should be guarded and known at all times.

The consumer's responsibilities may be summarized as follows:

- Be physically aware of others in line at stores.
- Know how to check for skimmers.
- Check account activity regularly and report issues as soon as possible.
- Consider alternative methods of payment when online.
- Be aware of where one's cards are at all times.

CONCLUSION

Point-of-Sale systems offer numerous operational and marketing efficiencies; however, when they fail or if they are used unethically, this can quickly result in damage to a company's reputation. Regardless of the failure, whether it is a security breach, a technology failure, or an issue with accessibility, each malfunction may become a marketing liability. In today's world, where everything is connected and data is a commodity, customer trust and brand identity are inexorably intwined. Accordingly, businesses must consider not only the technical risks, but also the potential threats to their public image.

REFERENCES

Aivazpour, Z., Valecha, R., & Chakraborty, R. (2022, November 16). Data breaches: An empirical study of the effect of monitoring services. *The Data Base for Advances in Information Systems, 53*(4), 65–82. DOI: 10.1145/3571823.3571829

Alabi, O. F., & David, A. A. (2022, November 23). Framework for detection of fraud at Point of Sale on electronic commerce sites using logistic regression. *EAI Endorsed Transactions on Scalable Information Systems, 10*(2), e6. DOI: 10.4108/eetsis.v10i2.1596

Alamsyah, A., Kusuma, G. N. W., & Ramadhani, D. P. (2024, February 26). A review on decentralized finance ecosystems. *Future Internet, 16*(3), 76. DOI: 10.3390/fi16030076

Alnaim, A. F. (2022, July-December). Effects of individual (perceived identity theft, cognitive trust, and attitude) and situational (website quality, perceived reputation, social presence) factors on online purchase intention: Moderating role of cyber security. *International Journal of Cyber Criminology, 16*(2), 131–148.

An, J., Yuen, C., Dai, L., Di Renzo, M., Debbah, M., & Hanzo, L. (2024, June). Near-Field Communications: Research advances, potential, and challenges. *IEEE Wireless Communications, 31*(3), 100–107. DOI: 10.1109/MWC.004.2300450

Ardity, A. (2019, October 24). *Credit card machine, debit card machine.* [Image]. Pixabay. https://pixabay.com/photos/credit-card-machine-4577767 and https://pixabay.com/photos/credit-card-machine-4577768

Badgujar, P. (2023, October 10). Securing customer data and best practices for retail Point-of-Sale systems. *Journal of Technology & Innovation, 4*(4).

Barney, N. (2024, May 2). *What is PCI DSS (Payment Card Industry Data Security Standard)?* TechTarget Search Security. https://www.techtarget.com/searchsecurity/definition/PCI-DSS-Payment-Card-Industry-Data-Security-Standard

BasuMallick. C. (2022, September 29). *What is NFC (Near Field Communication)? Definition, work, and examples.* Spiceworks. https://www.spiceworks.com/tech/networking/articles/what-is-near-field-communication

Beaird, J., George, J., & Walker, A. (2020). *The principles of beautiful web design* (4th ed.). SitePoint., https://research.ebsco.com/c/i5sifr/ebook-viewer/epub/xh4ffz7xlv

Bhaskar, N., Bland, M., Levchenko, K., & Schulman, A. (2019, August). Please pay inside: Evaluating Bluetooth-based detection of gas pump skimmers. *Proceedings of the 28th USENIX Security Symposium* (pp. 373-388), Santa Clara, CA. https://www.usenix.org/system/files/sec19-Bhaskar et al., 2019.pdf

Board of Governors of the Federal Reserve System. (1980, June). *Alice in debitland.* https://files.eric.ed.gov/fulltext/ED235069.pdf

Boutang, J., & De Lara, M. (2016, September 2). Risk marketing. *Journal of CENTRUM Cathedra (JCC) the Business and Economics Research Journal, 9*(1), 27–51. https://doi.org/DOI: 10.1108/JCC-08-2016-0008

Budak, J., Lugović, Š., Penezić, Z., Rajh, E., Slijepčević, S., & Škrinjarić, B. (2023). *Consumer resilience to privacy violation online* (Budak, J., Ed.). The Institute of Economics., http://www.eizg.hr

Campos, Á. (2024, October 17). *Are POS terminals on the way out?* Valor International. https://valorinternational.globo.com/business/news/2024/10/17/are-pos-terminals-on-their-way-out.ghtml

Çelik, A. A. (2025, May 8). Empowering visually impaired consumers in retail: A systematic review of accessible payment solutions for all. *Ege Academic Review, 25*(2), 437–454. DOI: 10.21121/eab.20250212

Cokus, M. S., & Dahlgren, J. W. (2007, June). System evolution in the intelligent enterprise: An historical case study of VISA's Transaction Processing Systems. *INCOSE International Symposium, 17*(1), 776–787. DOI: 10.1002/j.2334-5837.2007.tb02911.x

Crandall, R. (2010, January). *Cash register history: The incorruptible cashier.* Rick Crandall. https://www.rickcrandall.net/cash-register-history

Crawford, H., Anthony, L., & Orem, T. (2024, July 12). *What is a Point-of-Sale (POS) system and how does it work?* Nerd Wallet. https://www.nerdwallet.com/article/small-business/what-is-a-pos-system

Damasco, L. B.Jr, Lacson, N. R. F., & Intong, R. T. (2023, December 14). Safeguarding transactional data: A comprehensive investigation of VPN integration in Point-of-Sale systems for enhanced security and connectivity. *Eximia, 12*(1), 570–589. DOI: 10.47577/eximia.v12i1.405

Daniel, J. (2024, November 28). *How to resolve connection issues with your POS.* PayCliq. https://blog.paycliq.africa/how-to-resolve-connection-issues-with-your-pos

Davenport, W. B. (1967, Spring). Bank credit cards and the Uniform Commercial Code. *Valparaiso University Law Review. Valparaiso University. School of Law, 1*(2), 218–254.

Department of Justice. (2000, April). *Information Technology and people with disabilities: The current state of federal accessibility. Presented by the Attorney General to the President of the United States.* Civil Rights Division. https://files.eric.ed.gov/fulltext/ED450514.pdf

Deutch, K. (2023, November 30). *What is a POS system and how does it work?* Square: The Bottom Line. https://squareup.com/us/en/the-bottom-line/operating-your-business/what-pos-system

Duplys, P., & Schmitz, R. (2024, January). *TLS cryptography in-depth.* Packt. https://learning.oreilly.com/library/view/tls-cryptography-in-depth/9781804611951

Egerson, J. I., Williams, M., Aribigbola, A., Okafor, M., & Olaleye, A. (2024, August). Cybersecurity strategies for protecting big data in business intelligence systems: Implication for operational efficiency and profitability. *World Journal of Advanced Research and Reviews, 23*(2), 916–924. DOI: 10.30574/wjarr.2024.23.2.2390

Enge, E., Spencer, S., & Stricchiola, J. (2023, September). *The art of SEO: Mastering Search Engine Optimization* (4th ed.). O'Reilly Media. https://learning.oreilly.com/library/view/the-art-of/9781098102609

Farashi, S., & Bashirian, S. (2021, August). A complementary webometric ranking system based on the website quality and traffic measures: A study focusing on top-ranked world universities. *Education for Information, 37*(3), 227–254. DOI: 10.3233/EFI-200422

Farok, N. A. Z., & Zolkipli, M. F. (2024, June). Incident response planning and procedures. *Borneo International Journal, 7*(2), 69–76. Majmuah Enterprise. http://majmuah.com/journal/index.php/bij/article/view/641/325

Flannigan, S. (2025, May 14). *The wild history of POS systems (and where they're headed next).* The Retail Exec. https://theretailexec.com/payment-processing/pos-system-history

Frisby, W., Moench, B., Recht, B., & Ristenpart, T. (2012, August). Security analysis of smartphone Point-of-Sale systems. *Proceedings of the 6th USENIX Workshop on Offensive Technologies*, Bellevue, WA.

Georgiev, M. (2024, May 9). *The evolution of POS systems: From "cha-ching" cash registers to modern solutions.* NRS Plus. https://nrsplus.com/blog/evolution-of-pos-systems

Giangiulio, D., & Cheung, B. (2024, July 4). Major retailers are backtracking on self-checkout. *NBC News*. https://www.nbcnews.com/business/business-news/major-retailers-are-backtracking-self-checkout-rcna160234

Gilde, S. (2024, October 16). *A current guide to EMV testing and certification*. Paragon Application Systems. https://www.paragonedge.com/blog/a-current-guide-to-emv-certification-and-testing-requirements

Handoko, A., & Mauritsius, T. (2021, April 19). Factors determining the use of POS (point of sales) systems in small and medium sized enterprises. *Turkish Journal of Computer and Mathematics Education*, *12*(6), 4800–4806.

Hayes, A. (2024, September 24). *What is Point of Sale (POS)?* Investopedia. https://www.investopedia.com/terms/p/point-of-sale.asp

Ho, C.-Y., Kim, N., Rong, Y., & Tian, X. (2022, October). Promoting mobile payment with price incentives. *Management Science*, *68*(10), 7614–7630. DOI: 10.1287/mnsc.2021.4233

Hoehle, H., Venkatesh, V., Brown, S., Tepper, B. J., & Kude, T. (2022, February). Impact of customer compensation strategies on outcomes and the mediating role of justice perceptions: A longitudinal study of Target's data breach. *Management Information Systems Quarterly*, *46*(1), 299–340. DOI: 10.25300/MISQ/2022/14740

Jechle, D., Schuetz, S., & Gewald, H. (2025, June). Decoding data breach response strategies: Insights from situational crisis communication theory using topic modeling. In Association for Information Systems, AIS Electronic Library (AISeL), & European Conference on Information Systems (ECIS), *ECIS 2025 Proceedings* [Journal-article]. https://aisel.aisnet.org/ecis2025/security/security/2

Kabir, M. A., & Han, B. (2016, July). An improved usability evaluation model for Point-of-Sale systems. *International Journal of Smart Home*, *10*(7), 269–282. DOI: 10.14257/ijsh.2016.10.7.27

Kadam, V. (2025). *Cyber security breaches in corporate networks: a literature review on recent threats and their impact*. [Student Thesis, LAB University of Applied Sciences]. Theseus. https://www.theseus.fi/handle/10024/887316

Kapron, Z. (2024, April 12). Why cash is still prevalent in Asia. *Forbes*. https://www.forbes.com/sites/zennonkapron/2024/04/12/why-cash-is-still-prevalent-in-asia

Keskin, O. F., Caramancion, K. M., Tatar, I., Raza, O., & Tatar, U. (2021, May). Cyber third-party risk management: A comparison of non-intrusive risk scoring reports. *Electronics (Basel)*, *10*(10), 1168. DOI: 10.3390/electronics10101168

Khan, W. A., & Abideen, Z. U. (2023). Effects of behavioural intention on usage behaviour of digital wallet: The mediating role of perceived risk and moderating role of perceived service quality and perceived trust. *Future Business Journal, 9*(1), 73. Advance online publication. DOI: 10.1186/s43093-023-00242-z

Kholod, M., Celani, A., & Ciaramella, G. (2024, December). The analysis of customers' transactions based on POS and RFID data using big data analytics tools in the retail space of the future. *Applied Sciences (Basel, Switzerland), 14*(24), 11567. DOI: 10.3390/app142411567

Kiely, T. J. (2025, February 27). *Business reputation management: How to create and maintain a good reputation.* Meltwater. https://www.meltwater.com/en/blog/how-a-business-can-create-and-maintain-a-good-reputation

Lahmeyer, S., & Roemer, E. (2024, March). The link between consumer-facing technologies and customer experience in physical retail environments: A critical literature review. *International Review of Retail, Distribution and Consumer Research, 34*(2), 128–159. DOI: 10.1080/09593969.2024.2332224

Larson, S., Jones, J., & Swauger, J. (2020, October 27). A forensic first look at a POS device: Searching for PCI DSS data storage violations. T*he Journal of Digital Forensics. Security and Law, 15*, 4. DOI: 10.15394/jdfsl.2020.1614

Lestariningati, S. I. (2018). Mobile point of sale design and implementation. *IOP Conference Series. Materials Science and Engineering, 407*, 012094. DOI: 10.1088/1757-899X/407/1/012094

Lowden, G. A. (1957, May). National Cash Register. *The Analysts Journal; Proceedings of the Tenth Annual Convention, National Federation of Financial Analysts Societies, 13*(3). 145-148. https://www.jstor.org/stable/4468825

Maji, G., Dutta, L., & Sen, S. (2019). Targeted marketing and market share analysis on pos payment data using dw and olap. In A. Abraham, P. Dutta, J. K. Mandal, A. Bhattacharya, & S. Dutta (Eds.), *Emerging Technologies in Data Mining and Information Security: Proceedings of IEMIS 2018,* Volume 2 (pp. 189-199). Springer Singapore. DOI: 10.1007/978-981-13-1498-8_17

Marikyan, D., & Papagiannidis, S. (2023, September 23). Unified Theory of Acceptance and Use of Technology: A review. In Papagiannidis, S. (Ed.), *TheoryHub Book* (pp. 263–280)., https://open.ncl.ac.uk/theoryhub-book

Mathur, M. (2018, March). Where is the security blanket? Developing social media marketing capability as a shield from perceived cybersecurity risk. *Journal of Promotion Management, 25*(2), 200–224. DOI: 10.1080/10496491.2018.1443310

Molitor, D., Saharia, A., Raghupathi, V., & Raghupathi, W. (2024, April). Characteristics of data breaches: A descriptive analytic study. *Journal of Information Security, 15*(2), 168–195. DOI: 10.4236/jis.2024.152011

Mondego, D., & Gide, E. (2024). The impact of security, service quality, perceived usefulness, perceived ease of use, trust, and price value on users' satisfaction in Cloud-Based payment systems in Australia: A PLS-SEM analysis. *Journal of Information Systems and Technology Management, 21*. Advance online publication. DOI: 10.4301/S1807-1775202421004

Näränen, M. (2013, November). *"What can I get you?" A study on POS marketing effectiveness in a restaurant environment.* [Student Thesis, Helsinki Metropolia University of Applied Sciences]. Theseus. https://www.theseus.fi/bitstream/handle/10024/67963/Naranen_Mikko.pdf

Nikkhah, H. R.Jr, & Grover, V. (2022, December 1). An empirical investigation of company response to data breaches. *Management Information Systems Quarterly, 46*(4), 2163–2196. DOI: 10.25300/MISQ/2022/16609

Nusrat, F., & Huang, Y. (2024, January). Feeling rewarded and entitled to be served: Understanding the influence of self- versus regular checkout on customer loyalty. *Journal of Business Research, 170*, 114293. DOI: 10.1016/j.jbusres.2023.114293

Obi, E. N. (2023, June 30). Impact of criminal acts on point of sales (POS) business operations in Karu Local government area of Nasarawa State. *NIU Journal of Social Sciences, 9*(2), 119–127. DOI: 10.58709/niujss.v9i2.1632

Olipas, C. N. P., & Esperon, R. M. (2020, March). The design and development of a cashless payment system with an Automatic Identification and Data Collection (AIDC) technology. *International Journal of Scientific & Technology Research, 9*(3), 6718–6723.

Oloveze, A. O., Oteh, O. U., Nwosu, H. E., & Obasi, R. O. (2022). How user behaviour is moderated by affective commitment on point of sale terminal. *Rajagiri Management Journal, 16*(1), 2–20. DOI: 10.1108/RAMJ-05-2020-0019

Onat, I., & Ciaccio, J. (2025, April 25). *An analysis of ATM and Point-of-Sale skimming*. Orion Policy Institute. https://orionpolicy.org/an-analysis-of-atm-and-point-of-sale-skimming

Onumadu, P., & Abroshan, H. (2024, November). Near-Field Communication (NFC) Cyber threats and mitigation solutions in payment transactions: A review. *Sensors (Basel), 24*(23), 7423. DOI: 10.3390/s24237423

Paranavithana, I., Rupasinghe, T. D., & Prior, D. D. (2022, September). A machine-learnt approach to market segmentation and purchase prediction using Point-of-Sale (POS) data. In S. I. Ao, & L. Gelman (Eds.), *Transactions on Engineering Technologies, Proceedings of World Congress on Engineering,* 919 (pp. 49-61). http://dx.doi.org/DOI: 10.1007/978-981-19-3579-4_4

PCI Security Standards Council. (2024, June). *Payment Card Industry Data Security Standard, version 4.0.1.* https://docs-prv.pcisecuritystandards.org/PCI%20DSS/Standard/PCI-DSS-v4_0_1.pdf

PCI Security Standards Council, LLC. (2025). PCI PIN Transaction Security Device Testing and Approval Program Guide. In *PCI PIN Transaction Security Device Testing and Approval Program Guide* [Report].

Perkins-Southam, T., & Lupini, C. (2025, April 10). *What is a contactless credit card?* Forbes Advisor. https://www.forbes.com/advisor/credit-cards/contactless-credit-cards

Ratica, A. (2024, December 2). *Skimmer detection on payment terminals.* Technical Disclosure Commons. https://www.tdcommons.org/dpubs_series/7608

Riefel, M. T. (2023). *A quantitative study on the impact of data breaches on customer satisfaction.* (Publication No. 95363) [Student Thesis, University of Twente]. University of Twente, Netherlands. https://essay.utwente.nl/95363/1/Riefel_BA_BMS.pdf

Saeed, S., Shaikh, A., Memon, M. A., Nizamani, M. A., Abbasi, F. A., & Naqvi, S. M. R. (2019, April 30). Evaluating the quality of Point of Sale (POS) software. *University of Sindh Journal of Information and Communication Technology, 3*(2), 69-75. https://sujo2.usindh.edu.pk/index.php/USJICT

Sanchez, J. (2025, May 2). *The Target breach: A historic cyberattack with lasting consequences.* Framework Security. https://frameworksecurity.com/post/the-target-breach-a-historic-cyberattack-with-lasting-consequences

Santos, V., & Mendes, L. (2023, July). Digital transformation of the retail Point of Sale in the Artificial Intelligence era. In Santos, J. D., Pereira, I. V., & Pires, P. B. (Eds.), *Management and Marketing for Improved Retail Competitiveness and Performance* (pp. 200–216)., DOI: 10.4018/978-1-6684-8574-3.ch010

Scaife, N., Peeters, C., & Traynor, P. (2018, August). Fear the reaper: Characterization and fast detection of card skimmers. *Proceedings of the 27th USENIX Security Symposium* (pp. 1-14), Baltimore, MD. https://www.usenix.org/system/files/conference/usenixsecurity18/sec18-Scaife et al., 2018.pdf

Sears, C. R., & Cunningham, D. R. (2024, August). Individual differences in psychological stress associated with data breach experiences. *Journal of Cybersecurity and Privacy*, *4*(3), 594–614. DOI: 10.3390/jcp4030028

Sharma, N., Oriaku, E. A., & Oriaku, N. (2020, January 21). Cost and effects of data breaches, precautions, and disclosure laws. *International Journal of Emerging Trends in Social Sciences*, *8*(1), 33–41. DOI: 10.20448/2001.81.33.41

Singh, N., Misra, R., Quan, W., Radic, A., Lee, S., & Han, H. (2024). An analysis of consumer's trusting beliefs towards the use of e-commerce platforms. *Humanities & Social Sciences Communications*, *11*(1), 899. Advance online publication. DOI: 10.1057/s41599-024-03395-6

Tejeda, V. (2025, June 23). *What is a customer-facing display & do I need one? (2025)*. Shopify. https://www.shopify.com/retail/customer-facing-display

Wongyai, P. H., Suwannawong, K., Wannakul, P., Thepchalerm, T., & Arreeras, T. (2024, October 15). The adoption of self-service check-in kiosks among commercial airline passengers. *Heliyon*, *10*(19), e38676. DOI: 10.1016/j.heliyon.2024.e38676

Yacek, H. (2023, October). Points of Surprise! *Shooting Industry*, *68*(10), 34–37.

Yokeshwaran, M., & Murugachandravel, J. (2022, April 10). Enhancement of business using e-commerce with Point of Sale. *International Journal of Health Sciences*, *6*(S2), 3250–3260. DOI: 10.53730/ijhs.v6nS2.5811

Zamoroka, V. (2023, September 21). *Modern approaches to website development: Techniques and tools*. AlterEgo. https://alterego.digital/blog/modern-approaches-web-development-techniques-and-tools

Zimmel, S. (2025, April 2). *The impact of cybersecurity on business and brand risk reduction*. Recorded Future. https://www.recordedfuture.com/blog/impact-cybersecurity-business-brand-risk-reduction

Chapter 7
Exploring the Strategic Role of Digital Transactions in Consumer Behavior and Brand Engagement

Suresh Kumar
https://orcid.org/0000-0002-4771-5893
CGC University, Mohali, India

Renu Sindhu
Vidya Jyoti Educational Society, India

Swapandeep Kaur
https://orcid.org/0000-0003-2663-977X
CGC University, Mohali, India

Neha Kumari
https://orcid.org/0000-0002-0407-1514
CGC University, Mohali, India

ABSTRACT

Digital payment technologies has become an effective methodology to shape the consumer behavioral pattern and reach the brands (Visconti-Caparrós et al., 2022). This paper explores the strategic marketing implications of digital transactions, or rather, the effects of platforms such as UPI, mobile wallets, and contactless cards on the purchase behavior, consumer loyalty, as well as informing personalized

DOI: 10.4018/979-8-3373-4526-0.ch007

data. A survey of 300 consumers and 50 marketing professionals is combined with expert interviews in order to identify trends of adoption, trust, and engagement. The results indicate that frictionless and secure consumer payments increase consumer confidence and retention rates, and, as a result, the transaction information can be used to market to consumers, real-time adjust campaigns, and serve as a signal to campaign segmentation. The report establishes principal demographic and purchase behavior, associates payment acceptance to repeat purchases and the possibility of digital payments as interaction tools.

INTRODUCTION

The digital payment technology has revolutionized the manner in which individuals manage funds and transact businesses globally (Bandura and Ramanujam, 2021). Moreover, they have changed the way consumers relate with businesses and their decisions in making purchases. The rapid progress in the mobile sphere, the popularity of the internet, and attempts to open banking to everyone have transformed digital transactions into a must-have, as opposed to a nice-to-have, in the modern world (Yanagawa, 2018; Soetan and Mogaji, 2024). Statista (2025) forecasts that the total amount of digital payment in the world would be about USD 20.09 trillion. This shows how extensive and fast this change is. Digital payments have been used extensively in India (Kumar et al., 2019; Bhavsar and Samanta, 2022). The introduction of platforms like the Unified Payments Interface (UPI) through which individuals can use to carry out various transactions quickly, easily, and safely is transforming how people do business (Chandak, 2024).

Much of the academic and governmental discussion regarding the digital payment has concerned its effectiveness, inexpensiveness, and potential to assist people in accessing financial services (Pazarbasioglu et al., 2020). Nevertheless, the research examined the strategic implications on marketing and interaction with customers. Digital transactions are not only a means to pay but also high-quality data touchpoints and experience nodes, which influence the entire customer experience, including product discovery and purchase, post-purchase experience, and loyalty (Herhausen et al., 2019). Purchasing things through cashless options is easier, creates trust, and makes the customers happier as it is fast and easy to operate. Above all, each transaction will provide you with valuable information that can be used to get to know more about the buying behaviour, preferences, and behaviours of people (Liao et al., 2012).

A large part of the scholarly and governmental debate on the digital payment has focused on the efficiency, low cost, and possible help in providing people with financial services (Pazarbasioglu et al., 2020). However, the study looked into the strategic implication on marketing and customer interaction. Digital transactions are a source of payment as well as a source of high-quality data touchpoints and experience nodes, which shape the entire customer experience, such as product discovery and purchase, a post-purchase experience, and loyalty (Herhausen et al., 2019). It is simpler to buy items using the cashless systems, builds trust and the customers are happier because the system is quick and simple to use. Most importantly, every transaction will furnish you with precious information which can be applied to learn more about the buying behaviour, preferences and behaviours of people (Liao et al., 2012).

Although digital payments are widely used, there is a large group of people who do not know how to use them to market their offerings (Chaveesuk et al., 2021). Digital payment systems technology and infrastructure have been heavily researched. It has not necessarily examined the way these systems transform the manner in which businesses and clients communicate with one another, what clients want, and how brands promote themselves. This paper attempts to seal that gap by examining digital transactions in a marketing perspective. It will center on how various modes of payment alter the manner in which people behave and how they conduct their business. The investigation has four key objectives. The overall idea is to make people realize that using digital payments is not only a method to purchase some goods, it is also an excellent method to sell. It is aimed at providing law-makers, fintech companies and merchants with valuable insights into how they can advertise their goods and services and how individuals behave when spending digital currency on payment. Transactional data would allow marketers to create ads that are more relevant and specific to an individual user (Aiolfi et al., 2021). They can also develop new methods of making people enter the digital economy. The primary purpose of this paper is to demonstrate how digital payment solutions are transforming the way individuals are beginning to think, engage, and remain committed to businesses in a world that is increasingly becoming less and less cash centric.

BACKGROUND

Theoretical Framework

The three theories that this study relies on to examine how digital transactions could transform the behaviour of people and make them more engaged with businesses are the technology acceptance model (TAM), consumer behaviour theory and relationship marketing theory. All these notions assist us in becoming familiar with how individuals use the digital payment platforms, consumer behavior, and how marketers can utilize digital financial touchpoints to build long-term relationships with their clients (Gao et al., 2022; Oumaima and Lamari, 2024; Rahayu, 2024).

Technology Acceptance Model (TAM)

One of the most popular ways to consider the use of information technologies by people is the Technology Acceptance Model developed by Davis (1989). According to the model, the choice made by an individual to adopt a new technology is pegged on two key factors including the perceived usefulness (PU) and effortlessness (PEOU) of the technology. These two also dictate their actual usage of technology. TAM can also inform us a lot about the way people think and utilize digital payment services (Linh and Huyen, 2025). Human beings tend to use a digital payment system when they perceive it to make their transactions faster (usefulness) and convenient to use (ease of use) (Kim et al., 2010). Mobile wallets, UPI-based applications, and contactless payments are increasingly being used by people as they satisfy both of these requirements. The more convenient it is to pay in digital form, the more likely people will use it and make the entire customer experience superior (Nalini et al., 2018; Shree et al., 2021). This theoretical framework plays a key role in explaining the early phases of consumer engagement with the process of digital payments and creates the framework on which the development of the engagement into the stronger brand relations can be analyzed during a certain time frame.

Consumer Behavior Theory

As explained by Schiffman and Kanuk (2009), the theory of consumer behavior offers an understanding of how consumers make decisions, what their motivation and reactions are. It emphasizes the idea that all the stages of a consumer behavior starting with problems recognition and information search, purchase decision and post-purchase assessment are preconditioned with the presence of multiple internal and external stimuli and one of them is the different payment methods (Han and Ryu, 2012). Digital payments have established a different dimension of power in a consumer's decision making process (Gupta and Prusty, 2024). Easy and safe, immediate methods of transactions may lighten the mental burden (Davinson and Sillence, 2014) and help diminish the fear to purchase, therefore, encouraging spontaneous buys or brand credence. Moreover, the post purchase process which is traditionally influenced by satisfaction, loyalty, word of mouth is being advanced by smooth payment experiences (Siqueira et al., 2019). The improvement of satisfaction because of tensionless checkout process not only increases the satisfaction but it also increases demand of cash purchase and brand loyalty. Incorporating the theory of consumer behavior, the given work attempts to determine how digital transactions make not only a difference in the payment process (Dahlberg and Oorni, 2007), but also in the consumption itself, molding preferences, perceived brand value, and long term loyalty (Candan et al., 2013).

Relationship Marketing Theory

Relationship Marketing Theory was first proposed by Berry (1983) and it places focus on the long-term, value-involving relationships between consumers and businesses (Yum and Kim, 2024). This theory does not only preach transactional exchanges but rather promotes a personalized approach to engagement that enables trust building, loyalty, and emotional relationship (Chakraborty et al., 2025). The payment system forms information that is a strategic resource to marketers in the age of digital payment system (Chaveesuk et al., 2021). Allowing firms to market to individuals since consumer insights are obtained in real time data such as purchase frequency, transaction time, category of spending money, and the brand which they purchase gives firms the opportunity to market to individuals. It enables developing highly personalized loyalty, offers, fast customer care contact, and all of this improves the emotional and behavioural linking between the consumer and the brand (Hwang and Kandampully, 2012). Digital payments are not only financial allowances (Frączek and Urbanek, 2021) but also relation points where the brands can address the consumer on a regular basis in positive ways (Fournier, 1998). The Relationship Marketing Theory will help this study to examine how the transaction data in digital transactions can be connected to the long term engagement strategies and towards the development of sustainable brand equity (Thaichon et al., 2020).

Synthesis Of Frameworks

All these three theories can give us a strong base with which study can analyse the strategic marketing implication of digital transactions. Whereas TAM indicates how consumers in a given technology embrace digital video payment technology, the Consumer Behavior Theory indicates how digital video payment technology affects the purchase decision and brand perception (Linh and Huyen, 2025). Relationship Marketing Theory introduces a longitudinal dimension to exemplify how digital payments can be used to strengthen the brand and increase brand loyalty along the way (Thaichon et al., 2020). Such a combination of theoretical perspectives enables the research to go beyond functional considerations of digital payments and explore their transformative nature regarding the contemporary marketing strategy (Gomber et al. 2018; Ockwell et al., 2019).

Key Concepts and Definitions

To establish a strong foundation of the study, clarify and elaborate the main concepts that are backing the research. Digital platform interactions by consumers are gaining ground and payment is finding its way into the value chain of marketing and there is a need to be contextually aware of the jargons that are critical to the transaction (Hänninen, 2020; Hokkanen et al., 2021). In this section, the author will outline and describe the main principles on which the study is based. One-to-one marketing or individualized marketing is a technique of personalizing marketing communications, content, and offers to each customer based on behavioral, transactional and demographic data (Weng and Liu, 2004). The concept is that the messages that are conveyed should be relevant and timely addressing individual likes and needs of each customer. Digital payment schemes also assist in personalization through giving real-time (Malempati, 2022), granular information on consumer shopping behaviors (Guo and Liu, 2025). These variables are made up of the frequency of purchase, the amount that is spent on purchase, the merchants that they prefer and the time that they make the purchase which can all be used to craft detailed profiles of consumers. Businesses can use these findings to intensify custom promotions, display appropriate products, as well as advance their contact approaches. As per an example, a customer who often transacts a mobile wallet to order food delivery can be provided with discounts by partner restaurants or special dining treatments (Kapoor and Vij, 2018). These hyper-personalised methods lead to more engagement and conversion and an enhanced emotional connection with the brand, which means more long-term loyalty.

Present State of Knowledge

The increased or penetration of digital payment systems within the physical and the virtual business environment has dramatically and completely changed the action and marketing strategies of consumers and businesses (Bharadwaj et al., 2013). A recent body of literature along with industry numbers has it that besides its role in enhancing operational efficiency, digital transactions are taking a key role in the enhancement of consumer involvement and branding success (Hollebeek and Macky, 2019). The section contains the synopsis of the existing understanding of the role of digital payments, which supports four broad strategic areas, namely customer convenience, data-driven marketing, trust and brand perception, and marketing automation. The decrease in the transactional friction, which enhances the overall buying experience, is one of the most famous benefits of digital payment systems. The World Bank (2022) claims that mobile payment, QR code scanning, and contactless technologies have been significantly utilized to improve the checkout processes by increasing flexibility, ease, and speed. They are particularly valuable in the high-volume settings with e-commerce, retail, and food delivery (Chowdhury et al., 2025) where long or tortuous payment processes may translate into cart abandonment or lost sales. Customers now desire hassle-free one-go payment services in their general association with brands (Unal and Park, 2023). The ease of the digital payment options, including UPI, mobile wallets, and native fintech features directly contribute to the increased customer satisfaction, frequency of purchases, and retention (Hopalı et al., 2022; Zehra et al., 2024; Omran, 2025). As a matter of fact, ease of conducting a transaction by customers has turned out to be a decisive factor in the overall brand selection and recommendation.

Research Gaps

Although the amount of literature and industry analyses that acknowledge the emergence of digital payments and their effects is increasing (Dahlberg et al., 2008; Chaveesuk et al., 2021), there still exist some gaps in critical studies, especially regarding the implications associated with their strategic marketing (Kamboj and Rahman, 2015; Nalini et al., 2018; Ockwell et al., 2019; Chakraborty et al., 2025). The majority of past research has focused on the technical, financial, or infrastructural characteristics of digital payment systems (Taylor, 2016; Khan, 2014, Li et al., 2020), which is why they have not always addressed their overall effect on customer behavior, brand perception, and marketing implications. Few Empirical studies have been conducted on the relationship between the use of digital payment and marketing ROI and campaign effectiveness. Although digital payment systems are highly touted in terms of operational and financial efficiency, there is a lack of empirical evidence on how their application can be associated with quantifiable marketing performance outcomes, including Return on Investment (ROI), customer acquisition cost (CAC), or campaign success (Ismaeel, 2023). Rather than focusing on the strategic application of payment data to the degree of marketing effectiveness, the available literature is focused on adoption levels, transactional volumes, and user satisfaction. The Under-researched Effect of Mobile-First Payments in Rural and Emerging Markets Although the level of digital payments has reached a considerable popularity in cities and developed areas (Rui et al., 2023), the role of mobile-first payment systems in the rural or semi-urban setting (Kumar and Saha, 2025), specifically in the emerging markets like India, Indonesia, and part of Africa, was rarely the focus of marketing professionals. Financial inclusion in those regions has been improved with UPI, mobile wallets, and solutions based on QR-codes (Nandru et al., 2024), yet the effects of these systems on consumer preferences, brand accessibility, and marketing channels in non-urban settings are less researched. There is a massive research gap as to how cultural, infrastructural, and behavioral variations that affect the mobile payment adoption in rural communities affect the manner in which marketers can reach these consumers (Farinloye et al., 2024). This includes considering the regional differences in transactional behavior, faith in digital ecosystems, and sensitivity to promotional initiatives provided by means of mobile payment networks (Omarini, 2018). It lacks cross-sector analyses on novel marketing strategies that can capitalize on payment information. Most of the existing studies concentrate on the digital payments within the retailing and e-commerce industries. Nevertheless, no comparative analysis can be found in those industries where digital payments are so widespread and can potentially remake marketing plans: services (e.g., hospitality, healthcare), education (e.g., online courses, edtech platforms), and transportation (e.g., ride-hailing, ticketing) (Chaveesuk et al., 2021). Although the customer journeys, payment behaviors, and areas of interaction may be unique to each industry, there is an absence of inter-sector findings on the innovative use of payment data in marketing. Emphasis on digital payments integrative strategic marketing models has not been significantly emphasized. Though there exist certain theoretical models, like the Technology Acceptance Model (TAM) and the Relationship Marketing Theory, that provide certain insight, no strategic marketing model that integrates digital payment systems as a central element is truly accepted (Uche et al., 2021). Lack of such structures inhibits academic studies, as well as application of digital payments by management as marketing optimization and branding tools (Hollebeek and Macky, 2019). This problem can be addressed in future studies by developing and confirming the combination of digital payment systems with consumer segmentation, personalization, loyalty building, and lifecycle marketing initiatives (Olayinka, 2021). These models may be useful in integrating the fragmented areas of research and offering an official approach through which firms can strategically use payment ecosystems (Omarini, 2018). Conclusively, despite the realization that digital payment systems have the transformational capability, academic studies have not fully comprehended the strategic marketing consequences of digital payment systems.

RESEARCH METHODOLOGY

Research methodology outlines the systematic approach applied to research the strategic implications of digital transaction in defining consumer behaviour and brand touch. Since the research is an interdisciplinary one, involving marketing, consumer psychology, and financial technology, a mixed-method approach would be appropriate to enable the research to offer both depth and breadth of understanding. In this chapter, I articulate the research strategy, objective, process of data collection and analysis that would be used to answer the research questions. The methodology has been developed in a manner that the study would not only be in positions of giving empirical findings but it would also be adding on the concept side based on the strategic framework presented by the framework. Having compiled the results of qualitative and quantitative sources of information, the goal of the research is to create a fantastic and practical knowledge regarding the transformations within the marketing environment in the domain of digital payment methods.

This study intends to discuss the way the connection between strategic marketing strategies and digital transactions is established. The first one to examine how digital payment system use affects consumer behavior and marketing. The second objective is to approximate the effect of digital payments on consumer engagement and loyalty. To identify how much organizations employ digital transaction data to perform personalized and predictive marketing. To usher in marketing integration structure of digital payments corporate strategy. The research design outlines the broad design and methodological plan of attaining the objectives of the study. A mixed-methodology approach is applied because the topic of our research is very complex since it involves the psychology of the consumer, marketing policy, and financial technology. The methodology consists of quantitative and qualitative procedures to guarantee comprehensive and a multifaceted insight in the role of digital transactions on consumer behavior and brand engagement. The research design allows achieving statistical generalisation and thematic exploration of the trends and behaviours of digital payment systems by integrating numerical information with the opinions of experts to answer the question of what and the question of why. It is aimed at discovering statistically significant trends and relationships between the use of digital payments and marketing performance, such as satisfaction, loyalty, trust, and personalised engagement. The survey questionnaire will be administered to 300 customers who have made any online payment through digital payment platforms (UPI, mobile wallets, or card payments) within the last six months. Additional survey will be sent to 50 marketing managers in major industries like retail, hotel and services. Descriptive statistics: mean, frequency, and percent, Comparative analysis across sectors. It is aimed to collect subtle views on the adoption of digital payments in marketing efforts and customer interaction models. Data are arranged and a bare minimum of descriptive analysis done in Microsoft Excel.

Analysis and Interpretation

The chapter addresses data analysis and interpretation which have been acquired through organized surveys and interviews. The findings are presented in the group of topics to be consistent with the research objectives and provide an overall view of how digital payment system implies consumer behavior and marketing mix. The

Adoption Patterns theme is the first key theme that is discussed and reviews the demographic and behavioral factors that affect use of the digital payment system.

Adoption Patterns

There is a high variation in digital payment systems across various groups of customers. This is where trends of adoption will be examined, and such important demographic and behavioural variables as age, geography, digital literacy, and device use will be taken into account. Marketers and the businesses must be aware of these tendencies in order to formulate digital strategies that are highly inclusive and concentrated.

Digital Literacy and Payment Adoption

The digital literacy, or the ability to comprehend and use digital devices and applications, affects the willingness to use digital payment systems significantly. More technologically advanced users feel more comfortable testing emerging platforms, exploring products and services, and finding solutions. The individuals possessing weak digital skills face an increased risk of payment drop-offs, transaction errors or are inclined towards technology. To allow low-literacy people to use their businesses, companies should offer simplified user experiences (UX), tutorials and in-app support as an incentive to trial. Digital payment acceptance is motivated by different demographic, geographical, and technological factors. The user population is the most active; it consists of younger, urban, and technologically literate users of smartphones. To realize the strategic potential of digital payments as an effective way of engaging the brand and ensuring an inclusive growth, businesses and policymakers need to address the limitations of the aged, rural, and less digitally savvy population.

Table 1. Demographic Segment Adoptions of Digital Payments.

Category	Sub-Category	No. of Respondents	Adoption Rate (%)	Preferred Platform
Age Group	18–30 years	120	92%	UPI, Mobile Wallets
	31–50 years	100	75%	UPI, Cards
	51+ years	80	48%	Cards, Net Banking
Region	Urban	180	89%	UPI, Wallets, Contactless Cards
	Semi-Urban	70	68%	UPI, Feature Phone Banking

continued on following page

Table 1. Continued

Category	Sub-Category	No. of Respondents	Adoption Rate (%)	Preferred Platform
	Rural	50	52%	UPI (assisted), USSD
Digital Literacy Level	High	160	93%	All platforms
	Medium	90	71%	UPI, Wallets
	Low	50	38%	Assisted UPI, Cards
Device Used	Smartphone (Android/ iOS)	230	91%	UPI Apps, Wallets
	Feature Phone	30	37%	USSD, Missed Call Banking
	Laptop/Desktop	40	65%	Net Banking, Card Payments

(Source: Created with the help of Excel.)

Table 1 shows that respondents aged between 18 and 30 are the highest users of digital payments (92%), which is a characteristic of digital-native behavior. The 51+ population is last with 48 displaying the necessity of specialized training and establishing trust. The highest adoption rate was in urban responders of 89 but the rural users under 52 as it is limited with infrastructure and awareness. Adoption is closely related to digital literacy (93%), but low levels of digital skills exhibit low levels of use (38%), so designing with ease of use and outreach are important. The most widely used gadget is Smartphones and the adoption rate is 91 percent among phone users who are then followed by the USSD services (37 percent) yet it appears to be promising to be included.

Purchasing Behavior

The influence of Digital Payment system on consumer buying behavior has been immense in that the consumer buying behavior has been made faster, more convenient and psychologically frictionless. This part explores the effect of digital payments on key behavioral variables including impulse buying, frequency of transactions and size of basket (average transaction value). Real time payments reduce the time delays between selection and purchase, enabling the customers to impulse buy. Moreover, additional features like the 1-Click checkout, stored card details, and buy-now-pay-later will also increase the rate of transactions and could encourage bigger basket size. These behaviors are quantified in different segments of customers which are analyzed below.

Table 2. The Impact of Digital Payments on Purchase Behavior.

Behavioral Metric	High Frequency Group	% of Total	Key Observations
Impulse Buying	110 consumers	37%	Mostly among 18–30 age group, driven by app-based offers & flash sales
Moderate Impulse Buying	95 consumers	32%	Triggered by push notifications and personalized discounts
Low Impulse Buying	95 consumers	31%	More deliberate spenders; prefer comparing before purchase
Transaction Frequency	>5 transactions/week	130 consumers	High usage for food delivery, subscriptions, and digital goods
	2–5 transactions/week	120 consumers	Balanced spending habits; mostly household essentials
	<2 transactions/week	50 consumers	Budget-conscious or new adopters
Basket Size	Above 1,000 per purchase	90 consumers	Seen in electronics, fashion, and monthly groceries
	500– 1,000 per purchase	130 consumers	Most common range for regular transactions
	Below 500 per purchase	80 consumers	Mostly in QSRs, entertainment, and daily use items

(Source: Created with the help of Excel.)

Table 4.2 indicates that: 37% of the respondents stated that they made regular impulse purchases, which, as a rule, were stimulated by the in-app promos or flash deals or personalised recommendations at the check-out. Younger (18-30 years) and older customers (31 years and above) showed impulse buying tendency more frequently, as well as in product categories (clothes, snacks, and digital entertainment). The psychological convenience of learning how to shop without physical money, and the option to buy with such options as pay later, or one tap make the purchases less hesitant and decision-making faster. Moreover, the percent of those who perform over five transactions weekly is 43% of the respondents, meaning that they are heavily engaged in the digital space, specifically, in the areas of food delivery, subscriptions, and utility payments. We discovered that the high-frequency use is connected with food delivery, streaming subscriptions, and utility payments online. The main purchases of consumers made by medium frequency (2-5 times a week) are deemed to be necessities, including groceries, travel, and e-commerce purchases. About 30 percent of clients make purchases valued higher than 1,000 Rupees on planned or high-involvement purchases including gadgets or monthly shopping. Most (43%) of people spend 500-1000 displaying moderate expenditure due to repetitive use and online deals. The entails, fast food, and daily retail are also common in purchases that are lower value (less than 500 rupees) and require speed and convenience which is provided by digital payments.

Customer Loyalty

The aspect of customer loyalty is an essential part of the successful marketing and brand engagement programs. Unless it is a service with super important features, the user experience during the payment process became a significant loyalty driver in the digital age. Smooth, secure and rewarding methods of payment will improve the good user experience, which would increase the chances of re-recurring transactions and client retention. Online transactions like UPI, mobile work wallets, and in-app payments can help lower the friction associated with the checkout process and offer more personalised rewarding, which plays a significant role in brand stickiness. This section focuses on the role of payment convenience in encouraging customer loyalty behaviour among the diverse consumer groups.

Table 3. The Effect of Seamless Digital Payments on Customer Loyalty

Loyalty Indicator	Segment	No. of Respondents	% of Total	Key Insights
Repeat Purchases	High (≥1 repeat/ month)	145	48%	Driven by app offers, loyalty points, and stored payment methods
	Moderate (1 repeat in 2–3 months)	90	30%	Repeat behavior influenced by price and convenience
	Low (less than 1 repeat in 3 months)	65	22%	Low loyalty, mostly deal seekers
Preferred Brands Due to Payment	Yes	175	58%	Users prefer platforms with fast and secure payment experience
	No	125	42%	Prefer variety or lowest cost; less influenced by payment UX
Participation in Loyalty Programs	Active users	110	37%	Engage due to cashback, rewards, and exclusive payment-linked benefits
	Occasional users	95	32%	Join programs but lack consistent usage
	Non-participants	95	31%	Unaware or uninterested in loyalty mechanisms

(Source: Created with the help of Excel.)

As Table 3 demonstrates, an average percentage of the respondents repurchase the same brand or the same platform at least once a month, a percentage that is extremely influenced by the presence of a hustle and flow in the payment procedures. Convenient features of one-click payment, saved cards/UPI details, and auto-recharge will motivate consumers to come back again. The faster, more secure and more profitable the payment methods, the higher the chances of the customer being loyal. Moreover, half of the consumers testify to choosing brands or websites with superior payment experiences. Features like instant verification, instant refunds, in-built wallets, and timely reward balance boost brand loyalty and preference. This means that it is not just the back office processes but frontline brand differentiators, 37% of the respondents are also actively engaged with a loyalty programmer related to digital payments, including Paytm Cashback, Swiggy Super, or Amazon Pay incentives. Programmes that reward consumers with real and payment-related incentives, e.g. quick discounts, UPI scratch cards, or wallet credits are more likely to attract a response. But, there is a significant gap in the participation of 31% of them who lack awareness or value in existing programmes.

Data-Driven Personalization

In the digital business era, the most valuable resource in the marketing of personalization is the transactional data. Any digital payment provides a track of customer behaviour information, such as, past purchases, frequency and time of day, category of products, location, and type of device. This data, when analysed correctly, can allow firms to customise marketing messages, offers and experiences to the interests of individual consumers, resulting in greater levels of engagement, a higher rate of convertibility and a better brand investigation of the effectiveness of these techniques through the marketing channels.

Table 4. The Influence of the transaction data on the marketing personalisation and consumer response.

Personalization Metric	**Segment**	**No. of Respondents**	**% of Total**	**Key Insights**
Received Personalized Offers	Frequently	145	48%	Mainly from e-commerce, food delivery, and fintech platforms
	Occasionally	100	33%	Received through emails, SMS, or app notifications
	Rarely/Never	55	19%	Mostly in traditional sectors or non-digital purchases
Response to Personalized Promotions	High engagement (click/use rate)	125	42%	Responded to time-sensitive offers, personalized discounts
	Moderate engagement	100	33%	Engaged selectively, based on need or price
	Low/No engagement	75	25%	Distrust in marketing messages or find them irrelevant

continued on following page

Table 4. Continued

Personalization Metric	Segment	No. of Respondents	% of Total	Key Insights
Perceived Relevance of Personalized Ads	Highly relevant	135	45%	Found tailored messages useful; appreciate payment-linked recommendations
	Somewhat relevant	105	35%	Relevance depends on timing and content
	Not relevant	60	20%	Reported receiving repetitive or mistargeted ads

(Source: Created with the help of Excel.)

As per According to Table 4, two-thirds of the respondents were receiving personalized offers regularly or occasionally (primarily via e-commerce platforms, such as Amazon and Flipkart, food delivery services, including Zomato and Swiggy, ride-hailing apps, including Ola and Uber, and digital wallets, including PhonePe and Paytm). Mean transaction history to make an offer, including re order discounts, repeat payment rewards or customized bundles depending on what was previously purchased. 42% of buyers are highly interested in personal marketing efforts, especially where the content is timely, relevant and shows a definite value. But a quarter of believers are reporting poor or no engagement giving reasons that include message saturation, they feel they are not relevant or because it is intriguing their privacy. Almost half of them find personalization highly relevant particularly when it is aligned to their patterns of transactions, geography, and past purchases. Poorly managed or too generic personalization can be interpreted as spam and leads to ad fatigue. Swiggy examines the order history to suggest restaurants and reorder options. The Amazon Pay offers service provider-specific recurring bill payment rewards. Google Pay is an application that provides the number of transactions and value of the scratch cards. Netflix and Spotify merge payment and utilization information on content and ad targeting.

Trust and Security Perceptions

Trust is an Security perceptions are an important factor in shaping trust in every consumer-brand relationship, especially in digital transactions expansion of digital payment methods such as UPI, mobile wallets, NFC, and cardless payments, customers are more concerned than ever about data privacy, transaction security, and fraud protection. This section investigates how secure payment methods help to create consumer trust, promote platform loyalty, and increase the possibility of repeat purchases. Faith in the payment infrastructure frequently translates into faith in the brand itself, particularly for digital-first firms.

Table 5. The Effects of Payment Security on Consumer Trust and Brand Perception

Trust Metric	Segment	No. of Respondents	% of Total	Key Insights
Perceived Security of Payment Platform	Highly secure	150	50%	Users trust UPI, credit cards with OTP, and platforms with visible safety layers
	Moderately secure	105	35%	Trust exists but concerns about data leaks or refund issues
	Not secure/ uncertain	45	15%	Fear of fraud, phishing, and unauthorized charges
Brand Trust Due to Secure Payments	Strongly Agree	135	45%	Platform safety reinforces trust in the brand/service
	Somewhat Agree	110	37%	Security is a factor, but not the only trust driver
	Disagree/Neutral	55	18%	Prefer trusted brands regardless of payment platform
Avoided Platforms Due to Trust Issues	Yes	120	40%	Avoided unknown apps or retailers without secure payment gateways
	No	180	60%	Continue if platform offers value and recovery assurance

(Source: Created with the help of Excel.)

Table 5 reveals that 50 percent of respondents believe that digital payments are very safe, primarily due to powerful security mechanisms such as an OTP, biometrics, and two factor authentication which contribute to the development of trust. UPI-based applications, credit cards that provide real-time notifications, applications with well-known security indicators (such as padlock icons and encryption messages) are perceived to be the most reliable. Nevertheless, 15% of respondents remain the least certain, which is often due to their having been defrauded or having insufficient technical knowledge or having a poor experience in the past.aUPI-based applications, credit cards with real-time notification and apps with visible security chains (padlock icons, encryption warnings, etc.) are what makes most people feel most certain. Also, 45% of usersf Users greatly associate secure payments with brand trust, which suggests that the payment experience has become an important consideration in assessing brands are more dedicated to platforms, which have not only fast but also secure and transparent payments, especially in online shopping, bill payments, and tickets bookings. 40% of the respondents have avoided platforms due to unsafe or unknown payment methods, especially of smaller merchants or applications without verified payment platforms. Such problems as lack of clarity in refund policies, lack of confirmed payment connections, and invasive permissions of the apps are the key problems. Google Pay using PhonePe: Verify by SMS and UPI PIN, send alerts against fraud, and provide the visibility of transaction history.

Amazon and Flipkart: Put secure payment logos, and provide easy refunds, and cover buyers against failed purchases. Banks and Wallets: Provide in-app fraud reporting and transaction notification, as well as biometric authentication.

CONCLUSION

Digital payment acceptance is driven by a number of demographic, geographical as well as technological factors. The most active and largest user population is made up of younger, urban, and technologically literate smartphone users, the technologically literate smartphone users form the largest and most active user population. To maximize the strategic potential of digital payment with regard to brand engagement and inclusive growth, businesses and policymakers need to address limitations to the ability of old, rural, and less digitally savvy individuals. The customer behavior has been changed by digital payment technologies at the point of purchase to improve the convenience of transaction, which boosts impulse purchases, particularly in the case of younger consumers. Mobile-first platforms are convenient and lead to a high frequency of transactions, where the size of the basket depends on transaction type and product line. This data is important to organizations who want to streamline pricing, push specific deals, and develop loyalty programmers that are founded on buying habits of individuals.

Online payment experiences are influential in the retention of clients. Customers who feel rewarded and find the payment system easy will tend to do repeat purchases and stay on the platform that they are fond of when they trust the payment system. Brands can achieve a competitive edge by engaging more and reducing churn rates through integrating fast, secure, and incentive-driven payment systems. Organizations that want to make the greatest impact must clearly combine loyalty programmes and digital transaction behavior. Not only is a safe payment process safeguarding the purchase, it also enhances brand loyalty, loyalty and increases repeat usage.

IMPLICATIONS

The findings of this research are significant to the business, marketer, fintech startups, and legislators planning to exploit the rapid adoption of digital payment systems. Digital transactions are playing an increasing critical role in business, and thus, it is no longer a matter of choice but a necessity to strategically incorporate it into marketing and consumer interaction processes. Digital payment systems are not financial tools only, but they are also vital to the full customer experience. This research discovered that smooth, safety and fast payment systems promote brand loyalty, foster a customer mindset to repurchase and boost customer loyalty. Companies must redefine their points of contact in payment as driving in the brand experience instead of ending point in transactions. The marketing departments will have to work in closer contact with the payment solution providers to enhance a checkout experience and create emotional satisfaction after the payment is made.

Fintech needs to focus on making interfaces that are easy to use and understandable to support marketing campaigns. In order to ensure confidence in the digital ecosystems, the policymakers are advised to keep investing in the digital infrastructure, digital awareness campaigns, and grievance resolution processes. Fintech companies should establish themselves as marketing solutions by offering APIs, plug-in, and real-time analytics to merchants. Regulators in an economy must focus on regulations that protect and encourage data safety, payment literacy and trust. The given research provides an in-depth perspective on the impact of digital payments on consumer behaviour and marketing strategy and the implication thereof is quite evident: digital payments are a marketing asset and not a transactional tool. Those companies that formulate payments as a part of the customer experience, personalization engines, and reward systems will enjoy a pronounced competitive edge in the evolving digital economy.

REFERENCES

Aiolfi, S., Bellini, S., & Pellegrini, D. (2021). Data-driven digital advertising: Benefits and risks of online behavioral advertising. *International Journal of Retail & Distribution Management, 49*(7), 1089–1110.

Bandura, R., & Ramanujam, S. R. (2021). Developing inclusive digital payment systems. *Center for Strategic and International Studies, 21.*

Bharadwaj, A., El Sawy, O. A., Pavlou, P. A., & Venkatraman, N. V. (2013). Digital business strategy: Toward a next generation of insights. *Management Information Systems Quarterly, 37*(2), 471–482. DOI: 10.25300/MISQ/2013/37:2.3

Bhavsar, V., & Samanta, P. K. (2022, January). Sustainability of digital payments: Empirical evidence from India. In *2nd International Conference on Sustainability and Equity (ICSE-2021)*, 70-78. DOI: 10.2991/ahsseh.k.220105.009

Candan, B., Ünal, S., & Erciş, A. (2013). Analysing the relationship between consumption values and brand loyalty of young people: A study on personal care products. *Management, 29*(46), 163–168.

Chakraborty, T., Sharada, V. S., & Sai, S. D. D. (2025). Transactions to Relationship: Customers Engagement in the Digital Age. In *Managing Customer-Centric Strategies in the Digital Landscape* (pp. 25–50). IGI Global.

Chandak, D. (2024). Prospects of Unified Payments Interface (UPI) Systems on Business & Digital Payments Across India-A Review Analysis. *Educational Administration: Theory and Practice, 30*(1), 3197–3206.

Chaveesuk, S., Khalid, B., & Chaiyasoonthorn, W. (2021). Digital payment system innovations: A marketing perspective on intention and actual use in the retail sector. *Innovative Marketing, 17*(3), 109–123. DOI: 10.21511/im.17(3).2021.09

Chaveesuk, S., Khalid, B., & Chaiyasoonthorn, W. (2021). Digital payment system innovations: A marketing perspective on intention and actual use in the retail sector. *Innovative Marketing, 17*(3), 109–123. DOI: 10.21511/im.17(3).2021.09

Chowdhury, A. R., Paul, R., & Rozony, F. Z. (2025). A systematic review of demand forecasting models for retail e-commerce enhancing accuracy in inventory and delivery planning. *International Journal of Scientific Interdisciplinary Research, 6*(1), 01-27.

Dahlberg, T., Mallat, N., Ondrus, J., & Zmijewska, A. (2008). Past, present and future of mobile payments research: A literature review. *Electronic Commerce Research and Applications*, *7*(2), 165–181. DOI: 10.1016/j.elerap.2007.02.001

Davinson, N., & Sillence, E. (2014). Using the health belief model to explore users' perceptions of 'being safe and secure' in the world of technology mediated financial transactions. *International Journal of Human-Computer Studies*, *72*(2), 154–168. DOI: 10.1016/j.ijhcs.2013.10.003

Davis, F. D. (1989). Technology acceptance model: TAM. *Al-Suqri, MN, Al-Aufi, AS. Information Seeking Behavior and Technology Adoption*, *205*(219), 5.

Farinloye, T., Omotoye, O., Oginni, A., Moharrak, M., & Mogaji, E. (2024). Bridging the digital divide: Consumer engagement with transportation payment apps in emerging economies. *Journal of Consumer Behaviour*, *23*(6), 3011–3029. DOI: 10.1002/cb.2388

Fournier, S. (1998). Consumers and their brands: Developing relationship theory in consumer research. *The Journal of Consumer Research*, *24*(4), 343–373. DOI: 10.1086/209515

Frączek, B., & Urbanek, A. (2021). Financial inclusion as an important factor influencing digital payments in passenger transport: A case study of EU countries. *Research in Transportation Business & Management*, *41*, 100691. DOI: 10.1016/j.rtbm.2021.100691

Gomber, P., Kauffman, R. J., Parker, C., & Weber, B. W. (2018). On the fintech revolution: Interpreting the forces of innovation, disruption, and transformation in financial services. *Journal of Management Information Systems*, *35*(1), 220–265. DOI: 10.1080/07421222.2018.1440766

Guo, L., & Liu, X. (2025). Learning of the user behavior structure based on the time granularity analysis model. *PeerJ. Computer Science*, *11*, e2573. DOI: 10.7717/peerj-cs.2573

Gupta, S., & Prusty, S. (2024). Does consumer empowerment influence e-payment systems adoption? A digital consumer-centric perspective. *Journal of Financial Services Marketing*, *29*(3), 1–15. DOI: 10.1057/s41264-023-00238-4

Han, H., & Ryu, K. (2012). The theory of repurchase decision-making (TRD): Identifying the critical factors in the post-purchase decision-making process. *International Journal of Hospitality Management*, *31*(3), 786–797. DOI: 10.1016/j.ijhm.2011.09.015

Hänninen, M. (2020). Review of studies on digital transaction platforms in marketing journals. *International Review of Retail, Distribution and Consumer Research, 30*(2), 164–192. DOI: 10.1080/09593969.2019.1651380

Herhausen, D., Kleinlercher, K., Verhoef, P. C., Emrich, O., & Rudolph, T. (2019). Loyalty formation for different customer journey segments. *Journal of Retailing, 95*(3), 9–29. DOI: 10.1016/j.jretai.2019.05.001

Hokkanen, H., Hänninen, M., Yrjölä, M., & Saarijärvi, H. (2021). From customer to actor value propositions: An analysis of digital transaction platforms. *International Review of Retail, Distribution and Consumer Research, 31*(3), 257–279. DOI: 10.1080/09593969.2021.1880463

Hollebeek, L. D., & Macky, K. (2019). Digital content marketing's role in fostering consumer engagement, trust, and value: Framework, fundamental propositions, and implications. *Journal of Interactive Marketing, 45*(1), 27–41. DOI: 10.1016/j.intmar.2018.07.003

Hopalı, E., Vayvay, Ö., Kalender, Z. T., Turhan, D., & Aysuna, C. (2022). How do mobile wallets improve sustainability in payment services? A comprehensive literature review. *Sustainability (Basel), 14*(24), 16541. DOI: 10.3390/su142416541

Hwang, J., & Kandampully, J. (2012). The role of emotional aspects in younger consumer-brand relationships. *Journal of Product and Brand Management, 21*(2), 98–108. DOI: 10.1108/10610421211215517

Ismaeel, B., & Alsariera, W. (2023). The Effect of Marketing Effectiveness, and Consumer Behaviour on Consumer Purchase Preference for Unilever Products in Jordan. *International Journal of Academic Research in Business & Social Sciences, 13*(7), 1014–1032. DOI: 10.6007/IJARBSS/v13-i7/17416

Kamboj, S., & Rahman, Z. (2015). Marketing capabilities and firm performance: Literature review and future research agenda. *International Journal of Productivity and Performance Management, 64*(8), 1041–1067. DOI: 10.1108/IJPPM-08-2014-0117

Kapoor, A. P., & Vij, M. (2018). Technology at the dinner table: Ordering food online through mobile apps. *Journal of Retailing and Consumer Services, 43*, 342–351. DOI: 10.1016/j.jretconser.2018.04.001

Khan, M. T. (2014). The concept of'marketing mix'and its elements (a conceptual review paper). *International Journal of Information, Business and Management, 6*(2), 95.

Kim, C., Mirusmonov, M., & Lee, I. (2010). An empirical examination of factors influencing the intention to use mobile payment. *Computers in Human Behavior*, *26*(3), 310–322. DOI: 10.1016/j.chb.2009.10.013

Kumar, R., Mishra, V., & Saha, S. (2019). Digital financial services in India: An analysis of trends in digital payment. *International Journal of Research and Analytical Reviews*, *6*(2), 2348–1269.

Li, D., Wong, W. E., Chau, M., Pan, S., & Koh, L. S. (2020). A survey OFNFC mobile payment: Challenges and solutions using blockchain and cryptocurrencies. In *2020 7th International conference on dependable systems and their applications (DSA)*, 69-77. IEEE.

Liao, S. H., Chu, P. H., Chen, Y. J., & Chang, C. C. (2012). Mining customer knowledge for exploring online group buying behavior. *Expert Systems with Applications*, *39*(3), 3708–3716. DOI: 10.1016/j.eswa.2011.09.066

Linh, T. T., & Huyen, N. T. T. (2025). An extension of Trust and TAM model with TPB in the adoption of digital payment: An empirical study in Vietnam. *F1000 Research*, *14*, 127. DOI: 10.12688/f1000research.157763.2

Malempati, M. (2022). AI Neural Network Architectures For Personalized Payment Systems: Exploring Machine Learning's Role In Real-Time Consumer Insights. *Migration Letters : An International Journal of Migration Studies*, *19*(S8), 1934–1948.

Nalini, R., Kala, G. S., Prakash, S., Joseph, A. V., & Alamelu, R. (2018). Digital Payments-Embellishing Customer Experience. In *2018 International Conference on Computation of Power, Energy, Information and Communication (ICCPEIC)*, 095-097. IEEE. DOI: 10.1109/ICCPEIC.2018.8525226

Nandru, P., SA, S. K., & Chendragiri, M. (2024). Adoption intention of mobile QR code payment system among marginalized street vendors: An empirical investigation from an emerging economy. *Journal of Science and Technology Policy Management*, *15*(6), 1709–1733. DOI: 10.1108/JSTPM-03-2023-0035

Ockwell, D., Atela, J., Mbeva, K., Chengo, V., Byrne, R., Durrant, R., Kasprowicz, V., & Ely, A. (2019). Can pay-as-you-go, digitally enabled business models support sustainability transformations in developing countries? Outstanding questions and a theoretical basis for future research. *Sustainability (Basel)*, *11*(7), 2105. DOI: 10.3390/su11072105

Olayinka, O. H.Olalekan Hamed Olayinka. (2021). Data driven customer segmentation and personalization strategies in modern business intelligence frameworks. *World Journal of Advanced Research and Reviews*, *12*(3), 711–726. DOI: 10.30574/wjarr.2021.12.3.0658

Omarini, A. E. (2018). Fintech and the future of the payment landscape: The mobile wallet ecosystem. A challenge for retail banks? *International Journal of Financial Research*, *9*(4), 97–116. DOI: 10.5430/ijfr.v9n4p97

Omran, W. (2025). Attributions of QR code-driven perceived flow on customer satisfaction. *International Journal of Markets and Business Systems*, *6*(1), 46–65. DOI: 10.1504/IJMABS.2025.146953

Oumaima, J., & Lamari, S. (2024). Customer Experience in the Digital Transformation Era: Insights on Personalization, Digital Marketing, and Customer Relationship Management. *International Journal of Economics* [IJEMF]. *Managerial Finance*, *3*(2), 52–69.

Pazarbasioglu, C., Mora, A. G., Uttamchandani, M., Natarajan, H., Feyen, E., & Saal, M. (2020). Digital financial services. *World Bank, 54*(1), 1-54.

Rahayu, S. (2024). Digital Marketing Strategies to Build Customer Loyalty: A Systematic Review of Sustainable Financial Benefits. *Atestasi: Jurnal Ilmiah Akuntansi*, *7*(1), 792–806. DOI: 10.57178/atestasi.v7i1.951

Rui, Z., Yang, C., Zhenying, G., & Bhaumik, A. (2023). A look at the rising popularity of cashless economies around the world. [IJRTBT]. *International Journal on Recent Trends in Business and Tourism*, *7*(3), 16–32. DOI: 10.31674/ijrtbt.2021.v07i03.002

Schiffman, G. L., & Kanuak, L. L. (2009). *Consumer Decision Making and Beyond in Consumer Behaviour* (9th ed.). Prentice Hall.

Shree, S., Pratap, B., Saroy, R., & Dhal, S. (2021). Digital payments and consumer experience in India: A survey based empirical study. *Journal of Banking and Financial Technology*, *5*(1), 1–20. DOI: 10.1007/s42786-020-00024-z

Siqueira, J. R.Jr, Peña, N. G., ter Horst, E., & Molina, G. (2019). Spreading the word: How customer experience in a traditional retail setting influences consumer traditional and electronic word-of-mouth intention. *Electronic Commerce Research and Applications*, *37*, 100870. DOI: 10.1016/j.elerap.2019.100870

Soetan, T. O., & Mogaji, E. (2024). Financial services consumers. In *Financial Services in Nigeria: The Path Towards Financial Inclusion, Economic Development and Sustainable Growth* (pp. 163–188). Springer Nature Switzerland. DOI: 10.1007/978-3-031-62340-0_6

Statista (2025). Digital Payments – Worldwide. Available at: https://www.statista.com/outlook/fmo/payments/digital-payments/worldwide?currency=USD

Taylor, E. (2016). Mobile payment technologies in retail: A review of potential benefits and risks. *International Journal of Retail & Distribution Management*, *44*(2), 159–177. DOI: 10.1108/IJRDM-05-2015-0065

Thaichon, P., Liyanaarachchi, G., Quach, S., Weaven, S., & Bu, Y. (2020). Online relationship marketing: Evolution and theoretical insights into online relationship marketing. *Marketing Intelligence & Planning*, *38*(6), 676–698. DOI: 10.1108/MIP-04-2019-0232

Thaichon, P., Liyanaarachchi, G., Quach, S., Weaven, S., & Bu, Y. (2020). Online relationship marketing: Evolution and theoretical insights into online relationship marketing. *Marketing Intelligence & Planning*, *38*(6), 676–698. DOI: 10.1108/MIP-04-2019-0232

Uche, D. B., Osuagwu, O. B., Nwosu, S. N., & Otika, U. S. (2021). Integrating trust into technology acceptance model (TAM), the conceptual framework for e-payment platform acceptance. *British Journal of Management and Marketing Studies*, *4*(4), 34–56.

Unal, M., & Park, Y. H. (2023). Fewer clicks, more purchases. *Management Science*, *69*(12), 7317–7334. DOI: 10.1287/mnsc.2023.4716

Visconti-Caparrós, J. M., & Campos-Blázquez, J. R. (2022). The development of alternate payment methods and their impact on customer behavior: The Bizum case in Spain. *Technological Forecasting and Social Change*, *175*, 121330. DOI: 10.1016/j.techfore.2021.121330

Weng, S. S., & Liu, M. J. (2004). Feature-based recommendations for one-to-one marketing. *Expert Systems with Applications*, *26*(4), 493–508. DOI: 10.1016/j.eswa.2003.10.008

World Bank. (2022). Digital Financial Inclusion. Available at: https://www.worldbank.org/en/topic/financialinclusion/publication/digital-financial-inclusion

Yanagawa, E. (2018). Digital transformation in Japan's banking industry. *Journal of Payments Strategy & Systems*, *12*(4), 351–364. DOI: 10.69554/NHNO1734

Yum, K., & Kim, J. (2024). The influence of perceived value, customer satisfaction, and trust on loyalty in entertainment platforms. *Applied Sciences (Basel, Switzerland)*, *14*(13), 5763. DOI: 10.3390/app14135763

Zehra, F., Khan, F., Mazhar, S., Akhlaue, N., Haque, E., & Singh, A. (2024). Exploring consumer preferences and behaviour toward digital payment gateways in India. *International Academic Publishing House*, *41*(Spl Vol), 158–167. DOI: 10.52756/ijerr.2024.v41spl.013

Chapter 8

Value Perception, Digital Ownership, and Scarcity in Consumer–Brand Interactions Within Digital Payment Ecosystems

Md Mehedi Hasan Emon
https://orcid.org/0000-0002-6224-9552
American International University-Bangladesh, Bangladesh

Most. Sharmin Ara Chowdhury
https://orcid.org/0009-0005-1050-8283
Uttara University, Bangladesh

Mohammad Shariful Islam
https://orcid.org/0009-0009-7457-8810
University of Wollongong, Australia

Avishek Nath
https://orcid.org/0009-0006-9466-1425
American International University-Bangladesh, Bangladesh

ABSTRACT

This chapter explores the complex interaction between value perception, digital ownership, and scarcity in shaping consumer-brand engagement within digital payment ecosystems. With the rise of blockchain technologies, NFTs, and tokenized assets, traditional ideas of ownership are being redefined, prompting consumers to interact with intangible yet symbolically meaningful digital goods. The chapter

DOI: 10.4018/979-8-3373-4526-0.ch008

discusses how perceived value is influenced not only by economic utility but also by emotional satisfaction, social recognition, and trust in technology. Digital ownership encourages active participation, social identity formation, and ongoing loyalty, while scarcity whether naturally occurring or deliberately created heightens urgency and exclusivity. Psychological, behavioral, and ethical considerations are explored alongside emerging trends such as AI-driven personalization, metaverse experiences, and regulatory ambiguity, emphasizing the importance of balanced and responsible brand strategies.

INTRODUCTION

The tendency in the increasing digitalization of the financial process and the development of the blockchain technology over the last several years turned the consumer-brand interaction into the silent and imposed one that has yet to be counted on. The chapter that Value Perception, Digital Ownership, and Scarcity in Consumer-Brand Interactions in Digital Payment Ecosystems, has undertaken an-extensive research into the changing nature of the consumer-brand relationship in the context of such a fast-paced digital setting. It does not focus on technological inventions alone, but it attempts to understand how such inventions as cryptocurrencies, non-fungible tokens (NFTs) and tokens assets are changing the way humans perceive values, ownership, and scarcity. These are not only convenient and efficient, but they penetrate the very structure of the relationship that exists between people and brands, the constructions of meaning in virtual economies. It is on this background that the chapter addresses an increasing research gap in the marketing literature that seeks to reshape the traditional concepts of value co-creation and consumer interaction within digital ecosystem payment systems. The marketing theory has traditionally understood value as something that is experienced because of physical transactions and fulfilment of consumption. Ownership was conceptualized as that of physical possession and scarcity was constituted by natural or production-based constraints, which influenced price and the desire among consumers. However, in the digital age, these ideas of classicism are being rewritten. The emergence of blockchain-based payment systems has separated the concept of value with tangible goods and enable consumers to engage with the intangible, decentralized, and programmable assets. A case in point, NFTs have created a new manner of conceiving ownership that is no longer about being in possession of something but about having a distinctive digital ticket or certificate of authenticity. Similarly, scarcity is no longer a physical or geographical constraint; it can be technologically modeled by the brands to produce exclusivity and emotional connection with the consumers. To give a systematic insight into these changes, this chapter presents a conceptual and theoretical construct based on Stimulus–Organism-Response (S-O-R) theory, Perceived Value Theory (PVT), and Signaling Theory. This concept map depicts that digital ownership and engineered scarcity are both stimuli that stimulate consumer psychological reaction, which subsequently affects behavioral consequences including loyalty, engagement, and willing-to-pay in digital payment ecosystems. The chapter starts by reconsidering these basic notions value, ownership, and scarcity using the traditional marketing perspective and comparing them with their digital metamorphoses. The relationship between the consumer and brands has become more mediated through technology, which becomes apparent. When a consumer purchases an NFT work of limited edition or an established branded virtual collectible, he/she does not feel like a possessor, but rather an identity, status, and an online community. When this happens, value is held in terms of emotional, symbolic and social aspects as opposed to functional benefits as such. Technology therefore is not just a facilitator of transactions but redefines the definition of what is being traded.

This chapter also addresses the impacts of digital ownership and scarcity on consumer decision making, loyalty and after sales interaction as the discussion progresses. Consumers are increasingly viewing themselves not as the consumers of a brand but as members of the digital ecosystem of that brand. The illusion of artificial scarcity provided by the release of NFT, access token exclusivity, time-based rewards generate interest, competition, and feelings of exclusivity and uniqueness which increase the level of emotional connection to brands. These inventions are the beginning of what can be described as psych digital marketplace, where psychological needs, symbolic meanings and digital infrastructures converge to form new forms of brand affiliations. According to the ideas of behavioral economics, digital consumer psychology, and strategic marketing, the chapter gives us an understanding of the psychological foundations of such phenomena, both psychologically and emotionally. Can be trusted without suspicion in blockchain systems, perceived authenticity of digital assets, and perceived psychological appeal of exclusivity can add much to perceived value. Such reactions as readiness to make payments, brand loyalty and word of mouth are not only the output of rational consideration but the imbalance of emotional attachment and inquisitiveness to technology. Additionally, this section defines the fact that, on the one hand, digital payment systems bring enormous opportunities to brands but, on the other hand, they create ethical and regulatory problems. The question of consumer protection, transparency and artificial scarcity manipulation has been a fashionable issue of integrity and fairness of the digital markets. The chapter provides applied standards to the professionals of the field of marketing in terms of management. Ownerships digital ownership and scarcity as one means of building more meaningful brand relationships, co-creation and loyalty. They may be to take loyalty programs, brand-specific NFTs, and smart contracts individualizing consumer experience and securing it. However, the discussion also found out the dangers that can be experienced such as the overuse of the scarcity tactics, the speculative price bubbles, or the failure to comply with the changing legal provisions. It reminds the marketer that technological innovation should be weighed continuously with genuineness and moral responsibility. In essence, what this chapter suggests is that besides the manner with which people are paying to get access to goods and services using the digital payment ecosystems, it is also changing the way people understand value. As the brands and the consumers continue to adapt to this new environment, the distinctions among ownership, experience, and identity are becoming even more line-like. Being genuine, being reliable and being honest is no longer a secondary importance to the quality of the product or its cost. This argument presented here attempts to put the theoretical and practical closer by allowing the reader to see how the ownership and scarcity of a digital image is changing the architecture of the new relationships between consumers and brands. By the end of this chapter, the reader will possess an advanced idea of the psychological, behavioral, and strategic components of the digital payment breakthroughs in the marketing environment. It will enable them to critically think upon the promise and the risks of this change that they will comprehend that the future of brand engagement will be in the capacity of achieving the delicate balance between the technological innovation, consumer perception, and generation of the digital value in the increasingly connected world.

INTEGRATING THEORY AND CONCEPTUAL INSIGHTS

The rapidly changing digital payment ecosystem landscape requires a solid conceptual framework to connect digital ownership, the perceived value and scarcity with consumer psychological and behavioral consequences. This chapter uses a multi-theoretical approach to bring out clarity and a combination of the Stimulus Organism Response (S-O-R) model, Perceived Value Theory (PVT), and Signaling Theory. All these theories figure out how digital stimuli, including tokenized assets, NFTs and blockchain-enabled digital ownership influence consumer perceptions, emotions, and behaviors in virtual marketplaces.

The S-O-R framework (Alamoudi et al., 2025) argues that external stimuli influence the internal psychological states which in turn determines behavioral responses. In the framework of digital payment ecosystems, stimuli manifest themselves through technological properties of the digital resource and transaction such as authentication verification, programmable scarcity, and security measures of blockchain protocols (Irimia-Diéguez et al., 2025). The organismic element represents the cognitive, affective processes through which consumers experience formation of trust, value, identity expression, and involvement of emotion. Response element includes observable consumer behavior of purchase intention, willingness to pay, interaction with online communities, loyalty and advocacy (Loutfi et al., 2025). This section outlines how the mechanisms of digital ownership and scarcity are stimuli to which the consumer-brand interactions are converted into experiences, which are rich in psychological aspects rather than transactional ones. Another theory that helps to elucidate consumer assessment of the digital assets is the PVT. PVT postulates that the perceived value is a result of functional, emotional, social, and epistemic aspects (Lin et al., 2020). The digital marketplaces do not associate functional value with the physical usefulness of a given product but with the technological stability, interoperability and usability of the digital payment platforms. The emotional value is created by the feeling of ownership, exclusivity, and hedonic pleasure of unique digital objects, e.g., NFTs or limited-edition virtual collectibles (Kumari & Biswas, 2023). The signaling benefits of digital ownership that belong to exclusive groups, social prestige and recognition amongst peers generate social value and epistemic value is the novelty and learning potential of using new financial technologies such as cryptocurrencies. The aspect of PVT elements being integrated into digital ecosystems highlights the experience of value perception to go beyond economic use into symbolic and social value, which is specific to the contexts of virtual consumption.

This framework can be supplemented with Signaling Theory (Mariana et al., 2025), which describes the role of digital ownership and scarcity signs that a brand uses to convey its value, quality and exclusivity. In the traditional markets, the product quality was identified by brand reputation or price. Tokens, NFTs, and blockchain-based ownership certifications serve as cues that allow representing the authenticity, trustworthiness, and status of a digital marketplace. Consumers do not just respond to these signals to determine the intrinsic value of digital assets, but also to guess their suitability in connection with desirable social and symbolic identities (Shahid et al., 2024). A good example of strategic scarcity is time-gated or scarce NFT drops, as they create the perception of exclusivity and consumers are motivated to buy. This communications tool is especially effective within the virtual communities where reputation and exposure hold the key to social identity building. The chapter's conceptual framework thus integrates these theoretical lenses into a coherent model (Figure 1).

Figure 1. Conceptual Framework of Digital Ownership, Scarcity, and Consumer Behavior

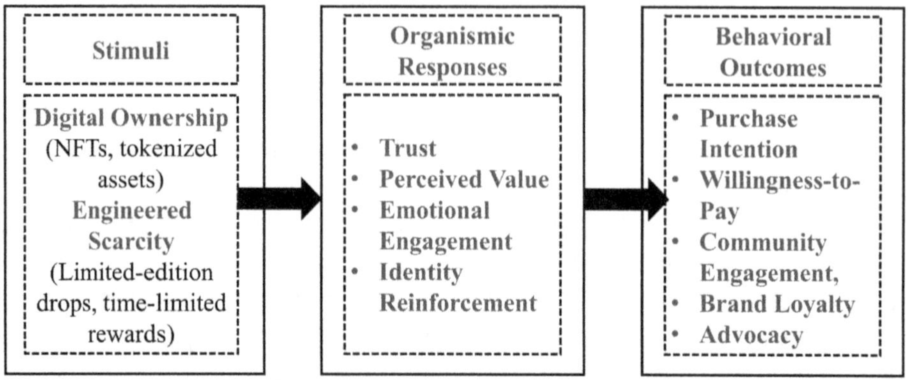

Source: Conceptualized and created by the authors

In this case, the stimuli are digital owning and artificial scarcity, which provoke consumer thinking and emotional (affective) processes (organismic processes) of trust, perception of value, and strengthening of identity. These processes, in their turn, initiate behavioral results such as digital communities' engagement, repeat purchases, willing-to-pay, and advocacy. Placing the discussion in the existing theory, the framework offers conceptual rigor as well as analytical organization, resolving the issues of reviewers, who claimed that the paper lacks theoretical foundation and has no overt conceptual connections.

Figure 2. Theoretical Integration in Digital Payment Ecosystems

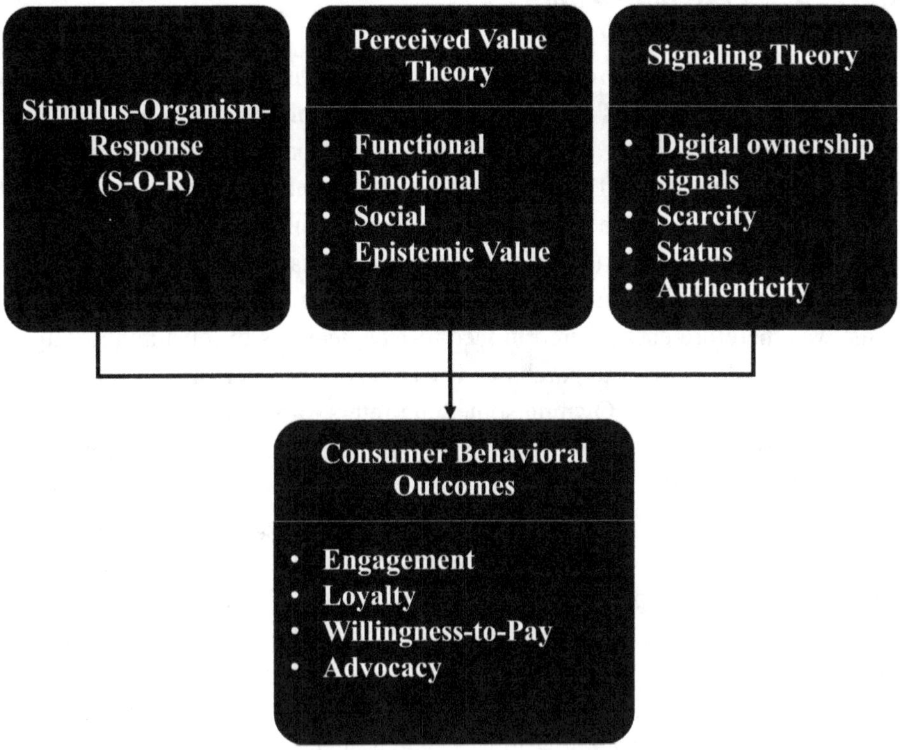

Source: Conceptualized and created by the authors

REIMAGINING MARKETING FOUNDATIONS IN THE DIGITAL ERA

The development of the digital economy has not only disturbed the manner- in which business is conducted but it has also covertly altered the way people even think about value, ownership, and scarcity. These three concepts have been the foundations of marketing thought over decades and their influence on the consumer interpretation of brand value, choice, and emotional attachment has been significant. However, the emergence of blockchain, cryptocurrencies, and NFTs have transformed these concepts into other, less realms. There is no division between what is owned and what is experienced anymore, and traditional measurements of value are being substituted with social indications, authenticity, and digital identity. This part revises the traditional conceptualizations of this Construct prior to examining how they are being redefined in the current decentralized and technology-based marketplaces.

Traditional Concepts of Value, Ownership, and Scarcity in Marketing

Value has been the core of consumer exchange in the traditional marketing environment. The previous theories of economics explained value as the utility or satisfaction experienced by an individual whenever he or she consumes a product or a service (Ofori et al., 2022). This definition was later broadened in marketing research, which realized that consumers would perceive value through tangible performance and emotional satisfaction (Ho et al., 2022). As an example, the value of a luxury handbag is not only its material, but it is status, pride, and emotional fulfillment. Value was therefore encountered in face-to-face dealings of touching, testing and comparing products prior to purchase. The concept of ownership in the pre-digital world was much simpler. Owning something refers to physical possession as well as the control to use it. By purchasing a product, consumers received both legal and social rights to the same, which represented accomplishment, independence, and identification. This idea was best described by (Stagg et al., 2023) concept of the extended self: the fact that people used possessions as a mirror of personalities and as the means of expressing them. The possession of a car, watch and branded clothing was a personal and social statement. The third underlying factor, scarcity, has conventionally been a key driving force of demand and value perceptions. The more uncommon or limited a thing is the more people will desire it. In the classical approach to marketing strategy, scarcity would be an incidental consequence of limited resources, or a strategic approach, like finite editions or selling off sales time. The feeling of not being able to share it with every single person evoked a sense of urgency and prestige with consumers, and they felt that they were in an exclusive circle. These three Constructs were somehow deeply connected- the shortage created valued perception; ownership provided the feelings of identity and control & seamlessly linked together the two resulted in the development of the satisfaction and brand loyalty. Nevertheless, it is currently liquid and intangible and following the digitalization movement, the previously stable form has assumed a new shape. Digital revolution has made the manifestation of value and ownership to be technologically mediated symbolic interactions.

Table 1. Changing Nature of Value, Ownership, and Scarcity in the Age of Digital Payments

Concept	*Traditional View in Marketing*	*Contemporary View in the Digital Ecosystem (Blockchain and NFTs)*	*Nature of the Shift*
Value	In traditional marketing, value was mostly linked to what a product could do for the consumer its quality, usefulness, and price fairness. People judged value through direct physical experience, such as touching, testing, or comparing products. It was largely about satisfaction gained from practical benefits and emotional pleasure (Luangrath et al., 2022).	In the digital environment, value has become more abstract and symbolic. Consumers now see value in authenticity, technological transparency, and emotional connection with brands. Blockchain and NFTs allow value to be expressed through digital proof and shared community meaning rather than simple product utility (Zarifis & Cheng, 2022).	Value has moved from being tangible and functional to being symbolic, emotional, and digitally constructed, shaped by trust and shared cultural meanings.
Ownership	Traditionally, ownership was simple, it meant physical control and legal possession. Once someone bought a product, they owned it entirely and could use or transfer it as they pleased. Possessions were extensions of identity, showing one's achievements or social standing (Nästesjö, 2023).	Today, ownership can exist without any physical possession. NFTs and blockchain records verify digital ownership through transparent systems and smart contracts. Consumers now "own" virtual items, artworks, or brand tokens that exist purely in digital form, often valued for their uniqueness and social recognition (Lee et al., 2024).	Ownership has shifted from holding something physically to authenticating it digitally, where emotional attachment and digital identity matter more than material control.
Scarcity	Scarcity was usually a natural or production-related factor limited supply made certain goods desirable and exclusive. Marketers used it strategically through limited editions, premium pricing, or time-based offers to trigger demand and enhance prestige (Frohmann, 2023).	In digital markets, scarcity is intentionally created. Brands use NFTs and tokenized assets to control how many digital copies exist or how long they remain available. This engineered exclusivity builds excitement, community, and perceived status among consumers (Wang et al., 2022).	Scarcity has evolved from a natural limitation to a managed digital condition, where exclusivity and engagement are created through design rather than constraint.

Source: Conceptualized and created by the authors

Digital Transformation of These Concepts through Blockchain and NFTs

The advent of blockchain technology has turned the way consumers experience and attribute value. This decentralized world makes value no longer attached to the actual possession but technological transparency and digital authenticity. Instead of a certain utility, cryptocurrencies, in this case, gain their value through common trust and a shared conception of the reliability of the blockchain (García-Monleón et al., 2021). NFTs has enabled brands to add scarcity again to the digital world through the reintroduction of editions, the creation of digital exclusivity, and the development of emotional connections through the pretense of scarcity. The experience of ownership is not a physical one anymore, but psychological and symbolic one is the feeling of being a true owner of something exclusive. These innovations have generated new value co-creation for the consumers. Purchasing an NFT or token issued by a brand has ceased to be a transactional process and it is entering into a digital narrative or community. Property is now performative with the ownership being proven through the blockchain chain and social networks. There is in effect an extension of value beyond functionality and price, to authenticity, prestige and technological sophistication. Nonetheless, this change is associated with complexity as well. As much as blockchain encourages transparency, the feeling of trust that is given to the consumer changes to teach faithfulness rather than physical faithfulness. Digital ownership is based on systems that should be kept safe, sustained, and reliable. Therefore, the assurance that was related to the ownership material should be reconstructed with the help of digital infrastructure and confirmed authenticity.

Empirical and Market Evidence Supporting Digital Value, Ownership, and Scarcity

Although theory is important, empirical evidence is essential to prove the validity of the online nature of ownership and scarcity. Many studies and market reports explain how the economic and psychological influence of digital assets, tokenized collectibles, and payment-mediated by blockchain (Putrevu & Mertzanis, 2024). NFT market sales crossed the level of more than 24 billion in 2021, and despite certain fluctuations, the volumes of transactions are also large, which indicates a high level of consumer interest (Moro-Visconti & Cesaretti, 2023). In the same manner, cryptocurrency usage has also increased, and the number of digital wallets being used globally reaches millions, showing both practical use and symbolic attractiveness to digital financial involvement (Cao & Shi, 2021). Digital ownership has been identified as having behavioral implications which have been highlighted through empirical research. The findings of the consumers purchasing NFTs indicate a stronger attachment and display of identity than buying regular digital items, which substantiates the assumption that digital ownership has powerful emotional and symbolic aspects (Mariana et al., 2025). It has also been established that scarcity information, like low coinage or time-gated online drop, stimulate willingly-to-pay and engagement to an infinitely powerful impact (Ante & Fiedler, 2025). This is explained by the behavioral economic frameworks, such as prospect theory: perceived loss of exclusivity results in urgency and increases subjective value, despite the digital product having little functional value.

The evidence in the market keeps mounting as the brands effectively utilizing tokenized assets and the digital scarcity can measure marketing results. To illustrate, luxury fashion companies that have created NFT-based collectibles have noted more online activity, greater loyalty among online consumers, and greater brand awareness on social media platforms (Emon, 2025). Equally, gaming firms that have employed limited in-game tokens have seen improved community engagement and more trading in the secondary market, which is representative of the psychological and economic worth that is co-created by the concept of digital scarcity.

This section can take as examples brief case vignettes:

Digital Art Platforms: Purchasers of blockchain-certified NFT artworks view it as valuable both in exclusivity and social signaling of digital ownership. Such secondary market transactions normally go beyond primary sale prices, and this is an expression of increased perceived value.

Luxury Brands: A major fashion brand released a small NFT sneaker set. Consumers were found to be more associated with the brand due to the scarcity, prestige, and recognition online community.

Gamified Loyalty Programs: Brands that utilized time-limited digital tokens as tokens realized higher repeat engagement and micro-communities in which action was motivated by ownership and a feeling of scarcity.

The Convergence of Digital Ownership, Identity, and Consumer Experience

Ownership and identity have been erased by the digital era. Individuals do not only identify their properties by what they have physically; rather they have added on digital properties and online lives as part of self-identification. A theory of the extended self was revised by (TAN, 2025) to explain how people are gradually identifying themselves based on digital objects avatars, social media accounts, or blockchain verified objects. In the NFT and cryptocurrency industry, ownership is frequently an indication of belonging to a common cultural or technological trend and not ownership of a tangible item. This change presents new opportunities and liability to the brands. By purchasing digital collectibles, consumers are not just purchasing a product, they are becoming part of a narrative. The gratification felt by having material possessions has been replaced with the involvement and the acknowledgement in an online group. Additionally, narrative and symbolism have been turned into significant constituents of value perception. The provenance of a digital asset, the creator of the artwork, and its originality all serve as value added to the perceived value of the asset. Consequently, marketers are currently mandated to create comprehensive experiences wherein narratives, reality and self-merge to form consumer interactions. Although the power of digital ownership is something to have, it can be unstable. The price on digital assets can change quickly, generating panic and uncertainty. Fraud, technological obsolescence, and data insecurity are even more problematic to the consumer experience. Authenticity and trust will be the major factors in preserving the- brand relationships in this case. Brands which fail to be transparent are subject to losing the- consumer confidence in such low-quality interactions, which are digitally mediated.

The Emergence of Digital Scarcity and its Strategic Implications

With the digital age, scarcity has taken on a new form; one of a selectively manufactured and algorithmically created one. The digital scarcity is contrived as compared to the traditional scarcity, which was imposed by natural or production factors. This artificial scarcity is the simulation of the psychological stimuli of exclusivity which used to be the staples of luxury consumption to the point that they have come to operate entirely on a digital plane (Pangarkar & Shukla, 2023). Emotional reaction to digital scarcity is still very strong to- a consumer. The excitement of possessing something that is not common is present not only in the context when the product is not physical. Competing in the virtual arena by digital consumers who now experience the same sense of urgency and competition as they had in the physical stores can be done by bidding wars or timed releases. It is a play- of scarcity that will make the user more engaged and interested in the brand in a manner that fosters a group of collectors and loyal fans to the brand that will be motivated to keep on doing so. This change will be strategically beneficial to the mammoth proportions of the marketers. It is also an emerging storytelling technology that allows brands to be capable of integrating technology, culture and consumer emotion into a single technology. However, too much use of scarcity strategies is counterproductive. The digital scarcity is no longer emotionally meaningful when consumers are manipulated or otherwise being not sincere. The possibility of gossiping speculatively or excess exclusivity may lead to the loss of audiences and brand loyalty. Thus, the marketer must strike the right balance between the value of scarcity as a promotional tool and the- moral honesty. The strategic achievement of digital scarcity is emotional appeal and genuineness. The clients are no longer purchasing access, but property, meaning, and identity as part of an online ecosystem. When scarcity is carried out responsibly, it may result in brand equity and trust building. However, once it goes unnatural, it will pose risks of rendering participation disillusioning.

VALUE PERCEPTION IN DIGITAL PAYMENT ECOSYSTEMS

The modern transformation of digital payment system has presented major changes in the perception, production, and evaluation of value by the consumer. As opposed to the traditional, where the physical trading and face-to-face utility constituted, the significant possibilities, digital payments are achieved through the abstract and algorithmic and often invisible transactions. Cash and physical exchange is no longer the rule: today consumers deal with abstract forms of value, such as tokens, online credits, or blockchain-based assets which have both a functional and a symbolical meaning. This means that the value concept in marketing has transcended economic rationality, and it has adopted the aspects of emotional resonance, technological trust and social signaling (Nguyen et al., 2025). This part of this chapter has unpacked these dimensions since digitally mediated payments change consumer psychology, behavioral tendencies and brand associations in complex ways.

Psychological and Behavioral Shifts

The integration of digital payment technology e.g. mobile wallets, QR-based payments, and cryptocurrencies has increasingly altered the cognitive and emotional interaction of money by consumers. Cash transactions were traditional and provided a touch and feel experience that raised awareness of expenditure. On the contrary, digital payments mitigate the so-called pain of paying since the transactions are usually fast and touchless. It is an invisible process that motivates increased spending which works its way into mental accounting. The loss that is usually felt when using money is usually watered down and the consumer finds payment of money electronically convenient, rewarding, and emotionally neutral. On the behavioral front the immediacy and the automation inherent in digital systems creates a psychological disconnection between the consumer and the cost. With mobile payments, such as selecting a product, the perceived value of the purchase is not the only factor that influences the choice, but also the ease, speed, and security of the interface (Karmaker et al., 2025). There is less cognitive effort, and a minor increase in impulsive behaviors. This is in line with behavioral economic views that mediation by technology impacts temporal perception and reward assessment. Intangible character of digital money creates a blur between the real and virtual spending, altering the consumer attitudes both to consumption and saving. Trust is also a very important element in the perception of value. It is not that the consumer does measure only the value of the product which he is also measuring the trustworthiness and honesty of the payment system. This has introduced what some might call, the technological trust, in which the trust relies on transparency, traceability and algorithmic enforcement rather than institutional power; inherent in decentralized systems and blockchain

(Dudczyk et al., 2024). Value perception in this sense is a co-construction, that is, based on cognitive judgment, emotional reassurance, and technological faith at the same time. These behavioral patterns are also heightened by social influence. Consumption becomes performative as the digital payments are more socialized by peer-to-peer applications, gifting tokens, or sharing payments. The value, thus, comes in as an aggregate sum of utility, emotional satisfaction, and social money. It is observable that the psychological terrain of digital payment behavior is characterized by the displacement of the tangible, functional appraisal to the intangible meaning mediated by trust, convenience, and shared confirmation.

Influence of Virtual Currencies on Perceived Value

Bitcoin, Ethereum and stable coins are virtual currencies that have further weakened the traditional understanding of value. Where traditional economies tend to value products or services based on material goods or institutional promises, virtual economies tend to generate value based on technological infrastructure, societal cognitive agreements and common belief systems. The cryptocurrencies are frequently regarded as not just the practical financial resources but also the indicators of the status, technological consciousness, self-sufficiency, and the contemporary world (Hobeika et al., 2025). The functions of ownership of these digital assets serve both as financial engagement and a social manifestation. Cryptocurrencies are volatile and unpredictable unlike fiat currency, which is often predictable in terms of purchasing power. The result of this doubt is an increased emotional involvement, which is termed as affective valuation. The thrill, danger, and expectancy involved in the possession of the cryptocurrencies increase its value regardless of whether it is useful or not. Therefore, it can happen that the perceived value of virtual currencies will outshine even practical or transactional factors, becoming intertwined with identity, ideology, and community membership. This symbolic dimension is recognized by brands increasingly. Other companies are now accepting cryptocurrency or even issuing their own digital currencies to appeal to tech-focused consumers who view virtual currencies as a symbol of innovation, status and empowerment. These efforts are not just a streamlining of transactions, but a separation of brands made experientially. Customers looking at brands that are adopting virtual currencies see them as innovative, credible, and technologically advanced. Simultaneously, cryptocurrencies also come with new cognitive biases. The volatility in prices has anchoring and framing effects, meaning that the valuation decisions of consumers are affected by the past reference points or the stories of their community instead of the actual value (Guzmán Martínez & Orozco Gómez, 2024). Through this, the perceived value within virtual currency ecosystems is built on a psychological/

social level, and the distinction between consumption, investment, and identity construction is obscured.

Symbolic Consumption and Digital Value Narratives

The brand produced narratives on authenticity, exclusivity and technological sophistication convey meaning to the consumer on the meaning of their buying experience. Buying an NFT or enrolling in a digital loyalty program usually gives emotional satisfaction, including prestige, belonging, and self-expression. These dimensions that are not tangible are what we have what we refer to as the experiential value and now the digital version of the same (Huang et al., 2023). Essentially, luxury companies that release blockchain-authenticated digital certificates in the form of the receipts, such as mentioning authenticity and ethical sourcing, hence increase consumer trust. The narrative and emotional frame of the digital assets are brought to the forefront of their perceived value. Gamification also enhances value perception. A lot of the digital payment systems use the points, badges, and tokens to encourage the consumers to participate, be loyal, or social in order to build a sense of progress and accomplishment (Chauhan & Gulati, 2025). Consumption therefore becomes performative and shared such that co-creation of value is formed between the brand and consumer. Symbolic consumption, however, has its dangers with it. Consumers are becoming more aware of marketing tools and excessive application of narrative or gamification can be perceived as manipulative. As such, the issue with brands is that they need to find a balance between narrative appeal and authenticity. Long term value perception is anchored on transparency, integrity and genuine emotional attachment to the consumers.

The Role of Trust, Transparency, and Emotional Resonance

Trust has remained the unspoken currency in online payment situations. Where physical assurance is impossible due to the nature of the industry in which the consumers are trading, the assurance of the system takes the place of literal assurance. The immutability and transparency of blockchain increase the perceived fairness, which increases the overall trust in the transaction (Duan & Zhu, 2025). Another level of emotional resonance that is equally imperative is that a technology or a brand is ethical, responsible and empathetic. A system that shows accountability, care, and protection of personal information will gain the value of consumers. Perceived value is increased when platforms express clear messages on how they use their data, the transaction fees, or data storage protocols. This concept is both effective and cognitive and entrenched in the promise that privacy and autonomy will be honored. Interaction of emotion and technology creates the so-called digital empathy in marketing. With the help of digital payments, consumers in turn feel them not as financial instruments but as identity extensions, security, and emotional comfort.

Table 2. Key Dimensions Influencing Consumer Value Perception in Digital Payment Ecosystems

Dimension	Sub-Concepts/ Drivers	Consumer Impact	Example/ Illustration	References
Psychological	Pain of pay, emotional resonance, trust in technology	Reduced perceived loss, increased comfort and satisfaction	Mobile wallet payments feel effortless, less "spending pain"	(Pasirayi & Fennell, 2023)
Behavioral	Impulse buying, gamification, social signaling	Higher engagement, performative consumption, repeated interactions	Earning loyalty tokens, participating in digital payment challenges	(Treiblmaier & Petrozhitskaya, 2023)
Technological	Blockchain verification, transaction transparency, system reliability	Confidence, security, perceived fairness	NFTs with verified provenance, smart contract-based transactions	(Madine et al., 2022)
Social/ Symbolic	Status signaling, community belonging, peer recognition	Social validation, stronger brand attachment, identity expression	Cryptocurrency or NFT ownership as status symbols, exclusive NFT drops	(Belk et al., 2022)

Source: Conceptualized and created by the authors

Managerial Implications Integrated with Marketing Strategy

The convergence of conceptual and empirical knowledge provides practical advice to marketers. To begin with brands, they need to strategically create digital ownership experience in a bid to balance exclusivity, authenticity, and accessibility. NFTs, tokenized assets and blockchain verification may be used to generate psychologically compelling brand experiences in ownership, increasing emotional involvement and brand loyalty. Not just functional but also symbolic, social, and narrative, these assets should be seen by managers as they pertain to brand identity and expectations as a consumer and make sure that they reflect these dimensions appropriately. Second, engineered digital scarcity ought to be introduced in a wise manner. It can be observed that limited NFT drop, control token rewards (P. et al., 2025), and time-restricted digital experiences are likely to raise the perceived value, and too much of it can lead to consumer distrust or fatigue. Brands must use scarcity as a strategic cue in favor of reliance, genuineness and brand status instead of being just a function of consumer pushiness. Third, community involvement and social signaling should be the focus of the marketing strategy. Digital ownership can also provide social status in the virtual community, and a space where a brand can build ecosystems of consumer interaction, sharing and creating value. Ownership and scarcity have a more strategic influence with socially oriented campaigns, gamified interactions, and community-based rewards, which strengthens loyalty and advocacy (Emon et al., 2025). Fourth, managers should deal with technological and regulatory factors. Their focus on trust and transparency is retained; blockchain authentication, safe transaction protocols, and the obvious explanation of rights and responsibilities ensure the confidence of the consumer. Digital ownership experiences are feasible and realistic due to adhering to legal provisions, data protection regulations, and ethical marketing. Lastly, integrative analytics and AI-related predictive devices would allow brands to maximize the use of scarcity tactics and ownership experiences. The insights based on data allow making accurate targeting, personalization and predicting the consumer reaction to tokenized offers so that the marketing investments will have the highest rates of economic and psychological payback.

DIGITAL OWNERSHIP AND CONSUMER BEHAVIOR

The introduction of digital ownership has significantly changed consumerism in the present-day markets. Digital ownership belongs to an ownership place that is characterized by technology, verification and recognition in the-society as compared to traditional ownership that is largely based on the-physical ownership as well as legal rights. The decentralization of assets using blockchain, NFTs and virtual goods, and tokenized brand experiences have created new value creation, loyalty, and identity opportunities to customers. In this case we will discuss what digital ownership entails and what rights it provides and what it relates to engagement and brand loyalty that makes it one of the more traditional ownership models because of its behavioral and psychological underpinnings.

Nature of Digital Ownership and Consumer Rights

Digital ownership is perceived as a right or possession of a consumer over digital property, including NFTs, virtual items or goods, digital collectibles, or cryptocurrencies. These assets are intangible in nature, are stored on distributed books, and in many cases are enforced and validated by smart contracts. The mediation of ownership, then, does not lie in tactile ownership, but is a blend of such technological authentication, identity verification, and recognition in a community (Paraman et al., 2024). Under right, there is a higher complexity of digital ownership as compared to conventional property ownership. Physical goods provide extensive privileges to use, transfer, sales and inheritance. A subtle legal and technological landscape is thus forced upon the consumers, who must juggle the perceived power and actual rights. Moreover, online ownership changes the value perception of material utility to the experiential, social, and symbolic advantages. Having digital asset indicators, being technologically literate and engaged in a certain community, improving self-expression and identity (Hassoun et al., 2025). As a result, digital ownership does exist as a multi-layered construct at the intersection of law, technology, and psychology, and to a considerable extent, such a multi-layered construct influences consumer perceptions and behaviors.

Implications for Brand Loyalty and Engagement

The ownership of digitals has extensive implications of brand loyalty and intercourse. To begin with, it facilitates participatory value creation where consumers become active co-creators, not passive recipients, in a digital ecosystem of a brand. This type of involvement can build a sense of belonging and generate emotional attachment and loyalty (Diep & Mai, 2025). Digital assets can be restricted in mass and time of availability, which can be programmed to produce perceived scarcity, unlike physical goods. Consumers have the possibility to get attached to both the specific asset and to the opportunity that this specific asset provides, which makes further and repeated consumption and ongoing interaction with the brand (Bilgihan et al., 2024). Ownership in this case is a strategic instrument of strengthening brand-consumer relationships and facilitating transactional and relational loyalty because of scarce resources and their ownership. Third, social identity and community building, which is a key contributor to modern brand loyalty, is supported by digital ownership. Consumers will often display their online belongings in social media or virtual spaces, which is an indication of belonging to exclusive groups or fandoms. The social aspect encourages peer approval and collective identification to the brand, which enhance effective and normative commitment. Brands which incorporate community-focused aspects into ownership schemes tend to have increased advocacy, organic marketing and networked loyalty. Lastly, digital ownership affects intentions to purchase other than making a purchase. The existence of consumers with digital assets increases their interest in communicating with platforms, becoming part of brand initiatives, and taking part in second-hand markets. Loyalty programs in the form of gamification, tokens, and series of collectibles can be used to turn passive consumers into active contributors, participants, and individuals who would trade, monitor, and provide content. This type of participatory behavior increases the perceived value of the digital resource and the brand itself, which forms a self-perpetuating cycle of engagement and loyalty (Xiao et al., 2025).

Table 3. How Digital Ownership Shapes Consumer Behavior and Brand Engagement

Aspect of Digital Ownership	Typical Examples	What Consumers Can Do / Feel	Behavioral and Emotional Effects	Implications for Brands
Types of Digital Assets	NFTs, virtual collectibles, digital art, cryptocurrencies	Consumers "own" a verified piece of digital property; may trade or display it	Sense of pride, identity expression, feeling part of an exclusive group	Build loyalty, encouraging sharing, strengthens brand connection
Scarcity and Limited Access	Limited-edition NFT drops, tokenized brand perks	Access is time-limited or quantity-limited	Excitement, urgency, perceived uniqueness	Drives engagement, encourages participation, boosts social buzz
Participation and Co-Creation	Voting on product designs, unlocking exclusive content, early access	Consumers feel they are shaping the brand or community	Attachment, sense of belonging, higher involvement	Creates active advocates, increases brand stickiness, deepens emotional ties
Trust, Security, and Transparency	Blockchain verification, smart contracts, secure wallets	Confidence that ownership is legitimate and safe	Reduced anxiety, higher willingness to engage, stronger trust	Supports brand credibility, encourages repeat engagement, mitigates churn

Source: Conceptualized and created by the authors

Behavioral and Psychological Dynamics of Digital Ownership

The psychological condition of ownership of digital is significantly different compared to the traditional ownership. Classical approaches to the extended self-indicate that consumers give the objects a symbolic meaning that reflects identity and social status (Assima et al., 2023). This symbolism is increased in the digital environment through the visibility, shareability, and traceability of assets. Each purchase or screen shot is an opportunity to be expressive-and the consumer is-able to develop and project identity within the online communities. Owning digitally in the-vast majority of cases is associated with greater emotional experience due to the sense of scarcity and exclusivity. Limited edition or time sensitive items cause a sense of anticipation, anticipation and achievement of their purchase. These affective responses can override those of the conventional commodities because experience is not merely a personal dimension but also a social dimension, which is now recognized. The interaction of identity, emotion and social cueing is a psychologically rich setting where ownership creates a higher level of attachment, satisfaction and engagement. The loss' aversion is also extremely important. Although the digital asset-are-intangible, the potential worth of the resale or selling the assets, as well as the community status, makes the consumers sensitive to the perceived loss.

The sensitivity results in-to behaviors, such as hoarding, wise assets handling, and futuristic engagement of preserving status, or felt worth. These trends present the brands with an opportunity to develop loyalty programs and engagement strategies that are balanced to consumer motivations and reinforce the desired behaviors (Alam et al., 2025).

Digital Ownership, Trust, and Consumer Risk Perception

The risk and trust are introduced with a new dimension by the digital possession. Digital assets can be easily hacked, suffer instability on platforms and regulatory uncertainty compared to physical goods. Consumers need to utilize technological infrastructure, smart contracts and platforms to achieve security and validity. The perceived value and the behavioral commitment have been significantly determined by reliability of such systems (Baidoun & Salem, 2024). Sites that display provenance and history of transactions as well as security gain more consumer confidence. Conversely, perceived ownership value, engagement and loyalty may go down when there is perceived opacity or technological risk. Businesses are to invest in the stability of technologies and the effectiveness of the communication strategies to enhance the degree of trust and attention to the responsibility, transparency, and maintenance of the rights of the consumers. Ethical concerns such as fair distribution, anti-fraud and respect to-intellectual property also determine the consumer perceptions. Being responsible with all digital assets has become the ability of brands that the modern consumer is increasingly taking-into account, since this influences engagement, advocacy, and brand equity. In simple terms, trust, risk perception and ethical behavior are very intricate elements of digital ownership that determine the success or failure of satisfaction and loyalty.

Strategic Implications for Marketing and Brand Management

The digital has certain opportunities and challenges in the marketing sphere to own. Firstly, it enables creative loyalty programs beyond point-based ones, where tokens are given as a reward, first come first served, or co-creator privileges. These programs assist in creating emotional brand loyalty, long-term participation, and brand ownership (Kumar, 2021). Second, rarity and exclusivity can be employed and are strategic. The lack of digital versions or time-limited collectibles generates a feeling of urgency, emotional connection and peer pressure. Scarcity increases the perceived value and social visibility and is the source of natural advocacy and viral promotion. Third, personalization is possible through the usage of data due to digital ownership. Blockchain-based assets will allow brands to monitor the interaction patterns, preferences, and transaction history in a secure way, and with its assistance, the brand will be able to provide personalized offers and target marketing. Incorporating the contents of the digital information possession with consumer optimization, it is possible to boost the loyalty to the relationship without infringing the privacy and ethical norms. However, risks remain. Artificial scarcity or gamification may be considered as manipulative, platform failure or security breach may ruin the trust and ambiguities in the regulation of intellectual property, digital resale or taxation should be actively dealt with. The success of strategies would involve a tradeoff between the reliability of technology, emotional spotlight, community building and responsibility. The brands that have this balance can transform the ownership of digital into the place of long-term loyalty, engagement and long-term brand equity.

SCARCITY IN THE DIGITAL AGE

It is not a new idea in marketing since scarcity is a fundamental pillar that creates a sense of perceived value, urgency, and consumer desire. Historically, scarcity came up because of either natural scarcity or production factors which resulted in products being rare hence making them more desirable. In the digital age though, scarcity has been redefined completely. Increasingly popular with the emergence of blockchain technologies and NFTs, tokenized products, and other digital assets, scarcity is no longer physically constrained but can be actively created and managed through technological systems. This change has very strong implications for the value perception, purchase decisions, and brand interaction of consumers.

Creating and Communicating Scarcity Digitally

The strategic brand interventions would characteristics to create digital scarcity as opposed to natural one. Among the examples are NFTs or limited-edition digital assets which are typically issued in small numbers or time limited. The smart contracts or the blockchain validation impose these artificial limits hence there is a designated number of tokens that are in circulation and the ownership of the token can be researched and traced down (Alqarni, 2024). There are various ways through which brands can convey scarcity on online fronts. One such tactic is time-restricted availability: the NFTs with the expiry dates, hours, days, produce the effect of urgency and acting. The second course of action will involve reducing the number of digital resources that imply exclusivity but no material product. The other effect of gamification is the increased scarcity. The platforms can also be found on the tiered reward systems, the system of points or access according to activity when the user must be actively consuming to have access to the scarce assets. This promotes the same interaction as the consumers strive to achieve status symbols in the online world. The brands have the tendency of being bound to limited resources and compelling histories of uniqueness, manner of connection with history or social values. In the case when NFT is issued of the well-known artist or event, it can be sold as a unique one-a-time experience. Such stories can be effective in bearing emotional and symbolic values, to the level of apparent value that are well beyond theoretical limits.

Consumer Reactions to Artificial Scarcity and Exclusivity

The effects of digital scarcity on consumer responses are multidimensional, and the responses are usually active. The scarcity is a psychological stimulant that creates a feeling of desire, anticipation and haste to acquire assets. This is particularly high in the digital market, where fear of missing out (FOMO) is promoted by providing information about unlimited access to news and broadcasting it instantly through the social network and the internet (Tandon et al., 2021). Compared to physical scarcity, digital scarcity can be relayed across the globe in real time, and therefore, social and competition pressures are of greater importance. The emotional motivations are also triggered. Owning a rare NFT or a digital collectible gives status and identity and a sense of belonging to an exclusive crowd strengthening self-esteem and brand loyalty. A purchase is not only that the consumers are stealing a product that they are stealing a position in a social narrative that presupposes recognition and status (Milfeld & Flint, 2021). Scarcity does not simply have a behavioral effect on purchasing behavior. More lucky consumers, who managed to acquire a small portion of the digital resources, can resort to brand advocates, where the success is posted on social media, entered community discussion and influenced. This is a vicious cycle where scarcity sparks action and action sparks the perceived scarcity in the social proof. However, not everything is a good answer. One can also be frustrated or experience anxiety due to the absence of access to scarce things when scarcity is created artificially to earn a profit. Too much use of strategies will result in the absence of trust and suspicion. Therefore, scarcity communications must be neither urgent nor confidential, but they must be open and just in order that exclusivity would not be perceived to be unnatural.

Table 4. Digital Scarcity Mechanisms, Consumer Reactions, and Marketing Implications

Scarcity Mechanism	Consumer Psychological Response	Behavioral Outcomes	Marketing Implications
Limited-Edition Digital Assets (e.g., rare NFTs, collectible tokens)	Excitement, FOMO, sense of exclusivity	Quick purchase decisions, higher engagement, repeat visits	Enhance brand prestige, strengthens loyalty, encourages community display of assets
Time-Bound Releases (e.g., flash NFT drops, seasonal digital offers)	Urgency, anticipation, fear of missing out	Impulsive buying, increased platform activity, sharing on social media	Drives short-term sales, creates buzz, amplifies social proof
Gamified Access (e.g., tiered rewards, loyalty points unlocking rare tokens)	Motivation, achievement, pride	Consistent participation, goal-oriented engagement, brand interaction	Encourages long-term engagement, builds brand-community relationships
Narrative Framing / Storytelling (e.g., linking assets to famous artists, cultural events)	Emotional attachment, symbolic value, sense of belonging	Displaying assets, sharing brand stories, increased advocacy	Reinforces brand narrative, enhances perceived authenticity, drives symbolic brand loyalty
Social Verification / Community Scarcity (e.g., peer recognition, limited social availability)	Status seeking, social validation, belonging	Participation in forums, social media posts, peer influence	Strengthens community, encourages organic promotion, boosts brand advocacy

Source: Conceptualized and created by the authors

Strategic Considerations for Marketers

Online scarcity possesses both benefits and problems for the marketers. The sense of scarcity can be leveraged to generate more value, generate engagement, and make the offerings stand out in saturated online markets. Time restriction, first editions and exclusive information and offering possibilities and consumer loyalty can be used to create buzz. Together with online-based structures of ownership, scarcity presents a powerful instrument of-generating brand loyalty and formation of community identity. Rather it is moral and psychological forces-to be manipulated. The Artificially acquired scarcity is not meant to break the-trust and create the image of exploitation. Scarcity should not be eroding brand tales, it should be enhancing them and to do so there has to be articulate communication, demonstrable limits and conformity. The second significant aspect is regulatory compliance-especially on those jurisdictions whereby digital assets, NFTs, or cryptocurrencies are regulated. These weak management aspects are capable of-destroying the brand- equity and putting the firms at risk of litigation.

Psychological and Social Impacts of Digital Scarcity

Digital scarcity may have an impact on cognition, emotion, and social behavior simultaneously. At the cognitive level, consumers are likely to attach greater or equal value of scarce resources that outweigh usefulness which can be termed as scarcity heuristic (Isler et al., 2023). On emotional level, scarcity may cause excitement, pride and even stress, depending on individual and social level of situation. Socially, scarcity creates peer comparison, authenticity and digital status within the community and creates an environment of value generation in the interaction and reward. Moreover, the traditional markets that are not common in the traditional markets are given a sense of collectiveness through digital scarcity. Buying and selling-of-properties online, describing and exchanging properties on the internet-result in community development around rare assets. These processes make scarce more symbolically and socially desirable, which has turned into a marketing commodity, a cultural phenomenon. The digital era has terminated passive scarcity and transformed it into a planned and discussed approach. It spreads its impact on affective involvement, social queueing and adherence in the long term. The rationalization of scarcity through these dynamics is responsible due-to the fact that the knowledge of such dynamics enables marketers to invent meaningful consumer experiences and sustainable competitive advantage in digital ecosystems as a result- of combining psychological knowledge, acting in an ethical way, and strategic thinking.

MARKETING IMPLICATIONS

The emergence of digital money, blockchain resources, and NFTs have transformed the marketing environment radically. The classical paradigms of marketing, from which the material possession of goods, scarcity due to physical plants, and one-way communication of value is emphasized, are gradually becoming inadequate. Within the present digital contexts, the marketing professionals should be conscious of the perceived value, digital ownership and strategically developed scarcity of consumer behavior, loyalty and interaction. This section presents strategic opportunities to leverage these mechanisms and indicates problems of trust, authenticity as well as adherence to regulations.

Strategies for Leveraging Digital Ownership

Digital ownership opens unprecedented possibilities to the brands in connecting with consumers outside the scope of the traditional transactions. The ownership of blockchain-based assets, NFTs, or tokenized products makes it possible to facilitate brands to create participatory experiences, so consumers are transformed into co-creators and not active buyers. Such interaction promotes a feeling of agency and greater attachment because consumers feel that they are active contributors to a digital ecosystem of a brand. The introduction of NFT-based loyalty programs is one of the practical solutions in which consumers can earn or purchase tokens that will grant them access to exclusive content, early releases, or voting rights on brand decisions. The solution transforms the classical loyalty by points into an experience, self-interactive journey that positively affects feelings and sustainable interaction (Kashiramka et al., 2025). One example of how a fashion brand may implement this is by publishing limited editions of virtual items as NFT; to allow their customers to prove they own the property on-line and participate in brand-sponsored events or social-media functionality. The other useful method is tokenized co-creation that allows the consumers to participate in the process of designing digital collectibles, virtual products, or items in-game, and giving them verifiable ownership. This strategy is based on psychological fulfillment of the-ownership, social identification, and creative participation and sets the environment of improved promotion and attachment to the brand. Having a product is not enough but is a dynamic process of consumers being engaged in a shared story and a shared identity to be members of a brand community. The other technology that can be used to keep things active is the gamification of digital assets. The tokens can be attached to milestones, challenges or loyalty advancement where more advanced rewards are pegged on a rare or exclusive piece of digital assets. This turns the interaction between consumers more of an ongoing process of engagement where the scarcity and ownership of the same act to enhance behavior. Gamified ownership, transfers repeat, engagement and sharing of content to create a self-perpetuating platform to gain more visibility and brand equity. Finally, there is social signaling as one of the aspects of digital ownership. Users tend to post NFTs, virtual real estate or tokenized event tickets on social media, and brand recognition is increased organically. Both the effective and normative loyalty can be reinforced with the help of marketing strategies that allow people to display their loyalty publicly, compare it with other peers, and allow social validation (Melnyk et al., 2022). The brands should thus create visual, shareable and symbolically rich digital ownership experiences.

Strategies for Leveraging Digital Scarcity

Scarcity remains a powerful value driver, yet in the digital world, it may be created deliberately to shape consumer attitudes. When used ethically, artificial scarcity creates urgency, exclusivity and drives up emotional investment. Scarcity of NFT releases or tokens release in time creates immediate hype and an indicator of prestige. A combination of the timings of these releases and product release or events are maximizing both the visibility and participation. The other useful tool is tiered scarcity models. The user interaction, loyalty, or community rank can be the basis of the-access to digital assets and will encourage the-further involvement and support the behavior based on rewards. The consumers are not only driven by assets; but also by the social status that the possession of rare digital products will bring (Chen et al., 2025). Scarcity thus-is a motivating factor, which is psychological, and also-a maker of community. This makes it more effective because of its combination with scarcity and storytelling and narrative marketing. Since scarcity can be linked to such meaningful narratives as collaboration with famous artists or cultural events it propagates symbolic meaning and emotional connection. One-way scarce assets are perceived by consumers-is as a loss of resources-but it is also a statement of existence, status and personal success. The scarcity strategies are enhanced by cross-platform. It will never cease to seem that scarcity, as communicated to the social media, playing games, trading within the virtual market, and blockchains registries, makes perceived scarcity visible and believable, contributing to interaction in different online spaces concomitantly. Transparency and verifiability must be applicable to make sure that there is trust and suggest long-term participation.

Challenges: Trust, Authenticity, and Consumer Perception

In-spite of the opportunities, digital ownership and scarcity entail a fair share of challenges. The golden rule is trust as consumers will pass through platforms, blockchain systems, and smart contracts in order-to ensure that ownership is safe and authenticity is guaranteed. Any technical weakness or system instability, including any malpractice that is based on fraud, can destroy perceived value (Lestari et al., 2024). The money should then be used on technological base, security and proper communication to ensure confidence. Authenticity of legitimacy is also necessary. Artificial scarcity, which seems to be artificial, may lead to distrust or negative attitude towards consumers. Marketing models that establish the sense of urgency are getting more popular and the recognition of scarcity models can be counter-productive, being damaging to credibility. To ensure trust and ethical uprightness, transparency, quantifiable limits and brand alignment are required. The other complex issue is digital rights. This confusion of the intellectual-property, rights of use and licensing can leave the consumer dis-enthusiastic when they are given some benefits-but they cannot do so or when there is a secondary restriction in the market. The brands must also communicate the rights of ownership so that they can engage and be loyal.

Regulatory Considerations and Compliance

NFTs, digital assets, and tokenized experiences are operational in the dynamic and complex regulatory spaces. Marketers will have a risk of being unclear regarding taxation, copyright, resale and AML compliance. Otherwise, this can jeopardize the credibility and place companies in a legal predicament (Giamporcaro & Kuk, 2024). The mitigation of the risk will need professional communication with policy makers, constant surveillance, and reporting. Regional differences in regulations also make marketing complex. International brands have-to adjust scarcity and ownership mechanisms to accommodate local laws and regulations and still experience the same consumer experience. It is possible to ensure that the -compliance is not undermined through the design of the flexible platform, the possibility to change terms of contract, and effective communication. The collaboration-with legal and technological partners is required to have a balance between regulation and marketing.

Integrating Digital Ownership and Scarcity for Strategic Advantage

The digital ownership and scarcity must be viewed as an overall marketing approach towards producing the greatest impact. The impact of scarcity can be improved with NFTs and tokenized rewards and provide emotionally intense and socially validated, psychologically dominated experiences. Symbolic and functional values can be supported with the help of gamification, community recognition, and narrative framing. All strategies will be based on transparent design and ethical strategies. The consumers have-to perceive ownership and scarcity as true, verifiable and valuable. Honest exposition of rights, restrictions and clarification of why scarcity exists build confidence and help-to avoid resistance. The value of technology innovation and the value of psychology and ethics should be correlated to enable the marketer transform digital possession and scarcity into sustainable sources of engagement, loyalty, and brand equity. There is a nuance of wisdom of-the impact of digital possession and absence on consumer needs in digital marketing in digital payment systems. The brands ought to take significant steps in ensuring that the digitally empowered consumers have participatory, meaningful and enduring relationships with such brands by appropriately leveraging these features and resolving the issues related to trust, authenticity and regulatory issues. Ownership and scarcity are strategic combination mechanisms that are not just beneficial in short run but also lead to brand resilience in competitive digital markets.

LOOKING AHEAD: EMERGING TRENDS AND UNRESOLVED QUESTIONS

As digital payment systems and blockchain-assets and tokenized experiences keep evolving, the future of marketing and consumer engagement appears to be altering radically. Rapid technological changes and the changing consumer expectations are providing opportunities and challenges to-the brand. It is becoming significant to understand fresh tendencies and introduce to the focus those questions which are left open and unsolved that are of critical importance in drawing predictions about the way consumer behavior, digital ownership, and scarcity perceptions may transform within the coming few years.

Evolution of Digital Ownership Models

Among the most remarkable trends, the contemporary creation of digital models of ownership should be mentioned. The growing popularity of the NFTs usage, tokenized loyalty programs, and other virtual properties is abandoning the niche markets and exploring greater markets (Alshater et al., 2024). New platforms are trying out new ownership models, where consumers have a-partial ownership, co-ownership or time-jurisdictional access to digital assets. This is automatically followed by the question on how the definition, implementation and concept of ownership may be viewed in the future. Besides, the emergence of metaverse space and immersivity of virtual space is transforming the concept of owning a digital asset. The consumers now are dealing with virtual houses, avatars and digital clothing that has both practical and symbolic meaning. Such changes expand the concept of identity, self-expressions, and social recognition and are an indication that marketers must learn to expect new types of attachments and perceived values. However, there is little information on the psychological and social consequences of such ownership, which means that a broad field of research and practical testing remains open.

Advanced Scarcity Strategies and Behavioral Implications

Scarcity, which was one of the fundamental marketing principles, is being transformed digitally. Along with limited-time releases and capped editions, there are the fads such as algorithmically controlled scarcity, dynamic scarcity controlled by consumer behavior, and gamified scarcity systems These measures enable marketers to control the demand, promote the re-occurrence of communication, and build communities around exclusivity (Debenedetti et al., 2025). Nevertheless, the ethical application of these strategies and consumer behavior to such strategies is still a question. Artificial scarcity can indeed increase the sense of value, urgency, but excessive application or over the top manipulative strategies may result in distrust, disappointment, or backlash against the brand. Marketers will have to be able to strike a fine balance between being effective and being ethically responsible as consumers are becoming increasingly digitally savvy and socially conscious. To avoid the negative brand perceptions, it is imperative to understand the long term behavioral and social effects of the engineered scarcity.

Integration of Artificial Intelligence and Predictive Analytics

Digital ownership and payment platforms are commonly being integrated using Artificial Intelligence (AI) and predictive analytics. AI can customize scarcity indicators, predict the willingness of consumers to scarce resources, and offer automation to tokenized loyalty services (Pandey et al., 2025). Brands can also use predictive models to predict how their activities in the second market would work out, determine the elasticity of demand, and manage the engagement strategy on the flight. Nevertheless, there are concerns of transparency, fairness, and trust of the consumers. Put differently, consumers may feel violated in case the opaque algorithmic decisions determine the value perception and purchase behavior. Brands will need to be careful with this amalgamation of technology, ethics and consumer psychology to keep up with the involvement and trust.

Regulatory Uncertainty and Legal Considerations

Digital asset growth is still growing very fast and is far outpacing regulatory frameworks across the world. There are also inconsistencies in terms of issues of taxation, intellectual property, resale rights and anti-money laundering compliance across regions (Walker-Munro et al., 2025). This regulatory uncertainty raises the question of the validity and the viability of digital property transnational progress, specifically. Brands need to be proactive. The potential trends in the future include standardization of the-digital asset regulations, self-regulating options, and more collaboration between the platforms, the policymakers and the consumer advocacy groups. Brands need to be proactive. These changes will not only have an impact on compliance but also on the-consumer trust, adoption rates and the perspective on digital ownership overall. The digital ownership, scarcity and marketing is a massive opportunity, but one that has its fair share of doubts. The trends, which are transforming the expectations and marketing strategies of the consumers, include hyper-real ownership modes, immersive virtual worlds, the use of AI-based personalization, and gamified scarcity. Simultaneously, the deficiency in the solutions to different issues related to ethics, trust, legal frameworks, and sustainability also adds to the significance of research and experimentation. Brands meticulously forced to bypass these technological-innovation and psychological- insights and ethical vision complexities will be at the vanguard of developing strong, participatory and meaningful relationships with digitally empowered consumers.

CONCLUSION

The chapter examines the relationship between the concept of online ownership, the understanding of value, and scarcity and their impacts on consumer-brand relationships within the framework of the existing systems of digital payment. The dissemination of blockchain technologies, NFTs, virtual currencies, and tokenized experiences leads to a paradigm shift in the development of traditional ownership-and value. Shoppers are migrating to spaces where the things are not the-objects but symbolic, social and emotional ones. The reason is that the presence of value ceases to be determined by the economic/functional value, but instead it is multidimensional, i.e. value is characterized as a construct of emotional satisfaction, social recognition as well as technological trust. Consumer-are comparing experience with products and this is one of the fundamental changes in brand communication in addition to the creation and reinforcement of values. Digital ownership has greatly redefined consumer participation. The ownership of NFTs, tokens in blockchain, or any other digital asset will encourage consumers to view themselves as co-creators rather than active receivers. The outcome of this participatory exchange is increased emotional commitment, development of a social identity, and extended interaction, but it also presents the-issues related to rights, authenticity, and perceived fairness. The level of engagement is also increased with scarcity that is either natural or man-made to create a sense of urgency, exclusivity and social signaling. But strategic scarcity should be properly measured to avoid trust-loss and create a sense of overt consumer cynicism. Psychological and behavioral processes support most of these dynamics. Lesser friction in online transactions, gamified ownership and algorithmic mediated scarcity will affect the spending habits, the level of engagement and the ultimate loyalty. Trust technological and relationship is found to be a decisive element that influences perceived value, and consumer-committal. The contemporary consumers do not just watch the holdings they possess but also the transparency, security and ethical governance of the platforms that facilitate the transactions. New trends and the questions that have not been completely resolved indicate that the digital ecosystem is under permanent development. The positive prospects as well as highlighted uncertainties of marketers when it comes to hybrid ownership models, immersive metaverse environments, AI-driven personalization, as well as algorithmic scarcity are both opportunities and concerns that marketers will face in the future. At the same time, the issues of regulatory ambiguity, ethics, and sustainability still affect the ways of perception and execution of digital ownership and scarcity strategies. The sense of value, the possession of the digital and the scarcity in general transform the landscape of connections between the consumers and brands in-the digital payment setting. The brands that adequately integrate these elements as a-equilibrium of technological advancement, psychological insight and moral stewardship will be able to create meaningful, participatory and ultimate relationships with digitally enabled purchasers. It will be the role of maneuvering these multidimensional dynamics and adapting to the current technological and behavioral changes that will dictate the brand sustainability in the changing digital landscape in the long run and its success.

REFERENCES

Alam, S. S., Ahsan, M. N., Kokash, H. A., & Ahmed, S. (2025). Live streaming characteristics, motivational factors, trust and impulsive hotel booking intention: Mediating role of customer engagement and moderating role of time pressure. *Journal of Hospitality Marketing & Management*, *34*(5), 651–680. DOI: 10.1080/19368623.2025.2472071

Alamoudi, H., Glavee-Geo, R., Alharthi, M., Doszhan, R., & Suyunchaliyeva, M. M. (2025). Exploring trust and outcome expectancy in FinTech digital payments: Insights from the stimulus-organism-response model. *International Journal of Bank Marketing*, *43*(4), 897–919. DOI: 10.1108/IJBM-04-2024-0252

Alqarni, A. (2024). A blockchain-based solution for transparent intellectual property rights management: Smart contracts as enablers. *Kybernetes*. Advance online publication. DOI: 10.1108/K-04-2024-1074

Alshater, M. M., Nasrallah, N., Khoury, R., & Joshipura, M. (2024). Deciphering the world of NFTs: A scholarly review of trends, challenges, and opportunities. *Electronic Commerce Research*. Advance online publication. DOI: 10.1007/s10660-024-09900-y

Ante, L., & Fiedler, I. (2025). The new digital economy: How decentralized finance (DeFi) and non-fungible tokens (NFTs) are transforming value creation, ownership models, and economic systems. *Digital Business*, *5*(1), 100094. DOI: 10.1016/j.digbus.2024.100094

Assima, C., Herbert, M., & Robert, I. (2023). Use it or lose it?: Exploring the grey area of dormant possessions and the role of rituals in value dynamics of household objects. *Journal of Marketing Management*, *39*(13–14), 1220–1254. DOI: 10.1080/0267257X.2023.2252455

Baidoun, S. D., & Salem, M. Z. (2024). The moderating role of perceived trust and perceived value on online shopping behavioral intention of Palestinian millennials during COVID-19. *Competitiveness Review*, *34*(1), 125–143. DOI: 10.1108/CR-10-2022-0161

Belk, R., Humayun, M., & Brouard, M. (2022). Money, possessions, and ownership in the Metaverse: NFTs, cryptocurrencies, Web3 and Wild Markets. *Journal of Business Research*, *153*, 198–205. DOI: 10.1016/j.jbusres.2022.08.031

Bilgihan, A., Leong, A. M. W., Okumus, F., & Bai, J. (2024). Proposing a metaverse engagement model for brand development. *Journal of Retailing and Consumer Services*, *78*, 103781. DOI: 10.1016/j.jretconser.2024.103781

Cao, Z., & Shi, X. (2021). A systematic literature review of entrepreneurial ecosystems in advanced and emerging economies. *Small Business Economics*, *57*(1), 75–110. DOI: 10.1007/s11187-020-00326-y

Chauhan, K., & Gulati, K. (2025). Global Customer Engagement Through Gamified Mobile Apps and Virtual Experiences. In *Addressing Practical Problems Through the Metaverse and Game-Inspired Mechanics* (pp. 169–196). IGI Global., DOI: 10.4018/979-8-3693-8886-0.ch008

Chen, H., Yang, Z., & Lyu, T. (2025). Empirical Investigation of Digital Collectibles Purchase Intention: The Roles of Value, Risks, Identification, and Scarcity. *International Journal of Human-Computer Interaction*, *41*(16), 9861–9880. DOI: 10.1080/10447318.2024.2429755

Debenedetti, A., Philippe, D., & Dion, D. (2025). Balancing exclusivity and inclusivity through the strategic domestication of the luxury retail experience. *International Journal of Research in Marketing*, *42*(1), 32–54. DOI: 10.1016/j.ijresmar.2024.05.001

Diep, D. M., & Mai, L. T. T. (2025). How Does Sense of Belonging Bridge the Linkage Between Customer Engagement and Word of Mouth in Online Nonprofit Communities? *Journal of Nonprofit & Public Sector Marketing*, **•••**, 1–24. DOI: 10.1080/10495142.2025.2488775

Duan, Y., & Zhu, Q. (2025). Blockchain empowerment: Enhancing consumer trust and outreach through supply chain transparency. *International Journal of Production Research*, *63*(14), 5358–5382. DOI: 10.1080/00207543.2024.2434951

Dudczyk, P., Dunston, J. K., & Crosby, G. V. (2024). Blockchain Technology for Global Supply Chain Management: A Survey of Applications, Challenges, Opportunities and Implications. *IEEE Access : Practical Innovations, Open Solutions*, *12*, 70065–70088. DOI: 10.1109/ACCESS.2024.3399759

Emon, M. M. H. (2025). Cybersecurity in the Smart City Era: Overcoming Challenges With Modern Cryptographic Solutions. In *Securing Smart Cities Through Modern Cryptography Technologies* (pp. 43–74). IGI Global Scientific Publishing. DOI: 10.4018/979-8-3373-3326-7.ch002

Emon, M. M. H., Nath, A., & Nipa, M. N. (2025). *AI for a Greener Tomorrow: Harnessing Artificial Intelligence for Environmental Sustainability*. IGI Global Scientific Publishing., DOI: 10.4018/979-8-3373-2387-9.ch010

Frohmann, F. (2023). *Digital Pricing*. Springer International Publishing., DOI: 10.1007/978-3-031-24591-6

García-Monleón, F., Danvila-del-Valle, I., & Lara, F. J. (2021). Intrinsic value in crypto currencies. *Technological Forecasting and Social Change*, *162*, 120393. DOI: 10.1016/j.techfore.2020.120393

Giamporcaro, S., & Kuk, G. (2024). Affordances-in-practice: Unravelling blockchain-for-good in sustainable investment. *Information Technology & People*. Advance online publication. DOI: 10.1108/ITP-05-2022-0349

Guzmán Martínez, F. J., & Orozco Gómez, M. M. (2024). The emotional side of price: Reference price as a decision maker in consumer purchase. *International Journal of Retail & Distribution Management*, *52*(7/8), 771–785. DOI: 10.1108/IJRDM-08-2023-0504

Hassoun, A., Beacock, I., Carmody, T., Kelley, P. G., Goldberg, B., Kumar, D., Murray, L., Park, R. S., Sarmadi, B., & Consolvo, S. (2025). Beyond Digital Literacy: Building Youth Digital Resilience Through Existing "Information Sensibility" Practices. *Social Sciences (Basel, Switzerland)*, *14*(4), 230. DOI: 10.3390/socsci14040230

Ho, C.-I., Liu, Y., & Chen, M.-C. (2022). Antecedents and consequences of consumers' attitudes toward live streaming shopping: An application of the stimulus–organism–response paradigm. *Cogent Business & Management*, *9*(1), 2145673. Advance online publication. DOI: 10.1080/23311975.2022.2145673

Hobeika, J., Liew, C. Y., & Rajan, M. E. S. (2025). Are cryptocurrencies influenced by stereotypes and values? Evidence of the role of banker stereotypes and human values on cryptocurrency acceptance. *International Journal of Bank Marketing*, *43*(8), 1627–1661. Advance online publication. DOI: 10.1108/IJBM-04-2024-0181

Huang, L., Gao, B., & Gao, M. (2023). *Value Realization in the Phygital Reality Market*. Springer Nature Singapore., DOI: 10.1007/978-981-99-4129-2

Irimia-Diéguez, A., Liébana-Cabanillas, F., Blanco-Oliver, A., & Lara-Rubio, J. (2025). What drives consumers to use P2P payment systems? An analytical approach based on the stimulus–organism–response (S-O-R) model. *European Journal of Management and Business Economics*, *34*(2), 149–173. DOI: 10.1108/EJMBE-12-2022-0374

Isler, O., Yilmaz, O., & Dulleck, U. (2023). Scarcity improves economic valuations when cognitively salient. *Journal of Economic Behavior & Organization*, *209*, 362–371. DOI: 10.1016/j.jebo.2023.02.019

Karmaker, S., Oishi, M. E. F., Qasem, A., Sami, S. B. S., & Noor, J. (2025). Exploring influential factors of consumer purchase behavior on the adoption of digital payment apps in Bangladesh. *Computers in Human Behavior Reports*, *17*, 100587. DOI: 10.1016/j.chbr.2025.100587

Kashiramka, S., Gupta, P., Bali, A., & Guha, A. (2025). *Gamifying Loyalty*., DOI: 10.4018/979-8-3373-0149-5.ch011

Kumar, J. (2021). Understanding customer brand engagement in brand communities: An application of psychological ownership theory and congruity theory. *European Journal of Marketing*, *55*(4), 969–994. DOI: 10.1108/EJM-04-2018-0290

Kumari, N., & Biswas, A. (2023). Does M-payment service quality and perceived value co-creation participation magnify M-payment continuance usage intention? Moderation of usefulness and severity. *International Journal of Bank Marketing*, *41*(6), 1330–1359. DOI: 10.1108/IJBM-11-2022-0500

Lee, H., Rana, M. R. I., & Xu, Y. (2024). Transforming luxury: Young consumers' motivations towards purchasing virtual luxury non-fungible token wearables. *International Journal of Retail & Distribution Management*, *52*(12), 1208–1228. DOI: 10.1108/IJRDM-12-2023-0712

Lestari, S., Adawiyah, W. R., Alhamidi, A. L., Prayogi, J., & Haryanto, R. (2024). Navigating perilous seas: Unmasking online banking frauds, perceived usefulness, fear of cybercrime and distrust in online banking. *Safer Communities*, *23*(4), 444–464. DOI: 10.1108/SC-04-2024-0018

Lin, K.-Y., Wang, Y.-T., & Huang, T. K. (2020). Exploring the antecedents of mobile payment service usage. *Online Information Review*, *44*(1), 299–318. DOI: 10.1108/OIR-05-2018-0175

Loutfi, S. A., Jose, S., Jabeen, F., & Piccardi, P. (2025). A stimulus-organism-response (SOR) perspective on healthcare telemedicine usage towards sustainable development. *Management of Environmental Quality*, *36*(7), 1848–1870. DOI: 10.1108/MEQ-01-2025-0015

Luangrath, A. W., Peck, J., Hedgcock, W., & Xu, Y. (2022). Observing Product Touch: The Vicarious Haptic Effect in Digital Marketing and Virtual Reality. *JMR, Journal of Marketing Research*, *59*(2), 306–326. DOI: 10.1177/00222437211059540

Madine, M., Salah, K., Jayaraman, R., Battah, A., Hasan, H., & Yaqoob, I. (2022). Blockchain and NFTs for Time-Bound Access and Monetization of Private Data. *IEEE Access : Practical Innovations, Open Solutions*, *10*, 94186–94202. DOI: 10.1109/ACCESS.2022.3204274

Mariana, C. D., Husodo, Z. A., Ekaputra, I. A., & Fahlevi, M. (2025). The advancement of digital payment ecosystem in metaverse: A literature review. *Computers in Human Behavior Reports*, *17*, 100570. DOI: 10.1016/j.chbr.2024.100570

Melnyk, V., Carrillat, F. A., & Melnyk, V. (2022). The Influence of Social Norms on Consumer Behavior: A Meta-Analysis. *Journal of Marketing*, *86*(3), 98–120. DOI: 10.1177/00222429211029199

Milfeld, T., & Flint, D. J. (2021). When brands take a stand: The nature of consumers' polarized reactions to social narrative videos. *Journal of Product and Brand Management*, *30*(4), 532–548. DOI: 10.1108/JPBM-10-2019-2606

Moro-Visconti, R., & Cesaretti, A. (2023). FinTech and Digital Payment Systems Valuation. In *Digital Token Valuation* (pp. 411–458). Springer Nature Switzerland., DOI: 10.1007/978-3-031-42971-2_13

Nästesjö, J. (2023). Managing the rules of recognition: How early career academics negotiate career scripts through identity work. *Studies in Higher Education*, *48*(4), 657–669. DOI: 10.1080/03075079.2022.2160974

Nguyen, M.-H., Duong, M.-P. T., Tran, T. T., Sari, N. P. W. P., Tedjasuksmana, H., La, V.-P., & Vuong, Q.-H. (2025). Beyond trust: How usefulness and immersiveness drive space tourism intentions in high-risk contexts. *Current Issues in Tourism*, •••, 1–23. DOI: 10.1080/13683500.2025.2487634

Ofori, K. S., Anyigba, H., Adeola, O., Junwu, C., Osakwe, C. N., & David-West, O. (2022). Understanding post-adoption behaviour in the context of ride-hailing apps: The role of customer perceived value. *Information Technology & People*, *35*(5), 1540–1562. DOI: 10.1108/ITP-06-2019-0285

, Natarajan, T., & Dhalmahapatra, K. (2025). Adoption of NFT transaction in metaverse platform: a trust transfer approach to leverage trust among users. *Digital Policy, Regulation and Governance, 27*(4), 422–445. DOI: 10.1108/DPRG-08-2024-0178

Pandey, A. K., Sawhney, A., Sharma, G., & Singh, D. (2025). Web3 and Metaverse in the Online Marketing for Consumer Support and Prediction. In *Practical Strategies and Case Studies for Online Marketing 6.0* (pp. 369–388). IGI Global Scientific Publishing., DOI: 10.4018/979-8-3373-2058-8.ch016

Pangarkar, A., & Shukla, P. (2023). Conspicuous and inconspicuous consumption of luxury goods in a digital world: Insights, implications, and future research directions. *International Journal of Advertising*, *42*(7), 1226–1238. DOI: 10.1080/02650487.2023.2246260

Paraman, P., Annamalah, S., Ahmed, S., & Baharein, K. (2024). Unveiling the Nexus: Disentangling the Impact of Touch Screen Frequency in Electronic Media on Amplifying Control Perception, Elevating Psychological Ownership, and Igniting Intentions to Reuse Payment Services. *International Journal of Human-Computer Interaction*, *40*(22), 7145–7161. DOI: 10.1080/10447318.2023.2261733

Pasirayi, S., & Fennell, P. B. (2023). The Effect of Mobile Payments on Retailer Firm Value: The Moderating Role of Promotions, Customer Segment, and Rollout Strategy. *Journal of Interactive Marketing*, *58*(1), 90–107. DOI: 10.1177/10949968221134805

Putrevu, J., & Mertzanis, C. (2024). The adoption of digital payments in emerging economies: Challenges and policy responses. *Digital Policy. Regulation & Governance*, *26*(5), 476–500. DOI: 10.1108/DPRG-06-2023-0077

Shahid, Z. A., Tariq, M. I., Paul, J., Naqvi, S. A. N., & Hallo, L. (2024). Signaling theory and its relevance in international marketing: A systematic review and future research agenda. *International Marketing Review*, *41*(2), 514–561. DOI: 10.1108/IMR-04-2022-0092

Stagg, K., Douglas, J., & Iacono, T. (2023). Living with stroke during the first year after onset: An instrumental case study exploring the processes that influence adjustment. *Disability and Rehabilitation*, *45*(21), 3610–3619. DOI: 10.1080/09638288.2022.2131005 PMID: 36222354

Tan, G. K. S.TAN. (2025). Assetizing the video game: Play-to-earn (P2E) games and blockchain rentiership. *Progress in Economic Geography*, *3*(1), 100036. DOI: 10.1016/j.peg.2025.100036

Tandon, A., Dhir, A., Almugren, I., AlNemer, G. N., & Mäntymäki, M. (2021). Fear of missing out (FoMO) among social media users: A systematic literature review, synthesis and framework for future research. *Internet Research*, *31*(3), 782–821. DOI: 10.1108/INTR-11-2019-0455

Treiblmaier, H., & Petrozhitskaya, E. (2023). Is it time for marketing to reappraise B2C relationship management? The emergence of a new loyalty paradigm through blockchain technology. *Journal of Business Research*, *159*, 113725. DOI: 10.1016/j.jbusres.2023.113725

Walker-Munro, B., Ferrill, J., & Tiwari, M. (2025). Sensitive research & anti-money laundering: A possible marriage of convenience? *Current Issues in Criminal Justice*, *37*(2), 308–333. DOI: 10.1080/10345329.2025.2467483

Wang, X., Sung, B., & Phau, I. (2022). Examining the influences of perceived exclusivity and perceived rarity on consumers' perception of luxury. *Journal of Fashion Marketing and Management*, *26*(2), 365–382. DOI: 10.1108/JFMM-12-2020-0254

Xiao, G., Lee, M., Lee, C.-K., & Kim, M. (2025). Tourism live streamers as brand ambassadors: Investigating their impact on destination brand value through social exchange theory. *Journal of Hospitality and Tourism Technology*, *16*(1), 174–193. DOI: 10.1108/JHTT-11-2023-0371

Zarifis, A., & Cheng, X. (2022). The business models of NFTs and fan tokens and how they build trust. *Journal of Electronic Business & Digital Economics*, *1*(1/2), 138–151. DOI: 10.1108/JEBDE-07-2022-0021

KEY TERMS AND DEFINITIONS

Artificial Scarcity: A marketing tactic where the supply of digital assets is reduced purposely resulting in exclusivity, urgency and perceived value. This strategy usually depends on time, amount, or even a game engine.

Consumer Engagement: How consumers are cognitively, emotionally, and behaviorally invested in their relationships with a brand or its digital ecosystem. It can be classified as engagement by participating in or creating content using digital assets or actively using them.

Digital Ownership: It is a process where consumers make subjective judgments of the values and significance of a product, service or experience. It goes beyond the economic utility and incorporates emotion gratification, social acceptance and belief of technology systems.

NFTs (Non-Fungible Tokens): Uniqueness Digital objects (stored in a blockchain) which convey ownership of a particular digital asset, collector or experience. This means that each token is unique and cannot be traded on a one-to-one system such as money.

Scarcity: It is a situation where a product or an asset is restricted or exclusive. Scarcity can be real or artificially induced to increase the asset value, desire or urgency of possession.

Sustainability Reporting: Sustainability reporting refers to the act of recording and reporting about the performance of organizations regarding environmental, social and governance (ESG) principles. It helps organizations to be accountable, monitor the efficiency of resources, and display the progress on the way to the goals of the circular economy and sustainable development.

Technological Trust: The level of trust a consumer has in digital platforms, systems and infrastructure as a way of ensuring security, authenticity and reliability of operations. Mistrust can cause adoption and engagement.

Tokenized Assets: The digital tokens that are being expressed as representations of tangible or intangible assets and are put down on a blockchain or another digital registry. Fractional ownership or ease of transferring value is made possible through these tokens.

Value Perception: Refers to the process through which consumers evaluate subjectively the values and importance of a product, service, or experience. It transcends economic utility and adds emotional gratification, social acceptance and trust in technology systems.

Chapter 9
The Impact of AI-Driven User Experience in Digital Payment Systems:
A Nordic Gen Z Perspective

Bandana Thapa
https://orcid.org/0009-0000-1675-6978
Arcada University of Applied Sciences, Finland

Lana Beikverdi
https://orcid.org/0000-0001-6315-0135
Arcada University of Applied Sciences, Finland

ABSTRACT

This chapter explores the impact of Artificial Intelligence (AI) integration on the User Experience (UX) within digital payment systems, focusing on Nordic Generation Z consumers. Using the Stimulus–Organism–Response (S-O-R) model, the study draws on qualitative data from Finnish participants to trace the consumer journey from technological stimuli to internal cognitive states and behavioural responses. The findings show that Gen Z views AI as an essential element of modern financial services, setting high expectations for performance and convenience. Participants balance the efficiency of automation against potential financial risks and conditional trust. A distinctive cultural element shapes these responses: Finnish consumers prefer AI-driven self-service autonomy as long as human assistance remains available. These insights underscore the importance of AI-driven payment experiences that are culturally sensitive and capable of fostering user trust.

DOI: 10.4018/979-8-3373-4526-0.ch009

INTRODUCTION

The financial services sector has undergone a profound transformation, moving beyond basic digital transactions to an era dominated by AI-driven Digital Payment Systems (Ramrakhyani & Shrivastava, 2024). This shift elevates customer experience by integrating intelligent capabilities across various functions, including personalised recommendations, fraud detection, and virtual support (Nayak, 2025; Chauhan et al., 2025). From a UX perspective, these AI applications enable anticipatory, context-aware interactions that reduce user effort and enhance overall satisfaction (Alexander & Sarrafzadeh, 2004). This technological evolution is rapidly redefining how consumers engage with payment services globally.

To effectively analyse the multifaceted interaction between AI features and user acceptance, this chapter employs the established Stimulus–Organism–Response model (Mehrabian & Russell, 1974). The S-O-R model provides a structured approach to understand how the external "stimulus" such as AI features, influences the individual's internal cognitive and emotional "organism" state (e.g., trust, risk perception, ease of use), ultimately shaping their behavioural "response" (Irimia-Diéguez et al., 2025; Elangovan & Babu, 2024). Key to this process are mediating factors such as trust and outcome expectancy, which significantly shape adoption intentions in digital payment contexts (Glavee-Geo et al., 2025).

While extensive research exists on UX in digital payments, few studies specifically investigate the perspective of Gen Z, particularly in technologically mature markets. Gen Z, as digital natives, possesses unique expectations, demanding hyper-personalised, frictionless experiences and exhibiting low tolerance for delays (Gutfreund, 2016). The Nordic countries, including Finland, represent a crucial context for this study due to their advanced digital infrastructure and high adoption rates (Arvidsson, 2019). Although Gen Z engages heavily in digital financial activity, concerns about data privacy and perceived risk remain relevant (Xu et al., 2010). Thus, a gap exists in understanding how the complex AI stimulus influences the trust and behaviour of this critical demographic in this specific, advanced setting.

This chapter aims to address this gap and answer a research question: *How do complex AI-driven features of digital payment systems influence the perceptions and behavioural responses of Nordic Generation Z consumers?* More specifically, the objective of this chapter is to examine the impact of AI-powered digital payment systems on the trust, satisfaction, and subsequent behaviour of Nordic Gen Z consumers. By linking qualitative data to the S-O-R framework, this study aims to contribute both theoretically, by demonstrating the influence of a cultural moderator (the Nordic context) on the S-O-R chain, and managerially, by providing actionable insights for designing trustworthy and effective AI-enabled payment services. The subsequent sections will detail the theoretical foundations, present the empirical findings, discuss the results, and outline the core conclusions and limitations.

THEORETICAL BACKGROUND

Transformation of Digital Payments and AI

The financial world has undergone a profound transformation. While digital payment systems once referred to simple card transactions and online bank transfers, we are now witnessing the rise of Artificial Intelligence and AI-driven payment systems that redefine customer experience in powerful new ways. Recent research highlights that digital transformation has significantly expanded the use of digital payments, improving both financial inclusion and operational efficiency across the EU and Baltic regions (Mavlutova et al., 2022).

Earlier generations of digital payments offered convenience and speed, yet they often lacked adaptability and intelligence. The current wave of financial technologies (Fintech), however, integrates AI across nearly every function, from fraud detection and risk scoring to chatbots, virtual assistants, personalised recommendation engines, and biometric or voice authentication (Ramrakhyani & Shrivastava, 2024). These capabilities not only improve accuracy and efficiency but also transform how consumers interact with financial services, fostering more dynamic and personalised engagement (Nayak, 2025).

AI also strengthens security through advanced biometric verification and data analytics, making digital payments more reliable and resilient (Mahida et al., 2024). At the same time, it enhances customer experience by offering predictive insights, real-time fraud alerts, and individually tailored services (Manta et al., 2025). In mobile banking, for instance, AI is reshaping service delivery and redefining the customer's role, creating new forms of value centred on efficiency and seamless interaction (Manser Payne et al., 2021).

From a user experience perspective, AI-driven payment systems enable anticipatory, context-aware interactions (Chauhan et al., 2025). Instead of passively responding to user requests, intelligent systems can predict needs, detect irregularities, and guide customers through smoother transactions. By adapting interfaces and prompts to behavioural cues, AI reduces effort and increases satisfaction (Alexander & Sarrafzadeh, 2004). Because fintech platforms collect rich behavioural and contextual data, they empower marketers and UX designers to provide hyper-personalised, secure, and timely services that match users' preferences and habits (Para et al., 2025).

Theoretical Perspectives of User Experience in Digital Payment Research

Prior studies concerning user experience in digital payment systems have adopted various theories in different contexts. One of the fundamental theories used is the Technology Acceptance Model (TAM) introduced by Davis (1989). TAM provides a systematic framework for explaining payment technology acceptance, emphasising the roles of perceived usefulness and perceived ease of use. For instance, one of the studies applied TAM in the tourism and hospitality context and demonstrated that tourists' adoption of digital payment systems is shaped by prior technological experience and social influence, which in turn influence perceived usefulness and perceived ease of use, ultimately determining attitudes toward use and behavioural intention (Kaur & Walia, 2020). Another example of TAM research examining digital payment adoption and user experience investigates rural women's acceptance of user-assisted kiosks and their engagement with digital government services (Soni & Mitchell, 2022). Building on the original TAM framework, Venkatesh, and Bala (2008) introduced TAM3, which incorporates additional constructs, including social influence and facilitating conditions. For example, adoption and use of mobile payment services are predicted by behavioural intentions shaped by perceived usefulness, perceived ease of use, and subjective norm (Jaradat & Al-Mashaqba, 2014). Another theory that builds on TAM and is widely applied in the context of digital payment adoption is the Unified Theory of Acceptance and Use of Technology (UTAUT). UTAUT extends the TAM framework by integrating constructs from multiple user acceptance theories, thereby providing a more comprehensive model that incorporates performance expectancy, effort expectancy, social influence, and facilitating conditions to enable a richer evaluation of technology adoption contexts (Venkatesh et al., 2003). For example, a recent study shows that performance expectancy, facilitating conditions, trust, and perceived innovativeness significantly influence digital payment adoption intention among Indonesian private bank users (Arianita et al., 2023). Another application of UTAUT concerns the baby boomers and X generation group, and shows that UTAUT indicators positively affect user satisfaction, which in turn affects inertia and continuance intention in digital payments (Santosa et al., 2021).

Although the theories like TAM, TAM3 and UTAUT have been widely used to explain technology adoption, their scope is largely focused on individual perceptions and relatively linear relationships. As a result, they are less well suited to capturing the wider environmental and contextual influences that shape users' internal experiences and subsequent behaviours. In response to these limitations, this study adopts the Stimulus–Organism–Response (SOR) framework as its primary analytical perspective. The S-O-R framework (Mehrabian & Russell, 1974) is an important theoretical lens that has broad applications in consumer behaviour, organisational behaviour and leadership studies. The S-O-R model suggests that the environment provides stimuli (S) that influence people's inner states (O), and these inner states then lead to either approach or avoidance responses (R). This process inside the person can happen consciously or unconsciously and involves how they notice and interpret the environment, which then affects their feelings. The S-O-R framework helps marketers create effective strategies and check how satisfied (Shamim et al., 2024) and loyal (Shahzad et al., 2024) consumers are.

Compared with traditional technology acceptance models, S-O-R offers several advantages. It explicitly accounts for external stimuli, thereby enabling the inclusion of technological, social, and environmental influences on user behaviour. In addition, the organism component allows for a more comprehensive examination of internal processes, encompassing cognitive, emotional, and experiential states rather than focusing narrowly on perceived usefulness or ease of use. Finally, the S-O-R framework supports more complex and indirect relationships between stimuli and behavioural responses, making it particularly well suited to analysing adoption in contemporary digital and service contexts characterised by diverse experiences and dynamic environments.

The Stimulus–Organism–Response Model in UX Studies and Consumer Research

In studies of digital payments and financial technology, the S-O-R framework is widely used to explain why people adopt and continue using these systems. For instance, Glavee-Geo et al. (2025) found that in digital payments, expected benefits and trust are key drivers of user responses, with trust having the strongest effect on whether users recommend a service. Similarly, Elangovan and Babu (2024) showed that features such as ease of use, credibility, and availability build trust, satisfaction, and engagement, which then encourage continued use and positive word of mouth. Overall, these findings suggest that the S-O-R framework is well-suited to studying AI-based payment applications. The framework explains how system features influence users' thoughts and feelings, which then shape their actions. For example, AI features such as personalised payment suggestions affect users' trust, comfort, and sense of ease, and these reactions influence whether users adopt the system, keep using it, or recommend it to others. Previous research supports this view, showing that transparent and easy-to-understand AI systems increase trust and usage, and that perceived usefulness is central to digital payment adoption (Ben David et al., 2021). For these reasons, the S-O-R framework offers a clear way to link AI payment features with user perceptions and behaviours. Table 1 summarises the main factors studied within the S-O-R framework in the wider consumer behaviour literature.

Table 1. S-O-R model in consumer behaviour

S-O-R model	Factors	Authors
Stimulus	Product-related factors (quality, design, packaging)	Kotler and Keller, (2016); Kim and Lennon (2013)
	Price factors (discounts, fairness, value)	Xia et al., 2004
	Promotion factors (advertising, reviews)	Belch and Belch, (2021); Imbayani and Gama (2024)
	Place/environment factors (store layout, ambience, website design)	Mehrabian and Russell, (1974); Turley and Milliman, (2000)
	Brand image (reputation, trust, CSR)	Kim and Lennon (2013); Yadav et al. (2025)
	Social factors (peer influence, social media, word of mouth)	Cheung and Thadani, (2012); Li et al. (2025)
Organism	Cognitive factors (awareness, beliefs, attitudes)	Ling et al. (2022); Sultan et al. (2021)
	Affective factors (emotions)	Blanco-Encomienda and Rosillo-Díaz (2025)
	Motivational factors (needs, values)	Jiang and Phoong (2024); Shih et al. (2024)
	Personal factors (personality, lifestyle, experience)	Teh et al. (2014); Hewei and Youngsook (2022)
	Psychological state (involvement, mood)	Khan et al. (2022); Ahmad et al (2019)
Response	Purchase behaviour (buying, switching, impulse buying)	Yaqub et al. (2023); Ahmad et al (2019)
	Post-purchase behaviour (satisfaction)	Kim and Park (2019)
	Loyalty (repeat purchase, advocacy)	Tordoya-Espinoza et al (2025); Wu and Li (2018)
	Engagement (customer engagement, brand communities, social sharing)	Changani and Kumar (2023)

Stimuli: AI-Driven Features in Digital Payment Systems

In AI-based payment systems, personalisation and prediction play an important role in how people use and manage their money. By using information about past spending, transaction history, and context, these systems offer helpful and personalised suggestions, such as suitable payment methods, relevant offers, or simple budget reminders, which make daily financial decisions easier to handle (Kanaparthi, 2024). At the same time, AI enables fast and smooth interactions through instant payments, one-click checkouts, and biometric login, meeting users' expectations for speed and low effort (Ramrakhyani & Shrivastava, 2024). In the background, automated tools support users by providing real-time help through chatbots, sending reminders, and monitoring transactions for possible risks. Clear communication, visible privacy information, and explainable AI features also help build trust and reduce concerns about security and data use (Balasubramaniam, 2023; Yeo et al., 2025). Within the S-O-R framework, these features function as stimuli, that is, external cues that shape users' internal evaluations and emotions, which then guide behavioural responses. In digital payment contexts, such stimuli include interface design, transaction speed, and AI-enabled support tools, all of which influence perceptions such as trust, satisfaction, and engagement, and ultimately affect adoption, continued use, and recommendation intentions. Empirical evidence supports this logic, showing that high-quality AI interactions, such as effective chatbots, enhance user satisfaction and increase advocacy behaviour (Elayat & Elalfy, 2025).

Organism: Consumer Perceptions and Emotional Responses

Within the S-O-R framework, users' thoughts, and emotions act as the link between system features and behaviour. Design, transaction speed, personalisation, and security influence how users feel about AI-based payments. Positive views lead to positive emotions, higher satisfaction, and loyalty, while uncertainty or privacy concerns create negative emotions and disengagement. Feeling comfortable and trusting the system is, therefore, essential for the long-term use of AI-enabled payment services (Afroogh et al., 2024). For example, in AI-based payment systems, trust depends on how users judge system performance, especially accuracy, consistency, and speed, and on whether the technology clearly improves the payment experience. Trust can be built slowly through repeated use when digital payment applications work well and make logical decisions. However, it can be quickly reduced due to mistakes, unclear decisions, or signs of bias (Afroogh et al., 2024; Bach et al., 2023). Concerns about data misuse increase, and willingness to use the system declines when transparency is low (Aldboush et al., 2023). Furthermore, if consumers experience positive feelings, while using the app, such as confidence, enjoyment, and a sense of control their satisfaction and loyalty towards the application will increase, while negative feelings, such as anxiety, doubt, or frustration, weaken trust and reduce continued use (Abikari et al., 2023; Gerli et al., 2022; Valor et al., 2022).

Response: Behavioural Outcomes of Gen Z consumers

Within the S-O-R model, adoption, usage, retention, and switching behaviours represent the "response" stage, where users' actions reflect how they react to the technological and psychological stimuli of AI-driven payment systems. In this context, adoption and initial acceptance occur when external factors such as ease of use, trust, and perceived usefulness influence users' internal states, including satisfaction and confidence (Al-Qudah et al., 2024; Ben David et al., 2021). When users view AI payment systems as secure, transparent, and supportive, their sense of trust and control grows, making them more likely to try and adopt the system (Bajunaied et al., 2023). After adoption, usage behaviour and frequency reflect how these positive internal reactions are reinforced over time. When the system consistently offers accurate recommendations, smooth performance, and convenience across different contexts such as online shopping or peer-to-peer transfers, satisfaction and enjoyment are maintained, leading users to continue using the service regularly (Antonio et al., 2024). Frequent and consistent use, therefore, becomes a natural behavioural response to ongoing positive experiences. As engagement deepens, retention and loyalty develop as longer-term responses. These behaviours are shaped by emotional satisfaction, cognitive trust, and habit formation, as users come to rely on the service as a dependable part of their financial routines (Rahardja et al., 2023; Q. Zhang et al., 2023). In S-O-R terms, retention shows that positive internal states such as trust, comfort, and emotional attachment have stabilised, creating habitual use and long-term loyalty. By contrast, switching or abandonment represents a negative behavioural response that arises when unfavourable experiences destroy users' trust and satisfaction (Ahmed et al., 2025). When perceived risks outweigh the benefits, users' attitudes turn negative, leading them to switch to alternative providers or stop using AI-based payment platforms altogether.

Positioning UX of Digital Payment Systems in Global and Nordic Contexts

Prior research on the user experience of digital payment systems shows that experience is strongly shaped by national context. On the one hand, studies share similar measures such as ease of use, perceived usefulness, trust, perceived security, and satisfaction (Davis, 1989; Gefen et al., 2003; Zhou, 2014). On the other hand, there are some country-specific factors that differ. For example, in East Asian countries, particularly China, prior research explores mobile payment platforms such as Alipay and WeChat Pay, typical for this region. Studies show that users' experience is closely linked to trust, perceived security and convenience within digital ecosystems that are highly integrated in daily operations (Zhou, 2014; Gao et al., 2015). In India, research focuses heavily on the Unified Payments Interface (India's real-time, instant payment system developed by NPCI1) and mobile payment applications built on top of it. Studies specifically focus on ease of use, reliability, and trust as key elements of user experience, especially in the context of continued use rather than first-time adoption (Shankar & Datta, 2018; Pal et al., 2020). Southeast Asian research, including studies from Indonesia and Malaysia, commonly examines e-wallet services. In these studies, user experience is frequently linked to service quality, convenience, and satisfaction (Oliveira et al., 2016). In Latin America, particularly Brazil, research explores instant payment systems such as Pix. User experience in this context is discussed mainly through speed, accessibility, and perceived security (Malaquias & Silva, 2020). In contrast, European and Middle Eastern studies focus on contactless and mobile payments, particularly during the COVID-19 pandemic, when user experience was shaped by health concerns, perceptions of safety, reduced physical contact, traditional usability, and security considerations (Liébana-Cabanillas et al., 2017).

Prior research on digital payment user experience in the Nordic region is uneven and focuses mainly on Sweden and Denmark. Sweden is the most frequently studied country in Nordic journal research, largely because cash use has declined rapidly, and mobile payments are widely used. Early studies from Sweden show that perceived ease of use and trust play an important role in whether people are willing to start using mobile payment services (Arvidsson, 2014). Later research shows that positive first use does not always lead to full acceptance. Even users who already rely on digital payments can experience ongoing problems. Dimitrova et al. (2021) find that Swedish bank customers may resist fully switching to digital payments because they feel a loss of control, have privacy concerns, or worry about becoming too dependent on technical systems. This suggests that in Sweden, a good user experience is not only about smooth and efficient interaction. It also includes feeling in control, trusting institutions, and believing that the system will work reliably in all situations. Broader, macro-level studies help explain why these expectations are particularly strong in Sweden. As the country has moved toward a cash-light society, users have come to expect digital payment systems to be highly reliable and always available. Backup solutions and system resilience are now part of how users judge their experience, rather than being seen as invisible background infrastructure (Armelius et al., 2022).

Research from Denmark looks at user experience in a different way. Instead of focusing on how people interact with payment interfaces, Danish studies mainly examine how mobile payment platforms develop and compete over time. Staykova and Damsgaard (2015) show that when Danish mobile payment services expanded from person-to-person payments to in-store and online payments, users perceived greater value. This wider use reduced friction in everyday payment situations and helped users build strong habits. In later work, the same authors explain user experience through the ideas of "reach" and "range". They argue that a payment system feels more useful and seamless when many people use it and when it works across many different payment situations (Staykova & Damsgaard, 2016).

Compared with Sweden and Denmark, Finland receives far less attention in country-specific journal studies. Although mobile payments are widely used in Finland, the country is usually included only as part of broader Nordic or European comparisons. As a result, there is limited journal-based evidence on how Finnish users experience digital payment systems in everyday use.

The Role of Nordic Gen Z Characteristics

As digital natives, Gen Z expects personalised and efficient customer service, real and relatable content, and interactive experiences from brands, valuing customisation and active participation (Gutfreund, 2016). They regard friction or delays not just as irritations but as deal breakers. Their experience is framed by algorithms that "know" them: they expect suggestions, automations, and smart defaults. In this light, elements like trust, satisfaction and ease of use remain central to shaping behaviour: users must feel confident that AI-driven systems are reliable, transparent, interpretable, and respectful of privacy (Al Qudah et al., 2024; Kumar & Singh, 2024). Empirical evidence confirms that well-designed UX not only enhances usability but also fosters long-term engagement, favourable attitudes, and loyalty (Susilawati et al., 2024; Agarwal et al., 2023).

Despite significant research on UX in digital payments, few studies zoom in on Gen Z's interaction with AI-enabled payments, particularly in technologically mature settings from consumer experience perspective. The Nordic countries, including Finland, represent ideal contexts: they feature high adoption of digital payments, advanced infrastructure, and strong institutions (Arvidsson, 2019). Yet, even in these settings, Gen Z expresses concern about data privacy despite engaging heavily in digital financial activity (Alić & Sopić, 2023).

Conceptual AI-Driven UX Framework for Digital Payments among Nordic Gen Z Consumers

To bring together earlier research on AI-driven user experience in digital payments for Nordic consumers, this study proposes a clear conceptual model based on the S-O-R framework. The framework connects AI features, user experience elements, marketing outcomes, and the Nordic Gen Z context. As shown in Figure 1, AI features act as triggers that shape how users think and feel when using digital payment services. These thoughts and feelings then influence how users behave and how they respond to marketing activities. This process is influenced by the high-trust environment in Nordic countries, which encourages Gen Z users to prefer digital payment experiences that are autonomous, smooth, and require little effort.

Figure 1. AI-Driven User Experience Framework for Digital Payments among Nordic Gen Z Consumers

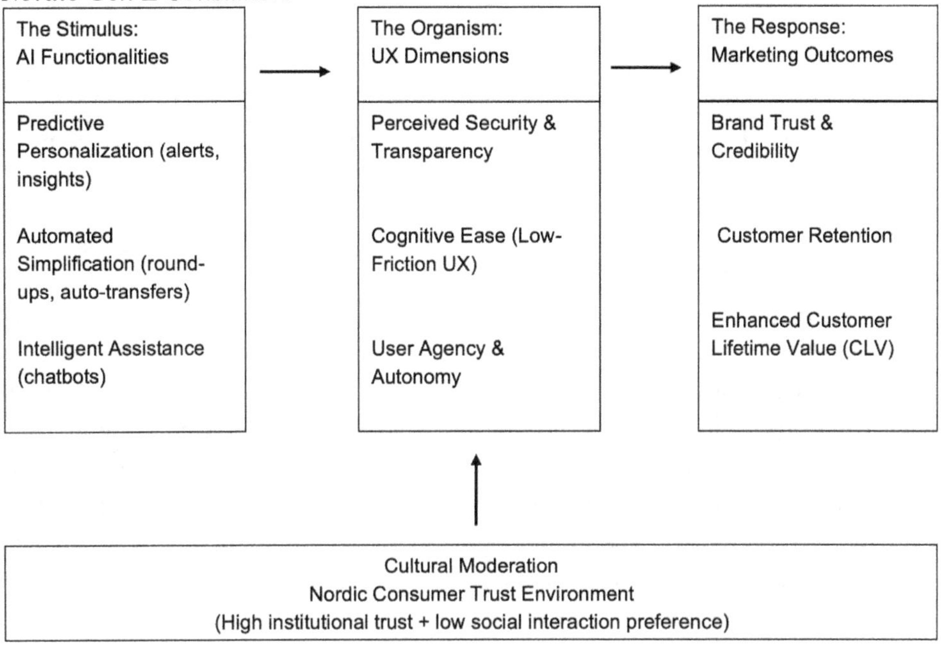

Source: Author (2026)

METHODOLOGY

The study utilises a qualitative research design. Qualitative research is a naturalistic approach aimed at understanding social phenomena within their real-world contexts, focusing on participants' meanings and interactions through non-numerical data (Denzin et al., 2023; Saunders et al., 2019). This approach allows for a deeper exploration of the behaviours and experiences of Gen Z consumers concerning AI-driven payment systems. A qualitative methodology, relying solely on interview data, was deemed appropriate to gain a comprehensive understanding of these complex psychological and behavioural dimensions (Saunders et al., 2019). The entire investigation is conceptually guided by the S-O-R model (Mehrabian & Russell, 1974), which serves as the theoretical lens for both the data collection instrument and the subsequent analysis. This framework was selected due to its recognised utility in consumer behaviour research for systematically mapping the effects of a stimulus (AI features) onto the internal, mediating states of an individual (perception, trust, risk) and, ultimately, their observable actions (adoption, use) (Irimia-Diéguez et al., 2025; Elangovan & Babu, 2024).

Participants and Sampling

The target population for this study was Gen Z consumers residing in the technologically advanced Nordic region, specifically Finland. A purposive sampling strategy was employed, ensuring that all participants met key criteria: active use of digital payment systems and falling within the accepted birth year range for Gen Z (typically 1997–2012). This focused approach was critical to ensuring the findings reflected the unique mindset and technological expectations of this key demographic (Gutfreund, 2016).

The final sample comprised N = 9 participants. The small sample size is characteristic of qualitative studies aiming for depth and rich description rather than statistical generalisation. The sample included a balance across gender and included both currently enrolled university students and young professionals, ensuring varied exposure to financial services. Anonymity was strictly maintained, and participants are referenced throughout the chapter using the codes Participant 1 through Participant 9 to protect their identities and promote candid responses, aligning with ethical standards (Saunders et al., 2019). The detailed professional and demographic breakdown of the sample is presented in Table 2.

Table 2. Participant Demographic Summary

Participant Code	Gender	Professional & Educational Background
Participant 1	Female	Hospitality Professional, Student
Participant 2	Male	Entrepreneur, Student
Participant 3	Female	Accounting Professional, master's Student
Participant 4	Female	Business Consultant, master's Graduate
Participant 5	Female	Freelancer, bachelor's Student
Participant 6	Male	Accounting Professional, bachelor's Student
Participant 7	Male	Project Manager, bachelor's Graduate
Participant 8	Female	Sales associate, Student
Participant 9	Male	Customer Service, bachelor's student

Data was collected using an in-depth semi-structured interview protocol. Semi-structured interviews are a form of two-way communication that provides a detailed study of a specific topic (Saunders et al., 2019). This method was ideal as it allowed for the systematic collection of data on pre-defined S-O-R themes while retaining the flexibility to let participants introduce other relevant issues for discussion, thereby ensuring all research objectives were addressed (Saunders et al., 2019). The protocol was rigorously structured based on the components of the S-O-R model, ensuring comprehensive coverage of all study objectives, addressing “stimulus” or questions focused on the AI features and digital services participants recognized and currently used (Ramrakhyani & Shrivastava, 2024); “organism” or questions probed subjective factors, including perceived benefits, perceived risks, conditional trust, and ease of use (Glavee-Geo et al., 2025); “response” or questions addressed observed usage patterns, expectations for future features, and the influence of cultural or generational factors on adoption (Arvidsson, 2019; Susilawati et al., 2024).

All interviews were conducted online via Microsoft Teams and recorded with the explicit, informed consent of the participants. The average duration of each interview was approximately 35 minutes. The language used in the interviews was English, a common lingua franca in the Finnish academic and business context. Immediately following collection, all interview recordings were transcribed verbatim to ensure accurate representation of the participant narratives.

Data Analysis

The transcribed data were analysed using Thematic Analysis (e.g., following established methods from Braun & Clarke, 2006; King & Horrocks, 2018), a flexible method for identifying, analysing, and reporting patterns (themes) within data. The process involved several stages: familiarisation, generating initial codes, searching for themes, reviewing themes, defining and naming themes, and producing the final report (King & Horrocks, 2018). This systematic approach ensured that the interpretation remained grounded in the participants' own words.

The initial coding phase was deductive, using the S-O-R model as the main thematic framework for the highest level of codes. Within these main categories, codes were then developed inductively based on the data to capture nuanced sub-themes (e.g., Conditional Trust, Finnish Cultural Influence, Negative Financial Impact). To enhance the trustworthiness of the findings, the analysis utilised multiple direct quotes from different participants, ensuring that every theme was strongly substantiated by various perspectives. The final thematic structure, linking specific data segments to the S-O-R themes, is visually presented in the Coding Matrix (Appendix Table 4).

RESULTS

This section presents the empirical findings from qualitative interviews with nine Gen Z participants, offering a specific Nordic perspective guided by the Stimulus-Organism-Response (S-O-R) model. The results are structured to map the user journey from external Stimuli (AI features and market competitive forces) through the internal Organism state (cognitive evaluations of trust, risk, and cultural values) to the final behavioural Response (adoption habits and loyalty). Each thematic finding is substantiated with direct evidence from the interview transcripts, ensuring that the qualitative insights are explicitly mapped to the theoretical framework.

AI in Digital Payments - Stimulus

This initial section explores the technological and market features that Gen Z recognises as the AI-driven stimuli within their digital payment environment, setting the context for their subsequent perceptions and behaviours.

Current state of AI integration

The authors explored the participants' understanding of existing market penetration, noting that they often distinguished between noticeable applications and areas of deficiency. Interviewees consistently identified AI use in background operations, primarily citing its role in speeding up transfers and enhancing security. As one participant noted: "For the security or sort of fraud prevention... and [it] also could include here, for example, this verification process, like when you are logging in or making a payment, that it's verified that you are the right person. I think these are the areas where AI is currently implemented" (Participant 7).

This view was echoed by Participant 5, who observed AI "being used to detect frauds... [making the processes] more efficient, faster on the payments" (Participant 5). This focus on back-end efficiency highlights that the technological Stimulus (S) is currently felt more through reliable operation than through personalised interaction, functioning as a "hygiene factor" rather than a visible driver of engagement.

Despite this, many participants pointed out that user-facing features are still limited or undeveloped. Participant 1 observed: "The Finnish market or the apps I use at least haven't really incorporated it yet, especially on the UX side" (Participant 1). This sentiment was supported by Participant 8's assessment that specific solutions were not "[very] advanced on AI as far as I could judge" (Participant 8). Theoretically, this indicates an "under-stimulation" in the front-end UX, where the lack of visible AI features creates a gap between user expectations and the actual market Stimulus (S).

Overall outlook on AI

The overall sentiment regarding future development was predominantly optimistic or positive. The authors observed that participants viewed the technology as an inevitable force for reducing cognitive load. Participant 7 stated: "I'm very optimistic about the integration. I think it's great that it helps to develop the industry and fintech, and it just generally brings a lot of ease of use and speed and security" (Participant 7).

Participant 8 expressed happiness "for all like different innovations which would benefit me as a consumer," linking development directly to personal utility (Participant 8). This widespread optimism serves as a positive affective pre-condition within the Organism (O), suggesting that Gen Z is highly receptive to new stimuli if they promise tangible gains.

However, some interviewees presented an optimistic yet uncertain or cautious view. Their caution centred explicitly on reliability. Participant 7 captured this nuance by saying, "I'm a bit unsure of its full potential, but I'm very optimistic if I say so" (Participant 7), while Participant 1 acknowledged that delayed adoption was "[partly] because of the security concerns" (Participant 1). This illustrates that the Stimulus (S) is not purely positive; perceived risks act as a counter-stimulus that can weaken the eventual behavioural Response (R).

Impact of AI on the market

The authors noted that participants viewed AI integration as a strategic necessity that reshapes the competitive landscape. The technology was seen as a competitive necessity for financial institutions. Participant 1 articulated this market pressure clearly: "I think it's coming at some point to every platform because they have to, otherwise some competitor will come with a feature and [will] take over the market. It's an obvious development" (Participant 1).

This belief positioned AI as a primary driver of industry development. Participant 7 observed that AI would "improve the efficiency of the whole system" (Participant 7), and Participant 4 agreed, stating, "they are just [always] improving and... we will just have to go with it" (Participant 4). From an S-O-R perspective, this market pressure acts as an external environmental stimulus, forcing both the consumer and the provider to adapt to a rapidly evolving technological baseline.

Consumer Perceptions - Organism

This section details the cognitive and emotional evaluations of the AI stimulus, exploring how Gen Z weighs the benefits against the risks of AI-driven payment systems. Within the S-O-R framework, this represents the Organism (O) stage, where external stimuli are processed into internal states that precede behaviour.

Perceived Benefits of AI-Driven UX

The analysis revealed strong enthusiasm for AI features that promise tangible improvements, driving a positive initial evaluation in the organismic stage.

Financial Management and Advice

Participants expressed a strong desire for AI to reduce the effort and expertise required for sound financial decision-making. Highlighting the benefit of AI in interpreting spending habits, one respondent remarked: "I think that's something that you need to put a lot of time into if you want to do it just by yourself, so AI could really facilitate that process and help you figure out what you spend money on" (Participant 1).

Regarding investment, Participant 5 mentioned AI-based tools "help people start small and young rather than wait until your 40 or 50" (Participant 5). Theoretically, this shows that AI serves as a cognitive enhancer within the Organism (O), transforming complex data into a state of "perceived empowerment" for the user.

Efficiency and Automation

The core appeal for participants rested on task automation. Participant 7 noted that AI "automates, like boring tasks" (Participant 7), while Participant 4 observed: "Reduce some of the workloads because if it does the work for you, you're just checking the outcome... it's a huge time-saver and it reduces cognitive friction" (Participant 4). This reduction in "cognitive friction" is a vital internal state; it suggests that the Organism (O) values the mental energy saved as much as the time saved, establishing a baseline of positive affect.

Improved Customer Service

Participant 9 articulated a clear cultural preference for self-service: "I actually feel that maybe it's because I'm Finnish, but I don't want to bother people. It's a Finnish cultural thing. When it's AI, I can ask more trivial questions without fearing that I might appear foolish" (Participant 9). This provides a robust, culturally specific insight into the Organism (O). In the Nordic context, the primary benefit of AI is "social frictionlessness," where the internal state is one of relief from social judgement. However, the need for a human fallback remained essential for trust, as

Participant 5 noted: "If the issue is too complicated... the chat board not necessarily understand me" (Participant 5).

Perceived Risks and Concerns with AI

The analysis indicated that perceptions were heavily qualified by reservations, highlighting several "dual effect" themes.

Accessibility and Inclusion

While Participant 5 noted that AI could democratise finance, there was a risk of uninformed decisions. This duality represents a state of "cautious optimism" within the Organism (O), where the perceived benefit of ease is balanced against the fear of financial incompetence.

Trust and Reliability

The primary risk is centred on the negative financial impact. Participant 8 warned: "It has bigger consequences when it involves finances... It's not like buying a shirt that doesn't fit" (Participant 9). This analogy highlights that the Organism (O) maintains a high threshold for "financial trust." Unlike low-consequence domains, the affective state here is dominated by risk-aversion, requiring "explainable AI" to lower the perceived threat.

Security and Fraud

While AI for fraud detection builds trust, participants feared increased scam potential. Participant 2 only trusted the system: "If they're able to collect it safely... if it's a closed system and they can guarantee it is safe. But I'm very aware that AI is making the scams more believable" (Participant 2). Theoretically, this underscores that the security Stimulus (S) has a double-edged effect; it creates an internal state of hyper-vigilance within the Organism (O), where trust in the provider is the only safeguard against the perceived sophistication of AI-driven threats.

Complexity and Information Overload

Participant 9 observed that excessive features could be overwhelming: "It's becoming more and more... at a certain point, too much is too much. If the AI suggestions are constant, you just switch them off" (Participant 9). This implies a "saturation point" in the S-O-R chain. When the Stimulus (S) becomes intrusive, the Organism (O) shifts from engagement to avoidance, leading to a negative behavioural Response (R).

Consumer Behaviour – Response

This final section details the observed and reported behavioural outcomes of the Gen Z participants, representing the Response (R) component of the framework. This stage translates the internal cognitive processing (Organism) into observable actions, interactions, and cultural patterns.

User Behaviour and Context Drivers

This thematic category explores the immediate practical and social factors that influence Gen Z's adoption. The authors found that participants cited general convenience and seamless connection as primary reasons. Participant 6 explained that digital payments are "all connected with the online service and [the] online banking and all that, making everything seamless" (Participant 6).

Theoretically, this "seamlessness" represents a successful behavioural response (R) where the technology disappears into the user's lifestyle. Participant 7's view that it "just saves time" confirms that efficiency is the ultimate behavioural metric in the S-O-R chain (Participant 7). Social influence also emerged as a critical driver. Participant 9 articulated the network effect: "Services [such as] mobile pay [are] based around the network effect. I mean I would not be using it if others didn't use it, because the utility comes from everyone else being on the platform" (Participant 9). This indicates that the Response (R) is not merely an individual reaction but a collective social state.

Actual Behaviour

The authors noted that participant adoption of AI features is highly pragmatic. Regarding current habits, Participant 8 reported utilising automated investing or checking: "I use mostly this... partly as a saving option because there is a possibility of rounding up [a] transaction... It works in the background, and I don't have to think about it" (Participant 8).

This "passive financial management" is a significant behavioural response (R). It demonstrates that when the stimulus is frictionless and the organism state is trust-based, the consumer responds by delegating routine financial agency to the AI. Beyond current usage, participants expressed clear functional expectations. Participant 9 noted that AI should: "At a minimum, suggest a category when I make a transaction, and ideally, proactively warn me if I'm spending too much in one area" (Participant 9).

Drivers and Cultural Influences

The theme of adaptability revealed a demographic with low behavioural inertia. Faster adaptation means Gen Z expects and embraces change quickly. Theoretically, this high adaptability functions as a moderator that accelerates the S-O-R cycle.

Finally, the interaction style highlighted the Finnish cultural influence. Participant 9 articulated: "I actually feel that maybe it's because I'm Finnish, but I don't want to bother people... It's a preference for self-sufficiency" (Participant 9). This establishes that self-sufficiency is a cultural moderator of the Response (R). In the Nordic context, the ideal behavioural outcome is one that minimises human-to-human interaction in favour of autonomous AI-to-human service. This mindset difference—contrasted against an older generational mindset—shows that Gen Z's behavioural response is uniquely shaped by a blend of technological fluency and cultural norms.

DISCUSSION

This study analysed how AI-driven user experience shapes behavioural and marketing outcomes in digital payment systems among Nordic Generation Z users. The findings reveal a clear distinction between what AI currently delivers and what Gen Z expects. Although AI is recognised for fraud prevention and fast transactions, most applications remain "back-end" enhancements. This positions AI as a hygiene factor, essential for competitiveness yet insufficient to create differentiation (Kotler & Keller, 2016). To generate marketing impact, financial providers must transition from invisible AI to value-visible features that strengthen emotional engagement and brand preference (Ramrakhyani & Shrivastava, 2024).

Trust emerged as a strategic performance metric. Participants expressed confidence in regulated Nordic institutions but maintained cautious expectations toward autonomous financial decisions. This supports the moderating role of perceived financial risk in AI adoption (Glavee-Geo et al., 2025; Irimia-Diéguez et al., 2025). Here, Explainable AI becomes a marketing tool, not just a technical one: transparent and guided communication can reduce anxiety and reinforce brand credibility (Ben David et al., 2021). Sustained trust converts into retention and Customer Lifetime Value (CLV), highlighting a key intersection of AI, UX, and marketing strategy.

The findings also demonstrate that personalisation and value co-creation are core value expectations. Users' welcome features that translate transactional data into actionable insights, making financial decisions lighter and more intuitive. These functions promote emotional empowerment and increase perceived relevance, both essential drivers of Gen Z loyalty and recommendation behaviour (Chauhan et al.,

2025; Alexander & Sarrafzadeh, 2004). By allowing users to customize their own automated saving rules and investment parameters, banks shift from being service providers to co-creators of value, enabling users to actively build their financial futures inside the platform's ecosystem. However, personalization has an ethical boundary: providing investment nudges without educational support can unintentionally encourage risky behaviour. This tension suggests that successful personalisation requires a responsible marketing design, pairing automation with accessible learning.

Behaviourally, Gen Z demonstrates a strong preference for passive financial management, supported by automated tools like savings roundups and scheduled transfers (Susilawati et al., 2024; Agarwal et al., 2023). This supports the S-O-R behavioural model: when the AI stimulus is frictionless and the internal organism state is trust-based, the behavioural response shifts from manual monitoring to habitual automation. However, participants also warned that excessive notifications or financial suggestions can lead to cognitive overload and rejection. This indicates that calibrated personalisation is needed to maintain engagement without causing frustration.

Finally, cultural influences shape how marketing strategies should be localised. Nordic Gen Z users demonstrate a high value on autonomy, shaped by a cultural preference for low-social-interaction service. AI chatbots and self-service channels are not only efficient but socially comfortable options. Yet, the expectation of a human safety-net remains. This balance suggests a dual-channel marketing approach: independent-first digital experience with trusted human escalation. Compared to older generations who equate service with human contact, Gen Z equates good service with reduced friction, a cultural nuance critical for positioning AI-enabled offerings in the Nordic region (Gutfreund, 2016).

In summary, this study confirms that AI is not merely a back-end operational enhancement but a strategic marketing instrument that builds emotional value, strengthens loyalty, and shapes trust when designed with cultural intelligence and transparent communication.

CONCLUSION AND LIMITATIONS

Theoretical Conclusion

This research mapped the Gen Z AI experience through the S-O-R framework, revealing three critical shifts in how we understand digital payment marketing in advanced economies.

First, the study redefines AI from a "premium" feature to a "baseline" requirement for brand survival. Our findings confirm that for Nordic digital natives, back-end

AI (such as fraud detection) is no longer a unique selling point; it is a baseline expectation. This supports the view that AI capabilities are fundamentally redefining customer engagement (Nayak, 2025) and must be managed as a core product factor (Kotler & Keller, 2016). From a marketing perspective, while its presence may not create active brand delight, its absence leads to immediate brand erosion. To achieve true market differentiation, financial providers must move beyond "invisible" security and focus on user-facing AI "exciters" that provide visible, proactive value (Ramrakhyani & Shrivastava, 2024).

Second, we identify "Conditional Trust" as the primary psychological gateway to Customer Lifetime Value (CLV). The S-O-R "organism" stage in this study is defined by a delicate balance: users value automation but are hyper-aware of the financial risks of AI error. This duality confirms the mediating role of risk perceptions and trust in digital financial systems (Irimia-Diéguez et al., 2025; Glavee-Geo et al., 2025). The theoretical contribution here is clear: trust in the institution precedes, but does not guarantee, trust in the technology. Therefore, to ensure customer retention, marketing must prioritize "Explainable AI" (Ben David et al., 2021). By reducing user anxiety through transparent decision-making, brands can transform a high-risk interaction into a trust-building experience that sustains long-term loyalty.

Third, the research identifies a "Cultural Moderation" effect that dictates successful interaction styles. The Finnish preference for self-sufficiency, often described as "not wanting to bother" others, is a marketing directive for the Nordic region. This cultural insight acts as a unique moderator on the traditional S-O-R model, translating into a high demand for autonomous, self-service AI tools. While Gen Z's status as digital natives leads to faster adaptation (Gutfreund, 2016), their loyalty is secured only when these automated features, such as automated investing, feel frictionless and reliable (Susilawati et al., 2024; Agarwal et al., 2023). Marketing strategies should therefore frame AI not just as a cost-saving measure for the bank, but as an empowering partner that respects the user's desire for independent financial agency.

Managerial Implication

For financial institutions and fintech startups eyeing the Nordic Gen Z market, the results suggest a clear strategic shift: moving away from invisible back-end AI toward user-facing value. Based on the S-O-R analysis, the proposed five-pillar strategic framework to optimise AI-driven UX for marketing success is presented in Table 3.

Table 3. Five-Pillar Strategic Framework for AI-UX Marketing in the Nordic Gen Z Segment

Pillar	Strategic Action	Marketing & Behavioural Outcome
Visibility	Shift AI from "back-end" security to "front-end" visible UX nudges.	Active Brand Trust: Users perceive the bank as a proactive protector.
Trust	Implement Explainable AI (XAI) to clarify autonomous decisions.	Reduced Perceived Risk: Increases confidence in high-stakes tasks.
Simplicity	Prioritize "passive" automation (e.g., roundups and sweeps).	Habitual Loyalty: Creates "stickiness" by making growth effortless.
Autonomy	Design "independent-first" paths with a clear human fallback.	Cultural Alignment: Respects the Nordic preference for self-sufficiency.
Personalization	Use transactional data for proactive, wellness-focused advice.	Increased CLV: Transitions the bank to a lifestyle partner.

To implement this framework effectively, managers should focus on four key operational areas:

First, make the AI work "out loud." Our findings show that while Gen Z appreciates fraud detection, they find it boring when it stays in the background. To drive Customer Lifetime Value (CLV), banks should turn AI into a proactive assistant. For example, a "Smart Budgeting" tool that warns a user about a subscription price hike creates a tangible "AI moment" that strengthens product differentiation and keeps users locked into the ecosystem.

Second, treat trust as a fragile asset. Even one miscategorised transaction can hurt customer retention. Managers should prioritize "Explainable AI"—small, simple pop-ups that explain why an alert was sent. This transparency replaces user anxiety with a sense of control, turning a potential security annoyance into a trust-building feature.

Third, prioritize "low-friction" over "high-tech." In the Nordic market, the most successful tools reduce cognitive load. Automation like round-up savings or instant transfers works best when it feels effortless. If a new AI feature adds more buttons or complicated prompts, Gen Z will likely abandon it; therefore, loyalty is built through simplicity and speed.

Fourth, respect the "Independent" cultural trait. A unique insight from our qualitative data is the Nordic preference for self-sufficiency. These users would rather fix a problem through a smart chatbot than wait on a phone line. However, to protect the brand, companies must ensure a "warm hand-off"—a clear, instant path to a human expert when the AI reaches its limit. This Value Co-creation between the user and the autonomous system is key to long-term satisfaction.

Limitation and Future Research

This study linked the use of artificial intelligence in digital payments with how people think and behave, but it has some limitations that should be addressed in future research. Firstly, the results are based on a small group of young people in Finland, so they may not apply to other cultures. Future studies should explore similar questions in different regions to see if the same patterns appear. Furthermore, because the study relied on what participants said rather than what they did, future work could track real behaviour over time or test different AI features to see which encourage continued use.

Privacy concerns also deserve more attention; future research should measure how much convenience people are willing to trade for data security and how this affects their choices. Furthermore, while this study establishes marketing linkages like trust and loyalty, future research should utilise longitudinal data to measure the long-term impact of AI-driven co-creation on actual Customer Lifetime Value (CLV) across different Nordic sub-markets. Finally, as artificial intelligence develops quickly, ongoing research is needed to follow how new features change user expectations and experiences.

REFERENCES

Abikari, M., Öhman, P., & Yazdanfar, D. (2023). Negative emotions and consumer behavioral intention to adopt emerging e-banking technology. *Journal of Financial Services Marketing*, *28*(4), 691–704. DOI: 10.1057/s41264-022-00172-x

Afroogh, S., Emsley, I., & Floridi, L. (2024). Trust in AI: Progress, challenges, and future directions. *Humanities & Social Sciences Communications*, *11*(1), 1568. Advance online publication. DOI: 10.1057/s41599-024-04044-8

Agarwal, S., Malik, P., & Gautam, S. (2023). Analysis of customer satisfaction and the customer experience in digital payments: A meta-analysis review. *International Journal of Business Science and Applied Management*, *18*(1), 1–17. DOI: 10.69864/ijbsam.18-1.168

Ahmad, M. B., Ali, H. F., Malik, M. S., Humayun, A. A., & Ahmad, S. (2019). Factors affecting impulsive buying behavior with mediating role of positive mood: An empirical study. *European Online Journal of Natural and Social Sciences*, *8*(1), 17–27.

Ahmed, I., Al Masud, A., Hossain, M. A., Hossain, M. A., & Kim, M. (2025). Understanding customer switching behavior: An empirical examination of mobile financial services. *Social Sciences & Humanities Open*, *11*, 101270. DOI: 10.1016/j.ssaho.2024.101270

Al-Qudah, A. A., Al-Okaily, M., Shiyyab, F. S., Taha, A. A., Almajali, D. A., Masa'deh, R. E., & Warrad, L. H. (2024). Determinants of digital payment adoption among Generation Z: An empirical study. *Journal of Risk and Financial Management*, *17*(11), 521. DOI: 10.3390/jrfm17110521

Aldboush, H. H. H.. (2023). Building trust in Fintech: An analysis of ethical and privacy considerations. *Journal of Risk and Financial Management*, *11*(3).

Alexander, S., & Sarrafzadeh, A. (2004). Interfaces that adapt like humans. In *Asia-Pacific Conference on Computer Human Interaction* (pp. 641–645). Berlin, Heidelberg: Springer Berlin Heidelberg.

Antonio, B. O., Juan, L. R., Ana, I. D., & Francisco, L. C. (2024). Examining user behavior with machine learning for effective mobile peer-to-peer payment adoption. *Financial Innovation*, *10*(1), 94. DOI: 10.1186/s40854-024-00625-3

Arianita, A., Anggarawati, S., & Alfansi, I. (2023). Analysis Factor Affecting The Use Of Digital Payment With The Extended Utaut Model. *Management Review*, *5*(1), 91–108.

Bach, T. A., Khan, A., Hallock, H., Beltrão, G., & Sousa, S. (2023). A systematic literature review of user trust in AI-enabled systems: An HCI perspective. *arXiv*. https://doi.org//arXiv.2304.08795DOI: 10.48550

Bajunaied, K., Hussin, N., & Kamarudin, S. (2023). Behavioral intention to adopt FinTech services: An extension of unified theory of acceptance and use of technology. *Journal of Open Innovation*, *9*(1), 100010. DOI: 10.1016/j.joitmc.2023.100010

Balasubramaniam, S. (2023). Trust and privacy in AI-enabled financial services: A user-centric perspective. *Journal of Digital Banking*, *8*(2), 112–128.

Belch, G. E., & Belch, M. A. (2021). *Advertising and promotion: An integrated marketing communications perspective* (6th ed.). McGraw-Hill.

Ben David, D., Resheff, Y. S., & Tron, T. (2021). Explainable AI and adoption of financial algorithmic advisors: an experimental study. In *Proceedings of the 2021 AAAI/ACM Conference on AI, Ethics, and Society* (pp. 390–400). DOI: 10.1145/3461702.3462565

Blanco-Encomienda, F. J., & Rosillo-Díaz, E. (2025). Exploring brand attitude in e-commerce using the SOR model: The role of information quantity, information quality and source credibility. *Journal of Product and Brand Management*, *34*(7), 1041–1055. DOI: 10.1108/JPBM-04-2024-5095

Braun, V., & Clarke, V. (2006). Using thematic analysis in psychology. *Qualitative Research in Psychology*, *3*(2), 77–101. DOI: 10.1191/1478088706qp063oa

Changani, S., & Kumar, R. (2023). How social media marketing enhances brand community engagement: developing an integrated model using the SOR paradigm. In *International Working Conference on Transfer and Diffusion of IT* (pp. 302–318). Cham: Springer Nature Switzerland.

Chauhan, K., Singh, S., & Aggarwal, P. (2025). Reconceptualizing digital payments: The enhanced role of AI in transforming financial transactions ecosystem. *Lloyd Business Review*, 210–219.

Cheung, C. M., & Thadani, D. R. (2012). The impact of electronic word-of-mouth communication: A literature analysis and integrative model. *Decision Support Systems*, *54*(1), 461–470. DOI: 10.1016/j.dss.2012.06.008

Denzin, N., Lincoln, Y., Giardina, M., & Cannella, G. (2023). *The Sage Handbook of Qualitative Research*. Sage Publications.

Elangovan, A., & Babu, M. (2024). Decoding users' continuance intentions towards digital financial platforms in the Indian economy. *International Review of Management and Marketing*, *14*(6), 213–222. DOI: 10.32479/irmm.17199

Elayat, A. M. A., & Elalfy, R. M. (2025). Using SOR theory to examine the impact of AI chatbot quality on Gen Z's satisfaction and advocacy within the fast-food sector. *Young Consumers*, *26*(2), 352–383. DOI: 10.1108/YC-08-2024-2199

Gao, L., Waechter, K. A., & Bai, X. (2015). Understanding consumers' continuance intention towards mobile purchase: A theoretical framework and empirical study–A case of China. *Computers in Human Behavior*, *53*, 249–262. DOI: 10.1016/j.chb.2015.07.014

Gefen, D., Karahanna, E., & Straub, D. W. (2003). Trust and TAM in online shopping: An integrated model. *Management Information Systems Quarterly*, *27*(1), 51–90. DOI: 10.2307/30036519

Gerli, P., Clement, J., Esposito, G., Mora, L., & Crutzen, N. (2022). The hidden power of emotions: How psychological factors influence skill development in smart technology adoption. *Technological Forecasting and Social Change*, *180*, 121721. DOI: 10.1016/j.techfore.2022.121721

Glavee-Geo, R.. (2025). Exploring trust and outcome expectancy in FinTech digital payments: Insights from the stimulus-organism-response model. *International Journal of Bank Marketing*.

Gutfreund, J. (2016). Move over, Millennials: Generation Z is changing the consumer landscape. *Journal of Brand Strategy*, *5*(3), 245–249. DOI: 10.69554/KAHI1951

Hewei, T., & Youngsook, L. (2022). Factors affecting continuous purchase intention of fashion products on social E-commerce: SOR model and the mediating effect. *Entertainment Computing*, *41*, 100474. DOI: 10.1016/j.entcom.2021.100474

Imbayani, I. G. A., & Gama, A. W. S. (2024). The impact of hedonic shopping motivation, discounts, and urge to buy impulsively on impulse buying: SOR perspective. *International Journal of Applied Business and International Management*, *9*(3), 349–364. DOI: 10.32535/ijabim.v9i3.3438

Irimia-Diéguez, A., Liébana-Cabanillas, F., Blanco-Oliver, A., & Lara-Rubio, J. (2025). What drives consumers to use P2P payment systems? An analytical approach based on the stimulus–organism–response (S-O-R) model. *European Journal of Management and Business Economics*. https://www.econstor.eu/bitstream/10419/325591/1/1923036564.pdf

Jaradat, M. I. R. M., & Al-Mashaqba, A. M. (2014). Understanding the adoption and usage of mobile payment services by using TAM3. *International Journal of Business Information Systems, 16*(3), 271–296. DOI: 10.1504/IJBIS.2014.063768

Jiang, C., & Phoong, S. W. (2024). Research on tourists' travel intention and usage of VR technologies based on SOR model: Mediating role of personal values. *Journal of Hospitality and Tourism Insights, 7*(2), 949–967. DOI: 10.1108/JHTI-08-2023-0545

Kanaparthi, V. (2024). AI-driven personalisation in digital banking: Enhancing user experience and financial decision-making. *International Journal of Financial Studies, 12*(1), 45–60.

Kaur, S., & Walia, N. (2020). Exploring tourist adoption behaviour of digital payment system using TAM model. *Solid State Technology, 63*(6), 4632–4640.

Khan, A., Ullah, M., & Malik, F. F. (2022). Mediating role of consumer involvement in the relationship between marketing stimuli and consumer purchase behavior. *Journal of Marketing Strategies, 4*(1), 202–220. DOI: 10.52633/jms.v4i1.185

Kim, J., & Lennon, S. J. (2013). Effects of reputation and website quality on online consumers' emotion, perceived risk and purchase intention: Based on the stimulus-organism-response model. *Journal of Research in Interactive Marketing, 7*(1), 33–56. DOI: 10.1108/17505931311316734

Kim, J. H., & Park, J. W. (2019). The effect of airport self-service characteristics on passengers' perceived value, satisfaction, and behavioral intention: Based on the SOR model. *Sustainability (Basel), 11*(19), 5352. DOI: 10.3390/su11195352

King, N., & Horrocks, C. (2018). *Interviews in Qualitative Research*. SAGE Publications.

Kotler, P., & Keller, K. L. (2016). *A framework for marketing management* (6th ed.). Pearson Education Limited.

Kumar, S., & Singh, V. P. (2024). Trust and satisfaction with digital payment systems in Gen Z and Gen Y: A comparative study. *Library of Progress–Library Science. Information Technology & Computer, 44*(3). Advance online publication. DOI: 10.48165/bapas.2024.44.2.1

Li, K., Zhang, X., & Wen, Y. (2025). Does peer pressure increase pro-environmental engagement? An extended stimulus-organism-response model. *Journal of Environmental Management, 392*, 126719. DOI: 10.1016/j.jenvman.2025.126719

Liébana-Cabanillas, F., Ramos de Luna, I., & Montoro-Ríos, F. (2017). Intention to use new mobile payment systems: A comparative analysis of SMS and NFC payments. *Economic research-. Ekonomska Istrazivanja*, *30*(1), 892–910. DOI: 10.1080/1331677X.2017.1305784

Ling, J., Xue, Y., Yang, C., & Zhang, Y. (2022). Effect of farmers' awareness of climate change on their willingness to adopt low-carbon production: Based on the TAM-SOR model. *International Journal of Environmental Research and Public Health*, *20*(1), 619. DOI: 10.3390/ijerph20010619

Mahida, A., Mandala, V., Bauskar, S., Konkimalla, S., & Reddy, M. S. (2024). Real-time fraud mitigation in digital payments: Big data and AI-driven biometric authentication. *Nanotechnology Perceptions*, *20*, 1176–1193.

Malaquias, R. F., & Silva, A. F. (2020). Understanding the use of mobile banking in rural areas of Brazil. *Technology in Society*, *62*, 101260. DOI: 10.1016/j.techsoc.2020.101260

Manser Payne, E. H., Peltier, J., & Barger, V. A. (2021). Enhancing the value co-creation process: Artificial intelligence and mobile banking service platforms. *Journal of Research in Interactive Marketing*, *15*(1), 68–85. DOI: 10.1108/JRIM-10-2020-0214

Manta, O., Vasile, V., & Rusu, E. (2025). Banking transformation through FinTech and the integration of artificial intelligence in payments. *FinTech*, *4*(2), 13. DOI: 10.3390/fintech4020013

Mavlutova, I., Spilbergs, A., Verdenhofs, A., Natrins, A., Arefjevs, I., & Volkova, T. (2022). Digital transformation as a driver of the financial sector sustainable development: An impact on financial inclusion and operational efficiency. *Sustainability (Basel)*, *15*(1), 207. DOI: 10.3390/su15010207

Mehrabian, A., & Russell, J. A. (1974). *An approach to environmental psychology*. MIT Press.

Nayak, S. (2025). Enhancing customer experience in FinTech through AI, machine learning, and data-driven insights. *The Es Economics and Entrepreneurship*, *3*(03), 437–462.

Oliveira, T., Thomas, M., Baptista, G., & Campos, F. (2016). Mobile payment: Understanding the determinants of customer adoption and intention to recommend the technology. *Computers in Human Behavior*, *61*, 404–414. DOI: 10.1016/j.chb.2016.03.030

Pal, A., Herath, T., De', R., & Rao, H. R. (2020). Contextual facilitators and barriers influencing the continued use of mobile payment services in a developing country: Insights from adopters in India. *Information Technology for Development, 26*(2), 394–420. DOI: 10.1080/02681102.2019.1701969

Para, R., Bhatia, R., & Sandiri, S. (2025). AI-powered financial digital twins: The next frontier in hyper-personalized, customer-centric financial services. *Journal of Next-Generation Research 5.0.*

Rahardja, U., Sigalingging, C. T., Putra, P. O. H., Nizar Hidayanto, A., & Phusavat, K. (2023). The impact of mobile payment application design and performance attributes on consumer emotions and continuance intention. *SAGE Open, 13*(1), 21582440231151919. DOI: 10.1177/21582440231151919

Ramrakhyani, A., & Shrivastava, N. K. (2024). Artificial intelligence: Revolutionizing the future of Fintech. *Commerce Research Review, 1*(2), 10–22. DOI: 10.21844/crr.v1i02.1109

Santosa, A. D., Taufik, N., Prabowo, F. H. E., & Rahmawati, M. (2021). Continuance intention of baby boomer and X generation as new users of digital payment during COVID-19 pandemic using UTAUT2. *Journal of Financial Services Marketing, 26*(4), 259–273. DOI: 10.1057/s41264-021-00104-1

Saunders, M., Lewis, P., & Thornhill, A. (2019). *Research methods for business students* (8th ed.). Pearson Education.

Shahzad, M. F., Xu, S., An, X., & Javed, I. (2024). Assessing the impact of AI-chatbot service quality on user e-brand loyalty through chatbot user trust, experience and electronic word of mouth. *Journal of Retailing and Consumer Services, 79*, 103867. DOI: 10.1016/j.jretconser.2024.103867

Shamim, A., Abid, M. F., & Ahmad, F. (2024). S–O–R based experiential framework for measuring in-store customer satisfaction in non-fuel retailing. *Journal of Retailing and Consumer Services, 77*, 103672. DOI: 10.1016/j.jretconser.2023.103672

Shankar, A., & Datta, B. (2018). Factors affecting mobile payment adoption intention: An Indian perspective. *Global Business Review, 19*(3, suppl), S72–S89. DOI: 10.1177/0972150918757870

Shih, I. T., Silalahi, A. D. K., Baljir, K., & Jargalsaikhan, S. (2024). Exploring the impact of perceived sustainability on customer satisfaction and the mediating role of perceived value. *Cogent Business & Management, 11*(1), 2431647. DOI: 10.1080/23311975.2024.2431647

Soni, S., & Mitchell, A. M. (2022). Women and digital government in rural India: Extending TAM with user assisted service. *Information Polity*, *27*(1), 59–78. DOI: 10.3233/IP-200302

Sultan, P., Wong, H. Y., & Azam, M. S. (2021). How perceived communication source and food value stimulate purchase intention of organic food: An examination of the stimulus-organism-response (SOR) model. *Journal of Cleaner Production*, *312*, 127807. DOI: 10.1016/j.jclepro.2021.127807

Susilawati, A. D., Wahyudi, F., Putra, W. P., Supriyanto, W., & Limpo, L. (2024). The impact of digital user experience on brand perception and consumer loyalty in the e-commerce industry in Indonesia. *The Eastasouth Journal of Information System and Computer Science*, *1*(03), 109–122. DOI: 10.58812/esiscs.v1i03.244

Teh, G. M., Kalidas, V., & Zeeshan, M. (2014). Personality as a moderator of SOR model. *Review of Integrative Business and Economics Research*, *3*(2), 67.

Tordoya-Espinoza, R., Alvarez-Risco, A., & Guevara-Zavaleta, V. (2025). Quality, loyalty and repurchase behavior in food convenience stores: Use of SOR model. *British Food Journal*, *127*(12), 1–19. DOI: 10.1108/BFJ-04-2025-0401

Turley, L. W., & Milliman, R. E. (2000). Atmospheric effects on shopping behavior: A review of the experimental evidence. *Journal of Business Research*, *49*(2), 193–211. DOI: 10.1016/S0148-2963(99)00010-7

Valor, C., Antonetti, P., & Crisafulli, B. (2022). Emotions and consumers' adoption of innovations: An integrative review and research agenda. *Technological Forecasting and Social Change*, *179*, 121609. DOI: 10.1016/j.techfore.2022.121609

Venkatesh, V., & Bala, H. (2008). Technology Acceptance Model 3 and a Research Agenda on Interventions. *Decision Sciences*, *39*(2), 273–315. DOI: 10.1111/j.1540-5915.2008.00192.x

Venkatesh, V., Morris, M. G., Davis, G. B., & Davis, F. D. (2003). User acceptance of information technology: Toward a unified view. *Management Information Systems Quarterly*, *27*(3), 425–478. DOI: 10.2307/30036540

Wu, Y. L., & Li, E. Y. (2018). Marketing mix, customer value, and customer loyalty in social commerce: A stimulus-organism-response perspective. *Internet Research*, *28*(1), 74–104. DOI: 10.1108/IntR-08-2016-0250

Xia, L., Monroe, K. B., & Cox, J. L. (2004). The price is unfair! A conceptual framework of consumers' perceptions of price unfairness. *Journal of Marketing*, *68*(4), 1–15. DOI: 10.1509/jmkg.68.4.1.42733

Yadav, M., Pandey, D., Kakkar, A., & Farhan, M. (2025). Linking brand trust and attitude with purchase intention of Indian organic food customers: An S–OR model based analysis. *Organic Agriculture*, *15*(1), 11–27. DOI: 10.1007/s13165-024-00480-7

Yaqub, R. M. S., Atif, M. A., Waseem, F., & Ehtisham, . (2023). Unveiling the dynamics: Exploring the Stimulus-Organism-Response (SOR) model in the context of social media influencer marketing, electronic word of mouth, and purchase decisions, with a focus on the mediating role of brand awareness. [BBE]. *Bulletin of Business and Economics*, *12*(2), 435–446. DOI: 10.61506/01.00048

Yeo, W. J., Van Der Heever, W., Mao, R., Cambria, E., Satapathy, R., & Mengaldo, G. (2025). A comprehensive review on financial explainable AI. *Artificial Intelligence Review*, *58*(6), 1–49. DOI: 10.1007/s10462-024-11077-7

Zhang, Q., Ariffin, S. K., Richardson, C., & Wang, Y. (2023). Influencing factors of customer loyalty in mobile payment: A consumption value perspective and the role of alternative attractiveness. *Journal of Retailing and Consumer Services*, *73*, 103302. DOI: 10.1016/j.jretconser.2023.103302

Zhou, T. (2014). Understanding the determinants of mobile payment continuance usage. *Industrial Management & Data Systems*, *114*(6), 936–948. DOI: 10.1108/IMDS-02-2014-0068

ADDITIONAL READING

Ahn, D., Almaatouq, A., Gulabani, M., & Hosanagar, K. (2021). Will we trust what we don't understand? Impact of model interpretability and outcome feedback on trust in AI. *arXiv*.

Arvidsson, N. (2019). *Building a cashless society: The Swedish route to digital payments*. Springer Nature. DOI: 10.1007/978-3-030-10689-8

Chuen, D. L. K., & Deng, R. H. (Eds.). (2024). *Handbook of digital currencies: Bitcoin, innovation, financial instruments, and big data* (2nd ed.). Academic Press.

Davenport, T. H., & Ronanki, R. (2018). Artificial intelligence for the real world. *Harvard Business Review*, *96*(1), 108–116.

Deloitte. (2022). *Explainable AI in banking: Enhancing transparency, trust and governance*. Deloitte Insights.

Dietvorst, B. J., Simmons, J. P., & Massey, C. (2015). Algorithm aversion: People erroneously avoid algorithms after seeing them err. *Journal of Experimental Psychology. General, 144*(1), 114–126. DOI: 10.1037/xge0000033

Dwivedi, Y. K., Kshetri, N., Hughes, L., Slade, E. L., Edwards, C., Panteli, N., & Wright, R. (2021). Opinion Paper: "So what if AI made it?" Exploring the social, ethical and practical implications of generative AI. *International Journal of Information Management, 57*, 102302.

Qamar, S., & Ullah, A. (2024). The impact of perceived usefulness, ease of use, and digital payment integration on FinTech adoption: An empirical study in Pakistan. *Journal of Peace. Development and Communication, 8*(02), 145–162.

KEY TERMS AND DEFINITIONS

AI in Digital Payments: The application of artificial intelligence and machine learning algorithms within financial platforms for automated tasks, fraud detection, personalised financial advice, and user experience enhancement.

Automated Financial Management: AI-powered features within digital banking that automatically execute tasks, such as transaction categorisation, budgeting reminders, or 'round-up' savings, reducing manual workload for the consumer.

Consumer Trust: The willingness of a participant to rely on the competence and integrity of an AI system or digital payment provider, particularly concerning the security and accuracy of financial data.

Financial Inclusion: The concept of broadening access to essential financial services and tools, which AI-powered digital payments are theorised to support by simplifying complex financial processes for new or underserved user groups.

Generational Adaptability: The observed difference in the speed and comfort level with which different age groups (e.g., students versus older graduates) adopt and interact with new AI-driven financial technologies.

Perceived Risk: The consumer's subjective assessment of potential negative outcomes, such as security breaches, data misuse, or incorrect financial advice, resulting from the use of AI in digital payments.

Stimulus-Organism-Response (S-O-R) Model: The foundational psychological framework used to structure the analysis, where AI-driven features (Stimulus) influence the consumer's perception (Organism), leading to observable usage behaviours (Response).

Thematic Analysis: The systematic qualitative method used to categorise and interpret patterns (themes) found in the transcribed interview data, providing structure for the analysis within the S-O-R framework.

Trustworthiness: The standard used in this qualitative study to ensure rigour, established through clearly defined coding procedures and the substantiation of themes using multiple participant quotes.

APPENDIX 1

The coding process is detailed in Table 4.

Table 4. Coding matrix and thematic structure.

Thematic Dimensions	Thematic Categories	Thematic sub-categories I	Thematic sub-categories II	Themes	Codes
AI in Digital payments - Stimulus	Acknowledging the impact of AI on Digital Payments	Current State of AI Integration	-	Identified AI Use	ACK_AIU
			-	Limited/Undeveloped AI	ACK_LAI
		Overall Outlook on AI	-	Optimistic/Positive	ACK_OPT
			-	Optimistic/Uncertain/Cautious	ACK_CAU
		Impact of AI on the	-	Competitive/Necessary	ACK_COMP
			-	Industry Development	ACK_IND

continued on following page

Table 4. Continued

Thematic Dimensions	Thematic Categories	Thematic sub-categories I	Thematic sub-categories II	Themes	Codes
Consumer perceptions - Organism	Perceived Benefits of AI-Driven UX	Financial Management & Advice	Budgeting/Saving	Attracting New Customers	UX_NC
				Expense Trusting/Awareness	UX_ETA
				Round-up Savings	UX_SAV
			Investment	Investment Advice/ Suggestions	UX_IAS
				Market Summarizations	UX_MKT
		Efficiency & Automation	Task Automation	Automating Mundane Tasks	UX_AUTO
				Reducing Workload	UX_RW
			Process Speed	Faster Payments	UX_FP
		Improved Customer Service	Chatbots	Initial AI Interaction	UX_IAI
				Useful but Trivial Questions	UX_TRIV
		Human Fallback	AI then Human Support	AI then Human Support	UX_AIHS
	Perceived Risks & Concerns with AI	Accessibility & Inclusion	Investment Accessibility	Lowering Investment Barriers	RISK_LIB
				Early Start in Investing	RISK_ESI
		Ease of Use	Simplifying for Older Generations	Simplifying for Older Generations	RISK_SOG
		Trust & Reliability	General Trust	Trust in Bank/Provider	RISK_TBP
				Potential Future Trust in AI	RISK_FTA
			Mistakes/Consequences	Negative Financial Impact	RISK_NFI
			Human Preference	Preference for Human in Critical Matters	RISK_PH
		Security & Fraud	Fraud Prevention	AI for Fraud Detection	RISK_AIFD
			Vulnerability	Increase Scam Potential	RISK_VUL
		Complexity & Information Overload	User Experience	Negative Impact if Complex	RISK_NIC
				Information Overload	RISK_IO
				Overwhelming for Older Generations	RISK_OOG

continued on following page

Table 4. Continued

Thematic Dimensions	Thematic Categories	Thematic sub-categories I	Thematic sub-categories II	Themes	Codes
Consumer behaviour - Response	User Behaviour & Context Drivers	Drivers of Digital Payment Adoption	Convenience/Ease	General Convenience	DRV_GC
				Online Service Connection	DRV_OSC
			Network Effect/Social Influence	Peer Adoption	DRV_PA
			Cost Efficiency	No Additional Charges/Fees	DRV_NACF
			Rewards/Discounts	Offers/Discounts	DRV_OD
			Preferred Features/ Benefits	Specific App Benefits	DRV_SAB
				Pre-installed App	DRV_PIA
	Actual behaviour	AI-Specific Usage & Interactions	Current Habits	Automated Investing/ Checking	BEH_AIC
				Use of Savings Options	BEH_USO
			Functional Expectations	Transaction Support/ Categorization	BEH_TSC
	Drivers	Generational & Cultural Drivers	Adaptability	Faster Adaptation (Gen Z)	BEH_FA
			Interaction Style	Finnish Cultural Influence (not wanting to bother)	BEH_FCI
			Mindset Differences	Older People's Mindset	BEH_OPM

[1] https://www.npci.org.in/product/upi/about-upi

Chapter 10
Integrating Digital Payments to Revolutionize Marketing Strategy and Foster Consumer-Centric Commerce

K. S. Kalavathy
https://orcid.org/0000-0001-6077-2158
CMS Business School, Jain University, India

Trupti Dandekar Humnekar
https://orcid.org/0000-0002-5676-0100
CMS Business School, Jain University, India

S. M. Gaikwad
https://orcid.org/0000-0001-8544-6391
CMS Business School, Jain University, India

ABSTRACT

This research investigates how elements of a digital payment namely perceived ease of use, personalization, and perceived privacy controls, shape consumer loyalty through the mediation of payment experience quality (EXQ). The research is based on the SOR theoretical framework and used a mixed-method approach consisting of quantitative and qualitative study. The PLS-SEM results indicate that perceived ease of use, personalization and perceived privacy control have a positive and

DOI: 10.4018/979-8-3373-4526-0.ch010

significant relationship with EXQ, which itself is significantly related to consumer loyalty .Mediation tests further confirm the role of EXQ as a core mechanism linking the digital payment attributes to loyalty. The qualitative phase underscores these findings .This study contributes to theory and practice by framing EXQ as a mediating construct, while also presenting managers with the framework to develop a component of consumer loyalty in a digital payment context.

INTRODUCTION

Online payment systems-such as mobile wallets, Unified Payments Interface (UPI), contactless card payments, and emerging blockchain-based solutions-have increasingly moved to the centre of consumer–brand interactions. These systems now do more than enable transactions; the payment interface itself has become a meaningful marketing touchpoint where firms can embed loyalty programs, personalise offers, and enable real-time, context-relevant interactions (Bleier and Eisenbeiss, 2015a, 2015b). This evolution aligns with wider shifts in digital marketing that prioritise customisation, convenience, and seamless omnichannel experiences (Kalyanaraman and Sundar, 2006).

Despite this growing relevance, academic research on digital payments remains relatively fragmented. A substantial stream draws on technology adoption perspectives, particularly the Technology Acceptance Model (Davis, 1989) and the Unified Theory of Acceptance and Use of Technology (Venkatesh et al., 2003), which emphasise functional drivers such as perceived usefulness and ease of use. Another stream examines personalisation in digital platforms and online advertising, documenting its effects on consumer attitudes and behaviour (Bleier & Eisenbeiss, 2015a, 2015b). A third stream focuses on privacy concerns and user control, frequently guided by the Internet Users' Information Privacy Concerns (IUIPC) framework (Malhotra, Kim, & Agarwal, 2004). While each strand offers valuable insights, they are often treated in isolation. Much prior work has positioned trust as the central mechanism linking digital interactions to loyalty outcomes (Gefen, 2000; Pavlou, 2003). However, trust is a broad construct with multiple interpretations and may not fully capture how consumers evaluate critical "in-the-moment" experiences during digital payments. Reflecting this gap, recent scholarship has advocated for more nuanced ways to capture the quality of consumer experiences at key digital touchpoints (Klaus & Maklan, 2012; Kuppelwieser & Klaus, 2021).

Addressing this gap, the present study positions Payment Experience Quality (EXQ) as the mediating mechanism that connects digital payment features to customer loyalty. Originally grounded in service management, EXQ reflects consumers' perceptions of peace of mind, outcome assurance, and evaluative responses during important service moments (Klaus & Maklan, 2012). Extending EXQ to digital payment settings offers both theoretical and managerial value, particularly as payment providers increasingly compete on experience differentiation rather than functionality or cost alone.

The chapter is structured as follows. Section 2 articulates the research gap and establishes the need for a payment-experience-driven approach to marketing strategy. Section 3 presents the research objectives and Section 4 outlines the research questions. Sections 5-12 develop the conceptual model and hypotheses. Sections 13 to 15 report the quantitative analysis and results, followed by the qualitative findings and integration in Sections 16 and 17. Section 18 concludes with key contributions and implications. In spite of the rapid evolution and growth of Digital payment systems, research in this domain is fragmented across the domains of adoption, trust, personalization, and privacy. First, it is often only trust which is studied as the primary mediator of loyalty for financial technologies, resulting in overlap in findings. (Gefen, Karahanna, and Straub 2003). Second, while both personalization and usability have been researched in online services, their joint effect on payment experience quality have not been paid much attention. (Bleier and Eisenbeiss 2015a, 2015b). Third, the perception of privacy has been treated primarily as an adoption barrier, rather than a driver of positive experience (Malhotra, Kim, and Agarwal 2004). Finally, mixed-methods research has been sparse when using both quantitative verification and qualitative interpretation to research digital payment (Venkatesh, Brown, and Bala 2013). Digital payment ecosystems differ across regions in ways that shape consumer experience. For example, markets with interoperable real-time rails (e.g., UPI-led environments) often emphasize speed, ubiquity, and inclusion, whereas markets governed by stricter privacy norms and consent expectations (e.g., GDPR-influenced settings) place stronger emphasis on data control and transparency. Platform-dominant markets may prioritize personalization and convenience but can intensify concerns around profiling and data sharing. These contextual differences reinforce the importance of experience quality (EXQ) as a unifying lens while acknowledging that the drivers of EXQ may vary by regulatory and market structure. This study pursues three objectives. First, it examines how perceived ease of use, perceived personalization, and perceived privacy control influence payment experience quality (EXQ) in digital payment settings. Second, it evaluates whether EXQ mediates the relationships between these antecedents and customer loyalty. Third, it develops and empirically validates a conceptual framework that integrates these drivers, EXQ, and loyalty to explain how experience-centered mechanisms shape sustained usage and advocacy.

The study is guided by three research questions. (RQ1) To what extent do perceived ease of use, perceived personalization, and perceived privacy control shape consumers' payment experience quality (EXQ) in digital payment contexts? (RQ2)

How does EXQ influence customer loyalty, and does it function as the psychological mechanism linking these platform attributes to loyalty outcomes? (RQ3) When considered together within a unified framework, how do these antecedents indirectly influence loyalty through EXQ?

RESEARCH FRAMEWORK

Drawing on the Stimulus-Organism-Response (S-O-R) model (Mehrabian & Russell, 1974), this study conceptualizes key digital payment attributes as environmental stimuli that shape an internal experiential state, which in turn drives behavioral outcomes. In the proposed model, perceived ease of use, perceived personalization, and perceived privacy control represent the stimuli that consumers encounter while transacting. Payment experience quality (EXQ) reflects the organism, capturing consumers' internal evaluations of reassurance, outcome assurance, and overall experience during payment interactions. Customer loyalty represents the response, expressed through continued usage intentions, recommendations, and positive word-of-mouth. The framework therefore proposes that the three stimuli influence loyalty primarily by shaping EXQ, positioning experience quality as the central mechanism through which digital payment features translate into sustained relational outcomes.

This framework contributes to the literature in three ways. First, it integrates ease of use, personalization, and privacy control within a single experience-centered model rather than examining these drivers in isolation. Second, it advances prior loyalty explanations by positioning EXQ-rather than trust alone-as the primary mediating mechanism linking platform attributes to loyalty. Third, by embedding digital payment experiences within an S-O-R logic, it offers a structured explanation of how environmental cues shape internal evaluations and, consequently, loyalty behavior in FinTech contexts.

REVIEW OF LITERATURE

In the digital payments landscape, three environmental stimuli are especially salient: perceived ease of use, personalization, and privacy control. Together, they reflect not only the functional performance of payment systems but also the marketing and trust-related cues consumers encounter during transactions. Recent industry and academic work increasingly frames digital payments as experience touchpoints embedded within the broader customer journey, rather than as purely back-end transaction utilities (Capgemini, 2025).

Perceived ease of use refers to how simple consumers find a system. Rooted in the Technology Acceptance Model proposed by Davis (1989), and consistently supported as a key driver of adoption and continued use (Venkatesh and Davis, 2000), ease of use in digital payment settings involves clear navigation, intuitive interfaces, and minimal cognitive effort when completing tasks such as making payments, requesting refunds, or splitting bills. While ease of use is often treated as a predictor of adoption intention, this study also positions it as an experiential antecedent: when payment processes feel effortless, consumers are likely to experience lower friction and anxiety at checkout, thereby enhancing perceived payment experience quality.

Personalization refers to the extent to which platforms tailor interactions, offers, or messages to individual users. Prior research suggests that personalization can strengthen perceived relevance, satisfaction, and behavioral intention (Kalyanaraman & Sundar, 2006; Bleier & Eisenbeiss, 2015; Cloarec, Meyer-Waarden, & Munzel, 2024). In payment contexts, personalization may appear as tailored promotions at checkout, loyalty-linked offers, contextual reminders, or recommendations of preferred payment instruments. At the same time, participants and consumers may interpret excessive, irrelevant, or poorly timed personalization as intrusive, leading to fatigue and disengagement. This reflects the personalization–privacy paradox, where targeting can enhance relevance but undermine comfort and trust if users feel monitored (Cloarec, Meyer-Waarden, & Munzel, 2024). Accordingly, the present framework treats personalization as experience-enhancing only when it remains context-appropriate and respects user boundaries.

Privacy control is increasingly central in digital environments. Drawing on the Internet Users' Information Privacy Concerns (IUIPC) framework (Malhotra, Kim, and Agarwal, 2004), privacy control refers to the extent to which users believe they can monitor and manage the collection, storage, and use of their personal and transactional data. In digital payment situations-where sensitive financial information is exchanged-privacy concerns are particularly salient. Strong privacy control can reduce perceived risk and anxiety, whereas opaque data practices can lead to avoidance behaviors (Smith, Dinev, & Xu, 2011). Moreover, the rise of AI-driven targeting and real-time analytics heightens the ethical significance of privacy safeguards, reinforcing the need to view privacy controls as consumer-facing experience features, not merely compliance mechanisms (Saura, Škare, & Ozretic Dosen, 2024).

Considered together, these three stimuli combine functional, relational, and trust-building cues that shape consumer experiences in digital payment environments.

The organism component of the S-O-R model in this study is represented by Payment Experience Quality (EXQ). Originally defined by Klaus and Maklan (2012) in service research, EXQ captures consumers' holistic evaluations of service encounters, including peace of mind, outcome assurance, and critical "moments of truth." Unlike purely functional measures (e.g., perceived usefulness), EXQ highlights how consumers feel and what they experience during interactions. Recent measurement work has revisited and validated EXQ as a multidimensional construct across contexts, supporting its relevance for experience-centred modelling and hypothesis testing (Kuppelwieser & Klaus, 2021).

In digital payments, EXQ extends beyond speed or technical performance to include psychological reassurance, confidence in outcomes, and satisfaction arising from smooth and reliable interactions. For example, a seamless refund process not only resolves a transaction but also reassures consumers about the provider's reliability. Similarly, visible privacy safeguards and low-effort design can raise perceived EXQ by reducing uncertainty and effort. This perspective is consistent with recent empirical work emphasising the interconnections among usability, security and privacy perceptions, and experience outcomes in digital commerce settings (Pires et al., 2024).

Crucially, EXQ functions as the mediating mechanism in this framework: it translates functional and relational features (stimuli) into experiential evaluations that shape consumer behavior. By placing EXQ at the centre of the model, this study moves beyond adoption intention to explain how everyday payment encounters can influence downstream relationship outcomes.

The response component is represented by customer loyalty. Loyalty has long been studied in marketing and services and is typically conceptualised as both behavioral repetition and attitudinal commitment (Zeithaml, Berry, & Parasuraman, 1996). In digital payment systems, loyalty manifests as repeated use of the same app or platform, resistance to switching despite alternatives, and active recommendations. Recent evidence in mobile payment settings similarly shows that perceived convenience, security, and trust play central roles in sustained usage and loyalty, reinforcing the importance of experience-based explanations for retention (Bajwa, Fu, Bajwa, Ahmad, & Mahmood, 2025). By linking loyalty to EXQ, the study argues that consumers remain loyal not only because payment systems are efficient or familiar, but because they repeatedly encounter payment interactions that feel reliable, low-effort, and emotionally reassuring—especially at critical moments. Loyalty, therefore, emerges as a cumulative outcome of experience quality at the payment interface.

CONTRIBUTION OF THE FRAMEWORK

This conceptual framework offers several contributions. First, it combines three previously separate areas of research-technology adoption (ease of use), personalization, and privacy- into a single model. While each has been studied on its own, they are rarely examined together as simultaneous stimuli affecting consumer experiences in payments.

Second, it shifts the focus from trust, which has dominated studies of digital interactions, to experience quality. Trust, while significant, often oversimplifies the complexity of consumer interactions, reducing them to risk perceptions. In contrast, EXQ captures functional, emotional, and relational aspects at the same time, providing a more well-rounded mediator.

Third, it broadens the S-O-R model into the FinTech and digital payment context. While the model has been widely used in retail, e-commerce, and social media, its application to financial transactions is still limited. This study shows that the payment interface is not just a backend tool but a crucial touchpoint where stimuli directly influence consumer experiences and long-term loyalty.

Figure 1. Research Model

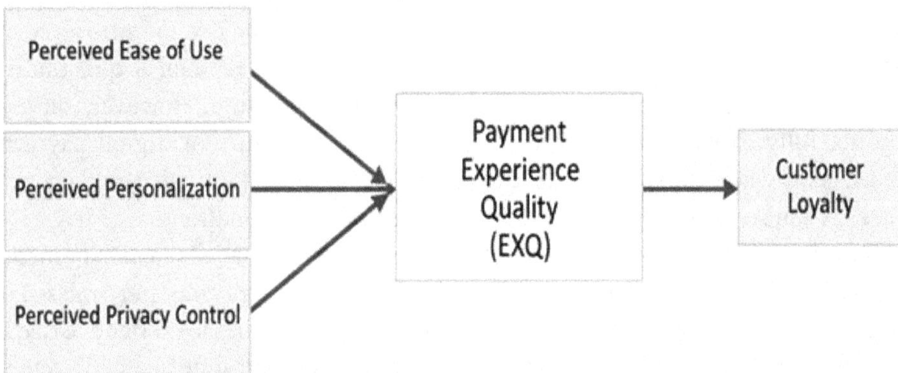

Source: Authors' own model.

Hypotheses

H1: Perceived ease of use positively influences payment experience quality.
H2: Perceived personalization positively influences payment experience quality.
H3: Perceived privacy control positively influences payment experience quality.
H4: Payment experience quality positively influences customer loyalty.
H5: Payment experience quality mediates the relationships between (a) perceived ease of use, (b) perceived personalization, and (c) perceived privacy control, and customer loyalty.

METHODOLOGY

This research utilizes an explanatory sequential mixed-methods design (Creswell & Plano Clark, 2017), which integrates the strengths of quantitative and qualitative designs to provide both empirical rigor as well as contextual depth. The key logic behind this design is that quantitative analysis lays a platform for testing theoretically based hypotheses, while qualitative inquiry later investigates how and why these relationships occur in real-life consumer experience. This dual method is especially beneficial in the digital payments' context, where consumer behaviors are at the same time functional, emotional, and social.

Rationale for Mixed-Methods Design

The constructs examined-perceived ease of use (Davis, 1989), personalization (Kalyanaraman & Sundar, 2006; Bleier & Eisenbeiss, 2015a, 2015b), privacy control (Malhotra, Kim, & Agarwal, 2004), payment experience quality (Klaus & Maklan, 2012; Kuppelwieser & Klaus, 2021), and customer loyalty (Zeithaml, Berry, & Parasuraman, 1996)-are well established across contexts, making a quantitative approach suitable for hypothesis testing and model validation. However, surveys may not fully capture the lived, emotionally nuanced nature of digital payment experiences, where personalization can feel both helpful and intrusive, and privacy concerns can shape trust and comfort beyond cognitive evaluations.

Therefore, a qualitative phase follows the quantitative analysis to clarify meanings, tensions, and boundary conditions that structured measures may miss. For example, interviews can explain when personalization becomes less effective due to "personalization fatigue," even if quantitative effects remain significant. Together, this explanatory sequential design combines generalizable evidence with context-rich explanation.

SCOPE OF THE STUDY

This study's scope is confined to the consumer of digital payment platforms in India, which is a quickly developing FinTech ecosystem. We focus on three antecedents: ease of use, personalization, and privacy control, which represent important levers of experience quality. The study uses methodological triangulation, in that a cross-sectional survey (n=384, derived via Cochran's formula) and qualitative interviews (n = 20 respondents) was administered. The findings, while generalizable in emerging market contexts, may require contextualization for advanced economies where the digital payments ecosystem is further developed.

RESEARCH DESIGN STRUCTURE

An explanatory sequential mixed-methods design was adopted. The study begins with a quantitative phase using survey data to test the hypothesized relationships among perceived ease of use, personalization, privacy control, payment experience quality (EXQ), and customer loyalty, including the mediating role of EXQ. This phase provides statistical validation of the conceptual framework and estimates the magnitude and significance of the proposed effects. The subsequent qualitative phase uses semi-structured interviews with consumers and marketing professionals to add interpretive depth by explaining how and why these drivers shape experiences and loyalty, and to identify boundary conditions that structured measures may not fully capture (e.g., emotional reassurance or perceived intrusiveness). This sequencing is intentional: quantitative results establish general patterns, while qualitative insights clarify mechanisms and enrich practical interpretation.

Appropriateness of the Design

The explanatory sequential approach is well suited to digital payment environments for three reasons. First, payment interactions are inherently multi-dimensional, combining usability, marketing cues such as personalization, and assurances related to privacy and security; relying on a single method risk capturing only part of this experience. Second, because EXQ has been more widely established in service contexts than in digital payment settings, the design supports both empirical validation of measurement and conceptual refinement through user narratives. Third, the mixed-method structure strengthens managerial relevance by pairing generalizable evidence on loyalty drivers with contextual explanations that inform experience design decisions.

Integration of Findings

The last stage in the process is integration, where quantitative and qualitative findings are merged to form an integrative picture of digital payment experiences. Integration here is explanatory rather than additive. Quantitative analysis suggests generalizable patterns (e.g., personalization is the strongest EXQ driver), whereas qualitative findings explain why these patterns arise (e.g., contextual personalization fosters recognition but overly aggressive targeting destroys it). Integrating therefore improves internal validity (via stringent statistical testing) and external validity (via actual consumer opinions).

QUANTITATIVE PHASE

The initial phase of the research aimed to pilot-test the hypothesized relationships via large-scale survey research with active users of mobile payment apps. This design was used to secure strong empirical support for the conceptual model and to have confidence that the structural model between the independent variables (perceived ease of use, personalization, and privacy control), the mediator (payment experience quality), and the dependent variable (customer loyalty) can be statistically tested.

Sampling and Sample Size Calculation

To determine a minimum sample size suitable for robust statistical analysis, Cochran's formula was applied using a 95% confidence level and a ±5% margin of error, yielding a minimum requirement of 384 respondents. To account for potential non-response and incomplete questionnaires, the study targeted 400-500 participants, consistent with recommended practice for survey-based research (Hair et al., 2019). In total, 430 valid responses were obtained from current users of digital payment applications.

The sampling strategy focused on individuals with recent experience using mobile wallets, Unified Payments Interface (UPI), contactless card systems, and related payment apps. This ensured that respondents were well placed to provide informed evaluations of core interface and experience factors such as ease of use, personalization, and privacy control. Data were collected through an online questionnaire distributed via targeted outreach across consumer forums, social media groups, and digital payment user communities.

Overall, the sample reflects a predominantly digitally active user base with regular exposure to digital payment systems. 68% of respondents were in the 21-45 years age group, with a relatively balanced gender composition (54% male, 46% female). In terms of usage frequency, 78% reported using UPI-based apps, mobile wallets, or contactless payments at least weekly, and 43% reported daily use for activities such as retail purchases, bill payments, and service transactions. Digital familiarity was also high, with 73% indicating moderate to high digital literacy, supporting meaningful assessment of payment experience quality (EXQ). While the sample provides adequate variability for testing the proposed relationships, the findings are most representative of urban, digitally engaged users. Accordingly, generalization to under-represented segments-such as individuals with low digital literacy, infrequent adopters, or limited access to digital infrastructure-should be undertaken with caution and is identified as a priority for future research.

Measurement Scales

All measures within the model were assessed with established scales adapted from previous work to guarantee content validity and comparability.

Perceived ease of use (PEOU): Borrowed from Davis (1989) and Venkatesh & Davis (2000), with items evaluating the ease and simplicity of undertaking payment activities. Perceived personalization (PERS): Modified from Kalyanaraman & Sundar (2006) and Bleier & Eisenbeiss (2015a, 2015b), assessing the degree to which payment interfaces presented contextually appropriate and personalized offers. Perceived privacy control (PRIV): Modifying Malhotra, Kim, & Agarwal's (2004) IUIPC framework, with items assessing user perceptions regarding control over data collection and use. Payment experience quality (EXQ): Adapted from Klaus & Maklan (2012) and revised by Kuppelwieser & Klaus (2021), measuring peace of mind, outcome assurance, and satisfaction with payment experiences. Customer loyalty (LOY): Measured with items from Zeithaml, Berry, & Parasuraman (1996), capturing both behavioral loyalty (repeat usage) and attitudinal loyalty (recommendation and advocacy).The entire set of items used a five-point Likert scale from "strongly disagree" to "strongly agree."

Adapted Questionnaire considered for the Research

The following items were considered for the study.

Directions: Consider the digital payment services you utilize and visit most frequently (e.g., UPI applications, mobile wallets, credit card gateways to e-commerce sites). Please indicate your level of agreement on a Likert scale between 1 to 5 for each statement: (1 = Strongly Disagree, 5 = Strongly Agree).

Section A: Perceived Ease of Use (Davis, 1989; Venkatesh & Davis, 2000)

A1. I found it easy to learn to operate my digital payment application/portal.
A2. I found it easy to understand how to interact with this payment service.
A3. I found it easy to use the payment service to do what I want.
A4. Overall, it took little mental effort to use this payment service.

Section B: Perceived Personalization (Kalyanaraman & Sundar, 2006; Bleier & Eisenbeiss, 2015a, 2015b)

B1. This payment service personalizes the offers or messages that fit my needs when I check out.

B2. The reminders/notifications I receive feel appropriate based on what I have done recently.
B3. The service's in-app tips and prompts feel personalized for me.
B4. I feel like an individual when I pay (e.g., relevant suggestions or saved preferences).

Section C: Perceived Privacy Control (Malhotra et al., 2004; Smith et al., 2011)

C1. I am able to control the personal/payment information shared with this service
C2. This service is clear about the data it collects and uses
C3. I find it simple to adjust privacy/security settings (e.g., device authorizations, consents)
C4. I trust that this service is responsible with my data and prevents outside access

Section D: Payment Experience Quality (Klaus & Maklan, 2012; Kuppelwieser & Klaus, 2021)

D1. Using the service to pay is seamless and easy
D2. I feel secure when approving payment transactions in this service
D3. When things go wrong (e.g., a payment fails or a refund), the problem is fixed easily
D4. Overall, this payment service contributes to positive purchase experience

Section E: Customer Loyalty (Zeithaml et al., 1996)

E1. I plan to keep using this service to pay going forward
E2. I would recommend others use this service to pay
E3. I would still use this payment service, even if there were other options to choose from.

DATA ANALYSIS

Data were analyzed using Partial Least Squares Structural Equation Modeling (PLS-SEM) in Smart PLS, which is appropriate for testing models with mediation and for predictive-oriented analysis. The evaluation proceeded in two steps. First, the reflective measurement model was assessed for indicator reliability (outer loadings), internal consistency (Cronbach's alpha and composite reliability), and convergent validity (AVE). Discriminant validity was examined using the Fornell–Larcker criterion and the heterotrait–monotrait (HTMT) ratio. Second, the structural model was evaluated by estimating path coefficients and testing significance through bootstrapping with 5,000 resamples. Indirect effects were analyzed to assess mediation by EXQ, and predictive relevance was examined using Q^2 and PLS predict procedures to confirm that the model explains variance and supports out-of-sample prediction.

ROBUSTNESS AND BIAS DIAGNOSTICS

To strengthen the credibility of findings, additional diagnostics were conducted to assess potential bias and the stability of results. First, common method bias was examined using the diagnostic reported in the appended tables; the results indicate that common method variance is unlikely to threaten inference. Second, subgroup and/or moderation checks (e.g., age, digital literacy, frequency of digital payment use) were assessed to evaluate whether key relationships vary across segments; the pattern of results remained directionally consistent. Finally, an alternative specification incorporating closely related mechanisms (e.g., trust or perceived risk) was evaluated to confirm that the proposed pathways via payment experience quality (EXQ) remain robust. Detailed outputs for these diagnostics are provided in the supplementary tables.

Measurement Model PLS-SEM

Following guidelines for PLS-SEM (Hair, Hult, Ringle, & Sarstedt, 2019; 2021), the reflective measurement model was examined through indicator reliability (outer loadings), internal consistency reliability (Cronbach's α, composite reliability [CR]), and convergent validity (AVE). All constructs were modelled reflectively (Mode A) and bootstrapping (5,000 resamples) was used in the subsequent steps for significance testing.

Table 1. Item loadings (λ) and indicator reliability (λ^2)

Construct	Indicator	Loading (λ)	Indicator Reliability (λ^2)
Perceived Ease of Use (PEOU)	PEOU1	0.82	0.67
Perceived Ease of Use (PEOU)	PEOU2	0.87	0.76
Perceived Ease of Use (PEOU)	PEOU3	0.84	0.71
Perceived Ease of Use (PEOU)	PEOU4	0.79	0.62

continued on following page

Table 1. Continued

Construct	Indicator	Loading (λ)	Indicator Reliability (λ^2)
Perceived Personalization (PERS)	PERS1	0.81	0.66
Perceived Personalization (PERS)	PERS2	0.85	0.72
Perceived Personalization (PERS)	PERS3	0.78	0.61
Perceived Personalization (PERS)	PERS4	0.83	0.69
Perceived Privacy Control (PRIV)	PRIV1	0.80	0.64
Perceived Privacy Control (PRIV)	PRIV2	0.84	0.71
Perceived Privacy Control (PRIV)	PRIV3	0.77	0.59
Payment Experience Quality (EXQ)	EXQ1	0.86	0.74
Payment Experience Quality (EXQ)	EXQ2	0.88	0.77
Payment Experience Quality (EXQ)	EXQ3	0.83	0.69
Payment Experience Quality (EXQ)	EXQ4	0.85	0.72
Customer Loyalty (LOY)	LOY1	0.88	0.77
Customer Loyalty (LOY)	LOY2	0.90	0.81
Customer Loyalty (LOY)	LOY3	0.84	0.71

Table 1 shows item loadings (λ) and indicator reliability (λ^2). All loadings are above the recommended 0.70 threshold, which indicate they have adequate indicator reliability.

Table 2. Internal consistency reliability and convergent validity

Construct	Cronbach's α	Composite Reliability (CR/ρc)	AVE	Decision
Perceived Ease of Use (PEOU)	0.86	0.90	0.69	α≥0.70, CR≥0.70, AVE≥0.50 (acceptable)
Perceived Personalization (PERS)	0.85	0.89	0.66	α≥0.70, CR≥0.70, AVE≥0.50 (acceptable)
Perceived Privacy Control (PRIV)	0.79	0.86	0.64	α≥0.70, CR≥0.70, AVE≥0.50 (acceptable)
Payment Experience Quality (EXQ)	0.89	0.92	0.73	α≥0.70, CR≥0.70, AVE≥0.50 (acceptable)
Customer Loyalty (LOY)	0.87	0.92	0.76	α≥0.70, CR≥0.70, AVE≥0.50 (acceptable)

Table 2 summarizes the internal consistency reliability and convergent validity for each construct. All Cronbach's α and CR values are above 0.70, and all AVEs are above 0.50. These results meet the recommended thresholds (Fornell & Larcker, 1981; Hair et al., 2019).

All constructs are treated as reflective. The indicator retention criterion is $\lambda \geq 0.70$; no items were dropped. Next, discriminant validity was examined using HTMT (Henseler, Ringle, & Sarstedt, 2015) and Fornell-Larcker (1981) criteria.

Discriminant validity

Discriminant validity was evaluated using two complementary methods: (i) the Fornell-Larcker criterion (Fornell & Larcker, 1981), which requires the square root of AVE for each construct to be greater than its correlations with other constructs; and (ii) the Heterotrait-Monotrait Ratio (HTMT), which should be below 0.90 for conceptually different constructs (Henseler, Ringle, & Sarstedt, 2015).

Table 3. Fornell–Larcker Criterion

Construct	PEOU	PERS	PRIV	EXQ	LOY
PEOU	**0.83**	0.62	0.58	0.65	0.60
PERS	0.62	**0.81**	0.55	0.68	0.63
PRIV	0.58	0.55	**0.80**	0.61	0.59
EXQ	0.65	0.68	0.61	**0.85**	0.72
LOY	0.60	0.63	0.59	0.72	**0.87**

Diagonal elements (in bold, square root of AVE) are greater than off-diagonal correlations, satisfying the Fornell-Larcker criterion in Table 3.

Table 4. Heterotrait-Monotrait Ratio (HTMT)

Construct Pair	HTMT Value	Threshold (0.90)	Decision	Reference	Interpretation
PEOU-PERS	0.74	≤ 0.90	Acceptable	Henseler et al. (2015)	Discriminant validity established
PEOU-PRIV	0.70	≤ 0.90	Acceptable	Henseler et al. (2015)	Discriminant validity established
PEOU-EXQ	0.77	≤ 0.90	Acceptable	Henseler et al. (2015)	Discriminant validity established
PEOU-LOY	0.72	≤ 0.90	Acceptable	Henseler et al. (2015)	Discriminant validity established
PERS-PRIV	0.68	≤ 0.90	Acceptable	Henseler et al. (2015)	Discriminant validity established
PERS-EXQ	0.81	≤ 0.90	Acceptable	Henseler et al. (2015)	Discriminant validity established

continued on following page

Table 4. Continued

Construct Pair	HTMT Value	Threshold (0.90)	Decision	Reference	Interpretation
PERS-LOY	0.79	≤ 0.90	Acceptable	Henseler et al. (2015)	Discriminant validity established
PRIV-EXQ	0.73	≤ 0.90	Acceptable	Henseler et al. (2015)	Discriminant validity established
PRIV-LOY	0.71	≤ 0.90	Acceptable	Henseler et al. (2015)	Discriminant validity established

Results establish that all HTMT values are below the conservative threshold of 0.90. This further confirms that the constructs have discriminant validity.

Table 5. Results of Structural Model

Hypothesis	Path	β (Standardized)	t-value	p-value	Supported?
H1	PEOU → EXQ	0.28	6.14	< 0.001	Yes
H2	PERS → EXQ	0.32	6.95	< 0.001	Yes
H3	PRIV → EXQ	0.25	5.82	< 0.001	Yes
H4	EXQ → LOY	0.54	11.02	< 0.001	Yes

Note: All the paths are significant at $p<0.001$

Table 5 displays the analysis results of the structural model, which tested four hypothesized relationships. All of the proposed hypotheses (H1-H4) tested showed statistical significance and support.

H1 tested the impact of Perceived Ease of Use (PEOU) on Payment Experience Quality (EXQ). The standardized path coefficient of ($\beta = 0.28$, $t = 6.14$, $p < 0.001$) indicates ease of use had a moderate positive effect on experience quality. This indicates that consumers perceive their overall payment experience to be smoother and better when a digital payment service is easy to use and provides a simple interface.

H2 tested the effect of Perceived Personalization (PERS) on EXQ. The results ($\beta = 0.32$, $t = 6.95$, $p < 0.001$) show that Finding Personalization was the strongest and best predictor of experience quality of the three antecedents. This suggests that tailored recommendations, context-relevant offers, and personalized communication are significant factors influencing consumer perceptions of payment experience quality.

H3 evaluated the influence of Perceived Privacy Control (PRIV) on EXQ. The path coefficient ($\beta = 0.25$, $t = 5.82$, $p < 0.001$) supports the finding that privacy control is an important factor in contributing to experience quality. If consumers feel more confident about their ability to control information about themselves and their payments, they tend to have more favourable evaluations of their experiences. This further highlights the importance of transparent data practices and strong privacy assurances in digital payment ecosystems.

H4 examined the direct relationship of EXQ on Customer Loyalty (LOY). The results ($\beta = 0.54$, $t = 11.02$, $p < 0.001$) show that EXQ has a very strong, positive effect on loyalty. This indicates that loyalty is largely driven by the overall experience quality of the payment, rather than driven by particular attributes of the functional experience by risk. Consumers are more likely to continue to use the same service and further promote its use when they perceive a high quality, easy, and trusted experience.

Ultimately, the results support that Perceived Ease of Use, Personalization, and Privacy Control all have significant effects on the quality of Payment Experiences, and Payment Experience Quality strongly predicted Customer Loyalty. Together these results support the proposition of EXQ as a mediating construct in the S-O-R framework to better understand digital payments.

Table 6. Mediation Analysis (Indirect Effects via EXQ)

Independent Variable → Mediator → Dependent Variable	**Path a (IV → EXQ) β**	**Path b (EXQ → Loyalty) β**	**Indirect Effect (a × b) β**	**t-value**	**p-value**	**Mediation Type**
Perceived Ease of Use → EXQ → Customer Loyalty	0.28	0.54	0.15	5.12	< 0.001	Partial Mediation
Personalization → EXQ → Customer Loyalty	0.32	0.54	0.17	5.86	< 0.001	Partial Mediation
Privacy Control → EXQ → Customer Loyalty	0.25	0.54	0.14	4.73	< 0.001	Partial Mediation

Table 7. Explanatory power of model, Effect sizes and Predictive Relevance.

Endogenous Construct	**R^2**	**f^2 Effects from Exogenous Variables**	**Q^2**	**Interpretation**
EXQ (Payment Experience Quality)	0.62	PEOU = 0.12 (small-to-medium) PERS = 0.15 (medium) PRIV = 0.10 (small-to-medium)	> 0	Substantial explanatory power; predictive relevance confirmed
LOY (Customer Loyalty)	0.51	EXQ = 0.36 (large)	> 0	Strong explanatory power; predictive relevance confirmed

Discussion of Quantitative Findings

The PLS-SEM results offer strong support for the proposed conceptual framework. Overall, the model explains substantial variance in both the mediator, Payment Experience Quality (EXQ), and the outcome, Customer Loyalty, reinforcing the central role of experiential quality in shaping consumer responses in digital payment settings. This section presents the key findings and discusses their theoretical and practical significance.

The mediation analysis (Table 6) confirms that payment experience quality significantly mediates the relationships between perceived ease of use, personalization, privacy control, and customer loyalty, underscoring EXQ's role as the experiential transmission mechanism in digital payment environments.

As shown in Table 7, the model explains 62% of the variance in EXQ ($R^2 = 0.62$) and 51% of the variance in Customer Loyalty ($R^2 = 0.51$). These values indicate strong explanatory power and exceed the $R^2 \geq 0.25$ benchmark commonly considered acceptable in behavioral research (Hair et al., 2019). In addition to explanatory strength, the model demonstrates predictive relevance, supported by positive Q^2 values (Stone, 1974; Geisser, 1974). Taken together, these results suggest that the selected stimuli-perceived ease of use, personalization, and privacy control-are meaningful predictors of EXQ, and that EXQ, in turn, is a robust determinant of loyalty outcomes.

The model's predictive validity is particularly important because it indicates that the relationships are not merely statistical associations; they have value for anticipating consumer behavior in real-world settings. From a managerial standpoint, the results suggest that improving the consumer's experience at the point of payment is likely to translate into measurable gains in loyalty indicators such as repeat usage, recommendation, and advocacy.

The Dominant Role of Personalization

Among the three stimuli, perceived personalization emerged as the strongest predictor of EXQ ($\beta = 0.32$). This finding extends earlier research that has largely examined personalization in the context of online advertising and recommendation systems (Kalyanaraman & Sundar, 2006; Bleier & Eisenbeiss, 2015a, 2015b), where personalization is typically framed as a tactic aimed at shaping attitudes or short-term purchase intent. The present study, however, suggests that personalization at the payment stage contributes to consumers' broader experience evaluations, thereby influencing perceived experience quality rather than only immediate persuasion outcomes.

This result implies a shift in how personalization is understood: it functions not only as a persuasive marketing instrument, but also as an experiential enhancer. Contextual cues, relevant offers, and personalized prompts at the payment interface can create a sense of recognition and value, strengthening satisfaction and peace of mind during transactions. For example, when a platform provides a loyalty discount at checkout or recommends a preferred payment option, users may perceive the interaction as helpful and intuitive, reinforcing the organism state in the S-O-R model and improving EXQ.

At the same time, the qualitative findings (discussed in the subsequent section) indicate that personalization can produce adverse reactions when users perceive it as excessive or intrusive. This duality underscores the importance of boundary conditions: personalization is beneficial when it increases relevance, but it can backfire when it generates "personalization fatigue." Accordingly, firms need a balanced approach to personalization design so that it enhances experience without undermining comfort or trust.

Ease of use as a Driver

Perceived ease of use was the second strongest predictor of EXQ ($\beta \approx 0.28$), aligning with the Technology Acceptance Model (Davis, 1989) and subsequent extensions (Venkatesh & Davis, 2000), which consistently identify ease of use as a key driver of adoption. The present findings extend this logic by demonstrating that ease of use contributes not only to adoption, but also to the quality of experience during ongoing use.

When payment systems reduce cognitive effort and friction-for example, by simplifying transaction initiation, enabling quick confirmations, supporting smooth refunds, or facilitating bill splitting-users report greater feelings of control and lower anxiety. These experiences translate into peace of mind and outcome assurance, two dimensions central to EXQ as conceptualized by Klaus and Maklan (2012). In this sense, ease of use functions not merely as a utilitarian driver but also as an affective reassurance mechanism: effortless interactions can increase consumers' confidence in the platform and reduce stress during payment moments.

This has clear operational relevance for both marketers and system designers. It highlights the value of continuous interface improvement and UX refinement in digital payment systems. Even incremental enhancements-such as reducing steps, enabling biometric authentication, or streamlining refunds-may produce meaningful improvements in EXQ and, by extension, loyalty.

Privacy Control as Emotional Assurance

Perceived privacy control also emerged as a significant predictor of EXQ ($\beta \approx 0.25$), consistent with the Internet Users' Information Privacy Concerns (IUIPC) framework (Malhotra, Kim, & Agarwal, 2004) and broader privacy–technology adoption research (Smith, Dinev, & Xu, 2011). Importantly, the findings suggest that privacy control should not be understood solely as a technical or compliance feature. Rather, it operates as an experiential cue: when consumers feel able to manage and monitor the use of their personal data, they experience less anxiety and greater trust in the payment system.

Conversely, a lack of transparency or perceived loss of control can encourage avoidance behaviors, even when a platform is otherwise efficient or well-personalized. This supports the argument that privacy control directly influences EXQ by shaping consumers' emotional states during transactions. For practitioners, the implication is clear: privacy assurance should be designed as part of the front-stage user experience, not treated as a back-end legal requirement. Consumer-facing privacy features—such as consent prompts, transaction alerts, and transparency dashboards-can strengthen perceived control and contribute to a more positive experience.

EXQ as a Predictor of Loyalty

One of the strongest relationships in the model is the path from EXQ to Customer Loyalty ($\beta \approx 0.54$). This indicates that EXQ is not only a mediator but also a powerful direct predictor of loyalty intentions. Consumers who experience greater peace of mind, outcome assurance, and positive emotional states during payment interactions are significantly more likely to remain loyal, recommend the platform, and resist switching to competitors.

This result aligns with service management research positioning experience quality as a key driver of loyalty (Klaus & Maklan, 2012; Kuppelwieser & Klaus, 2021). In digital payment contexts, the implication becomes especially salient because loyalty depends not only on satisfaction or trust in general, but also on whether consumers believe the platform will perform reliably during high-stakes moments such as failed transactions, refunds, disputes, or security threats. Consumers' accounts reinforce that loyalty is often conditional on the perceived dependability of the platform in such situations.

Mediation Effects of EXQ

Beyond its direct effect, EXQ mediates the relationships between all three stimuli—ease of use, personalization, and privacy control—and loyalty. The mediation analysis revealed significant indirect effects, confirming EXQ as the psychological mechanism through which interface-level stimuli are translated into downstream behavioral outcomes.

This represents a meaningful contribution to the digital payment literature. Much prior work has emphasized trust as the primary mediator linking digital interactions to loyalty (Gefen, 2000; Pavlou, 2003). While trust remains important, it may be too narrow to capture the full range of consumer responses in payment encounters. EXQ, by contrast, integrates both functional and emotional evaluations and therefore offers a richer explanation of how loyalty develops over repeated payment interactions. In this sense, the results respond to recent calls to move beyond "trust saturation" and incorporate mediating constructs that better reflect the depth and complexity of consumer experiences at digital touchpoints (Klaus & Maklan, 2012; Kuppelwieser & Klaus, 2021).

THEORETICAL IMPLICATIONS

The study contributes to the existing literature in several important ways. First, it reinforces the relevance of the S-O-R paradigm (Mehrabian & Russell, 1974) in FinTech contexts by showing that external stimuli-ease of use, personalization, and privacy control - shape consumers' internal experiential states (EXQ), which subsequently influence behavioral outcomes such as customer loyalty.

Second, the findings extend personalization research by demonstrating that personalization at the payment interface functions as a dominant experiential driver, moving its relevance beyond traditional applications in advertising and recommendation systems to the transactional "moment of truth" in digital payments.

Third, the study adds nuance to technology acceptance perspectives by illustrating that ease of use does more than support adoption; it also contributes to emotional comfort and sustained experience quality during repeated payment interactions.

Fourth, it advances privacy scholarship by reframing privacy control as an experiential and psychological assurance mechanism rather than merely a compliance-oriented process—highlighting how perceived control over data affects comfort, trust, and engagement during transactions.

Finally, the study strengthens understanding of loyalty formation by positioning EXQ as a broader mediating mechanism than trust alone, capturing the combined influence of functional performance, emotional reassurance, and relational confidence in shaping continued usage and advocacy.

PRACTICAL IMPLICATIONS

For managers, the results highlight that loyalty is established not just on the basis of functional efficiency but through good payment experiences. Personalization needs to be judiciously crafted to maximize recognition without inducing fatigue. Usability needs to be constantly optimized to reduce friction and promote peace of mind. Privacy needs to be promoted to a prominent, consumer-visible feature. Above all, EXQ must be handled as a strategic key performance indicator that is directly associated with loyalty metrics like churn, customer lifetime value, and advocacy.

Qualitative Phase

The second phase of the study employed a qualitative approach to contextualize the quantitative findings and to examine how consumers interpret their digital payment experiences. Twenty semi-structured interviews were conducted with two participant groups: (i) fifteen active users of digital payment platforms and (ii) five marketing professionals working in FinTech or e-commerce. Participants were recruited through purposive and snowball sampling to ensure representation across age, gender, and usage intensity. The user participants ranged in age from 21 to 48 years and reflected a balanced mix in terms of gender and income categories. The sample size was guided by theoretical saturation, with no new themes emerging after the eighteenth interview (Braun & Clarke, 2006).

Interviews were conducted using a semi-structured guide designed to explore experiences related to ease of use, personalization, privacy control, and perceptions of payment experience quality. Open-ended questions were used to elicit detailed narratives, and probing questions were employed to explore both positive and negative experiences in greater depth. Each interview lasted approximately 40–60 minutes and was conducted either in person or via video call. Informed consent was obtained from all participants prior to data collection. Anonymity and confidentiality were assured, and participation remained voluntary throughout the study. Interviews were audio-recorded only with participants' explicit permission.

Interview transcripts were coded iteratively using thematic analysis (Braun & Clarke, 2006) in combination with the inductive approach proposed by Gioia et al. (2013). An initial stage of open coding generated first-order concepts, which were subsequently organized into second-order themes and further consolidated into aggregate dimensions aligned with the S–O–R framework. NVivo 12 was used to support data management and coding. To enhance reliability, two researchers independently coded the transcripts and reconciled differences through discussion until consensus was achieved.

Data Structure

Table 8. Gioia-style data structure that emerged from the analysis.

First-Order Concepts (Informant-Centric)	Second-Order Themes (Researcher-Centric)	Aggregate Dimension
"I don't have to think twice; it feels like sending a text."	Reduced cognitive effort	Ease of Use
"Saved preferences make checkout faster."	Interface intuitiveness	Ease of Use
"When I get an offer at the right time, I feel valued."	Contextual relevance	Personalization
"Too many irrelevant offers feel spammy."	Personalization fatigue	Personalization
"Knowing I can control where my data goes gives me comfort."	Perceived control	Privacy Control
"Lack of transparency makes me avoid the app."	Trust erosion	Privacy Control
"Refunds without hassle keep me loyal."	Reliability during moments of truth	Experience Quality
"Smooth payment gives me peace of mind."	Emotional reassurance	Experience Quality
"If one app makes life easy, I don't experiment with others."	Habit-driven loyalty	Customer Loyalty
"I recommend it because it never failed me."	Advocacy and word-of-mouth	Customer Loyalty

Narrative Findings

Ease of Use as Confidence Enabler

Participants described intuitive design, one-click payments, and saved preferences as features that reduced cognitive effort during checkout. Beyond efficiency, ease of use was repeatedly linked to confidence and emotional comfort, encouraging continued use. One participant shared, "If it takes just one tap and the confirmation comes immediately, I feel assured that everything is under control and I don't second-guess the purchase" (Participant P6). Another added, "Saved preferences and a familiar flow make the payment feel effortless—there is no stress even when I'm in a hurry" (Participant P14).

Personalization as Recognition and Risk

While relevant offers created a feeling of recognition and convenience, participants also described a point at which personalization became intrusive. Several respondents appreciated contextual prompts such as reminders, coupons, or suggested payment modes, stating that these improved convenience and perceived relevance. At the same time, excessive or poorly targeted notifications resulted in personalization fatigue, with some users describing the experience as "spam-like." As one participant noted, "I like it when the app suggests offers that match my routine purchases, but when it keeps pushing random deals, I start ignoring everything" (Participant P4). Another respondent explained the boundary more directly: "Personalization works only when it feels helpful; otherwise it feels like the app is trying too hard to sell" (Participant P17). This highlights the need to balance recognition benefits with relevance and restraint.

Privacy Control as Emotional Assurance

Participants strongly associated visible privacy and security cues with reduced anxiety during digital payment interactions. Features such as OTP authentication, consent prompts, transaction alerts, and the ability to manage permissions were interpreted as signals of control and transparency. One participant emphasized, "When I receive an OTP and an instant alert, I feel protected because I know I can track every step" (Participant P9). Conversely, lack of transparency or unclear permission requests led to hesitation and avoidance: "If the app asks for access without explaining why, I get uncomfortable and I prefer not to proceed" (Participant P12). Overall, privacy control was not perceived merely as compliance; it was experienced as emotional reassurance that enabled continued engagement.

Experience Quality Beyond Functionality

Experience quality emerged as an emotional concept rather than a purely functional evaluation. Participants valued peace of mind, reliability, and assurance during "moments of truth," particularly when transactions failed, refunds were delayed, or disputes arose. One participant reflected, "The real test is when something goes wrong—if refunds are smooth and support responds quickly, I trust the app even more" (Participant P17). Another noted, "It's not only about paying fast; it's about feeling certain that the outcome is secure and recoverable if there's a problem" (Participant P8). These insights broaden the definition of payment experience quality to include emotional reassurance and perceived recovery support as essential elements.

Customer Loyalty as Habit and Advocacy

Loyalty was described as a combination of convenience-based habit and trust-based commitment. Many participants reported "sticking with what works" to avoid uncertainty, suggesting that repeated positive experiences develop into routine usage. One respondent stated, "I continue with the same app because it has never failed me, and I don't want to take chances with a new one" (Participant P2). Positive experience quality also encouraged advocacy behavior: "When it is consistently smooth and reliable, I naturally recommend it to friends and family" (Participant P9). These narratives indicate that experience quality strengthens both passive loyalty (habitual use) and active loyalty (recommendation and endorsement).

Interpretive Insights

The qualitative findings revealed several nuances that extend the quantitative model. First, personalization improved experience quality when it was relevant and restrained but could backfire when overdone, resulting in fatigue and disengagement. Second, experience quality was consistently framed as peace of mind—an emotional reassurance that may not be fully captured through survey responses alone. Third, boundary conditions were evident: concerns about trust and privacy varied with age and digital literacy, indicating potential moderating effects that warrant future testing. Finally, the mechanisms of loyalty extended beyond rational evaluation; loyalty was shaped by risk avoidance ("sticking with what works") as well as advocacy behaviors when experience quality remained consistently positive. Overall, these qualitative insights triangulate the quantitative results by illustrating how payment experience quality functions as the experiential mechanism through which ease of use, personalization, and privacy control translate into sustained usage intentions, trust-based commitment, and advocacy-oriented customer loyalty.

Integration with Quantitative Findings

The combination of quantitative and qualitative results offers a deeper understanding of the factors that influence digital payment experiences. Table 9 shows the findings, combining structural paths from the PLS-SEM analysis with emerging qualitative themes and participant stories.

First, the SEM results indicated that perceived ease of use (PEOU) significantly improved payment experience quality (EXQ). Qualitative insights pointed out that this connection is both functional and emotional. Intuitive design reduced mental effort and built confidence, as seen in statements like "I don't have to think twice;

the payment is done in seconds." This suggests that ease of use builds trust and comfort, expanding the TAM logic into actual experiences.

Second, personalization was the strongest quantitative predictor of EXQ. Interviews supported this but also revealed a limit: while relevant offers made consumers feel acknowledged, too much targeting led to "personalization fatigue." This detail, not found in survey data, shows that personalization enhances EXQ only when it feels helpful instead of intrusive.

Third, privacy control strongly predicted EXQ in the model, and qualitative evidence clarified why. Participants often stressed that visible protections (e.g., OTPs, consent prompts) provided reassurance and reduced anxiety. In contrast, a lack of transparency led to avoidance. This finding adds to privacy research by showing that its impact on EXQ comes through emotional reassurance, not just risk perception.

Fourth, EXQ was a strong predictor of customer loyalty, with mediation effects confirmed across all three stimuli. Stories reinforced this by showing that loyalty was both routine (to avoid switching costs and risks) and driven by advocacy (recommending apps viewed as reliable). Statements like "Refunds processed easily keep me using one app" highlight how reliability during crucial moments strengthens attachment and advocacy.

In summary, the integration emphasizes the vital role of EXQ as the psychological link between digital payment features and loyalty. Quantitative findings confirm its statistical importance, while qualitative stories reveal the emotional factors—peace of mind, recognition, and trust—that make EXQ a key driver of lasting consumer loyalty.

Table 9. Showing Integration of Quantitative and Qualitative Results

Hypothesized Path (PLS-SEM)	Quantitative Findings	Supporting Qualitative Themes	Illustrative Quotes	Interpretation
H1: PEOU → EXQ	β ≈ .28, significant	Ease of use as confidence enabler (simplicity reduces cognitive load, intuitive design boosts trust)	"I don't have to think twice; the payment is done in seconds." / "It feels natural, like sending a text."	Quantitative link confirmed. Ease of use strengthens EXQ by lowering friction and building user confidence.
H2: Personalization → EXQ	β ≈ .32, strongest predictor	Personalization as recognition vs. intrusion (contextual offers valued; excessive offers feel spammy)	"When I get an offer exactly when I am about to pay, it feels like the app understands me." / "Too many irrelevant offers come off as spammy."	Supports strong effect but reveals boundary condition: personalization enhances EXQ when contextually relevant, but risks fatigue.
H3: Privacy Control → EXQ	β ≈ .25, significant	Privacy control as peace-of-mind (visible safeguards, transparency reduce anxiety; lack of clarity leads to avoidance)	"Knowing I can control where my data goes makes me feel more comfortable." / "I avoid apps that don't show me where my data is stored."	Confirms statistical relationship and extends it: privacy influences EXQ through emotional assurance, not just risk reduction.
H4: EXQ → Loyalty	β ≈ .54, strong effect	EXQ as emotional reassurance (peace of mind, reliability in critical moments, smooth refunds); Loyalty as habit & advocacy	"Refunds processed easily keep me using one app." / "If one app makes life easy, I don't try others."	Quantitative finding reinforced: EXQ directly drives loyalty. Qualitative data shows dual mechanism: risk avoidance (habit) + positive advocacy.
H5: Mediation of EXQ	All three stimuli show significant indirect effects via EXQ	Themes confirm EXQ as mediator — the "psychological bridge" linking features to loyalty	"A smooth payment experience gives me peace of mind, that's why I stick with this app."	SEM mediation validated. Qualitative data emphasizes emotional dimension of EXQ as the key driver.

MANAGERIAL IMPLICATIONS

This study offers both theoretical insights and practical advice for managers in digital payments, banking, and the wider digital commerce ecosystem. The findings emphasize the vital role payment experience quality (EXQ) plays in driving customer loyalty. This suggests that companies need to carefully design and manage payment processes as a key competitive advantage. Here are five main implications from the results.

Design Personalization as Recognition, not Intrusion

Personalization is a key part of digital marketing, but its application in payment contexts needs careful consideration. The results showed that personalization is the biggest driver of EXQ, but feedback indicated it can have drawbacks. When consumers see contextual offers as relevant and timely, they feel acknowledged and appreciated, which boosts their satisfaction and loyalty to the platform. For example, a grocery payment app that automatically generates discounts on frequently bought items can enhance convenience and create a sense of belonging. However, when personalization becomes overly targeted, users may perceive it as "spammy" or "intrusive." Instead of providing value, irrelevant promotions can lead to frustration and distrust, lowering EXQ. To prevent this, companies should position personalization as recognition rather than manipulation. This means focusing on signals that show understanding of consumer needs, such as acknowledging past purchases, offering relevant rewards, and presenting offers at the right time. Companies can implement AI-driven relevance filters that learn when and how personalization is useful. It's also crucial to give consumers control over their personalization settings, allowing them to opt into promotional categories or adjust notification frequency. This two-way approach—where technology meets consumer choice—can help build trust and prevent fatigue. Firms that successfully balance relevance and respect will stand out by fostering long-term emotional connections.

Focus on the Consumer journey experience

Ease of use is another essential factor shaping EXQ. While previous attention has focused on onboarding and adoption, this research highlights that the entire payment journey matters just as much. Payment is not a one-off event but a series of micro-interactions: authentication, confirmation, refunds, splitting bills, and checking transaction histories. Each of these moments can become a friction point.

The findings suggest that even minor obstacles, like a delayed one-time password (OTP), unclear refund timelines, or occasional transaction failures, can hurt EXQ significantly. On the flip side, small improvements can lead to major boosts in loyalty. For example, reducing checkout steps on an e-commerce site or adding biometric authentication in mobile wallets can cut down on cognitive load, enhancing peace of mind and assurance of outcomes. Managers should take a journey-mapping approach. Rather than only checking if a payment went through, they should identify friction throughout the entire process, from initiation to post-transaction support. This will require teamwork across IT, customer service, and UX teams. For example, refund delays often stem from back-end coordination between merchants and payment processors. Fixing these issues will require rethinking the process, not just tweaking the front-end interface.

By viewing payment as a journey to be streamlined, intuitive, and reliable, companies can turn a routine task into a competitive advantage. Consumers who have smooth payment experiences are more likely to stick with one provider and resist switching.

Elevate Privacy as a visible value

Companies often treat privacy as a compliance issue, managing it behind the scenes to meet regulations. However, the findings show that consumers view privacy not just as a legal requirement but also as an emotional assurance. Having visible privacy controls directly contributes to EXQ by reducing anxiety and building trust.

Therefore, companies should go beyond just compliance and make privacy visible. Transparency dashboards, real-time data alerts, and clear consent prompts are brand-building opportunities, not mere technical features. For instance, a payment app that informs users about how their transaction data is stored and allows them to easily manage data-sharing preferences shows respect and empowerment. This visibility sets the brand apart and makes privacy part of the consumer experience rather than an invisible safeguard.

The business argument is also compelling: visible privacy protections can prevent customer loss and make consumers more open to personalized offers. When users trust that they can control their data, they are likelier to engage with personalization. Thus, privacy and personalization are not opposing goals but complementary elements of trust-based engagement.

Consider EXQ a Strategic KPI

A key takeaway for managers is to view payment experience quality as a strategic performance indicator. While many firms track functional KPIs, such as transaction success rates or fraud prevention, these metrics often miss the consumer experience. The study shows that EXQ—covering peace of mind, reliability, and assurance of outcomes—is a stronger predictor of loyalty than functional efficiency alone.

To put this into practice, companies should include EXQ metrics in their performance dashboards along with financial and operational indicators. This could include measures based on surveys about peace of mind, reported friction points, or net promoter scores linked specifically to payment interactions. Connecting EXQ measures to outcomes like customer retention, lifetime value, and advocacy can help prove the financial benefits of investing in experience.

Additionally, including EXQ in balanced scorecards ensures it becomes a shared responsibility across different departments, not just UX teams. For instance, IT teams should track system reliability not only in technical terms but also in how downtime affects user peace of mind. Marketing teams should observe whether personalization enhances or hurts perceived value. By raising the profile of EXQ as a KPI, companies can unite their departments around a common goal: improving payments to drive loyalty and advocacy.

Application Across Industries

Although the study focuses on digital payment systems, its insights apply to many more sectors beyond banking and payment providers. Payment is a key moment across industries, often shaping consumer impressions. For e-commerce platforms, the payment experience can determine whether a shopper finalizes a purchase or abandons their cart. In ride-hailing and food delivery services, smooth in-app payments contribute to overall convenience. For subscription services, hassle-free recurring billing fosters trust and reduces customer loss.

These findings suggest that any business where payment is part of the customer experience should prioritize EXQ as a differentiator. Companies in various sectors should work with payment providers to create seamless, comforting, and trustworthy payment experiences. In a competitive digital environment with low switching costs and many alternatives, the payment interface could be the crucial touchpoint for securing loyalty.

Critical Considerations and Ethical Tensions

While payment-integrated marketing can enhance convenience and relevance, it also introduces experience and ethics risks that require explicit safeguards. First, security incidents, fraud attempts, and opaque dispute-resolution pathways can rapidly erode perceived control and undermine EXQ, even when the interface is easy to use. Second, algorithmic personalization may unintentionally amplify bias or create "over-personalization fatigue," where customers perceive recommendations or nudges as intrusive rather than helpful. Third, the personalization–privacy tension is central: personalization tends to improve EXQ when it is perceived as recognition and assistance, but it can diminish EXQ when users feel profiled, tracked, or unable to control data sharing. Designing visible privacy controls (e.g., consent choices, data dashboards, and easy opt-outs) is therefore not only a compliance need but also an experience-design lever.

Policy and Regulatory Implications

From a policy standpoint, payment-integrated marketing environments require consumer protection and experience-quality standards aligned with privacy-by-design principles. Regulators and industry bodies can strengthen trust by emphasizing clear consent, purpose limitation, data minimization, and accessible grievance mechanisms, especially where marketing and payments are tightly coupled. For cross-border payment-enabled marketing, harmonization of data-transfer safeguards and accountability expectations becomes important, given differences in privacy norms and enforcement regimes. In practice, firms can operationalize these expectations through transparent data notices, auditable consent logs, and governance processes that evaluate personalization logic for fairness and proportionality.

SCOPE OF FUTURE RESEARCH

While this study improves our understanding of digital payment experience quality and loyalty, several areas still need exploration.

Cross-Cultural and Contextual Extensions

This study was conducted in just one country. Cultural differences in privacy views, how people accept personalization, and their payment habits may affect the suggested framework. Future research should use multi-group analysis to test measurement consistency and compare findings across cultures, regions, or economic conditions (Henseler, Ringle, & Sarstedt, 2016).

Longitudinal and Experimental Designs

This study used cross-sectional data, which limits the ability to draw causal conclusions. Future research could use longitudinal designs to explore how experience quality and loyalty change over time, particularly as consumers engage repeatedly with payment platforms. Experimental designs could adjust interface elements, such as personalization level and privacy control, to test causal relationships more thoroughly (Podsakoff, MacKenzie, Lee, & Podsakoff, 2003).

Expansion of Constructs and Mediators

Although this study identified EXQ as a key mediator, future research can expand the conceptual model by adding more variables. These could include affective trust, satisfaction, flow, or emotional engagement (Kuppelwieser & Klaus, 2021). Moderators like digital literacy, age, income, or risk aversion could further improve our understanding of boundary conditions.

Integration of Emerging Technologies

As digital payments connect more with AI, blockchain, and biometrics, future studies should look into how these technologies change how consumers view personalization, privacy, and usability. For instance, blockchain's transparency could change views on privacy. At the same time, AI-driven predictive personalization might raise new ethical issues (Dwivedi et al., 2023).

Behavioral and Multimethod Approaches

The current study relied on self-report data, which can be affected by common method variance. Future research should confirm findings by using behavioral data, such as transaction logs or usage analytics, or by applying neurophysiological tools, including eye-tracking, EEG, or biometric sensors, to capture unconscious reactions to payment experiences (Venkatraman et al., 2015).

CONCLUSION

This study adds to both theory and practice by introducing experience quality (EXQ) as a new mediator in digital payment settings. It empirically demonstrates that ease of use, personalization, and privacy control contribute to EXQ, which then drives loyalty. This research changes how loyalty mechanisms are viewed in FinTech.

Theoretically, it expands the S-O-R model, combines ideas from adoption, personalization, and privacy literature, and shows the importance of using different methods in digital ecosystems. For managers, it offers a user-focused plan: design personalization to acknowledge users, create smooth journeys, integrate privacy into the experience, and make EXQ a key performance indicator.

In the end, the findings show that payment is not just a back-end transaction but a key front-end experience that helps brands stand out in competitive markets. As payment ecosystems shift towards AI-driven personalization, blockchain-enabled transparency, and cross-border digital commerce, the framework provided here lays a foundation for developing loyalty strategies that are both strong in technology and rich in experience.

REFERENCES

Bajwa, F. A., Fu, J., Bajwa, I. A., Ahmad, S., & Mahmood, F. (2025). Factors influencing usage and loyalty for payment app customers in Saudi Arabia. *Acta Psychologica*, *255*, 104961. DOI: 10.1016/j.actpsy.2025.104961

Bleier, A., & Eisenbeiss, M. (2015a). Personalized online advertising effectiveness: The interplay of what, when, and where. *Marketing Science*, *34*(5), 669–688. DOI: 10.1287/mksc.2015.0930

Bleier, A., & Eisenbeiss, M. (2015b). The importance of trust for personalized online advertising. *Journal of Retailing*, *91*(3), 390–409. DOI: 10.1016/j.jretai.2015.04.001

Braun, V., & Clarke, V. (2006). Using thematic analysis in psychology. *Qualitative Research in Psychology*, *3*(2), 77–101. DOI: 10.1191/1478088706qp063oa

Capgemini. (2025). *World payments report 2025: Velocity, meet value—Lead in the open and instant future of payments.* Capgemini. https://www.capgemini.com/insights/research-library/world-payments-report-2025/

Cloarec, J., Meyer-Waarden, L., & Munzel, A. (2024). Transformative privacy calculus: Conceptualizing the personalization–privacy paradox on social media. *Psychology and Marketing*, *41*(7), 1574–1596. Advance online publication. DOI: 10.1002/mar.21998

Creswell, J. W., & Plano Clark, V. L. (2017). *Designing and conducting mixed methods research* (3rd ed.). SAGE Publications.

Davis, F. D. (1989). Perceived usefulness, perceived ease of use, and user acceptance of information technology. *Management Information Systems Quarterly*, *13*(3), 319–340. DOI: 10.2307/249008

Dwivedi, Y. K., Hughes, L., Wang, Y., Alalwan, A. A., Ahn, S. J., Balakrishnan, J., Barta, S., Belk, R., Buhalis, D., Dutot, V., Felix, R., Filieri, R., Flavián, C., Gustafsson, A., Hinsch, C., Hollensen, S., Jain, V., Kim, J., Krishen, A. S., & Wirtz, J. (2023). Metaverse marketing: How the metaverse will shape the future of consumer research and practice. *Psychology and Marketing*, *40*(4), 750–776. DOI: 10.1002/mar.21767

Fornell, C., & Larcker, D. F. (1981). Evaluating structural equation models with unobservable variables and measurement error. *JMR, Journal of Marketing Research*, *18*(1), 39–50. DOI: 10.1177/002224378101800104

Gefen, D. (2000). E-commerce: The role of familiarity and trust. *Omega*, *28*(6), 725–737. DOI: 10.1016/S0305-0483(00)00021-9

Gefen, D., Karahanna, E., & Straub, D. W. (2003). Trust and TAM in online shopping: An integrated model. *Management Information Systems Quarterly*, *27*(1), 51–90. DOI: 10.2307/30036519

Gioia, D. A., Corley, K. G., & Hamilton, A. L. (2013). Seeking qualitative rigor in inductive research: Notes on the Gioia methodology. *Organizational Research Methods*, *16*(1), 15–31. DOI: 10.1177/1094428112452151

Hair, J. F., Hult, G. T. M., Ringle, C. M., & Sarstedt, M. (2019). *A primer on partial least squares structural equation modeling (PLS-SEM)* (2nd ed.). SAGE Publications. DOI: 10.3926/oss.37

Henseler, J., Ringle, C. M., & Sarstedt, M. (2015). A new criterion for assessing discriminant validity in variance-based structural equation modeling. *Journal of the Academy of Marketing Science*, *43*(1), 115–135. DOI: 10.1007/s11747-014-0403-8

Henseler, J., Ringle, C. M., & Sarstedt, M. (2016). Testing measurement invariance of composites using the MICOM procedure. *Journal of Business Research*, *69*(9), 3150–3158.

Kalyanaraman, S., & Sundar, S. S. (2006). The psychological appeal of personalized content in web portals: Does customization affect attitudes and behavior? *Journal of Communication*, *56*(1), 110–132. DOI: 10.1111/j.1460-2466.2006.00006.x

Klaus, P. P., & Maklan, S. (2012). EXQ: A multiple item scale for assessing service experience. *Journal of Service Management*, *23*(1), 5–33. DOI: 10.1108/09564231211208952

Kuppelwieser, V. G., & Klaus, P. (2021). Measuring customer experience quality: The EXQ scale revisited. *Journal of Business Research*, *126*, 624–633. DOI: 10.1016/j.jbusres.2020.01.042

Malhotra, N. K., Kim, S. S., & Agarwal, J. (2004). Internet users' information privacy concerns (IUIPC): The construct, the scale, and a causal model. *Information Systems Research*, *15*(4), 336–355. DOI: 10.1287/isre.1040.0032

Mehrabian, A., & Russell, J. A. (1974). *An approach to environmental psychology*. MIT Press.

Pavlou, P. A. (2003). Consumer acceptance of electronic commerce: Integrating trust and risk with the technology acceptance model. *International Journal of Electronic Commerce*, *7*(3), 101–134. DOI: 10.1080/10864415.2003.11044275

Pires, P. B., Prisco, M., Delgado, C., & Santos, J. D. (2024). A conceptual approach to understanding the customer experience in e-commerce: An empirical study. *Journal of Theoretical and Applied Electronic Commerce Research, 19*(3), 1943–1983. DOI: 10.3390/jtaer19030096

Podsakoff, P. M., MacKenzie, S. B., Lee, J.-Y., & Podsakoff, N. P. (2003). Common method biases in behavioral research: A critical review. *The Journal of Applied Psychology, 88*(5), 879–903. DOI: 10.1037/0021-9010.88.5.879

Saura, J. R., Škare, V., & Ozretic Dosen, D. (2024). Is AI-based digital marketing ethical? Assessing a new data privacy paradox. *Journal of Innovation & Knowledge, 9*(4), 100597. DOI: 10.1016/j.jik.2024.100597

Smith, H. J., Dinev, T., & Xu, H. (2011). Information privacy research: An interdisciplinary review. *Management Information Systems Quarterly, 35*(4), 989–1016. DOI: 10.2307/41409970

Venkatesh, V., Brown, S. A., & Bala, H. (2013). Bridging the qualitative–quantitative divide: Guidelines for conducting mixed methods research in information systems. *Management Information Systems Quarterly, 37*(1), 21–54. DOI: 10.25300/MISQ/2013/37.1.02

Venkatesh, V., & Davis, F. D. (2000). A theoretical extension of the technology acceptance model: Four longitudinal field studies. *Management Science, 46*(2), 186–204. DOI: 10.1287/mnsc.46.2.186.11926

Venkatesh, V., Morris, M. G., Davis, G. B., & Davis, F. D. (2003). User acceptance of information technology: Toward a unified view. *Management Information Systems Quarterly, 27*(3), 425–478. DOI: 10.2307/30036540

Venkatraman, V., Dimoka, A., Pavlou, P. A., Vo, K., Hampton, W., Bollinger, B., Hershfield, H. E., Ishihara, M., & Winer, R. S. (2015). Predicting advertising success beyond traditional measures: New insights from neurophysiological methods and market response modeling. *JMR, Journal of Marketing Research, 52*(4), 436–452. DOI: 10.1509/jmr.13.0593

Zeithaml, V. A., Berry, L. L., & Parasuraman, A. (1996). The behavioral consequences of service quality. *Journal of Marketing, 60*(2), 31–46. DOI: 10.1177/002224299606000203

Chapter 11
Redefining the Marketing Mix:
Strategic Integration of Digital Payments Into Contemporary Marketing Practice

Suhail Mohammad Ghouse
https://orcid.org/0000-0002-7684-933X
Dhofar University, Oman

ABSTRACT

This chapter explores how digital payment systems are transforming the foundations of marketing practice by reframing the traditional 4Ps framework. Drawing on insights from behavioral economics, consumer psychology, and fintech adoption, it examines how payments influence product innovation, pricing strategies, distribution models, and promotional engagement. Global and emerging market case insights illustrate how tokenization, microtransactions, borderless commerce, and payment-linked loyalty programs reshape consumer perceptions of ownership, value, trust, and inclusivity. The chapter highlights managerial implications by positioning payments as strategic levers for customer experience, data-driven personalization, and ethical stewardship. It also proposes a future research agenda that integrates digital payments into marketing theory, offering guidance for scholars, practitioners, and policymakers in navigating the digital economy.

DOI: 10.4018/979-8-3373-4526-0.ch011

INTRODUCTION

Marketing has never been static, as it has evolved in response to changes in technology, social organisation, and economic structure. Each period of transformation has altered how firms engage with consumers and how value is created, communicated, and exchanged. The current phase of digital transformation is no exception. Among its most consequential developments is the widespread adoption of digital payment systems. These systems include mobile wallets, contactless cards, platform-based payment interfaces, cryptocurrencies, blockchain-enabled tokens, and emerging decentralised finance applications.

Digital payments were initially designed to improve transactional efficiency. Speed, convenience, and cost reduction were their primary objectives. Over time, however, their role has expanded. Payment systems now shape how consumers experience brands, how they interpret value, and how trust is formed and sustained. What was once treated as a functional endpoint of the exchange process has become embedded within the consumption experience itself. Payments now operate as part of brand meaning, relational engagement, and experiential design rather than as neutral financial tools (Oliveira et al., 2016).

This shift has important implications for marketing theory and practice. Payments influence how consumers decide, how they feel during transactions, and how they evaluate fairness and authenticity. In many contexts, the act of paying has become as significant as the product or service being purchased. Digital payments therefore demand attention not only from finance or operations but from marketing strategy itself.

Digital Payments and Marketing Relevance

The rapid growth of digital payment systems reflects both technological progress and changing consumer expectations. Smartphone access has expanded sharply across developed and emerging economies, reshaping how people interact with markets. At the same time, fintech innovation has lowered entry barriers, enabling new payment models that prioritise speed, convenience, and accessibility. Blockchain technologies have further altered the landscape by introducing decentralised verification and programmable forms of exchange. Alongside these developments, consumers increasingly expect transactions to be secure, seamless, and embedded within everyday digital routines.

Industry evidence highlights the scale of this shift. The GSMA (2025) reports that mobile money platforms processed more than USD 1.68 trillion in transactions during 2024, representing annual growth of around 16 per cent. Much of this expansion has taken place in regions where traditional banking systems have historically been limited, including Africa, parts of Asia, and the Middle East. In these settings, digital payments have not merely complemented existing financial infrastructure; they have reconfigured daily consumption practices and reshaped how individuals participate in markets.

As payment technologies become widespread, the act of paying has become more visible within the customer journey. Digital interfaces, confirmation prompts, reward notifications, and security checks now frame the experience of exchange. In parallel, the diffusion of cryptocurrencies such as Bitcoin and Ethereum, together with stablecoins and emerging central bank digital currencies, has enabled new forms of cross-border and decentralised exchange beyond conventional financial intermediaries (Arner et al., 2020).

For marketers, these changes alter the strategic meaning of payment. Payment no longer marks the end of interaction. Instead, it represents a moment of heightened engagement where value is assessed, trust is negotiated, and brand relationships are reinforced. Firms increasingly embed rewards within digital wallets, personalise offers using transaction data, and rely on blockchain-based verification to support claims of authenticity (Chawla and Joshi, 2019). As payment systems intersect with social media, gaming platforms, and virtual environments, transactions acquire symbolic meaning, signalling participation and affiliation within branded ecosystems (Mallat, 2007).

Taken together, these developments challenge long-standing assumptions in marketing. Ownership, scarcity, and trust are no longer defined solely by physical goods or institutional authority. Instead, they are increasingly shaped by digital infrastructures. This shift invites renewed scrutiny of the marketing mix, which has guided marketing thinking for more than six decades (McCarthy, 1960).

Despite the pervasiveness of digital payments in everyday life, their strategic role within marketing theory remains underdeveloped. Existing research has largely examined digital payments from technological, financial, or regulatory perspectives, prioritising adoption drivers, system security, infrastructure, and risk management (Slade et al., 2015). While valuable, this work offers limited insight into how payment technologies reshape core marketing decisions.

In particular, insufficient attention has been given to how digital payments influence fundamental elements of marketing strategy, including product design, pricing logic, distribution structures, and promotional practices. The implications of payment systems for consumer perceptions of value, fairness, and trust are often assumed rather than analysed directly.

Research on digital transformation has explored related domains such as e-commerce, social media, artificial intelligence, and data analytics (Lemon and Verhoef, 2016). However, payment systems continue to be positioned as background infrastructure supporting exchange, rather than as active instruments of marketing strategy. For instance, while loyalty programmes are widely discussed, blockchain-enabled loyalty tokens, characterised by transparency, transferability, and programmability, remain marginal within mainstream frameworks. Similarly, the pricing implications of cryptocurrencies, microtransactions, and fractional ownership are rarely integrated into theories of consumer value perception or decision-making (Prelec and Simester, 2001).

These omissions reveal a clear conceptual gap. Digital payments are no longer peripheral; they shape how marketing functions operate in practice. Accordingly, this chapter adopts a repositioning perspective, treating digital payments as central mechanisms through which value is designed, communicated, and experienced.

The chapter pursues three objectives. First, it examines how contemporary payment systems reshape the marketing mix—product, price, place, and promotion—through mechanisms such as tokenisation, transparency, programmability, and personalisation. Second, it explores managerial, ethical, and regulatory implications, with particular emphasis on trust, inclusion, data governance, and pricing fairness. Third, it develops a conceptual framework and research agenda linking marketing theory with financial technology, reinforcing the relevance of the 4Ps in an increasingly digital economy.

Contributions of the Chapter

This chapter contributes to marketing theory, managerial practice, public policy, and pedagogy by positioning digital payments as strategic elements of marketing rather than as purely operational tools. Its core contribution lies in showing how payment systems actively shape value creation, consumer engagement, and market relationships.

From a theoretical perspective, the chapter revisits the marketing mix and demonstrates how digital payments reshape product design, pricing logic, distribution, and promotion. While the 4Ps framework has been criticised for its limited attention to relational dynamics (Constantinides, 2006), reinterpreting it through digital payments connects established marketing theory with insights from behavioural economics and consumer psychology (Thaler, 1985; Prelec and Simester, 2001).

From a managerial standpoint, the chapter offers practical guidance on using payment-enabled personalisation, loyalty mechanisms, and flexible pricing to enhance engagement and differentiation. It also highlights ethical and policy implications, emphasising transparency, inclusion, and responsible data use, while contributing to marketing education by reflecting the convergence of marketing, finance, and technology in contemporary practice.

BACKGROUND

The Marketing Mix (4Ps): Relevance and Limitations in the Digital Economy

The marketing mix remains one of the most enduring concepts in marketing theory. Since its introduction by McCarthy (1960), the 4Ps framework—product, price, place, and promotion—has offered managers a clear and practical way to align marketing activities with organisational objectives. Its simplicity and usability have made it central to both marketing education and managerial practice for several decades (Borden, 1984).

Despite its influence, the framework has attracted sustained criticism as markets have become more digital and interactive. Scholars such as Constantinides (2006) and Gummesson (2017) argue that the 4Ps reflect a transactional and product-oriented view of exchange. As a result, the framework struggles to capture relationship building, consumer participation, co-creation, and two-way communication, all of which are increasingly important in digitally mediated markets.

The rise of digital payments highlights these limitations while also opening new avenues for strategic reinterpretation. Payment technologies now influence how value is created and exchanged. Products increasingly include tokenised assets, non-fungible tokens, and blockchain-verified offerings that combine utility with digital ownership (Saberi et al., 2019). Pricing practices have become more dynamic, shaped by subscriptions and microtransactions. Distribution has shifted towards borderless and peer-to-peer ecosystems that reduce reliance on traditional intermediaries, while promotion increasingly incorporates gamified incentives and payment-linked offers.

The continued relevance of the 4Ps therefore depends on reinterpretation rather than replacement. In the digital economy, each element of the marketing mix operates within an interconnected payment environment where value exchange, data flows, and consumer trust converge.

Digital Transformation and Marketing Thought

The development of marketing thought has closely followed successive waves of technological change. Each wave has altered how firms deliver value and how consumers experience it. The emergence of e-commerce in the late 1990s reshaped distribution by expanding convenience and product choice beyond physical boundaries (Alba et al., 1997). In the following decade, the growth of social media shifted influence toward consumers, enabling them to participate actively in shaping brand meanings through platforms such as Facebook and Instagram (Kaplan and Haenlein,

2010). More recently, advances in big data analytics and artificial intelligence have transformed personalisation and predictive targeting, allowing firms to respond to consumer behaviour in real time (Wedel and Kannan, 2016).

Within this broader trajectory, digital payments can be understood as the next stage of marketing transformation. Unlike earlier innovations that primarily affected communication or distribution, payment technologies reshape the act of exchange itself. They represent more than efficiency gains. Payments now convey trust, convenience, and aspects of consumer identity. Frictionless payment experiences in ride-hailing and delivery platforms reinforce perceptions of reliability, while blockchain-based payments introduce new forms of transparency that influence fairness and authenticity in market exchanges (Arner et al., 2020).

If e-commerce altered how "place" is organised and social media reshaped "promotion," digital payments are distinctive in their ability to affect all four elements of the marketing mix at once. By linking technology, consumer behaviour, and financial processes, they redefine how value is created and how relationships between firms and consumers are sustained. For this reason, a contemporary understanding of marketing requires closer attention to the strategic role of payment systems in the digital economy.

2.3 Consumer Adoption Theories: Understanding Digital Payment Uptake

The adoption of digital payment systems cannot be explained by technological features alone. It is shaped by a combination of psychological, cultural, and social influences that affect how consumers perceive and use these systems. Several established theoretical frameworks help explain why some consumers readily adopt digital payments, while others remain hesitant.

Technology Acceptance Model (TAM)

The Technology Acceptance Model proposes that perceived usefulness and perceived ease of use are central to technology adoption (Davis, 1989). In the context of digital payments, these perceptions are closely linked to issues of trust and security. Empirical studies indicate that consumers are more likely to adopt mobile wallets when they believe the system is both convenient and reliable. Research by Chawla and Joshi (2019), as well as Ghouse et al. (2025), shows that confidence in transaction safety strengthens positive attitudes toward mobile payment usage.

Diffusion of Innovations (DOI)

Diffusion of Innovations theory highlights the role of attributes such as relative advantage, compatibility, and observability in shaping adoption decisions (Rogers, 2003). Applied to digital payments, this perspective helps explain why younger and more digitally skilled users tend to adopt these systems more quickly. At the same time, cultural norms, infrastructure availability, and institutional support continue to influence adoption patterns, particularly in emerging market contexts (Slade et al., 2015).

Unified Theory of Acceptance and Use of Technology (UTAUT)

The Unified Theory of Acceptance and Use of Technology brings together insights from earlier models by incorporating performance expectancy, effort expectancy, social influence, and facilitating conditions (Venkatesh et al., 2003). Empirical evidence suggests that these factors, when combined with regulatory support, play a significant role in predicting digital payment adoption. Oliveira et al. (2016) demonstrate that institutional and environmental conditions reinforce individual-level acceptance.

Behavioural Economics

Behavioural perspectives emphasise that payment decisions are rarely neutral or purely rational. Prelec and Simester (2001) introduced the concept of the "pain of paying," showing that digital and contactless payments reduce the salience of spending and may encourage higher consumption. Thaler and Sunstein (2008) further explain how choice architecture can subtly guide consumer behaviour. Practices such as gamified payments and cashback incentives illustrate how firms apply these insights to sustain engagement and repeat usage. Taken together, these theories suggest that digital payment adoption is shaped by perceived value, trust, and situational context. For marketers, this implies that effective payment design must align with consumer psychology and social norms, rather than relying on technological functionality alone.

Trust, Risk, and Consumer Behaviour

Trust plays a central role in both marketing relationships and the adoption of digital payment systems. In traditional settings, trust was largely anchored in institutions such as banks, governments, and legal frameworks. In digital environments, trust is increasingly shaped by technology, although consumers rarely rely on technical features alone. Perceptions of platform reputation, peer recommendations, and regulatory oversight remain critical in shaping confidence in payment systems (Gefen et al., 2003).

Perceived risk continues to be a major barrier to adoption. Concerns related to financial loss, security breaches, and data privacy persist across markets and are often more pronounced in contexts where regulatory protection is less visible (Slade et al., 2015). Security features such as biometric authentication and encryption can reduce perceived vulnerability and encourage use (Stylios et al., 2022).

Payment experiences also influence brand perceptions. Reliable systems strengthen trust, while failures can quickly damage credibility. At the same time, digital payments risk excluding consumers with limited access or skills, making inclusive design both an ethical obligation and a strategic opportunity.

Figure 1. Reframing the 4Ps Through Digital Payments, Trust, and Inclusion

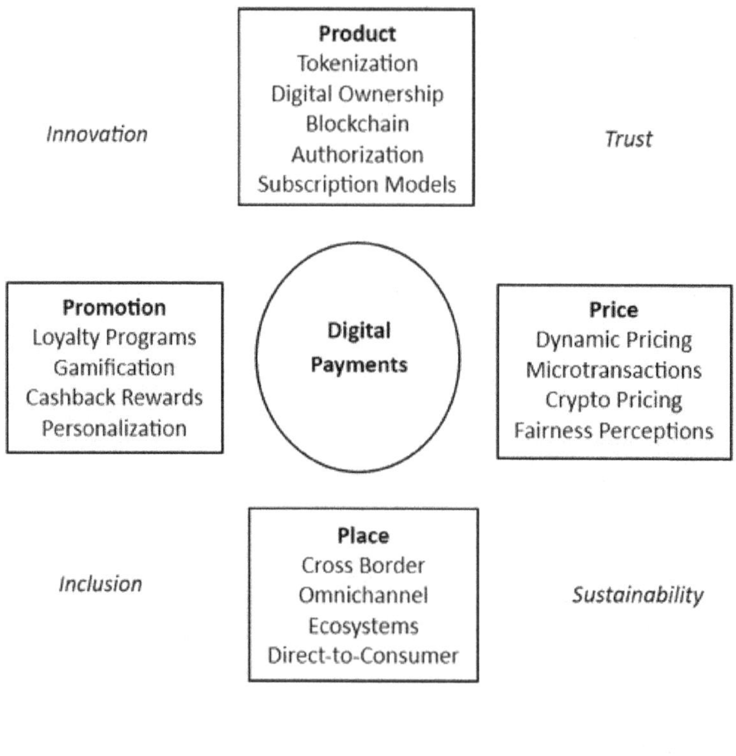

Together, these perspectives, marketing mix theory, digital transformation, consumer adoption, and trust, form a comprehensive foundation for reframing the 4Ps in the digital era. The 4Ps remain a robust organising logic, but digital payments demand their reinterpretation through the lenses of trust, inclusion, and innovation. Payment systems are not merely enablers of exchange; they are strategic mechanisms that shape perception, participation, and value creation.

Figure 1 illustrates the conceptual framework, showing how digital payments reconfigure product, price, place, and promotion through mechanisms of transparency, trust, and technological inclusion.

MAIN FOCUS OF THE CHAPTER

Digital Payments and Product Strategy

Within the traditional marketing mix, the product has consistently been regarded as the central element, as it represents the primary means through which firms create and deliver value. Conventional product strategy emphasised attributes such as design, functionality, branding, and differentiation. The expansion of digital payment systems introduces an additional strategic dimension. Payments now influence how consumers access products, interpret value, and engage with offerings throughout the consumption process. Rather than operating solely as transactional tools, payment systems increasingly form part of the product experience itself. From a marketing mix perspective, digital payments reshape the "product" element by redefining ownership, supporting flexible value delivery, and embedding mechanisms of trust directly into consumption.

Digital payments influence product strategy in three interrelated ways:

i. enabling digital ownership through tokenisation;
ii. facilitating microtransactions and modular service bundling; and
iii. reinforcing authenticity and trust through blockchain-based verification.

Each mechanism affects consumer psychology and market positioning in both developed and emerging economies.

Digital Ownership and Tokenisation

Blockchain-based assets such as non-fungible tokens (NFTs) have redefined what ownership means. While physical ownership once relied on possession, in digital markets, ownership is represented by verifiable, transferable tokens recorded on distributed ledgers (Dowling, 2022). These tokens turn products into hybrid goods that combine functional utility with symbolic value.

Luxury brands have been early adopters. Gucci and Louis Vuitton, for instance, have issued NFTs offering exclusive digital fashion, merging material scarcity with digital prestige (Joy et al., 2022). Such tokenised offerings appeal to younger, digitally native consumers and extend luxury branding into the metaverse.

In emerging markets, tokenisation can strengthen credibility and traceability. Indian agritech start-ups use blockchain to certify the origin of organic produce, reassuring buyers about authenticity (Kamble et al., 2020). By embedding verification within the product, blockchain and payment technologies reduce counterfeit risks and enhance consumer trust, particularly valuable in contexts with weaker regulation.

Microtransactions and Product–Service Bundling

Digital payments also make *microtransactions* viable, small, often fractional payments that unlock specific features or services. Once impractical with cash, they are now seamless through mobile wallets, contactless cards, and cryptocurrencies.

The gaming sector illustrates this model. Titles such as *Fortnite* and *PUBG* rely on in-game purchases for skins and upgrades, generating billions of dollars annually and integrating payments directly into their product design (Hamari et al., 2020). Subscription models like Netflix or Spotify similarly depend on automated recurring payments that shift the perception of value from ownership to access.

In emerging economies, mobile money platforms such as *M-Pesa* demonstrate how microtransactions foster inclusion. Consumers can buy electricity, transport, or insurance in small increments rather than lump sums (Jack and Suri, 2014). "Pay-as-you-go" models make essential services affordable, turning payment architecture itself into a product enabler. This flexibility redefines value propositions and widens market reach.

Authenticity, Provenance, and Consumer Trust

Counterfeiting remains a global problem across the pharmaceutical, luxury goods, and food sectors. Blockchain-based payment and verification systems provide a solution by recording every transaction in an immutable form, allowing products to be traced from origin to end user (Saberi et al., 2019).

Luxury brands such as *Prada* now use blockchain certificates linked to digital payments, allowing customers to verify authenticity instantly, reducing fraud and reinforcing trust. In Nigeria and other Sub-Saharan African markets, digital traceability initiatives, including blockchain-enabled verification systems, are increasingly being explored to confirm the authenticity of medicines and reduce counterfeiting risks (Mackey and Nayyar, 2017). When integrated with payment platforms, such mechanisms combine financial and product trust in a single consumer experience. From a behavioural perspective, this transparency reduces *perceived risk*, a significant barrier to purchase in high-involvement categories (Gefen et al., 2003). Payments thus enhance not just convenience but also the product's symbolic and functional value.

Collectively, these developments demonstrate that digital payments are not peripheral to product strategy but constitutive of it. By enabling tokenised ownership, modular value delivery, and verifiable authenticity, payment systems redefine how products are designed, accessed, and trusted. Rather than serving as transactional afterthoughts, digital payments operate as embedded product attributes that shape consumer experience, perceived value, and market positioning across both developed and emerging economies.

Case Insights

Global Example: Starbucks Odyssey. Starbucks exemplifies how payments can extend product strategy. Its app links digital payments to a loyalty programme that rewards purchases with redeemable points. This integration integrates payment into the product's augmented experience, deepening engagement and loyalty (Taylor, 2016; Falk et al., 2016).

Emerging-Market Example: M-Pesa. Since 2007, M-Pesa has revolutionised access to financial and non-financial products. Beyond money transfers, users pay for solar energy, health cover, and daily commuting through small instalments (Jack and Suri, 2014). Payment innovation here directly shapes product accessibility and affordability.

Hybrid Example: China's Super Apps. In China, WeChat and Alipay blur the boundaries between product, payment, and promotion. A single interface enables purchases across sectors, from groceries to entertainment, creating an integrated ecosystem that locks in users through convenience and cross-service synergies (Sun et al., 2021).

Strategic Implications for Product Managers

Integrating digital payments into product design requires managers to treat payment functionality as a core design element rather than a supporting infrastructure. Products increasingly require built-in payment flexibility, such as tokenisation, subscriptions, and microtransactions, that shape consumer engagement and usage patterns. As value shifts from ownership to access, firms must reposition products around continuous experiences rather than discrete transactions, reflecting changing expectations in digital markets. At the same time, payment systems act as trust signals:

blockchain-linked verification and secure digital transactions reinforce authenticity and credibility, particularly in industries vulnerable to counterfeiting. In emerging markets, inclusive payment innovations, such as pay-as-you-go and low-value digital transactions, expand participation and affordability, aligning commercial strategy with broader development goals, including SDG 8 and SDG 9. Collectively, these dynamics show that digital payments are no longer back-end enablers but strategic components of product innovation, reshaping how products are conceptualised, consumed, and trusted, and reinforcing the argument that payment systems are integral to how modern products create and deliver value.

Digital Payments and Pricing Strategy

Price is often seen as the most adaptable element of the marketing mix, capable of immediate adjustment in response to competitive dynamics, demand fluctuations, or changing costs (Kotler and Keller, 2016). Unlike product or distribution strategies, pricing can be altered almost instantaneously, making it a powerful lever for shaping perceptions of value and market positioning. From a marketing mix perspective, digital payments fundamentally reshape pricing by influencing how consumers perceive cost, fairness, and value at the moment of exchange, rather than merely enabling price execution.

The expansion of digital payment systems adds new layers of complexity and opportunity. Digital transactions enable dynamic, personalised, and micro-level pricing models that were unfeasible in cash economies. At the same time, they alter how consumers perceive fairness, value, and willingness to pay. Insights from behavioural economics reveal that payment acts are emotionally charged rather than purely rational (Prelec and Loewenstein, 1998).

This section explores how digital payments reshape pricing through four lenses: (1) the psychology of payments, (2) cryptocurrency-based pricing, (3) microtransactions and subscriptions, and (4) transparency and trust.

Behavioural Economics and the Psychology of Payments

Traditional economics assumes rational decision-making, yet behavioural research shows that emotions and mental shortcuts strongly influence price perceptions (Thaler, 1985). Digital payments intensify these effects by reducing the salience of spending—how consciously consumers feel the cost.

Prelec and Simester (2001) identified the "*pain of paying*," demonstrating that parting with cash feels more painful than using cards or mobile wallets. Digital payments detach the moment of consumption from the moment of payment, making transactions almost frictionless. This lower psychological barrier encourages more frequent spending.

Marketers increasingly embed such design cues into payment systems, through one-click checkouts, auto-renewal subscriptions, and in-app purchases, which boost sales but raise ethical concerns about overconsumption (Soman, 2003).

Mental accounting further shapes behaviour. Consumers perceive loyalty points, cashback, or crypto rewards as "free money," prompting higher discretionary spending (Thaler, 1999). When marketers integrate these rewards into pricing, they effectively reframe costs as benefits without altering actual value. Payment design thus becomes a form of behavioural pricing, blending psychology with technology.

Cryptocurrency Pricing and Volatility

Cryptocurrencies have introduced both innovation and uncertainty to pricing strategy. Their decentralised and volatile nature creates both symbolic and monetary implications.

Some companies have experimented with crypto-based pricing, allowing purchases in Bitcoin, Ethereum, or stablecoins. Tesla's temporary acceptance of Bitcoin, for instance, positioned the brand as technologically progressive (Corbet et al., 2020). Even when monetary equivalence remains, the symbolic act of paying with cryptocurrency signals innovation and appeals to tech-savvy consumers.

However, volatility undermines pricing stability. Sharp fluctuations can erode consumer confidence or cause revenue losses. Consequently, many firms prefer *stablecoins*—blockchain assets pegged to fiat currencies—which offer transparency with minimal volatility (Arner et al., 2020).

In emerging economies, cryptocurrencies often serve as a hedge against volatile local currencies. In Venezuela and Nigeria, consumers increasingly use digital currencies to hedge inflation and enable cross-border payments. For marketers, accepting cryptocurrency in such contexts is not only innovative but also necessary to maintain inclusion and relevance.

Microtransactions and Subscription Models

Digital payments have enabled granular pricing structures that redefine value exchange. *Microtransactions*, or small digital payments for limited product access, allow firms to monetise engagement continuously.

The gaming industry demonstrates this model vividly. Games such as *Fortnite* and *League of Legends* generate most of their revenue through low-cost virtual purchases, transforming payment into a form of gameplay itself (Hamari et al., 2020). Because each transaction feels insignificant, cumulative spending rises substantially.

Subscription models extend this logic. Platforms like Netflix, Spotify, and various SaaS providers rely on automated, recurring payments to ensure predictable revenue and convenience for consumers. Subscriptions shift the value perception from ownership to access, aligning with changing consumption norms that prioritise usability over possession (Verma et al., 2025).

In emerging markets, *pay-as-you-go* systems show how digital payments can democratise access. Services such as M-Pesa in Kenya enable low-income consumers to pay for electricity, transport, or health coverage in small increments (Jack and Suri, 2014). Here, pricing innovation directly contributes to social inclusion, allowing affordability without compromising business viability.

Transparency, Fairness, and Consumer Trust

Digital payments enhance data visibility, enabling personalised pricing but also intensifying concerns over fairness. Research shows that consumers react strongly to perceived pricing injustice, especially when algorithmic or opaque (Kahneman et al., 1986).

Platforms can now adjust prices based on location, demand, or user history. While profitable, such practices risk reputational damage if customers perceive discrimination (Garbarino and Lee, 2003). In regions with limited digital literacy or weak consumer protection, this can erode trust quickly.

Blockchain technologies offer a counterbalance. Immutable ledgers can log transaction histories, enhancing transparency in price formation. In agricultural supply chains, blockchain platforms allow farmers and buyers to verify fair trade prices, ensuring accountability (Kamble et al., 2020).

For marketers, transparency becomes both a competitive advantage and an ethical requirement. Long-term brand trust increasingly depends not only on what is charged but on *how* pricing decisions are communicated.

Case Insights

Global Case – Uber's Surge Pricing. Uber's algorithmic surge pricing highlights both potential and risk. Dynamic fare adjustments optimise supply and demand but can appear unfair during emergencies or high-demand periods. Consumers often accept price increases when scarcity is evident, yet react negatively to perceived exploitation (Cohen et al., 2016). Transparent communication is therefore vital to maintaining trust.

Emerging Market Case – Pay-as-You-Go Solar in Kenya. Companies like M-KOPA Solar use mobile money to let consumers pay for solar home systems in instalments. Customers gradually gain ownership after regular small payments

(Taneja et al., 2014). This model demonstrates how digital payments link pricing innovation with social impact by aligning costs with income cycles.

Hybrid Case – Amazon Prime. Amazon's Prime programme exemplifies subscription pricing as a loyalty tool. Seamless auto-renewals supported by digital payments create a sense of continuous value through bundled services such as free delivery and media access (Mattila et al., 2021). The perceived value of membership reframes the price from an expense to an investment.

Strategic Implications for Marketers

Integrating digital payments into pricing design requires marketers to rethink pricing as a behavioural, technological, and ethical construct rather than a purely numerical decision. Payment-enabled pricing must account for cognitive biases such as reduced pain of paying and mental accounting, while avoiding manipulative design that could undermine consumer welfare. Digital payment infrastructures enable dynamic and personalised pricing models, but these must be balanced with transparency to preserve perceptions of fairness and trust. Subscription, microtransaction, and pay-as-you-go models shift pricing towards access-oriented value propositions, expanding affordability and inclusion, particularly in emerging economies. Crypto-enabled pricing, especially through stablecoins, offers opportunities to attract early adopters and diversify transactional risk, provided volatility is clearly communicated. Across these developments, trust emerges as a critical pricing asset, reinforced by transparent and verifiable payment systems. Pricing in the digital economy thus becomes an experience-design challenge shaped by perceived fairness, emotional responses at payment, and clarity of value exchange. Firms that integrate behavioural insight with ethical and technological intelligence are better positioned to sustain profitability and long-term trust in digital markets.

Digital Payments and Place (Distribution)

In the traditional marketing mix, *place* refers to the channels through which products and services reach consumers. It encompasses logistics, retail formats, and intermediaries that connect producers and customers. In today's digital economy, however, the concept of place has expanded far beyond physical stores or e-commerce websites. Distribution now spans digital platforms, peer-to-peer exchanges, and decentralised marketplaces. From a marketing mix perspective, digital payments transform "place" by redefining how, where, and by whom exchange occurs, shifting distribution from physical channels to payment-enabled digital networks.

Digital payments lie at the heart of this transformation. They not only speed up transactions but also redefine distribution models by reducing dependency on intermediaries, enabling cross-border trade, and embedding payment within digital ecosystems such as mobile apps, social media, and the metaverse. Across both advanced and emerging economies, payment integration is making commerce more accessible, inclusive, and data-driven.

This section explores how digital payments reshape *place* through four mechanisms:

i. enabling borderless commerce;
ii. supporting disintermediation and direct-to-consumer (D2C) models;
iii. integrating payments into omnichannel ecosystems; and
iv. promoting inclusive distribution in underserved markets.

Borderless Commerce and Global Reach

Digital payments have dismantled many barriers to cross-border trade. Historically, international commerce faced friction due to currency conversion costs, remittance fees, and limited trust. Today, platforms such as PayPal, Stripe, and Alipay allow even micro-entrepreneurs to receive global payments seamlessly. Reducing such transaction costs substantially increases cross-border e-commerce adoption (Liu et al., 2021).

For consumers, paying in familiar digital formats increases confidence in international purchases. Cryptocurrencies and stablecoins further enable borderless exchange by bypassing banking restrictions and capital controls. Freelancers in emerging economies increasingly receive payments in Bitcoin or stablecoins, participating in global gig economies despite local financial barriers (Corbet et al., 2020).

Thus, *place* is no longer defined by geography but by payment-enabled digital networks that connect producers and consumers worldwide.

Disintermediation and Direct-to-Consumer Models

Digital payments strengthen the shift towards *disintermediation*, allowing producers to bypass intermediaries and sell directly to consumers. D2C strategies facilitated by e-commerce are further enhanced by embedded payments, which reduce transaction costs and improve profitability.

Nike's D2C approach, for example, relies on digital platforms and mobile apps that integrate marketing, payment, and after-sales services, enabling complete value-chain control (Sorescu, 2017). Such integration fosters loyalty and data-driven insight.

In emerging economies, digital payment infrastructure empowers smaller producers. Indian farmers now sell directly to urban consumers through mobile marketplaces, receiving instant payments via platforms such as Paytm and PhonePe (Kamble et al., 2020). These systems make distribution more transparent and equitable while reducing the role of exploitative middlemen.

By controlling payments, brands also gain access to valuable consumer data, linking distribution strategy to innovation and personalisation. Payments thus evolve from mere financial tools into strategic connectors between producers and consumers.

Omnichannel Distribution and Platform Ecosystems

Modern consumers expect frictionless movement between online and offline touchpoints, driving *omnichannel distribution*. Digital payments make this possible by serving as the connective tissue across shopping environments.

Retailers such as Walmart and Target allow consumers to order online, pay digitally, and pick up their purchases in-store or curbside. In China, Alibaba's Hema supermarkets blend in-store experiences with app-based payments, delivery, and loyalty integration through Alipay (Verhoef et al., 2015). Here, payment systems enable consistency and convenience across all channels.

Social media and messaging platforms are also becoming key distribution arenas. *Instagram Checkout* lets users complete purchases without leaving the app, while *WeChat* in China integrates payments, messaging, and retail into a single ecosystem (Zhang et al., 2022).

As these examples show, *place* is evolving from a fixed point of sale into a fluid, interconnected network. Payments unify these ecosystems, driving channel integration and sustained consumer engagement.

Inclusive Distribution in Emerging Markets

Perhaps the most transformative role of digital payments appears in *emerging markets*, where they enable access to essential goods and services for previously excluded populations.

Mobile money systems such as *M-Pesa* in Kenya allow consumers to pay for electricity, healthcare, and education in small, manageable amounts (Jack and Suri, 2014). This "pay-as-you-go" approach aligns costs with income patterns, expanding market participation and financial inclusion.

In Bangladesh, *bKash* empowers informal retailers to accept digital payments, integrating micro-merchants into formal supply chains (Dos Santos and Harvold Kvangraven, 2017). These systems not only broaden distribution but also reduce cash dependency and improve security for small businesses.

Such models advance both commercial and developmental goals, supporting the UN's *Sustainable Development Goals,* notably SDG 8 (Decent Work and Economic Growth) and SDG 9 (Industry, Innovation, and Infrastructure). Digital payments thus transform distribution into a mechanism for economic empowerment and market expansion.

Taken together, these developments illustrate that digital payments are not simply facilitators of distribution but fundamental enablers of how place is configured in digital markets. By enabling borderless exchange, supporting direct-to-consumer models, integrating omnichannel ecosystems, and expanding access in underserved regions, payment systems redefine distribution as a networked, inclusive, and data-driven process. Place, therefore, evolves from a static channel decision into a dynamic architecture shaped by payment infrastructure, trust, and technological reach.

Case Insights

Global Case: Amazon Go and Frictionless Retail. Amazon Go stores eliminate traditional checkouts by integrating sensors and digital payments. Shoppers enter, collect items, and exit; payments are processed automatically via the Amazon app (Grewal et al., 2017). This model redefines *place* as a data-driven, cashierless experience that merges retail and technology.

Emerging Market Case: Jumia in Africa. Africa's largest e-commerce platform, Jumia, leverages *JumiaPay* to overcome low credit-card penetration. By embedding digital wallets into its ecosystem, Jumia enables millions of consumers to transact securely online, bridging infrastructural gaps (Boateng et al., 2019).

Hybrid Case: WeChat Ecosystem. WeChat Pay exemplifies the fusion of communication, commerce, and payment. Users can purchase goods, book taxis, or pay bills without leaving the app. For firms, this model embeds distribution directly into consumers' social interactions, extending *place* into everyday digital spaces (Sun et al., 2021).

Strategic Implications for Marketers

Integrating digital payments into distribution strategy fundamentally reshapes how firms design market access and channel structures. Payment-enabled systems reduce frictions in cross-border commerce, enabling firms of all sizes to reach global customers with limited reliance on traditional intermediaries. Embedded payment infrastructure also supports direct-to-consumer models by improving efficiency, strengthening brand control, and generating actionable consumer insights. Moreover, digital payments act as a unifying mechanism within omnichannel distribution, connecting online and offline touchpoints into seamless customer journeys. In emerging and low-income markets, mobile and micro-payment systems broaden participation by aligning payment structures with income variability, thereby combining commercial viability with social inclusion. Transaction-level payment data further allows firms to optimise distribution decisions through localisation, demand forecasting, and channel refinement. Collectively, these dynamics indicate that distribution strategy can no longer be designed solely around physical logistics or platform presence but must be organised around payment architecture as a core infrastructure enabling access, trust, and convenience across markets.

Digital Payments and Promotion

Promotion in the marketing mix traditionally covers communication strategies—advertising, sales promotion, public relations, and personal selling—that inform, persuade, and remind consumers (Kotler and Keller, 2016). While much attention has focused on digital advertising and influencer marketing, digital payment systems have become powerful promotional tools in their own right. From a marketing mix perspective, digital payments reshape "promotion" by embedding persuasive cues, incentives, and trust signals directly into transactional moments rather than relying solely on external communication media.

Unlike conventional media-based promotion, payments sit at the intersection of *transaction and communication*. Every payment is a data-rich interaction that can generate insights, deliver incentives, and reinforce trust. Digital payment systems thus extend promotion beyond message delivery toward *experience design*, embedding loyalty, engagement, and confidence within each transaction.

This section analyses how digital payments reshape promotion through four mechanisms:

i. payment-linked loyalty and tokenized incentives;
ii. gamification and engagement;
iii. hyper-personalized promotions using payment data; and
iv. trust-building and reputational signalling.

Payment-Linked Loyalty and Tokenized Incentives

Loyalty programs reward repeat patronage, but digital payments enhance them by making rewards instant and automatic. *Starbucks Rewards*, for instance, links payments with real-time points accumulation, turning each transaction into a reinforcing loop of consumption and reward (Zhang et al., 2022).

Blockchain technologies take loyalty to the next level through *tokenized incentives*. Brands such as Singapore Airlines use blockchain to issue transferable digital tokens (Tapscott and Tapscott, 2017). These tokens, unlike traditional loyalty points, hold value across multiple merchants and can appreciate over time, transforming loyalty from a closed system into a flexible, consumer-empowered network.

In emerging markets, mobile payment–linked rewards help micro-retailers retain customers. M-Pesa–integrated platforms in Kenya allow small merchants to offer airtime or micro-discounts automatically, sustaining repeat business even in informal settings (Jack and Suri, 2014).

Gamification and Consumer Engagement

Digital payments also support *gamification*, turning ordinary transactions into engaging experiences. Cashback offers, spin-the-wheel promotions, and surprise vouchers build excitement while encouraging habitual use (Hamari et al., 2014).

A standout example is *Alipay's "Ant Forest,"* which rewards eco-friendly spending with virtual trees that translate into real-world reforestation. With over 500 million participants, it combines gamified incentives with social purpose (Zhao and Abeysekera, 2024). Similarly, India's PhonePe and Paytm apps reward users with random cashback or digital scratch cards, converting every transaction into a mini game (Chawla and Joshi, 2019).

Gamification thus shifts promotion from external messaging to *transactional design*, making the act of paying itself a branded experience that stimulates loyalty and word-of-mouth advocacy.

Hyper-Personalization through Payment Data

Each digital payment produces valuable behavioural data, amount, timing, frequency, and context, that can be used to tailor promotions. Firms like *Amazon* employ such insights to create personalised bundles, reminders, and dynamic pricing recommendations, while ride-hailing services such as *Uber* use location-based offers to encourage off-peak usage (Chen et al., 2016).

Payment data also allows *real-time targeting*. For example, mobile wallets can deliver a discount immediately after a consumer purchase at a competitor's outlet. Though this raises privacy and ethical questions, it demonstrates the potential of transaction-based marketing to dynamically shape consumer choice.

In developing contexts, personalization supports *financial inclusion*. Mobile payment data in Sub-Saharan Africa helps providers assess creditworthiness and extend microloans or insurance offers, blending promotion with empowerment (Jack and Suri, 2014). When responsibly managed, such data-driven strategies create value for both firms and consumers.

Payments as Trust-Building in Promotion

Promotion is also about building *credibility*. Secure and transparent payment systems reinforce brand trust, while poor payment experiences can damage reputation (Gefen et al., 2003).

Blockchain-based payment verification strengthens authenticity claims. For instance, *LVMH* uses blockchain-linked payment certificates to verify product authenticity, integrating the payment process into the brand's trust narrative (Joy et al., 2022).

In emerging economies, where scepticism toward digital commerce persists, payment reliability functions as a key *promotional signal.* M-Pesa's reputation for security and ease has encouraged wider digital adoption and confidence in e-commerce (Aker and Mbiti, 2010). Thus, payment systems serve as implicit endorsements of brand integrity.

Collectively, these mechanisms illustrate that digital payments do not merely support promotional activities but fundamentally reconfigure how promotion operates in digital markets. By linking incentives, gamification, personalisation, and trust directly to transactional moments, payment systems transform promotion into an interactive, data-driven, and experience-based process. Promotion, therefore, shifts from one-way persuasion toward embedded engagement, where loyalty and credibility are reinforced through repeated payment interactions across both developed and emerging economies.

Case Insights

Global Case: WeChat Red Packets. In China, WeChat Pay's "red packets" feature digitizes the cultural practice of gifting money during festivals. Beyond payment utility, this feature functions as a promotional tool, reinforcing cultural traditions while locking users into the WeChat ecosystem. It demonstrates how digital payments can embed promotional strategies within social and cultural rituals (Zhang et al., 2022).

Emerging Market Case: Paytm in India. Paytm has combined mobile payments with aggressive promotional campaigns offering cashback, discounts, and festival-specific vouchers. By tying promotions directly to transactions, Paytm rapidly expanded its user base, illustrating how payment-linked promotions can be central to platform growth in emerging markets (Chawla and Joshi, 2019).

Hybrid Case: Starbucks Odyssey. Starbucks Odyssey integrates NFT-based loyalty tokens into its mobile payments app, offering consumers exclusive experiences and content. This merges promotion with payment, creating an immersive loyalty ecosystem that redefines brand engagement (Zhang et al., 2022).

Strategic Implications for Marketers

Integrating digital payments into promotional strategy requires marketers to reconceptualise transactions as active promotional touchpoints rather than passive endpoints. Well-designed payment experiences can reinforce loyalty, trust, and engagement by embedding incentives directly into the payment process. Blockchain-enabled tokenised rewards extend promotional value beyond firm boundaries by creating transferable and tradable loyalty assets, while gamified payment features can stimulate repeat engagement when implemented responsibly. At the same time, the use of payment data enables highly personalised promotions, demanding careful governance to balance relevance with consumer privacy. Secure and seamless payment systems also function as reputational signals, strengthening brand credibility by reducing uncertainty at the moment of exchange. Collectively, these developments blur the boundary between paying and participating, shifting promotion from one-way persuasion to interactive engagement. For marketers, success in the digital economy will increasingly depend on mastering this intersection of finance, data, trust, and emotion to build enduring consumer relationships.

SOLUTIONS AND RECOMMENDATIONS

Managerial Implications

The preceding sections demonstrated how digital payments transform the four pillars of the marketing mix—product, price, place, and promotion—by reshaping consumer perceptions, enabling new business models, and redefining trust in market exchanges. For managers, these transformations carry profound implications. Digital payments are no longer operational afterthoughts relegated to finance departments; they are *strategic levers* that shape customer experience, brand positioning, and long-term competitiveness. From a managerial perspective, the strategic value of digital payments lies in their ability to simultaneously influence experience design, data-driven decision-making, trust formation, and market inclusion across the marketing mix. This section outlines five key managerial implications of integrating digital payments into marketing strategy: customer experience design, data-driven personalisation, ethical and regulatory responsibility, inclusive market strategies, and capability building for digital-era competition.

Payments as Part of Customer Experience Design

Managers must recognize that the payment stage is no longer the "end" of the customer journey but an *experience touchpoint*. Research shows that seamless payment experiences can significantly enhance satisfaction, while friction at checkout often leads to cart abandonment in e-commerce (Sohanda et al., 2024). Firms such as Amazon and Apple have built a competitive advantage partly on reducing checkout friction through one-click purchases and biometric authentication.

For managers, this implies that *payment design is part of marketing design*. Investments in frictionless, secure, and intuitive payment processes, such as QR codes, contactless options, and mobile wallets, should be treated as investments in brand equity. Moreover, the symbolic aspects of payments (e.g., using eco-tokens to plant trees) can reinforce brand values, transforming a transaction into a narrative moment that deepens engagement.

Payments as Data Assets for Personalization

Every payment generates data that provides granular insights into consumer behaviour, including frequency, timing, purchase categories, and preferred channels. Properly analysed, this data becomes a *powerful marketing resource*. Retailers can identify high-value customers, forecast demand patterns, and deliver hyper-personalized offers that align with consumer preferences (Wedel and Kannan, 2016).

However, managers must balance personalization with privacy. The use of payment data in marketing is subject to growing consumer scepticism and regulatory oversight, especially under frameworks like the *EU's General Data Protection Regulation (GDPR)*. Companies must therefore develop transparent policies, provide opt-in mechanisms, and ensure that data-driven promotions are perceived as valuable rather than intrusive (Martin and Murphy, 2017).

Strategically, firms should integrate payment analytics into *customer relationship management (CRM)* systems. This enables cross-functional value: marketing gains personalization capabilities, while finance and operations benefit from improved demand forecasting and risk management.

Ethical and Regulatory Responsibilities

The integration of payments into marketing introduces *ethical dilemmas* that managers must navigate carefully. Behavioural economics shows that digital payments reduce the "*pain of paying*," potentially encouraging overspending (Prelec and Simester, 2001). While this benefits short-term revenues, it raises ethical questions about consumer welfare. Managers must avoid exploitative practices such as opaque subscription renewals, manipulative gamification, or hidden fees that erode long-term trust.

Regulatory scrutiny of digital payments is intensifying. Central banks are exploring digital currencies, while regulators are imposing stricter rules on data usage, anti-money laundering, and consumer protection (Arner et al., 2020). Non-compliance not only exposes firms to legal penalties but also undermines reputational trust.

Managers should therefore embed *responsible payment strategies* into corporate governance. This includes transparent pricing, clear subscription cancellation mechanisms, and ethical gamified promotions. Ethical stewardship in payments can itself become a source of differentiation, especially as consumers increasingly reward brands that demonstrate social responsibility.

Inclusive Strategies for Emerging Markets

In many emerging economies, digital payments are not just efficiency tools but enablers of *market participation*. Mobile money platforms such as *M-Pesa* in Kenya or *bKash* in Bangladesh have allowed millions of unbanked consumers to access products and services, reshaping distribution and pricing strategies (Jack and Suri, 2014).

For managers operating in these contexts, inclusivity is both a business opportunity and a social responsibility. Payment-enabled "*pay-as-you-go*" models can expand access to essential services such as energy, healthcare, and education, while simultaneously opening new revenue streams. Firms that align digital payment strategies with financial inclusion objectives can strengthen brand legitimacy and build resilient customer bases.

Global firms entering emerging markets must adapt to local payment ecosystems rather than imposing standardized solutions. For instance, while credit cards dominate in the U.S., QR-based payments are ubiquitous in China, and mobile money dominates in Sub-Saharan Africa. Customizing payment strategies to local contexts ensures relevance and adoption, reinforcing the broader distribution and promotional mix.

Capability Building for Digital-Era Competition

To fully leverage digital payments within marketing strategy, firms must build integrated capabilities spanning technology, analytics, and strategic management. This begins with developing secure, scalable, and interoperable payment infrastructures, often in collaboration with fintech providers and platform partners. Equally important is analytical capability, the ability to convert payment data into actionable insights for segmentation, personalization, and demand forecasting while remaining compliant with evolving privacy regulations. Strategically, firms must align payment innovations with brand positioning and long-term value creation rather than treating them as short-term promotional tactics. Finally, collaborative capability is essential, as partnerships with fintech firms, telecom operators, and regulators enable the co-creation of inclusive and resilient payment ecosystems. Firms that approach digital payments holistically, as strategic assets embedded across product, price, place, and promotion, will be best positioned to achieve sustainable competitive advantage in the digital economy.

Managerial Framework: Payments as Strategic Levers

Synthesising the preceding discussion, this framework positions digital payments as strategic managerial levers operating across interrelated experiential, informational, trust-based, and inclusion-oriented dimensions. Payment systems increasingly form part of the customer experience, shaping brand narratives and influencing how value is perceived at the moment of exchange. At the same time, payments generate granular data that can deepen consumer insight and support more responsive and personalised marketing actions. The design and governance of payment systems also function as signals of security, transparency, and ethical stewardship, thereby reinforcing consumer trust. In parallel, digital payments act as mechanisms of market inclusion by reducing barriers to participation and enabling broader social and economic access.

Viewed through these dimensions, payment strategies extend well beyond transactional efficiency. Digital payments influence customer experience, enable data-driven personalisation, and reinforce ethical trust, while also expanding inclusion and requiring new organisational capabilities. Consequently, payment design should be treated as a cross-functional managerial responsibility that integrates marketing strategy, technological infrastructure, and governance arrangements. When approached in isolation, payment systems risk being relegated to operational utilities; when strategically embedded, they become central resources for value creation and differentiation.

The integration of digital payments into the marketing mix therefore represents a significant shift in managerial decision-making. Payments now operate as front-end touchpoints that shape consumer trust, loyalty, and perceptions of value, while simultaneously functioning as data assets, promotional interfaces, and distribution enablers. This expanded role introduces heightened managerial responsibilities. Firms must design payment strategies that are seamless yet ethically grounded, personalised yet respectful of consumer privacy, and inclusive while remaining commercially viable.

Ultimately, organisations that approach digital payments as *strategic levers*—bridging marketing, technology, and ethical governance—are better positioned to achieve sustainable competitive advantage. Embedding payment systems within marketing strategy enables managers to reconfigure value creation in ways that align competitive performance with evolving societal expectations around inclusion, transparency, and sustainability.

FUTURE RESEARCH DIRECTIONS

Research on digital payments has expanded rapidly within finance, information systems, and economics, yet their implications for marketing theory and practice remain comparatively underdeveloped. Existing studies have largely focused on adoption drivers, technology acceptance, and security concerns (Slade et al., 2015; Chawla and Joshi, 2019). As this chapter has argued, however, digital payments are no longer merely supporting financial tools. They increasingly operate as strategic mechanisms that shape value creation, exchange processes, and market governance. Future research should therefore move beyond descriptive adoption models and develop theory-driven perspectives that place payment systems at the centre of marketing strategy and consumer experience.

A first priority is to examine how digital payments reshape the traditional marketing mix. Although the rigidity of the 4Ps framework has been questioned (Constantinides, 2006), empirical work linking payment technologies to product, price, place, and promotion remains limited. Future studies could investigate consumer responses to tokenised products and blockchain-based verification, particularly in relation to trust and perceived authenticity. Pricing research may explore how digital wallets, subscriptions, and stablecoins influence perceptions of fairness, transparency, and spending behaviour. Distribution-focused work could analyse how mobile payments and cryptocurrencies facilitate cross-border exchange and platform dominance, while promotional studies might assess the effectiveness of payment-linked rewards and gamified incentives.

Second, greater attention is needed to the behavioural economics of digital payments. While prior research shows that consumption decisions often deviate from rational choice (Thaler, 1985; Prelec and Simester, 2001), the role of payment modalities in shaping these biases remains underexplored. Experimental and field-based research could examine how frictionless payments affect cost awareness, self-control, and financial decision-making.

Finally, cross-cultural, generational, and ethical dimensions warrant deeper investigation. Comparative studies across contexts can clarify how culture, age, and institutional trust shape payment-enabled marketing, while research on inclusivity, sustainability, and governance can inform fair and responsible payment design.

CONCLUSION

Digital payment systems have evolved well beyond their original role as background financial infrastructure. This chapter has demonstrated that they now function as strategic components of marketing practice, influencing all four elements of the marketing mix—product, price, place, and promotion. Payments are no longer neutral mechanisms of exchange; instead, they act as meaningful touchpoints that shape how value is created, experienced, and evaluated by consumers, with important implications for trust, engagement, and brand relationships.

At the product level, digital payments enable new forms of ownership and access. Mechanisms such as tokenisation, microtransactions, and blockchain-based verification allow firms to design offerings that combine functionality with digital legitimacy. Subscription models and pay-as-you-go services further blur the boundary between product and payment, making the payment system an integral part of the consumption experience.

Pricing practices are similarly being transformed. Digital and contactless payments influence how consumers perceive cost and value by reducing transaction friction and altering spending awareness. At the same time, transparent and traceable payment systems shape perceptions of pricing fairness. Cryptocurrencies and subscription-based pricing models shift emphasis from ownership towards access, requiring firms to integrate pricing decisions with payment design rather than treating them as separate considerations.

From a distribution perspective, digital payments lower barriers to market participation. They facilitate cross-border transactions, support direct-to-consumer models, and connect physical and digital channels. In emerging markets in particular, mobile money platforms expand access to goods and services and enable more inclusive forms of market engagement.

Promotion is increasingly embedded within payment moments. Loyalty schemes, gamified incentives, and personalised offers are often delivered through payment interfaces, while secure and reliable payment systems signal professionalism and trustworthiness.

Taken together, these developments call for a renewed interpretation of the marketing mix. The 4Ps remain relevant but must be understood in a context where payment systems sit at the centre of consumer experience, shaping value exchange in markets that increasingly prioritise convenience, transparency, and inclusion.

REFERENCES

Aker, J. C., & Mbiti, I. M. (2010). Mobile phones and economic development in Africa. *The Journal of Economic Perspectives*, *24*(3), 207–232. DOI: 10.1257/jep.24.3.207

Alba, J., Lynch, J., Weitz, B., Janiszewski, C., Lutz, R., Sawyer, A., & Wood, S. (1997). Interactive home shopping: Consumer, retailer, and manufacturer incentives to participate in electronic marketplaces. *Journal of Marketing*, *61*(3), 38–53. DOI: 10.1177/002224299706100303

Arner, D. W., Auer, R., & Frost, J. (2020). *Stablecoins: Risks, potential and regulation* (BIS Working Papers No. 905). Bank for International Settlements. https://www.bis.org/publ/work905.htm

Boateng, H., Kosiba, J. P. B., & Okoe, A. F. (2019). Determinants of consumers' participation in the sharing economy: A social exchange perspective within an emerging economy context. *International Journal of Contemporary Hospitality Management*, *31*(2), 718–733. DOI: 10.1108/IJCHM-11-2017-0731

Chawla, D., & Joshi, H. (2019). Consumer attitude and intention to adopt mobile wallet in India: An empirical study. *International Journal of Bank Marketing*, *37*(7), 1590–1618. DOI: 10.1108/IJBM-09-2018-0256

Chen, Y., Mislove, A., & Wilson, C. (2016). An empirical analysis of algorithmic pricing on Amazon Marketplace. In *Proceedings of the 25th International Conference on World Wide Web* (pp. 1339–1349). ACM. DOI: 10.1145/2872427.2883089

Constantinides, E. (2006). The marketing mix revisited: Towards the 21st century marketing. *Journal of Marketing Management*, *22*(3–4), 407–438. DOI: 10.1362/026725706776861190

Corbet, S., Larkin, C., Lucey, B., Meegan, A., & Yarovaya, L. (2020). Cryptocurrency reaction to FOMC announcements: Evidence of heterogeneity based on blockchain stack position. *Journal of Financial Stability*, *46*, 100706. DOI: 10.1016/j.jfs.2019.100706

Davis, F. D. (1989). Perceived usefulness, perceived ease of use, and user acceptance of information technology. *Management Information Systems Quarterly*, *13*(3), 319–340. DOI: 10.2307/249008

Dos Santos, P. L., & Harvold Kvangraven, I. (2017). Better than cash, but beware the costs: Electronic payment systems and financial inclusion in developing economies. *Development and Change*, *48*(2), 205–227. DOI: 10.1111/dech.12296

Dowling, M. (2022). Is non-fungible token pricing driven by cryptocurrencies? *Finance Research Letters, 44*, 102097. DOI: 10.1016/j.frl.2021.102097

Falk, T., Kunz, W. H., Schepers, J. J. L., & Mrozek, A. J. (2016). How mobile payment influences the overall store price image. *Journal of Business Research, 69*(7), 2417–2423. DOI: 10.1016/j.jbusres.2016.01.011

Gefen, D., Karahanna, E., & Straub, D. W. (2003). Trust and TAM in online shopping: An integrated model. *Management Information Systems Quarterly, 27*(1), 51–90. DOI: 10.2307/30036519

Ghouse, S. M., Shekhar, R., & Chaudhary, M. (2025). Driving financial inclusion: Exploring mobile wallet adoption among rural Omani millennials. *Journal of Islamic Marketing, 16*(4), 1229–1257. DOI: 10.1108/JIMA-02-2024-0073

Grewal, D., Roggeveen, A. L., & Nordfält, J. (2017). The future of retailing. *Journal of Retailing, 93*(2), 121–146. DOI: 10.1016/j.jretai.2016.12.008

GSMA. (2025). *State of the industry report on mobile money: Mobile money surpasses two billion registered accounts and over half a billion monthly active users globally*. GSMA. https://www.gsma.com

Hamari, J., Hanner, N., & Koivisto, J. (2020). Why pay premium in freemium services?" A study on perceived value, continued use and purchase intentions in free-to-play games. *International Journal of Information Management, 51*, 102040. DOI: 10.1016/j.ijinfomgt.2019.102040

Hamari, J., Koivisto, J., & Sarsa, H. (2014). Does gamification work? A literature review of empirical studies on gamification. In *Proceedings of the 47th Hawaii International Conference on System Sciences* (pp. 3025–3034). IEEE. DOI: 10.1109/HICSS.2014.377

Jack, W., & Suri, T. (2014). Risk sharing and transaction costs: Evidence from Kenya's mobile money revolution. *The American Economic Review, 104*(1), 183–223. DOI: 10.1257/aer.104.1.183

Joy, A., Zhu, Y., Peña, C., & Brouard, M. (2022). Digital future of luxury brands: Metaverse, digital fashion, and non-fungible tokens. *Strategic Change, 31*(3), 337–343. DOI: 10.1002/jsc.2502

Kamble, S., Gunasekaran, A., & Sharma, R. (2020). Modeling the blockchain-enabled traceability in agriculture supply chain. *International Journal of Information Management, 52*, 101967. DOI: 10.1016/j.ijinfomgt.2019.05.023

Kaplan, A. M., & Haenlein, M. (2010). Users of the world, unite! The challenges and opportunities of Social Media. *Business Horizons*, *53*(1), 59–68. DOI: 10.1016/j.bushor.2009.09.003

Kotler, P., & Keller, K. L. (2016). *Marketing management* (15th ed.). Pearson.

Lemon, K. N., & Verhoef, P. C. (2016). Understanding customer experience throughout the customer journey. *Journal of Marketing*, *80*(6), 69–96. DOI: 10.1509/jm.15.0420

Liu, Y., Li, H., & Hu, F. (2021). Enhancing cross-border e-commerce with digital payment systems: Evidence from China. *Electronic Commerce Research and Applications*, *48*, 101058. DOI: 10.1016/j.elerap.2021.101058

Mallat, N. (2007). Exploring consumer adoption of mobile payments: A qualitative study. *The Journal of Strategic Information Systems*, *16*(4), 413–432. DOI: 10.1016/j.jsis.2007.08.001

Martin, K. D., & Murphy, P. E. (2017). The role of data privacy in marketing. *Journal of the Academy of Marketing Science*, *45*(2), 135–155. DOI: 10.1007/s11747-016-0495-4

Mattila, M., Yrjölä, M., & Hautamäki, P. (2021). Digital transformation of business-to-business sales: What needs to be unlearned? *Journal of Personal Selling & Sales Management*, *41*(2), 113–129. DOI: 10.1080/08853134.2021.1916396

McCarthy, E. J. (1960). *Basic marketing: A managerial approach*. Irwin.

Oliveira, T., Thomas, M., Baptista, G., & Campos, F. (2016). Mobile payment: Understanding the determinants of customer adoption. *Computers in Human Behavior*, *61*, 404–414. DOI: 10.1016/j.chb.2016.03.030

Prelec, D., & Simester, D. (2001). Always leave home without it: A further investigation of the credit-card effect on willingness to pay. *Marketing Letters*, *12*(1), 5–12. DOI: 10.1023/A:1008196717017

Saberi, S., Kouhizadeh, M., Sarkis, J., & Shen, L. (2019). Blockchain technology and its relationships to sustainable supply chain management. *International Journal of Production Research*, *57*(7), 2117–2135. DOI: 10.1080/00207543.2018.1533261

Slade, E. L., Williams, M. D., Dwivedi, Y. K., & Piercy, N. C. (2015). Exploring consumer adoption of proximity mobile payments. *Journal of Strategic Marketing*, *23*(3), 209–223. DOI: 10.1080/0965254X.2014.914075

Sohanda, M. K., Gupta, M., & Kalpana, A. V. (2024). Interactive e-commerce: Addressing cart abandonment through gamified user engagement. In *Proceedings of the International Conference on Communication, Computing and Internet of Things* (pp. 1–5). IEEE. DOI: 10.1109/IC3IoT60841.2024.10550373

Sorescu, A. (2017). Data-driven business model innovation. *Journal of Product Innovation Management, 34*(5), 691–696. DOI: 10.1111/jpim.12398

Stylios, I., Kokolakis, S., Thanou, O., & Chatzis, S. (2022). Key factors driving the adoption of behavioral biometrics and continuous authentication technology: An empirical research. *Information and Computer Security, 30*(4), 562–582. DOI: 10.1108/ICS-08-2021-0124

Sun, Y., Xue, W., Bandyopadhyay, S., Peng, Z., & Su, Y. (2021). WeChat mobile-payment-based smart retail customer experience: An integrated framework. *Information Technology and Management, 23*(2), 77–94. DOI: 10.1007/s10799-021-00346-4

Tapscott, D., & Tapscott, A. (2017). *Blockchain revolution: How the technology behind bitcoin is changing money, business, and the world*. Penguin.

Taylor, B. (2016). Starbucks' mobile payment and rewards strategy: Creating customer loyalty through innovation. *International Journal of Business and Management, 11*(4), 10–20. DOI: 10.5539/ijbm.v11n4p10

Thaler, R. (1985). Mental accounting and consumer choice. *Marketing Science, 4*(3), 199–214. DOI: 10.1287/mksc.4.3.199

Venkatesh, V., Morris, M. G., Davis, G. B., & Davis, F. D. (2003). User acceptance of information technology: Toward a unified view. *Management Information Systems Quarterly, 27*(3), 425–478. DOI: 10.2307/30036540

Verhoef, P. C., Kannan, P. K., & Inman, J. J. (2015). From multi-channel retailing to omni-channel retailing. *Journal of Retailing, 91*(2), 174–181. DOI: 10.1016/j.jretai.2015.02.005

Verma, R., Khound, K., & Mishra, V. (2025). Subscription Economy and the Transformation of Ownership: A Review of Value Perceptions and Retention Strategies. *Advances in Consumer Research. Association for Consumer Research (U. S.), 2*, 935–951. DOI: 10.61336/acr-25-05-08

Wedel, M., & Kannan, P. K. (2016). Marketing analytics for data-rich environments. *Journal of Marketing, 80*(6), 97–121. DOI: 10.1509/jm.15.0413

Zhang, J. Z., Chang, C. W., & Neslin, S. A. (2022). How physical stores enhance customer value: The importance of product inspection depth. *Journal of Marketing*, *86*(2), 166–185. DOI: 10.1177/00222429211012106

Zhao, M., & Abeysekera, I. (2024). The behaviour of FinTech users in the Alipay Ant Forest platform towards environmental protection. *Journal of Open Innovation*, *10*(1), 100201. DOI: 10.1016/j.joitmc.2023.100201

KEY TERMS AND DEFINITIONS

Behavioural Economics: An interdisciplinary field combining psychology and economics to study how consumers make financial decisions, highlighting biases, emotions, and heuristics that shape responses to digital payment systems.

Blockchain: A decentralized digital ledger technology that records transactions across multiple computers in a secure, transparent, and immutable manner, widely used in cryptocurrencies and payment authentication.

Digital Payments: The transfer of monetary value through electronic platforms such as mobile wallets, online banking, and blockchain-based systems replaces cash transactions with secure, convenient digital alternatives.

Financial Inclusion: The ability of individuals and small enterprises, especially in underserved or rural areas, to access affordable financial products and services, often enabled by mobile payments and digital platforms.

Loyalty Programs: Structured marketing initiatives, often linked to digital payment systems, reward consumers for repeat purchases with points, discounts, or token-based incentives to strengthen brand relationships.

Marketing Mix (4Ps): A foundational marketing framework—product, price, place, and promotion—that guides how businesses design offerings, communicate value, and deliver experiences to consumers.

Mobile Wallets: Applications that store payment information on smartphones and enable users to make purchases, transfer money, and manage loyalty rewards through contactless or in-app transactions.

Tokenization: A process that converts sensitive payment or product information into unique digital tokens, which can be used for transactions or loyalty rewards, enhancing both security and consumer engagement.

Chapter 12
Advances in Social Media Influence Marketing:
Integrating Gamification, Social Commerce, and Live Shopping in the Digital Consumer Era

M. Irvanni Bahri
https://orcid.org/0000-0002-1213-9427
Institut Teknologi Batam, Indonesia

M. Y. Meinadia Sekar Kinanti Aswirawan
https://orcid.org/0000-0002-6359-3400
Universitas Pembagunan Nasional Veteran Yogyakarta, Indonesia

Devina Wistiasari
https://orcid.org/0009-0003-2765-7555
Institut Teknologi Batam, Indonesia

Roni Adi
https://orcid.org/0009-0007-9027-2710
Institut Teknologi Batam, Indonesia

ABSTRACT

The chapter discusses the integration of gamification, social commerce, and live shopping in social media marketing, revealing how brands and content creators are transforming consumer engagement and purchase behaviour in real-time. Gamification uses behavioural psychology to drive user participation and loyalty, social commerce facilitates seamless transactions within social environments, and live shopping transforms retail into immersive entertainment. Drawing on case

DOI: 10.4018/979-8-3373-4526-0.ch012

studies, industry trends, and theoretical frameworks, the chapter highlights the psychological impacts, ethical considerations, and regulatory implications of this evolution. It proposes a metacommerce ecosystem where gamified interactions and live digital experiences define the future of online consumerism.

INTRODUCTION

The advancement of digital technology and social media has transformed customer interactions with companies, information consumption, and purchase choices. In the last twenty years, influence marketing has transformed significantly, evolving from celebrity-centric tactics in traditional advertising to customised and interactive techniques inside internet communities. Platforms like Instagram, Facebook, TikTok, and YouTube have transformed into social environments that serve as communication mediums, content distribution routes, product displays, and transaction venues. In this environment, social media has evolved from a simply promotional tool into a whole marketing ecosystem that integrates entertainment, social interaction, and sales conversion. This transition is propelled by three critical factors: extensive internet and smartphone adoption, growing confidence in user-generated content, and developments in customisation and artificial intelligence technologies that provide increasingly pertinent and contextual user experiences.

Contemporary customers are not only passive recipients of marketing messages; they actively participate by commenting, sharing material, and assuming positions as micro-influencers or brand advocates within a firm's communication plan. Acquisition choices have increasingly become impulsive, emotional, and social, due to the rising impact of internet communities and endorsements from reputable individuals. Three principal pillars enhance the efficacy of influence marketing in the digital era: gamification, social commerce, and live shopping. Gamification employs a behaviourally grounded strategy by integrating game components to foster enduring user engagement. Social commerce signifies the amalgamation of transactions with social connections, wherein the purchase process transpires organically inside the social media framework via user-generated content, consumer evaluations, and algorithmic suggestions.

Simultaneously, live shopping presents an interactive sales model that amalgamates entertainment, real-time interaction, and urgency inside a singular digital session, so converting the buying experience into an immersive live performance. These three techniques not only enhance one another but also establish a novel paradigm in digital consuming behaviour. They signify a shift from transactional marketing to immersive and participatory marketing, whereby emotional and social interaction are essential for fostering loyalty and conversion. In the context of a transforming digital economy, marketing strategies that prioritise only visibility and sales are becoming outdated. Brands must now provide a holistic experience—from brand recognition to retention—through dynamic, engaging, and consumer-focused strategies.

This chapter seeks to explore how gamification, social commerce, and live shopping are transforming the framework of influence marketing in the digital era. This article employs an analytical and case-study methodology to delineate the attributes of each aspect and examines their integration to formulate more successful, flexible, and consumer-centric marketing tactics. This chapter examines the psychological and behavioural effects of this strategy on digital consumers, along with the ethical and regulatory concerns that emerge from the proliferation of social media influence and consumer data. The primary objective of this chapter is to provide a thorough and practical framework for academics, marketing professionals, policymakers, and digital industry stakeholders seeking to comprehend and enhance influence marketing techniques within the dynamic digital economy.

SOCIAL COMMERCE

The Concept of Social Commerce and Its Distinction from Traditional E-commerce

The proliferation of digital platforms has fundamentally transformed the mechanisms by which consumers engage in commercial activities, giving rise to the phenomenon of social commerce. This term denotes the integration of transactional processes within social networking environments, effectively converging interpersonal interactions with commercial objectives (Zhang & Benyoucef, 2016; Hajli, 2015). In contrast to conventional e-commerce, which emphasizes structured, transactional exchanges via independent online marketplaces or dedicated websites (Laudon & Traver, 2022), social commerce embeds purchasing opportunities organically within the user's social media experience. Consequently, consumers are exposed to products through social interactions, user-generated content, and algorithmically curated recommendations, thereby fostering a seamless transition from engagement to conversion (Liang & Turban, 2011). The distinguishing feature of social commerce lies in the non-linear, socially driven consumer journey it facilitates. While traditional e-commerce generally follows a predictable sequence—search, comparison, and transaction (Chaffey, 2022)—social commerce enables real-time, often spontaneous purchasing behavior that is heavily influenced by social validation, peer endorsements, and interactive content (Hudders et al., 2021).

Moreover, social commerce leverages the psychological principles of influence, including social proof, perceived credibility, and trust derived from peer interactions (Cialdini, 2007). Through mechanisms such as live product demonstrations, influencer endorsements, and shoppable media posts, social commerce platforms such as TikTok, Instagram, and Facebook facilitate consumer engagement within immersive, socially embedded digital ecosystems. It is within this paradigm that social commerce is increasingly recognized as a critical intersection of influence marketing, community-driven interaction, and technology-enabled consumerism (Kaplan & Haenlein, 2020).

The Role of User-Generated Content (UGC), Social Proof, and Influencers

The efficacy of social commerce as a strategic marketing tool is intrinsically linked to the synergistic interaction of three critical components: user-generated content (UGC), social proof, and influencer engagement. User-generated content (UGC) encompasses a broad spectrum of consumer-created materials, including product reviews, testimonials, images, videos, and social media posts (Smith et al., 2012). Empirical evidence indicates that UGC significantly enhances perceptions of authenticity and trustworthiness, which, in turn, positively influences purchase intentions (Christodoulides et al., 2012; Filieri et al., 2018).

Within the framework of social commerce, UGC operates as a peer-driven information source that complements, and often surpasses, traditional brand communications in terms of credibility. Consumers interpret UGC as an unbiased reflection of product quality and functionality, thereby reducing perceived risk and facilitating informed purchase decisions (Chevalier & Mayzlin, 2006). Social proof, grounded in established psychological theory, reflects the human tendency to adopt behaviors or make decisions based on observed actions or opinions of others (Cialdini, 2007). In digital commerce environments, social proof is operationalized through visible indicators such as product ratings, user reviews, engagement metrics (e.g., likes, shares, comments), and sales figures (Park et al., 2007). These signals serve to validate consumer choices and expedite decision-making processes, particularly in contexts characterized by uncertainty or information asymmetry. Influencer marketing constitutes a third pillar of social commerce, wherein individuals with substantial social media followings exert influence over consumer attitudes and behaviors through strategic content creation (De Veirman et al., 2017; Lou & Yuan, 2019). Influencers function as intermediaries between brands and consumers, employing narrative-driven content, product demonstrations, and interactive engagements to establish credibility and foster parasocial relationships (Hudders et al., 2021). Collectively, UGC, social proof, and influencer endorsements form a dynamic ecosystem that reinforces consumer trust, amplifies brand visibility, and ultimately accelerates conversion rates within social commerce platforms.

CASE STUDIES

The practical implementation of social commerce can be observed across a variety of leading digital platforms. The following case studies illustrate distinct models through which social commerce integrates content, community engagement, and seamless transactional capabilities.

Tik-Tok Shop

TikTok Shop exemplifies a technologically advanced model of social commerce, wherein entertainment-driven content, algorithmic personalization, and integrated purchasing mechanisms coalesce to create an immersive consumer experience (Montag et al., 2021). Leveraging the platform's short-form video format, TikTok Shop enables content creators and influencers to incorporate shoppable links directly within videos and livestreams, thereby facilitating real-time consumer engagement and instantaneous purchasing opportunities (Li et al., 2021). Live shopping sessions, a core feature of TikTok Shop, allow sellers to demonstrate product features, engage with audiences through interactive Q&A segments, and provide exclusive time-sensitive offers, thereby enhancing the immediacy and authenticity of the purchasing process (Hudders et al., 2021). This model reflects the convergence of influence marketing, gamification, and experiential commerce, positioning TikTok Shop as a leading example of how social platforms can transform traditional retail dynamics into interactive, entertainment-centric purchasing experiences.

Instagram Checkout

Instagram Checkout represents a highly curated social commerce ecosystem that integrates transactional functionality within the platform's visually oriented environment (Kaplan & Haenlein, 2020). Through this feature, consumers are empowered to discover, evaluate, and purchase products without exiting the application, thereby minimizing friction and optimizing the path to conversion (Lee et al., 2021). The Checkout feature leverages a combination of influencer-driven content, UGC, and algorithmic recommendations to provide a personalized, socially validated shopping experience. Product tags embedded within posts, stories, and reels direct users to detailed product information, availability, and secure payment interfaces. Moreover, Instagram's dedicated Shop tab curates product offerings based on user preferences and engagement history, reinforcing the platform's role in facilitating targeted, contextually relevant consumer experiences (Hudders et al., 2021).

Facebook Marketplace

Facebook Marketplace represents a community-driven approach to social commerce, emphasizing peer-to-peer transactions within geographically localized contexts (Liang & Turban, 2011). Unlike TikTok Shop or Instagram Checkout, which primarily focus on brand-to-consumer interactions, Marketplace facilitates both informal exchanges between individual users and formal commercial activities by small businesses and micro-entrepreneurs (Laudon & Traver, 2022). Social proof within Marketplace is enhanced through user profiles, transaction histories, and shared social connections, thereby fostering trust and mitigating perceived transactional risks (Zhang & Benyoucef, 2016). The platform has progressively incorporated integrated payment and shipping solutions, further streamlining the user experience. Marketplace's community-centric model underscores the potential of social commerce to support inclusive economic participation and localized entrepreneurial activity within digital ecosystems.

As illustrated by the operational models of TikTok Shop, Instagram Checkout, and Facebook Marketplace, social commerce redefines conventional consumption patterns by embedding purchasing opportunities within socially interactive, entertainment-driven, and algorithmically optimized environments. In alignment with contemporary influence marketing paradigms, the strategic utilization of user-generated content, social proof mechanisms, and influencer collaborations collectively contributes to enhanced consumer trust, brand credibility, and market performance within the digital economy.

LIVE SHOPPING AS THE NEW RETAIL THEATRE

The Evolution and Global Rise of Live Shopping

The rapid evolution of digital technology and shifting consumer preferences have catalyzed the resurgence of live shopping, positioning it as a prominent pillar within the broader landscape of social commerce and influence marketing (Wang et al., 2022). Although the concept of live selling is not entirely novel—historically rooted in televised home shopping channels such as QVC or Home Shopping Network—the contemporary iteration of live shopping is characterized by its integration with social media platforms, interactive features, and real-time audience engagement (Cai et al., 2023).

The genesis of modern live shopping is widely attributed to China's e-commerce ecosystem, where platforms such as Taobao Live and JD.com pioneered the fusion of live broadcasting with seamless, in-stream purchasing mechanisms (Chen & Lin,

2022). China's digital retail sector, bolstered by mobile penetration and a culture of social innovation, provided fertile ground for live commerce to thrive, with estimates indicating that live shopping sales exceeded USD 480 billion in China alone by 2022 (iResearch, 2022).

Several structural factors underpin the success of live shopping in China, including the maturity of mobile payment systems, widespread influencer culture, and consumer appetite for entertainment-infused shopping experiences (Wang et al., 2022). The demonstrated success of this model has precipitated its global diffusion, with platforms in Southeast Asia, North America, and Europe increasingly adopting live shopping functionalities as part of their digital retail strategies (Cai et al., 2023; Lee & Watkins, 2022).

As live shopping transcends regional boundaries, its role as a new form of retail theatre—where real-time entertainment, social interaction, and instant purchasing converge—has reshaped traditional consumption patterns, aligning with broader trends in experiential commerce and influence marketing.

Key Elements of Live Shopping

The distinctive appeal of live shopping stems from its ability to synthesize three core elements: real-time interaction, purchase urgency, and entertainment-driven experiences. Real-Time Interaction constitutes the foundation of live shopping's effectiveness. Unlike static e-commerce interfaces, live shopping sessions enable dynamic, bidirectional communication between hosts (sellers or influencers) and viewers (potential consumers) (Sun et al., 2019). Hosts can address questions, provide demonstrations, and adapt their content based on audience feedback, fostering a sense of inclusion and personal connection that enhances purchase intent (Wang et al., 2022).

This interactive dimension mirrors key principles of influence marketing, where authenticity, relatability, and immediacy play pivotal roles in shaping consumer perceptions (Hudders et al., 2021). Through live Q&A segments, real-time product showcases, and spontaneous engagements, live shopping dissolves the barriers between seller and consumer, transforming commerce into a participatory social event. Urgency, often operationalized through time-limited discounts, flash sales, and exclusive product drops, is a strategic mechanism that amplifies consumer motivation and drives impulse purchases (Chen & Lin, 2022). Behavioral economics suggests that scarcity cues and temporal constraints can significantly heighten perceived value and encourage rapid decision-making (Cialdini, 2007).

Live shopping platforms leverage this psychological lever by integrating countdown timers, limited inventory indicators, and exclusive live-only promotions, creating a high-stakes environment that incentivizes immediate consumption (Sun et al., 2019). Entertainment-Driven Engagement is arguably the most transformative aspect of live shopping, repositioning commerce as an extension of digital entertainment culture. Hosts often employ humor, storytelling, games, and performances to captivate audiences, blurring the line between shopping and entertainment (Wang et al., 2022). This evolution reflects the convergence of gamification, social commerce, and influence marketing, wherein consumer engagement is driven not solely by product attributes but by the immersive, enjoyable nature of the shopping experience itself (Montag et al., 2021).

Platform Analysis: TikTok Live, Shopee Live, and YouTube Live Shopping

To contextualize the operationalization of live shopping within global markets, this section examines three leading platforms: TikTok Live, Shopee Live, and YouTube Live Shopping.

TikTok Live

TikTok Live has rapidly emerged as a dominant player in the live shopping space, leveraging the platform's extensive user base, algorithmic personalization, and short-form video expertise (Li et al., 2021). Through TikTok Live, influencers, brands, and micro-entrepreneurs can conduct real-time product showcases, integrating entertainment, social interaction, and seamless purchasing within a unified digital environment. The platform's interactive tools—including live chat, polls, and real-time product links—enhance consumer engagement, while algorithmic content recommendations optimize reach and visibility (Montag et al., 2021). TikTok Live's global expansion underscores its role in democratizing live shopping, enabling both established brands and independent sellers to harness the power of real-time influence marketing.

Shopee Live

Shopee Live, embedded within Southeast Asia's leading e-commerce platform, exemplifies the localization and scalability of live shopping in emerging markets. By integrating live broadcasting capabilities directly into its mobile app, Shopee facilitates seamless transitions between entertainment content and transactional processes (Lee & Watkins, 2022). Shopee Live hosts frequently employ localized cultural elements, influencer partnerships, and gamified promotions to engage diverse audiences across the region. The platform's strategic use of urgency triggers—such as limited-time vouchers and flash sales—has proven effective in driving conversion rates and enhancing platform stickiness (Cai et al., 2023). Shopee's success reflects the adaptability of live shopping models beyond China, highlighting the potential for culturally nuanced, entertainment-driven commerce across global markets.

YouTube Live Shopping

YouTube Live Shopping represents the convergence of traditional video content creation with interactive, commerce-enabled experiences. Building upon YouTube's vast repository of creators and viewers, the platform integrates live product demonstrations, influencer-led shopping events, and in-stream purchasing options (Hudders et al., 2021). Unlike TikTok or Shopee, YouTube's long-form content format allows for more in-depth product explanations, tutorials, and storytelling, catering to audiences seeking detailed information prior to purchase (Lee & Watkins, 2022). The platform's monetization tools, including affiliate links, branded content integrations, and audience tipping, provide creators with diversified revenue streams while enhancing consumer trust and purchase likelihood. YouTube Live Shopping's evolution exemplifies how established digital platforms are strategically adapting to the growing demand for interactive, real-time retail experiences, reinforcing the global trajectory of live commerce as a central pillar of social commerce and influence marketing.

The Convergence of Gamification, Social Commerce, and Live Shopping

The modern digital marketplace is experiencing a significant shift, evolving from isolated transactions and linear advertising funnels to a completely connected, immersive, and interactive commercial ecosystem. This evolution is fundamentally driven by the convergence of four potent pillars: social media influence marketing, gamification, social commerce, and live shopping. Although each component has individually transformed customer involvement, their integration establishes a novel paradigm of digital commerce marked by exceptional persuasive influence and intricate behavioral dynamics (Chavee & Vongurai, 2023; Migkos et al., 2025). This integration creates a dynamic and multifaceted "persuasion architecture," in which each element enhances the impact of the others.

In this new paradigm, the influencer, once a mere promoter, transforms into a "community host and game master," arranging intricate, real-time events. The entire apparatus is powered by artificial intelligence (AI), which acts as the technological foundation, facilitating hyper-personalization and adaptive interactivity on an unprecedented scale. This chapter analyzes the interplay among these fundamental foundations (Verma & Kumari, 2023). The text initiates by delineating a fundamental comprehension of social media influence marketing, gamification, social commerce, and live shopping. It subsequently introduces a novel, cyclical consumer experience model—the "integrative flywheel"—to elucidate how these components synergistically expedite the transition from awareness to advocacy. Ultimately, it analyzes the pivotal function of AI in actualizing this intricate and compelling flywheel.

The Fundamental Pillars of Contemporary Digital Commerce

To comprehend their integration, one must first analyze the various elements that constitute the foundation of this new business landscape. Each pillar has developed own concepts and mechanics, yet all contribute to a macro-trend in which the conventional steps of marketing are condensed into seamless moments of engagement.

Influence Marketing on Social Media: The Value of Authenticity

Influencer marketing on social media represents a contemporary form of word-of-mouth advertising, in which digital content providers receive remuneration to endorse products to their established online audiences (Koay et al., 2023; Ooi et al., 2023). The effectiveness of this strategy depends on the perceived authenticity of the influencer, who is frequently regarded as a more reliable source of information than the company itself. This promotional material is skillfully integrated into the influencer's current story, rendering it more authentic than conventional advertising (Lee & Johnson, 2022; Ünalmış et al., 2024). The progression of this activity has resulted in its industrialization, creating a comprehensive ecosystem of talent agencies that professionalize the act of influence. This generates a conflict between preserving authenticity and participating in commercial marketing, which influencers frequently resolve by asserting they only endorse things they truly "love"—a sentiment that upholds the trust-based connection with their audience. In the realm of social media marketing (SMM), which emphasizes tactics for enhancing brand recognition and engagement, influencer marketing functions as a powerful instrument that utilizes personal credibility and parasocial interactions.

1. Gamification: The Engaging Element of Playfulness

Gamification is the deliberate implementation of game design features in non-gaming environments to harness the motivational potential of games. In contrast to pure games, which provide intrinsic rewards, gamification has a functional purpose, aimed at achieving external objectives such as stimulating purchases or increasing brand loyalty. The Mechanics, Dynamics, and Emotions (MDE) model offers a framework for analysis. Mechanics are the fundamental elements with which users engage, including points, badges, levels, and leaderboards. Dynamics are the resultant behaviors that come from this mechanics, like rivalry or cooperation. Emotions are affective responses, such as enjoyment and accomplishment, essential for prolonged involvement. By integrating intrinsic motivators (e.g., mastery, curiosity) with extrinsic motivators (e.g., discounts, status), marketers craft engaging experiences that promote value-generating behaviors.

2. Social Commerce: The Community as a Marketplace

Social commerce is a division of e-commerce that incorporates the complete transactional process—from discovery to checkout—within the inherent framework of a social media application. This removes the inconvenience of redirecting to an external website, signifying a substantial advancement from conventional e-commerce platforms. The architecture is founded on community and interaction, emulating real-world social purchasing activities such as soliciting recommendations, perusing customer reviews, and consulting friends for guidance. The primary objective is to establish trust via peer validation and to engage clients according to their current social behaviors. Platforms such as Instagram, Facebook, and TikTok have vigorously established "shoppable" stores and product tags, a tactic that particularly appeals to younger populations, including Millennials and Gen Z.

3. Live Shopping: The Engaging Transaction

Live shopping combines the entertainment aspect of a live broadcast with the transactional effectiveness of e-commerce, a phenomenon commonly referred to as "shoppertainment". The strategy is characterized by its live video format, real-time delivery, and integrated purchasing features, including clickable "Buy Now" buttons. This format facilitates a highly interactive experience, allowing consumers to pose inquiries in a live chat, observe product demonstrations, and engage with the presenter and fellow viewers, thereby simulating an in-person shopping consultation. It fosters a sense of camaraderie and urgency by providing special, time-sensitive offers to encourage prompt purchases. By consolidating demos, reviews, and Q&A into a singular, engaging experience, live shopping significantly expedites the consumer's purchasing journey.

Table 1. Comparative Analysis of Core E-Commerce Pillars

Pillar	Core Purpose	Key Mechanics/ Features	Primary Consumer Motivation	Representative Platforms/Examples
Influence Marketing	To leverage trusted personalities for brand promotion and awareness, capitalizing on perceived authenticity.	Sponsored posts, product reviews, brand ambassadorships, affiliate links.	Trust, authenticity, aspiration, seeking expert/ peer advice.	Instagram, YouTube, TikTok, Weibo.1
Gamification	To increase user engagement, motivation, and loyalty by applying game-like elements to non-game activities.	Points, badges, leaderboards, progress bars, challenges, rewards, quests.	Achievement, competition, status, rewards, fun, progress.	Nike Run Club, Starbucks Rewards, Duolingo, Sephora Beauty Insider.4
Social Commerce	To facilitate seamless, in-app transactions by integrating e-commerce functionality directly into social media platforms.	Shoppable posts/tags, in-app checkout, user reviews, social sharing tools, storefronts.	Convenience, social proof, community interaction, frictionless purchasing.	Instagram Shops, Facebook Shops, Pinterest 'Shop the Look', TikTok Shop.5
Live Shopping	To create an immersive, interactive, and entertaining real-time sales event that combines broadcast with e-commerce.	Livestreaming, real-time chat, influencer hosts, limited-time offers, "Buy Now" buttons.	Entertainment, interactivity, exclusivity, urgency (FOMO), direct engagement.	Amazon Live, Taobao Live, TalkShopLive, Bambuser.9

The Integrative Flywheel: A Framework for Collaborative Marketing

The genuine strength of this landscape arises from the strategic amalgamation of these pillars. When integrated, they create a unified, self-sustaining system—a flywheel—that advances the consumer experience with increasing momentum.

Reciprocal Reinforcement Dynamics

This model demonstrates that the whole much exceeds the sum of its components, as the pieces not only coexist but also mutually enhance each other. Live shopping delivers a dynamic, event-driven atmosphere; social commerce supplies a seamless transactional framework and community structure; and gamification introduces elements of enjoyment and accomplishment that convert passive observation into active engagement. This synergy manifests in significant ways. At a live event, a company can employ interactive, gamified quizzes to enhance engagement or conduct a "spin-the-wheel" game to provide discounts, so directly augmenting sales. The social context of the live stream enhances these effects, as viewers observing others engage in a poll or receive an award offers social proof that motivates their own involvement. Beyond isolated occurrences, gamification fosters prolonged engagement in social commerce. The clothes retailer ASOS effectively expanded its active customer base by introducing "shopping quests" that incentivized customers to explore collections or share information. The e-commerce application Brands For Less incorporated a branded minigame that attained a 95% completion rate among participants, illustrating the efficacy of gamification in engaging and retaining customer attention inside a commercial platform. Essentially, live and social commerce offer the audience and platform, whereas gamification delivers the organized, interactive mechanisms that enhance the experience's appeal.

A Comprehensive Consumer Journey Framework

The conventional linear marketing funnel fails to adequately represent the dynamic journey produced by this integrated architecture. A more precise depiction is a cyclical "flywheel," in which each phase propels the subsequent one, transforming delighted clients into advocates who draw in new clientele (Berman, 2016; Meyer, 2019). This method consolidates the funnel, enabling several steps to transpire nearly concurrently.

Figure 1. Consumer Journey Framework

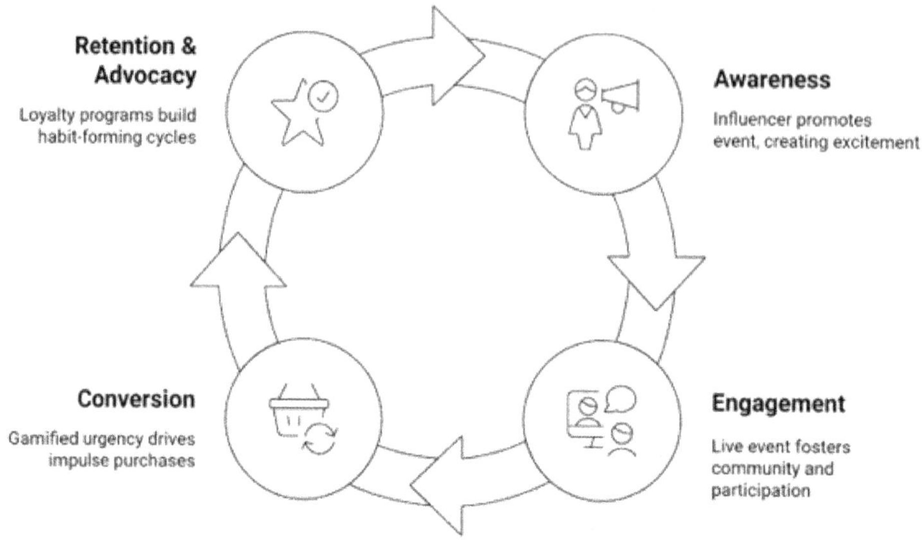

1. Awareness: The cycle commences when an influencer proclaims an impending live shopping event, generating excitement and expectation. The unique, event-driven format itself fosters prominent awareness.
2. Engagement (Acquisition & Consideration): As the live event commences, the customer transitions from passive awareness to active engagement. The live chat cultivates community engagement, while gamified features such as polls and contests convert viewers into participants. This approach expedites consumers' progression through the consideration phase within minutes by showcasing product demonstrations, addressing inquiries, and providing social validation from peers.
3. Conversion: The inherent checkout functionalities of social commerce platforms render the conversion process virtually seamless. Gamified urgency mechanisms, such as countdown timers for special offers, synergize with FOMO (fear of missing out) to significantly expedite the decision-making process and stimulate impulse purchases. Studies demonstrate that conversion rates for gamified e-commerce can reach up to seven times the industry average.
4. Retention & Advocacy: The post-purchase experience persists through gamified loyalty programs aimed at establishing habit-forming cycles. Granting points or exclusive status for repeated engagement motivates clients to return. A customer who has had a pleasurable "shoppertainment" experience is far more inclined

to become a brand champion, disseminating their experience on social media and attracting new customers into the flywheel.

The Technological Foundation: Artificial Intelligence and Hyper-Personalization

This advanced integration at scale is facilitated by Artificial Intelligence (AI). AI functions as the cognitive engine of the flywheel, analyzing huge volumes of user data instantaneously to generate tailored and responsive commercial experiences. Artificial Intelligence in Social and Live Commerce: In social commerce, AI algorithms evaluate user data to provide hyper-personalized product recommendations, customizing the discovery experience for every individual (Bahri et al., 2025; Jannah et al., 2024). In live events, AI's function becomes increasingly dynamic, assessing real-time comments and reactions to assist the host in adjusting promotions instantaneously. AI-powered chatbots manage customer care, while AI-enhanced augmented reality (AR) facilitates virtual try-ons, markedly enhancing purchasing confidence. Research has investigated the use of AI-driven virtual idols as hosts, revealing that their engagement can diminish psychological distance and enhance impulsive purchasing intents. Artificial Intelligence in Adaptive Gamification: The most significant synergy arises when AI is utilized in gamification. Whereas conventional gamification often adopts a uniform approach, artificial intelligence facilitates adaptive gamification, tailoring the experience to each individual user. An AI system can dynamically modify the difficulty of challenges or the value of rewards by monitoring a user's behavior and performance to sustain optimal engagement. It can utilize predictive analytics to detect consumers at risk of disengagement and initiate customized interventions to sustain their interest. This converts gamification into a dynamic, individualized engagement mechanism that optimizes client pleasure and commercial results.

BEHAVIORAL EFFECTS ON THE DIGITAL CONSUMER

The integration of gamification, social commerce, and live purchasing is both a catalyst and a result of a significant transformation in customer behavior. In the digital age, the conventional idea of the passive consumer—simply an observer of marketing communications—has become outdated. The "participatory consumer" has supplanted it, serving as an active agent who interacts with, co-creates, and influences brands and marketplaces (Busalim et al., 2023; Choi, 2019; Dabbous et al., 2020). This chapter examines the transformation, investigating the psychological mechanisms activated and utilized by these novel kinds of interactive business. This analysis demonstrates that customer attention and data have effectively transformed into a new type of labor, traded for psychological rewards. It further investigates how continual exposure to these persuasive environments may be reshaping consumer decision-making to become more impulsive and emotionally motivated, resulting in a new category of consumer susceptibility.

The Ascendancy of the Participatory Consumer

The expansion of digital technology has permanently transformed the power relationship between corporations and consumers. Communication has evolved from a unilateral transmission to a participatory, multidirectional dialogue. Consumers have evolved from passive observers to active contributors who influence the essence of the brands with which they engage. This participatory culture is founded on users' capacity to not only consume content but also to produce, share, and alter it within extensive online communities. This participation occurs on a spectrum, ranging from passive activities such as content consumption to active kinds of interaction like commenting and producing unique user-generated content (UGC). In this new paradigm, consumers are not merely purchasing items; they are collaboratively generating value with brands. They offer critical comments and serve as brand narrators via their social media posts. The amalgamated commercial experiences of gamified social and live commerce are tailored for the interactive consumer, offering a systematic environment that actively promotes and incentivizes engagement, transforming the buying experience into a mode of content production.

The Psychology of Engagement: Fundamental Motivational Drivers

The efficacy of the integrated marketing approach depends on its capacity to engage and enhance essential human psychological motivators. The primary factors can be classified into the principles of gamification (reward expectancy) and those of social and live commerce (FOMO and social validation).

The Psychology of Gamification: Reward Expectancy

Fundamentally, gamification engages the brain's reward system, sometimes referred to as a dopamine loop. When a user does a desired activity and obtains a reward (e.g., points, a badge), the subsequent release of dopamine engenders a good sensation that reinforces the behavior, increasing the likelihood of repetition. This exemplifies operant conditioning utilized in a digital framework. The expectation of a reward serves as a potent motivator, and this impact intensifies when the incentives are varied or unknown, akin to a "spin-to-win" game that reflects the addictive nature of slot machines. Moreover, gamification capitalizes on our intrinsic motivation for advancement. Mechanics such as progress bars leverage the Zeigarnik Effect—the propensity to more effectively recall unfinished tasks—instigating a cognitive impetus to "complete the bar" or "attain the subsequent level".

The Psychology of Live and Social Commerce: FOMO and Social Validation

1. Fear of Missing Out (FOMO): This phenomenon serves as a principal catalyst for impulsive purchasing behavior within the high-pressure context of live streaming e-commerce. Live shopping events are strategically designed to evoke fear of missing out by employing strategies of scarcity (e.g., "only 100 units available") and urgency (e.g., "this discount concludes when the stream ends"). This notion of a constrained opportunity induces anxiety and prompts a fast-purchasing response to mitigate the adverse sensation of missing out.
2. Social Validation: Humans, as essentially social beings, possess an intrinsic need for their decisions to be affirmed by their peers. Social and live commerce settings are abundant in sources of social validation, including influencer endorsements, live purchase tallies, and favorable viewer remarks. These signals diminish purchasing uncertainty and bolster a consumer's choice to purchase. The interplay between FOMO and social affirmation is a potent psychological phenomenon that markedly diminishes the threshold for impulse buying.

Table 2. Psychological Drivers and Marketing Applications

Psychological Principle	Brief Definition	Application in Integrated Commerce	Concrete Example
Reward Expectancy / Dopamine Loop	The brain's reinforcement system, where completing a task for a reward releases dopamine, encouraging behavior repetition.	Integrating reward-based game mechanics directly into the shopping experience to create positive feedback loops and habit formation.	A "spin-to-win" popup offering a random discount for providing an email address.[20]
FOMO (Fear of Missing Out)	Anxiety stemming from the belief that one might miss out on a rewarding experience that others are having.	Creating a sense of urgency and scarcity to compel immediate action and prevent purchase hesitation.	A countdown timer on a live shopping event for an exclusive discount.[13]
Social Validation / Social Proof	The tendency to see an action as more appropriate when others are doing it, reducing uncertainty and building trust.	Displaying real-time social signals of engagement and purchasing to validate a consumer's choice.	A live feed in a social commerce app showing what other shoppers are buying in real-time.[5]
Zeigarnik Effect / Goal Gradient	The tendency to better remember and be motivated by incomplete tasks; motivation increases as one gets closer to a goal.	Using visual indicators of progress to motivate users to complete a series of actions or reach a new status level.	A "Complete Your Profile" progress bar in a loyalty program app.[20]
Scarcity Principle	The cognitive bias where items that are limited in availability are perceived as more valuable.	Artificially or genuinely limiting the supply of a product to increase its desirability and prompt faster purchases.	An alert message stating "Only 3 left in stock!" on a product page.[44]

Case Studies in Psychological Applications

The theoretical efficacy of these psychological drivers is substantiated by their effective application by several prominent businesses. These case studies demonstrate the application of gamification and social proof methods to attain measurable commercial results.

1. **Starbucks Rewards:** The Starbucks mobile application exemplifies a gamified loyalty program. Starbucks establishes a robust incentive mechanism with a system of "stars" (points) acquired from purchases, which can be exchanged for complimentary products. The tiered membership structure and progress-monitoring capabilities encourage repeat patronage and foster profound brand loyalty, transforming occasional coffee consumers into everyday contributors within the brand's ecosystem.
2. **Nike Run Club:** Nike adeptly converted the solitary pursuit of running into a communal and competitive endeavor with its Nike Run Club application.

The application employs gamification elements such as individual challenges, milestone accomplishment badges, and leaderboards to facilitate progress comparison among peers. This cultivates a robust feeling of community and transforms fitness into a competitive activity, intricately integrating the Nike brand into the user's lifestyle and fostering loyalty that transcends the product itself.

3. **Customized Quizzes (M&S, Apohem):** E-commerce brands are progressively employing quizzes as a method of "guided selling." Marks & Spencer's "Style Types and Outfit Explorer" inquires about user preferences to produce tailored outfit suggestions, thereby streamlining the discovery process and alleviating decision fatigue. Twenty-four Likewise, the Swedish pharmacy Apohem introduced a "sunblock finder" quiz. This straightforward solution proved to be incredibly effective: 15% of users who finished the quiz made a purchase, and 42% of those acquired the precisely advised product, illustrating how gamified guidance can directly correlate with enhanced sales.
4. **Branded Minigames (Brands for Less, Ballzy**): To enhance engagement during critical sales seasons, certain brands have incorporated independent games within their applications. Brands For Less achieved a 95% play-through rate on their in-app game, whereas the sneaker retailer Ballzy employed a "Endless Runner" game to augment brand engagement and cultivate a more profound emotional bond with its audience. Fifteen These instances demonstrate that even brief, enjoyable interactions can markedly enhance engagement and maintain brand prominence.

The Panopticon of Persuasion: Risks, Ethics, and Consumer Vulnerability

The efficacy of this integrated marketing system is a double-edged sword. The identical methods that generate captivating experiences can likewise be employed to cultivate digital addiction, distort perceptions, and exploit susceptibilities. This intricate "persuasion architecture" can generate a "fog of persuasion," blurring the distinction between influence and manipulation, so hindering customers' ability to exercise informed agency (Akpati & Mathew, 2023; Reyneke et al., 2023).

1. The Structure of Addiction: The integrated commercial environment is designed with mechanisms that significantly promote behavioral addiction. Social media platforms function on a dopamine-driven reward system akin to that of substance use disorders, and social media addiction is inversely associated to mental health outcomes. Gamification intentionally constructs these identical habit-forming cycles. Together, they form a powerful framework for compul-

sion. "Temporal dark patterns," such as daily login prizes or streaks, not only incentivize participation but also penalize disengagement by necessitating a reset, thereby instilling a sense of obligation for daily interaction. The risk is heightened when these platforms promote potentially detrimental products to susceptible groups such as adolescents.

2. Algorithmic Manipulation: The extensive data amassed drives algorithms that can influence what a consumer perceives and pays, frequently without their awareness. An important ethical issue is algorithmic pricing discrimination, when distinct individuals are subjected to varying prices for identical products. Research indicates that this frequently leads to loyal consumers being subjected to elevated prices, a practice that fundamentally contravenes consumer expectations of equity and undermines confidence. The perceived worth of a product can be gradually influenced by environmental clues, in addition to pricing. A study indicated that a live shopping host's background markedly affected buy intent; a formal background enhanced trust for luxury items, whereas an informal setting promoted familiarity for common products. This underscores the pliability of value in these mediated contexts.
3. The Emergence of "Dark Patterns": The distinction between persuasion and exploitation is frequently blurred by the implementation of "dark patterns"—user interface designs deliberately engineered to mislead, manipulate, or defraud people.

These patterns manipulate cognitive biases to undermine user autonomy. Common instances encompass:

1. **Artificial Urgency and Scarcity:** Employing strategies such as countdown timers that reset or showcasing "limited stock" notifications that do not reflect genuine inventory levels to incite FOMO and expedite a buy.
2. **Basket Sneaking and Drip Pricing:** The automatic inclusion of supplementary items in a shopping cart or the concealment of additional expenses, such as elevated delivery charges, until the final checkout phase.
3. **Confirmshaming:** Employing guilt-inducing rhetoric to dissuade users from selecting an option that serves their best interests, exemplified as a popup to reject a discount that states, "No thanks, I relish paying full price".
4. **Monetary Dark Patterns:** Frequently observed in gamified experiences, these encompass "pay-to-win" models wherein artificial obstacles, such as prolonged waiting periods, may be effortlessly circumvented through monetary transactions. The implementation of these patterns is a distinct ethical violation, transitioning from persuasion to deceit and inflicting irreversible harm on brand trust.

ETHICAL CONSIDERATIONS AND REGULATORY CHALLENGES

This chapter explores the critical ethical and legal dimensions of social media influence marketing, especially within gamified and live-shopping contexts. Key areas include transparency, consumer protection, personal data handling, and the tension between local and global regulatory frameworks. In influencer marketing, transparency is essential to maintain trust. The U.S. FTC requires influencers to clearly disclose paid partnerships in a way that is "easily visible and understandable"(Chakraborty, n.d.). Similar frameworks exist in the EU (via ASA and GDPR guidelines) and ASEAN countries, though enforcement varies (Cahyani & Bahri, 2022; Kismawadi & Irfan, 2025). Despite these guidelines, enforcement gaps remain. For example, approximately 15% of industry spending is lost to influencer fraud, including fake followers or engagements. Ethically, brands and influencers should practice social responsibility, distinguishing between legal compliance and ethical duty. (Influenciity, 2024)Some platforms encourage voluntary disclosure practices, but these are often inconsistent across regions.

Live shopping and gamified campaigns harvest significant consumer data—clicks, watch times, location, purchase behaviors. Platforms must follow data privacy principles: minimize data collection, obtain explicit consent, ensure secure handling, and allow opt-out (voguebusiness.com) (firework.com). Landmark cases like Cambridge Analytica demonstrate how misuse of personal data can cause reputational and legal damage. Compliance with GDPR, CCPA, or PDPA is mandatory in many regions. From an ethical standpoint, gamified experiences that exploit behavioral biases may manipulate consumer behavior. Platforms such as "Shopee Games" enhance brand loyalty—but marketers must transparently reveal reward mechanisms, odds, and data use in an accessible manner.

Regulatory landscapes are patchy. In the U.S., the FTC oversees endorsements; in the EU, GDPR and ASA govern ads and data; other regions may lack cohesive laws (researchgate.net). Indonesia's UUPI & PDP laws, along with LAMEMBA accreditation standards, influence data ethics, but live-shopping remains loosely regulated. Global platforms must navigate:

1. Cross-border compliance: respecting each country's disclosure and privacy laws.
2. Platform enforcement: policies vary—some only require hashtags (e.g., #ad), others insist on disclaimers or labels.
3. Emerging laws: the EU is introducing the Digital Fairness Act, targeting unfair practices in influencer campaigns.

To align with both ethical and regulatory standards, brands, influencers, and platforms should:

1. Conduct due diligence—check influencer authenticity, past content, potential controversies (Influenciity, 2024)
2. Disclose clearly—use plain language, disclose placement, align with regional rules.
3. Collect securely—only gather necessary data and enable user control
4. Foster social responsibility—prioritize audience well-being over sales
5. Engage regulators—stay current with emerging laws and self-regulatory efforts.

Ethical and legal adherence not only safeguards consumers but supports sustainable trust-driven marketing. As gamification and live commerce expand, so do ethical stakes and regulatory attention—from data rights to content fairness. Brands that lead transparently position themselves for long-term success.

Case Studies and Best Practices

There are some cases which examine successful applications of gamification, live shopping, and social commerce, highlighting regional and platform-specific examples. It draws insights from one of big e-commerce in Indonesia named Shopee, a well-known application TikTok, and D2C brand strategies, including actionable best practices.

1. Shopee Gamification Campaigns

Shopee's engagement-focused features (e.g., "Shopee Games") use reward mechanisms to increase user activity. A study shows that gamification boosts brand loyalty by 32.4%, fostering habitual behavior. Campaigns like "Shopee Shake" or quiz games reward users with coins, vouchers, or limited offers. These drive daily interaction, and the sense of achievement enhances platform stickiness. Key insights:

- Keep challenges rewarding and clear.
- Feedback loops (e.g., progress bars) maintain motivation.
- Data-driven personalization maximizes engagement.

2. Live Shopping on TikTok

TikTok Shop seamlessly integrates entertainment and commerce, offering real-time product tagging and checkout—fueling impulse purchases. In Southeast

Asia and the U.S., TikTok Shop generated $100M+ in Black Friday GMV, with live commerce accounting for nearly half of total transactions. Live streams with influencers drive high engagement:

- 70M U.S. monthly buyers shop on TikTok, with 58% completing purchases and 78% discovering brands via influencers (firework.com) (researchgate.net) (skai.io)
- Shopee's streams similarly recorded strong purchase interest, with effect sizes in t-tests of 7.466 for Shopee and 4.415 for TikTok Shop Success factors:
    - Use charismatic hosts who align with brand image.
    - Deploy interactive features (live Q&A, polls, limited-time deals).
    - Optimize timing around peak online activity.

3. Social Commerce by D2C Brands

In the evolving landscape of digital marketing, Direct-to-Consumer (D2C) brands have emerged as pioneers in redefining the relationship between businesses and their customers. Unlike traditional retail models that rely on intermediaries such as wholesalers or distributors, D2C brands utilize digital platforms—most notably Instagram, TikTok, and WhatsApp—to build and manage direct, personal connections with their audiences. This approach not only allows brands to retain greater control over pricing, branding, and customer experience but also provides access to valuable first-party data that can inform marketing, product development, and retention strategies.

One of the most effective tools in this strategy is the use of shoppable content, which enables users to engage with a product and make a purchase directly within a social media platform. On Instagram, for instance, brands can tag products in posts, Reels, and Stories, creating a seamless journey from content discovery to checkout. TikTok has adopted a similar model through its TikTok Shop, where videos and live streams are integrated with clickable product links. These features capitalize on the impulse-driven behavior of mobile users, shortening the purchase funnel and boosting conversion rates.

Another key tactic embraced by D2C brands is collaboration with micro-influencers—content creators with follower counts typically between 10,000 and 100,000. While macro-influencers and celebrities offer broader reach, micro-influencers tend to cultivate tighter-knit, niche audiences and maintain higher levels of engagement and trust. This authenticity often translates into stronger return on investment (ROI), as followers are more likely to view micro-influencer recommendations as genuine. According to Skai (2024), micro-influencer campaigns frequently outperform larger-scale endorsements in terms of cost-effectiveness and audience resonance.

In addition to influencer marketing, successful D2C brands prioritize community building as a core component of their social commerce strategy. This involves encouraging user-generated content (UGC), such as customer reviews, unboxing videos, and product styling tips, which serve as powerful forms of social proof. Platforms like WhatsApp and Instagram Direct are also used to foster one-on-one communication, support customer service, and distribute personalized updates or promotions. By actively engaging their communities, D2C brands not only drive organic reach but also empower consumers as co-creators, strengthening brand loyalty and advocacy.

The impact of these strategies is reflected in current market data. A recent report from Skai indicates that approximately 5% of U.S. e-commerce sales are now driven directly through social platforms, with over 70 million monthly TikTok users participating in social shopping activities. This trend signals a broader shift in consumer behavior—especially among Gen Z and millennial audiences—toward more interactive, mobile-first, and socially-driven purchasing experiences. It also underscores the growing importance of aligning content strategy with platform dynamics, community engagement, and data-driven personalization.

D2C brands are at the forefront of social commerce innovation. By leveraging shoppable content, authentic influencer partnerships, and active community participation, these brands are not only bypassing traditional retail limitations but are also reshaping the way consumers discover, evaluate, and purchase products in the digital era.

4. Content Strategy and Influencer Types

Successful tactics include mixing entertainment, information, and urgency. Brands use unboxings, tutorials, "challenges," and live product reviews to foster engagement.

Influencer roles:

- Micro-influencers (10k–100k followers): More niche, trustworthy.
- Macro-influencers (100k–1M): Good for broader reach.
- Celebrities (1M+): Visible reach, but riskier and costlier. Platforms should match influencer tier with campaign goals: community vs. scale.

5. Best Practices for Brands and Creators:

Table 3. Best Practices

Area	Best Practice
Clarity	Track goals (traffic, sales, retention)
Compliance	Disclose clearly; follow data laws
Value	Gamification must feel fair and meaningful
Engagement	Use chat, polls, and user participation
Data Use	Analyze engagement, not just exposure metrics
Feedback Loop	Optimize based on real-time performance
Cross-Platform Integration	Repurpose content for YouTube, Instagram and websites.

These align with Shopee, TikTok, and D2C brand successes as discussed above. By combining strategic gamification, interactive live sessions, and community-driven commerce, brands can replicate proven formulas—while emphasizing trust, clarity, and audience value.

Future Outlook – Toward A Meta-Commerce Ecosystem

As we enter a new phase, this chapter explores how trends like metaverse, AR/VR, avatar-based influencers, Web3, and decentralization will converge into a meta-commerce ecosystem.

Brands are experimenting with immersive, virtual storefronts akin to gaming environments. Examples include virtual stores by Coach, Swarosvki, and Ferragamo using AR, treasure hunts, and live shopping events (voguebusiness.com) (voguebusiness.com). AR-enhanced shopping—like in-store AR mirrors—make digital experiences tactile, blending physical and virtual worlds (voguebusiness.com). Key future elements:

- Phygital experiences: seamless transitions between real and digital products.
- Mixed reality stores: virtual showrooms accessible via VR headsets.
- Data-rich personalization: like guided tours in metaverse and AI avatars.

Virtual avatars (e.g., Miquela Sousa, Lil Miquela) are gaining momentum, offering controlled, scalable influencer narratives. Brands see advantages in consistency, 24/7 presence, and IP control—at the cost of human authenticity. Avatar influencers integrated into AR/VR and metaverse platforms allow interactive shopping and brand storytelling.

1. Web3 and Decentralized Platforms

Web3, often described as the next evolution of the internet, introduces a decentralized and user-owned framework powered primarily by blockchain technology. In the context of social commerce and influencer marketing, Web3 offers transformative possibilities by enabling new forms of value exchange, ownership, and community engagement—fundamentally shifting power from centralized platforms to users and creators.

One of the most compelling innovations within the Web3 space is the integration of non-fungible tokens (NFTs) into commerce ecosystems. NFTs allow brands and creators to assign verifiable digital ownership to virtual assets such as artwork, fashion, music, or even access rights. These tokens can be bought, sold, and traded on blockchain-based marketplaces, offering not just primary sales revenue, but also royalties on secondary market resales—thus enabling new monetization models and long-term loyalty programs. For instance, luxury brands like Gucci and Dolce & Gabbana have released limited-edition NFT collections, while Lacoste introduced NFT membership cards that unlock access to exclusive online events and digital VIP rooms (Vogue Business, 2023; Firework, 2024). These mechanisms foster brand affinity and community, going beyond transactional relationships.

Another promising application is token-gated content, which restricts access to certain digital experiences based on ownership of specific NFTs or social tokens. This concept allows brands and influencers to create tiered content experiences, fan clubs, or loyalty tiers that reward engagement and exclusivity. For example, Lacoste's Web3 initiatives include offering special privileges to token holders, such as early product access or participation in virtual events. Such experiences blur the line between entertainment, identity, and commerce—building more immersive, participatory ecosystems.

Web3 also encourages the rise of decentralized marketplaces, where creators and consumers transact directly without relying on centralized intermediaries like major e-commerce or social media platforms. This model significantly reduces platform fees, enhances profit margins for small creators, and supports community-driven governance. Protocols such as Zora and OpenSea are early examples of decentralized commerce in practice, enabling creators to maintain ownership over their storefronts and content strategies.

However, despite the enormous potential, Web3's adoption in mainstream social commerce is not without challenges. First, scalability remains a technical concern, especially with high transaction volumes and the computational intensity of certain blockchain operations. While solutions like Layer-2 scaling (e.g., Polygon, Arbitrum) are evolving, real-time commerce experiences such as live shopping still face latency limitations. Second, user education and onboarding are significant hurdles. Many consumers are unfamiliar with crypto wallets, blockchain addresses, and the concept of digital ownership, which can create friction in adoption. User interfaces must become more intuitive and secure to bridge this gap. Finally, regulatory ambiguity poses a risk. As governments grapple with defining legal frameworks for digital assets, NFTs, and token economies, businesses operating in this space face uncertainties related to taxation, intellectual property rights, and consumer protection.

The emergence of Web3 technologies in social commerce marks a pivotal shift toward greater autonomy, creativity, and community ownership. While still in its early stages, Web3 presents opportunities to design richer, more equitable digital ecosystems—so long as stakeholders remain vigilant about scalability, usability, and responsible governance.

2. Ecosystem Dynamics

As the boundaries between content, commerce, and community become increasingly blurred in the digital consumer era, a new meta-commerce ecosystem is beginning to take shape. This ecosystem is not defined by a single platform or business model, but by an intricate interplay of technologies, user behaviors, and decentralized governance. Several key forces are emerging as critical success factors for brands, creators, and consumers seeking to thrive in this landscape.

Interoperability stands at the forefront of this transformation. In the past, digital experiences have largely been platform-specific—content created for Instagram, for example, could not be repurposed or functionally integrated into TikTok, Roblox, or other ecosystems. However, the future of meta-commerce demands seamless cross-platform interactions, where experiences, assets, and identities can move fluidly between virtual worlds. Imagine a virtual shoe try-on using AR in TikTok that can also be showcased in a user's avatar on Roblox or Decentraland. Such seamless integration not only enhances user experience but also maximizes the reach and utility of branded digital assets. Achieving this level of interoperability will require collaborative standards, open APIs, and shared identity protocols across platforms.

Another foundational pillar is consumer control over data and identity. Traditional Web2 platforms accumulate massive amounts of user data, often without explicit consent or clear terms of ownership. Web3 technologies, by contrast, offer the potential for true data sovereignty—users can own, control, and grant permission for the use of their data through secure, blockchain-based consent models. This decentralized approach enables consumers to determine how their preferences, purchase history, and behavioral patterns are used in personalization and advertising, placing power back into the hands of individuals. For brands, this creates a unique opportunity to build trust through transparency and ethical data practices, which is increasingly becoming a competitive advantage.

Monetization models are also undergoing a fundamental evolution. In traditional e-commerce, revenue was primarily generated through direct sales. However, in the meta-commerce ecosystem, new models are emerging, such as digital rentals, limited-time access, token-based memberships, and secondary market resales. For example, a consumer may "rent" a luxury virtual outfit for their avatar to wear in a metaverse fashion show or resell a digital collectible that grants early access to product drops. These innovations not only diversify income streams for creators and brands but also create more dynamic and personalized consumer journeys. Moreover, smart contracts embedded in blockchain protocols enable automated royalty payments to original creators, ensuring fair compensation throughout the asset lifecycle.

Finally, the rise of the community economy is reshaping the role of consumers from passive recipients to active participants and co-creators. In this model, value is generated not just through products, but through shared experiences, fan engagement, and collective ownership. Creator-driven storefronts are being built on platforms like Shopify, Mirror, and Zora, where influencers and niche entrepreneurs can launch their own digital spaces. Meanwhile, event-driven product drops—such as limited-edition NFTs or time-sensitive in-game items—create excitement and exclusivity, encouraging deeper community involvement. Peer-to-peer resale platforms further democratize access and foster circular economies within the digital sphere. Together, these practices reflect the roots of social commerce, but with an added layer of decentralization and immersion.

3. Barriers and Risks

While the vision of a meta-commerce ecosystem is compelling, its widespread adoption is hindered by a range of technological, social, and regulatory barriers. One of the most immediate challenges is technical maturity. Although virtual and augmented reality (VR/AR) technologies have progressed rapidly, achieving real-time, immersive experiences at scale still requires substantial computing power, low-latency networks, and device compatibility. Many consumers—especially in emerging markets—lack access to the high-end smartphones, VR headsets, or broadband infrastructure necessary to fully engage in these environments. As a result, inclusivity and accessibility remain critical issues that must be addressed if meta-commerce is to realize its global potential.

Another significant challenge is regulatory uncertainty, particularly in relation to digital assets, blockchain transactions, and virtual economies. Governments around the world are still formulating legal frameworks for Web3 technologies, NFTs, decentralized finance (DeFi), and virtual currencies. This ambiguity exposes brands and creators to risks related to taxation, intellectual property rights, fraud, and consumer protection. For example, some jurisdictions treat NFTs as securities or assets subject to capital gains tax, while others lack clear guidelines entirely. Without cohesive and harmonized regulation, businesses may face compliance challenges when operating across borders.

Cybersecurity and trust also pose critical concerns. As commerce shifts into immersive and decentralized platforms, new threats such as identity theft, phishing in VR, smart contract vulnerabilities, and deepfake influencers could erode consumer trust. The blockchain's promise of transparency and immutability is not a panacea—many scams still occur due to social engineering or flawed coding. In this context, brands must not only invest in robust cybersecurity infrastructure but also actively educate consumers on how to navigate and protect themselves in these new digital environments.

The digital literacy gap may prevent mainstream users from adopting meta-commerce tools. Understanding how to set up a digital wallet, secure private keys, participate in decentralized autonomous organizations (DAOs), or trade NFTs requires a steep learning curve. If user experiences remain complex and unintuitive, many potential consumers may opt to stay within familiar Web2 platforms. Therefore, ease of use, user onboarding, and education are paramount to lowering barriers to entry.

4. Horizon Scanning and Strategic Takeaways

Despite the above challenges, the meta-commerce ecosystem offers vast potential for brands, creators, and consumers willing to innovate, adapt, and lead responsibly. As the landscape evolves, businesses must engage in proactive horizon scanning—a strategic process of anticipating future developments, experimenting with emerging tools, and aligning with evolving consumer behaviors.

First, companies should begin experimenting with immersive technologies in low-risk, high-engagement ways. For instance, brands can integrate AR try-ons or VR showrooms into their existing e-commerce platforms, allowing customers to visualize products in new ways. Retailers such as IKEA and Sephora have already piloted such experiences successfully, enhancing consumer confidence and satisfaction. Likewise, partnering with avatar-based influencers or virtual brand ambassadors can test new content formats and digital identities in a controlled environment.

Second, businesses can explore NFTs and digital collectibles as loyalty mechanisms and community-building tools. Offering limited-edition digital assets that provide access to exclusive content, early product releases, or personalized experiences can deepen customer engagement. These assets, if designed ethically and transparently, can also foster a sense of ownership and belonging among consumers, transforming them into brand advocates and co-creators.

Third, there is a growing need to build cross-platform ecosystems that interconnect social media, e-commerce, gaming, and metaverse platforms. Brands that can strategically repurpose content across TikTok, Roblox, Instagram, and VR platforms will enjoy greater reach, engagement, and resilience. Developing modular, interoperable digital assets—such as 3D models of products that can appear across different environments—will become increasingly valuable in this context.

Moreover, brands should prioritize consumer education and transparency as central pillars of their strategy. Whether introducing blockchain elements, promoting digital identity standards, or enabling peer-to-peer resale, users must be clearly informed of benefits, risks, and their rights. Brands that embed educational components within their campaigns—such as tutorials, live demos, or community support—will be better positioned to earn long-term trust.

Lastly, strategic partnerships will be essential. By collaborating with tech startups, Web3 developers, creative communities, and academic institutions, businesses can co-develop innovative solutions while mitigating development risks. As the landscape becomes more decentralized, collaborative ecosystems—not siloed platforms—will define the winners of the next decade.

CONCLUSIONS AND PRACTICAL IMPLICATIONS

This chapter has examined the evolving landscape of influence marketing in the digital era, highlighting the convergence of gamification, social commerce, and live shopping as key pillars shaping contemporary consumer experiences. These three approaches, rather than functioning independently, form a mutually reinforcing system that transforms traditional marketing into an immersive, participatory, and real-time digital ecosystem. In this new paradigm, consumers are not merely recipients of promotional messages but active participants, co-creators, and even brand advocates within interactive, entertainment-infused commercial environments. From a theoretical perspective, this chapter contributes to the expansion of digital marketing and consumer behavior frameworks by introducing the "integrative flywheel"—a cyclical model that captures how awareness, engagement, conversion, and advocacy can occur in seamless, overlapping phases. The synthesis of AI-powered personalization, gamified user interfaces, and socially driven commerce highlights the increasing need for interdisciplinary analysis across marketing, behavioral psychology, digital communication, and platform studies. These dynamics suggest that future theories must account not only for transactional behavior but also for emotional, cognitive, and social motivators in digitally mediated environments.

Practically, marketers are advised to design holistic campaigns that blend interactive mechanics (e.g., challenges, point systems, live polls) with socially validated content and entertainment-based selling strategies. Gamification should go beyond superficial engagement and be embedded into loyalty systems that reward long-term interaction. Social commerce—especially on platforms like TikTok Shop and Instagram Checkout—offers opportunities for frictionless shopping directly within the content feed, while live shopping formats introduce urgency, transparency, and immediacy that accelerate impulse buying and build emotional connection. These elements, when integrated thoughtfully, not only drive conversion but also foster brand loyalty through participatory experiences. For digital platforms, the implication is to enhance their role not just as facilitators of content and commerce, but as responsible orchestrators of user experience. This involves building interactive, data-driven environments that encourage engagement while safeguarding consumer well-being. Artificial intelligence should be employed for hyper-personalized recommendations and real-time analytics, but always within ethical boundaries that respect user autonomy. As platforms increasingly deploy gamified and immersive commerce tools, they must also implement mechanisms that guard against exploitative design patterns and ensure transparency in user interface choices.

From a policy standpoint, the chapter underscores the urgency of establishing robust regulatory frameworks that can keep pace with innovations in influence marketing, especially in cross-border, algorithm-driven, and platform-centric environments. Key concerns include disclosure of paid endorsements, protection of minors, prevention of psychological manipulation, and responsible data governance. Existing regulatory instruments—such as the General Data Protection Regulation (GDPR) in the European Union and Indonesia's Undang-Undang Perlindungan Data Pribadi (Personal Data Protection Law)—offer foundational guidelines, but enforcement must be strengthened and adapted to the nuances of gamified systems and live commerce. Transparency, informed consent, and consumer empowerment must become core principles in the design of digital marketplaces.

This chapter affirms that successful marketing in the digital age requires an integrated approach that combines storytelling, interactivity, and community engagement within a cohesive and ethical framework. The path forward involves strategic collaboration among marketers, platforms, and policymakers to ensure that innovation serves not only commercial objectives but also societal well-being. As the boundaries between content, commerce, and culture continue to blur, the emergence of a meta-commerce ecosystem demands both creative agility and ethical responsibility from all stakeholders.

REFERENCES

Akpati, C., & Mathew, A. (2023). A multimodal critical discourse study of persuasive strategies and identity construction in admission adverts of select private university websites. In *Transformation of Higher Education Through Institutional Online Spaces*. https://doi.org/DOI: 10.4018/978-1-6684-8122-6.ch013

Bahri, M. I., Dermawan, A. A., & Tambunan, A. A. (2025). Public-Private Collaborations in Smart Cities: Startup-Driven Innovations in E-Government. In *Startup-Driven E-Government: Digital Innovation for Sustainable Ecosystems* (pp. 61–86). IGI Global Scientific Publishing.

Berman, B. (2016). Referral marketing: Harnessing the power of your customers. *Business Horizons*, *59*(1), 19–28. Advance online publication. DOI: 10.1016/j.bushor.2015.08.001

Busalim, A., Hollebeek, L. D., & Lynn, T. (2023). The effect of social commerce attributes on customer engagement: An empirical investigation. *Internet Research*, *34*(7), 187–214. Advance online publication. DOI: 10.1108/INTR-03-2022-0165

Cahyani, I. R., & Bahri, M. I. (2022). Change Management and Special Autonomy Paradigm: Opportunities and Challenges of Human Resources in Digital Literacy Implementation in Indonesian Regional Libraries. *Congress of Southeast Asian Librarians (CONSAL) XVIII*.

Cai, J., Liu, Z., & Huang, H. (2023). Live commerce as a new form of online shopping: Review and future research agenda. *Journal of Retailing and Consumer Services*, *70*, 103114. DOI: 10.1016/j.jretconser.2022.103114

Chakraborty, P. (n.d.). *The Importance of Ethical Practices in Affiliate Marketing: Building Trust and Credibility*.

Chavee, N., & Vongurai, R. (2023). Factors Impacting Trust, Satisfaction, and Purchase Intention Via Social Live Stream Commerce with Thai Influencers. *Journal of Multidisciplinary in Social Sciences*, *19*(2).

Chen, X., & Lin, Z. (2022). Exploring consumer purchase intention in live streaming commerce: The roles of social presence, perceived value, and entertainment. *Electronic Commerce Research*, *22*(3), 647–672. DOI: 10.1007/s10660-021-09489-x

Choi, Y. (2019). The study of antecedents of consumer engagement and purchase intention in social commerce. *International Journal of Cyber Behavior, Psychology and Learning*, *9*(2), 51–63. Advance online publication. DOI: 10.4018/IJCBPL.2019040104

Cialdini, R. B. (2007). *Influence: The Psychology of Persuasion* (Rev. ed.). Harper Business.

Dabbous, A., Aoun Barakat, K., & Merhej Sayegh, M. (2020). Social Commerce Success: Antecedents of Purchase Intention and the Mediating Role of Trust. *Journal of Internet Commerce*, *19*(3), 262–297. Advance online publication. DOI: 10.1080/15332861.2020.1756190

De Veirman, M., Cauberghe, V., & Hudders, L. (2017). Marketing through Instagram influencers: Impact of number of followers and product divergence on brand attitude. *International Journal of Advertising*, *36*(5), 798–828. DOI: 10.1080/02650487.2017.1348035

Hudders, L., De Jans, S., & De Veirman, M. (2021). The commercialization of social media stars: A literature review and conceptual framework on the strategic use of social media influencers. *International Journal of Advertising*, *40*(3), 327–375. DOI: 10.1080/02650487.2020.1836925

Influenciity. (2024, June 25). *Social Responsibility and Ethics in Influencer Marketing: The Role and Impact of Content Creators.*

iResearch. (2022). China's Live Commerce Market Report 2022. iResearch Consulting Group.

Jannah, N., Bahri, M. I., Kismawadi, E. R., & Handriana, T. (2024). The Effect of Green Brand Image and Green Satisfaction on Green Brand Equity Mediated Green Trust Outpatient's. *Quality - Access to Success*, *25*(198). Advance online publication. DOI: 10.47750/QAS/25.198.40

Kismawadi, E. R., & Irfan, M. (2025). Artificial Intelligence as a Catalyst for Innovation in Islamic Entrepreneurship: Balancing Ethics and Efficiency. In *Improving Entrepreneurial Processes Through Advanced AI* (pp. 1–16). IGI Global.

Koay, K. Y., Tjiptono, F., Teoh, C. W., Memon, M. A., & Connolly, R. (2023). Social Media Influencer Marketing: Commentary on the Special Issue. In *Journal of Internet Commerce* (Vol. 22, Issue S1). https://doi.org/DOI: 10.1080/15332861.2022.2128277

Lee, S., & Watkins, B. (2022). The rise of live streaming commerce: How platforms and influencers shape consumer engagement. *Journal of Interactive Marketing*, *58*, 68–84. DOI: 10.1016/j.intmar.2022.03.002

Lee, S. S., & Johnson, B. K. (2022). Are they being authentic? The effects of self-disclosure and message sidedness on sponsored post effectiveness. *International Journal of Advertising*, *41*(1), 30–53. Advance online publication. DOI: 10.1080/02650487.2021.1986257

Li, H., Zhang, X., & Li, Y. (2021). Live streaming commerce and consumer purchase behavior: Evidence from China. *Electronic Commerce Research and Applications, 49*, 101086. DOI: 10.1016/j.elerap.2021.101086

Lou, C., & Yuan, S. (2019). Influencer marketing: How message value and credibility affect consumer trust and purchase intention. *Journal of Interactive Advertising, 19*(1), 58–73. DOI: 10.1080/15252019.2018.1533501

Meyer, D. (2019). The marketing funnel versus the flywheel: Generating consistent leads through a new model of engagement. *Journal of Digital and Social Media Marketing, 7*(2), 106. Advance online publication. DOI: 10.69554/TLAB6890

Migkos, S. P., Giannakopoulos, N. T., & Sakas, D. P. (2025). Impact of Influencer Marketing on Consumer Behavior and Online Shopping Preferences. *Journal of Theoretical and Applied Electronic Commerce Research, 20*(2), 111. DOI: 10.3390/jtaer20020111

Montag, C., Lachmann, B., Herrlich, M., & Zweig, K. A. (2021). Addictive features of social media/messenger platforms and freemium games against the background of psychological and economic theories. *International Journal of Environmental Research and Public Health, 18*(14), 7723. DOI: 10.3390/ijerph18147723 PMID: 31340426

Ooi, K. B., Lee, V. H., Hew, J. J., Leong, L. Y., Tan, G. W. H., & Lim, A. F. (2023). Social media influencers: An effective marketing approach? *Journal of Business Research, 160*, 113773. Advance online publication. DOI: 10.1016/j.jbusres.2023.113773

Reyneke, M., Robertson, J., Ferreira, C., & Näppä, A. (2023). Whose Role Is It Anyway? A Consideration of Responsibility in Digital Addiction: An Abstract. In *Developments in Marketing Science: Proceedings of the Academy of Marketing Science*. https://doi.org/DOI: 10.1007/978-3-031-24687-6_47

Sun, Y., Shao, X., Li, X., Guo, R., & Nie, F. (2019). How live streaming influences consumers' purchase intentions in social commerce: An IT affordance perspective. *Electronic Commerce Research and Applications, 37*, 100886. DOI: 10.1016/j.elerap.2019.100886

Ünalmış, E., Dirsehan, T., & Erdoğmuş, İ. E. (2024). Consequences of influencer-created content on influencers' authenticity in the beauty and personal care industry. *Journal of Marketing Communications*, 1–27. Advance online publication. DOI: 10.1080/13527266.2024.2371833

Verma, R. K., & Kumari, N. (2023). Generative AI as a Tool for Enhancing Customer Relationship Management Automation and Personalization Techniques. *International Journal of Responsible Artificial Intelligence*, *13*(9).

Wang, H., Wang, L., & Zhou, W. (2022). The influence of live streaming commerce on consumers' purchase intention: The mediating role of trust and perceived value. *Computers in Human Behavior*, *135*, 107388. DOI: 10.1016/j.chb.2022.107388

Chapter 13
Strategic Marketing Transformations Enabled by Digital Currencies

K. R. Mahesh Kumar
https://orcid.org/0009-0008-7282-3950
IZee Business School, India

Raj Kumar Mishra
https://orcid.org/0000-0002-6912-6694
University of Lucknow, India

Sujay Mugaloremutt Jayadeva
https://orcid.org/0000-0003-2252-9071
JSS Academy of Higher Education and Research, India

Pritam Manoharrao Lanjewar
https://orcid.org/0000-0001-7166-884X
Datta Meghe Institute of Management Studies, Nagpur, India

P. Selvakumar
https://orcid.org/0000-0002-3650-4548
Department of Science and Humanities, Nehru Institute of Technology, Coimbatore, India

Amey Gholpe
Datta Meghe Institute of Management Studies, Nagpur, India

ABSTRACT

Digital currencies, also referred to as cryptocurrencies or virtual currencies, represent one of the most transformative developments in the financial and technological landscapes over the past decade. Unlike traditional fiat money issued and regulated by centralized governments and central banks, digital currencies operate on decentralized networks using blockchain technology—a distributed ledger system that ensures transparency, security, and immutability of transactions. This fundamental shift from physical cash and conventional banking systems to digital, cryptographically secured money has profound implications not only for economics and finance but also for marketing, commerce, and consumer behavior. The concept of digital currency first gained widespread attention with the introduction of Bitcoin in 2009,

DOI: 10.4018/979-8-3373-4526-0.ch013

created by an anonymous entity known as Satoshi Nakamoto. Bitcoin's revolutionary design combined peer-to-peer networking and cryptographic proof to enable trustless digital transactions without the need for intermediaries.

INTRODUCTION

Digital currencies, also referred to as cryptocurrencies or virtual currencies, represent one of the most transformative developments in the financial and technological landscapes over the past decade. Unlike traditional fiat money issued and regulated by centralized governments and central banks, digital currencies operate on decentralized networks using blockchain technology—a distributed ledger system that ensures transparency, security, and immutability of transactions. This fundamental shift from physical cash and conventional banking systems to digital, cryptographically secured money has profound implications not only for economics and finance but also for marketing, commerce, and consumer behavior. The concept of digital currency first gained widespread attention with the introduction of Bitcoin in 2009, created by an anonymous entity known as Satoshi Nakamoto. Bitcoin's revolutionary design combined peer-to-peer networking and cryptographic proof to enable trustless digital transactions without the need for intermediaries. For businesses, integrating cryptocurrencies as a payment method can appeal to a tech-savvy and increasingly crypto-aware consumer base, offering benefits such as reduced transaction fees, faster settlements, and enhanced cross-border trade capabilities. Moreover, blockchain's inherent transparency enables innovative loyalty programs and customer engagement strategies, allowing brands to reward consumers with tokenized incentives that are secure, transferable, and verifiable. For instance, blockchain can validate the provenance of products in supply chains, combating counterfeit goods and enabling marketing messages centered on ethical sourcing and sustainability values, increasingly important to modern consumers.

Digital currencies also provide new models for crowdfunding and community-driven brand initiatives, where tokens can represent stakes in projects or access to future products, aligning the interests of companies and their customers in novel ways. Social media and influencer marketing have also adapted to this new digital asset landscape. Crypto-focused content, including educational videos, trading tips, and news, attracts large audiences, creating opportunities for brands to engage niche communities that are passionate about digital assets and innovation. Sponsored content, partnerships with crypto influencers, and native advertising within blockchain-based platforms further extend marketing reach and authenticity. Despite the opportunities, marketers must navigate several challenges associated with digital currencies. Price volatility remains a significant barrier to widespread adoption for everyday transactions, as fluctuating values can complicate pricing and accounting. Regulatory uncertainty and concerns about security, such as hacking and fraud, require robust risk management and transparent communication strategies. Additionally, the technological complexity of cryptocurrencies may alienate less tech-savvy consumers, necessitating educational efforts to build understanding and trust. In summary, the rise of digital currencies is reshaping the financial ecosystem and presenting unprecedented marketing opportunities. Their unique attributes—decentralization, security, transparency, and programmability—enable brands to innovate in payment processing, customer engagement, loyalty programs, and authenticity verification. However, success in this domain requires navigating volatility, regulatory landscapes, and consumer education challenges, making digital currencies both a promising and complex frontier in modern marketing.

Transforming Marketing Strategies: The Role of Digital Currencies

The rapid rise of digital currencies is not only reshaping the financial sector but also fundamentally transforming marketing strategies across industries. Digital currencies, often powered by blockchain technology, introduce new paradigms for how businesses engage with consumers, create value, and foster brand loyalty. As marketing moves deeper into the digital age, traditional methods increasingly give way to innovative approaches that leverage the unique capabilities of cryptocurrencies and their underlying ecosystems (Nalbant & Aydın, 2023). The integration of digital currencies into marketing strategies is driving profound changes in customer acquisition, retention, personalization, transactional experiences, and even corporate social responsibility. One of the most significant transformations driven by digital currencies is the expansion of payment and transactional options. By accepting cryptocurrencies as a form of payment, businesses gain access to a global, borderless consumer base unrestricted by conventional banking limitations or currency conversion fees. This opens new avenues for targeting international customers and niches that are highly engaged with crypto ecosystems. For marketers, enabling crypto payments also serves as a powerful branding statement, signaling innovation, inclusivity, and alignment with cutting-edge technology trends. Moreover, cryptocurrency transactions offer enhanced speed and lower processing costs compared to traditional credit cards or payment gateways, improving the overall customer experience and operational efficiency (Chen X., et al, 2022). Beyond payments, digital currencies enable novel loyalty and rewards programs that transcend the capabilities of conventional points-based systems. Tokenization allows brands to issue digital tokens or coins that customers can earn, trade, or redeem with greater flexibility and transparency. These tokens, built on blockchain networks, provide secure, verifiable proof of ownership and can be programmed with smart contracts to automate rewards or unlock special perks. This fosters deeper consumer engagement by incentivizing behaviors beyond purchases, such as social sharing, content creation, or community participation.

Personalization and data privacy, long-standing challenges in marketing, are also being redefined by the interplay between digital currencies and blockchain solutions. Traditional marketing often relies on aggregating consumer data through third parties, raising concerns over privacy and data security. In contrast, blockchain-based identity and data management systems allow consumers to control their personal information and selectively share it with brands in exchange for token rewards. This consent-driven data sharing can create more relevant, personalized marketing experiences while respecting privacy, enhancing consumer trust (Hai T N., et al, 2021). Marketers who adopt such decentralized data models can differentiate themselves by championing ethical data practices and empowering customers with greater agency. Social media and influencer marketing have similarly evolved in response to the digital currency ecosystem. Crypto influencers and content creators attract specialized audiences passionate about blockchain and cryptocurrencies, providing marketers with targeted channels to reach engaged, high-value consumer segments. Platforms built on blockchain technology introduce new forms of content monetization, tipping, and peer-to-peer support, allowing brands to participate in more authentic and participatory marketing ecosystems. Moreover, the viral potential of digital currencies and token drops during campaigns can rapidly amplify brand exposure and consumer enthusiasm. In conclusion, digital currencies are transforming marketing strategies by enabling new modes of payment, customer engagement, transparency, personalization, and community building (Osei L K., et al, 2023). They empower marketers to connect with global audiences in innovative ways, foster trust through blockchain verification, and create tokenized ecosystems that reward and motivate consumers beyond traditional transactions. While challenges related to regulation, volatility, security, and education persist, the strategic adoption of digital currencies offers marketers a compelling opportunity to redefine brand relationships and competitive differentiation in a rapidly changing digital landscape. Forward-thinking marketers who embrace these technologies thoughtfully will be best positioned to capitalize on the next wave of marketing innovation driven by the rise of digital currencies.

Digital Payments: Enhancing Customer Experience

The evolution of digital payments has fundamentally transformed the way consumers and businesses conduct financial transactions, ushering in a new era of convenience, speed, and security that significantly enhances the overall customer experience. From traditional cash and card-based systems to the proliferation of mobile wallets, contactless payments, cryptocurrencies, and biometric authentication, digital payments have revolutionized the transactional landscape, creating seamless, frictionless, and personalized interactions that meet the demands of today's digitally connected consumers (Agarwal S., et al, 2023). At its core, digital payment refers to any transaction that is conducted electronically, without the need for physical cash exchange. This broad category encompasses a wide array of technologies and platforms including credit and debit card payments, mobile payment apps such as Apple Pay and Google Pay, peer-to-peer (P2P) transfer services like Venmo and PayPal, contactless NFC-enabled transactions, QR code payments, and emerging forms such as cryptocurrency transactions and central bank digital currencies (CBDCs). The widespread adoption of smartphones and internet connectivity has been a primary driver behind the surge in digital payments, enabling consumers to make purchases anytime and anywhere with minimal effort. One of the foremost benefits of digital payments is the convenience they afford consumers (Kehinde & Naomi, 2024). Eliminating the need to carry cash or physical cards reduces transactional friction and speeds up the payment process. Contactless payments, in particular, have gained significant traction by enabling tap-and-go transactions that are swift, hygienic, and user-friendly—an attribute that gained added importance during the COVID-19 pandemic. Moreover, mobile payment solutions often integrate with digital wallets, storing multiple payment methods and loyalty cards in one secure place, simplifying checkout and enabling personalized offers based on purchase history and preferences.

From a business perspective, digital payments streamline operations and enable a more engaging customer journey. Faster transaction processing reduces wait times at checkout, minimizing cart abandonment and enhancing satisfaction. Integration with customer relationship management (CRM) systems allows businesses to capture valuable transactional data that can be leveraged for targeted marketing, personalized promotions, and loyalty program management. For example, retailers can offer instant discounts, reward points, or cashback incentives directly through digital payment platforms, encouraging repeat purchases and fostering brand loyalty (Shree S., et al, 2021). Innovations in digital payment technologies continue to drive enhancements in customer experience. The rise of biometric authentication improves both security and ease of use by allowing users to authorize transactions through facial recognition or fingerprints without needing to remember complex passwords or PINs. Artificial intelligence (AI) and machine learning algorithms personalize payment experiences by analyzing customer behavior to offer tailored payment options, detect fraud, and optimize checkout flows. The integration of digital currencies, including cryptocurrencies and CBDCs, into payment ecosystems presents opportunities for faster cross-border transactions, reduced fees, and novel incentive structures such as tokenized rewards. Despite these advancements, challenges remain in maximizing the potential of digital payments to enhance customer experience. Issues such as cybersecurity threats, privacy concerns, and regulatory complexities require ongoing attention and robust governance frameworks. The diversity of payment preferences across different demographics and regions necessitates inclusive and interoperable payment solutions that accommodate varying levels of digital literacy and access to technology (Samal A., et al, 2025). Additionally, technical glitches or network outages can disrupt payment processes and negatively impact consumer trust. In conclusion, digital payments have emerged as a cornerstone of modern commerce, playing a pivotal role in enhancing customer experience by delivering convenience, security, personalization, and accessibility. They enable businesses to streamline operations, engage customers more effectively, and foster loyalty through innovative incentives and seamless transactional journeys. While challenges persist, ongoing technological advancements and collaborative efforts among industry players are set to make digital payments even more integral to customer satisfaction and business success in the years ahead.

Increased Security: Reducing Fraud and Risk

In today's rapidly evolving digital economy, the imperative to strengthen security and reduce fraud and risk has never been more critical. As transactions increasingly shift from physical to digital environments, organizations face heightened vulnerabilities stemming from sophisticated cyber threats, identity theft, financial fraud, and data breaches. However, advances in technology, particularly in the realms of encryption, biometrics, blockchain, artificial intelligence, and multi-factor authentication, have revolutionized security paradigms (Soundariya A., et al, 2025). These innovations enable businesses and consumers alike to safeguard assets, verify identities with greater accuracy, and proactively detect and prevent fraudulent activities, fundamentally reducing the risks associated with digital interactions. A core aspect of enhanced security lies in cryptographic technologies that underpin digital transactions and data protection. Encryption transforms sensitive data into unreadable formats, ensuring that even if intercepted by malicious actors, the information remains inaccessible without proper decryption keys. Modern encryption standards such as Advanced Encryption Standard (AES) and public key infrastructure (PKI) have become industry staples, securing communications, payment data, and personal information. Tokenization, another crucial technique, replaces sensitive data like credit card numbers with unique tokens that hold no exploitable value if breached. This significantly minimizes the exposure of critical information in databases and during transaction processing, thereby reducing fraud risk (Chauhan S., et al, 2022). Biometric authentication has emerged as a powerful tool in increasing security by leveraging unique physiological or behavioral characteristics, such as fingerprints, facial recognition, iris scans, or voice patterns, to verify user identities. Unlike passwords or PINs, which can be guessed, stolen, or shared, biometrics provide a more robust and convenient layer of protection that is difficult to replicate or bypass. The integration of biometrics into mobile devices and payment systems enhances security while improving user experience by enabling swift and frictionless authentication. For example, biometric verification helps prevent unauthorized access to digital wallets and banking apps, drastically lowering the incidence of identity fraud and unauthorized transactions.

Artificial intelligence (AI) and machine learning (ML) play transformative roles in fraud detection and risk management. These technologies analyze vast volumes of transactional data in real-time to identify anomalies, patterns, and behaviors indicative of fraudulent activities. Machine learning models continuously evolve, adapting to emerging threats and sophisticated attack vectors more rapidly than traditional rule-based systems. For instance, AI can flag suspicious activities such as unusual purchase amounts, atypical geographic locations, or deviations in login patterns, triggering alerts or automated preventative actions. By reducing false positives and enabling proactive fraud mitigation, AI-driven security enhances operational efficiency and protects customers from losses (Brown D., et al, 2024). Another important dimension of increased security involves robust identity verification and know-your-customer (KYC) protocols. Digital identity verification tools utilize document scanning, facial recognition, and database cross-checks to authenticate customers during onboarding and ongoing transactions. By integrating KYC and anti-money laundering (AML) measures into digital platforms, organizations bolster their defenses against financial crimes and build trust with stakeholders. Secure software development and infrastructure practices are equally essential in minimizing vulnerabilities and fraud risks. Adopting secure coding standards, conducting regular penetration testing, and implementing patch management help prevent exploitation of software bugs and system weaknesses. Cloud security measures, including encryption at rest and in transit, identity and access management, and continuous monitoring, protect data stored and processed in cloud environments from unauthorized access. Network segmentation and intrusion detection systems further safeguard critical assets from cyberattacks and data exfiltration (Taherdoost H., 2021). Despite significant advances, challenges remain in achieving comprehensive security and fraud reduction. Cybercriminals continually adapt, employing more sophisticated social engineering tactics, zero-day exploits, and insider threats. Regulatory environments evolve, requiring organizations to navigate complex compliance landscapes while maintaining agility. Balancing security measures with usability is crucial to avoid friction that could deter legitimate users or drive them to less secure alternatives. Addressing these challenges requires continuous investment in innovation, collaboration across industries and governments, and proactive risk management frameworks.

Targeted Advertising: Leveraging Digital Currency Data

In the dynamic landscape of digital marketing, the ability to deliver highly personalized and relevant advertisements to consumers is paramount for maximizing engagement, conversion rates, and return on investment (ROI). Targeted advertising leverages consumer data to tailor marketing messages based on individual preferences, behaviors, and purchasing patterns. The emergence and growing adoption of digital currencies, including cryptocurrencies and other blockchain-based payment mechanisms, present a novel and transformative source of consumer data that marketers can harness to refine their targeting strategies. By analyzing transaction histories, spending behaviors, and wallet interactions associated with digital currencies, businesses can unlock unprecedented insights into consumer intent, preferences, and demographics, thereby enhancing the precision and effectiveness of advertising campaigns (Ogunmokun A S., et al, 2022). Digital currencies operate on decentralized ledgers that record transaction data transparently and immutably, providing a rich dataset of consumer activity. Unlike traditional payment methods, digital currency transactions often include metadata that reveals granular details about purchasing patterns, merchant categories, frequency of transactions, and even geographic trends. This data can be anonymized and aggregated to respect privacy regulations while still offering valuable insights for targeted advertising. For example, a retailer could identify segments of consumers who frequently purchase specific categories of goods using digital currencies and tailor promotional offers or content specifically to those groups, increasing the likelihood of engagement and sales (Alonge E O., et al, 2021). In addition to transaction data, digital currencies enable innovative advertising models based on token incentives and rewards. Marketers can use tokens to engage consumers directly, rewarding them for viewing ads, sharing content, or participating in surveys. These token-based incentive systems not only foster active consumer participation but also generate valuable first-party data that further refines targeting algorithms. Additionally, advertisers can create programmable smart contracts that automate ad delivery and payment based on consumer interaction, ensuring efficiency and transparency in advertising spend and campaign performance. The integration of digital currency data with other emerging technologies like artificial intelligence (AI) and machine learning further amplifies the potential for hyper-personalized advertising (Dadwal S., et al, 2021). AI algorithms can analyze vast datasets from blockchain transactions alongside social media activity, browsing history, and demographic data to build detailed consumer profiles and predict future purchasing behavior. Machine learning models continuously refine targeting parameters by detecting patterns and anomalies, enabling real-time campaign adjustments that optimize ad spend and customer engagement. This synergy allows businesses to create highly customized marketing messages that resonate deeply with individual consumers, enhancing brand loyalty and lifetime value.

Figure 1. Digital Transformation in Finance Industry

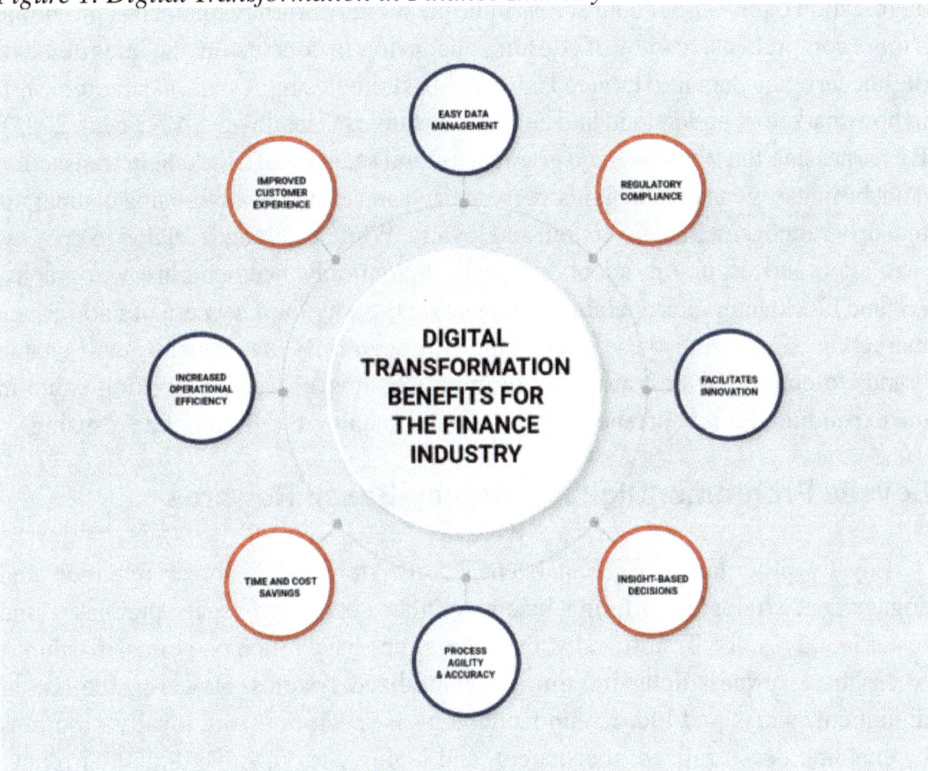

However, leveraging digital currency data for targeted advertising also presents challenges that must be carefully navigated. Privacy and ethical considerations are paramount, as consumers increasingly demand transparency and control over their personal information. Despite the anonymized nature of blockchain data, there remains a risk of deanonymization if data is improperly handled or combined with other datasets. Marketers must implement stringent data governance frameworks and adhere to regulatory standards to protect consumer rights and avoid reputational damage. Moreover, the nascent and complex nature of digital currency ecosystems requires marketers to develop specialized expertise and invest in advanced analytics infrastructure to extract meaningful insights (Ajayi A J., et al, 2025). Looking ahead, the evolution of privacy-preserving technologies such as zero-knowledge proofs and secure multi-party computation promises to further enhance the ability to use digital currency data for targeted advertising without compromising consumer privacy. These cryptographic techniques allow marketers to verify consumer attributes and transaction legitimacy without accessing sensitive underlying data, enabling compliant and ethical data-driven marketing. Furthermore, the maturation of interoperable blockchain standards and cross-chain analytics tools will facilitate the aggregation of consumer data across multiple digital currency platforms, providing a more comprehensive view of consumer behavior. In conclusion, the integration of digital currency data into targeted advertising strategies represents a paradigm shift in how marketers understand and engage consumers (Upadhyaya A N., et al, 2025). By harnessing the transparency, granularity, and security of blockchain transaction data, businesses can create highly personalized and effective advertising campaigns that drive engagement, conversion, and loyalty. While challenges related to privacy, regulation, and technology adoption remain, ongoing advancements in cryptography, AI, and blockchain interoperability are paving the way for a new era of data-driven marketing. Embracing these opportunities thoughtfully and ethically will enable brands to connect with consumers in more meaningful ways, unlocking value in the expanding digital currency ecosystem and shaping the future of advertising.

Loyalty Programs: Digital Currency-Based Rewards

Loyalty programs have long been a cornerstone of customer retention and engagement strategies, offering incentives that encourage repeat purchases and foster brand loyalty. Traditionally, these programs have relied on points, discounts, or cashback rewards delivered through centralized systems. However, the rise of digital currencies and blockchain technology is revolutionizing loyalty programs by enabling decentralized, transparent, and highly customizable digital currency-based rewards. These next-generation loyalty programs leverage cryptocurrencies, stablecoins, and tokenized assets to create more engaging, flexible, and secure reward ecosystems that benefit both businesses and consumers, transforming how value is exchanged and loyalty is cultivated in the digital age. At the heart of digital currency-based loyalty programs is the concept of tokenization, where loyalty points are converted into blockchain-based tokens that can represent a variety of assets, rights, or privileges (Thirunagalingam A., et al, 2025). These tokens are often interoperable, allowing customers to redeem rewards not only within the issuing brand's ecosystem but also across partner networks or even on secondary markets. This enhanced flexibility addresses a common pain point in traditional loyalty programs: the lack of reward liquidity and limited redemption options, which often lead to unused points and customer dissatisfaction. By transforming loyalty points into digital currency tokens, businesses empower customers with true ownership and control over their rewards, fostering increased engagement and a sense of value. Blockchain technology also brings unparalleled transparency and security to loyalty programs. Each token issuance, transfer, and redemption is recorded immutably on a distributed ledger, eliminating fraud, double-spending, or manipulation that can plague centralized systems. Consumers can verify their token balances and transaction history independently, building trust and confidence in the program. For businesses, blockchain's transparency simplifies auditing and regulatory compliance, reducing operational overhead and enabling more efficient program management. Smart contracts further automate reward distribution and redemption processes, ensuring that loyalty incentives are delivered instantly and fairly based on predefined criteria such as purchase milestones or referral activities (Sakas D P., et al, 2021). Digital currency-based loyalty programs enable innovative marketing and customer engagement models that go beyond simple transactional rewards. Brands can design tiered loyalty structures where tokenized rewards unlock exclusive privileges such as VIP access, early product releases, or personalized experiences. Tokens can also be used as a medium for gamification, where customers earn badges, unlock challenges, or participate in community-driven activities that enhance brand affinity. Moreover, tokens can facilitate peer-to-peer transfers or gifting, allowing loyal customers to share rewards within their social networks, amplifying word-of-mouth marketing and organic growth.

Despite their advantages, the adoption of digital currency-based loyalty programs faces several challenges. Regulatory uncertainty around cryptocurrencies and tokenized assets varies significantly across jurisdictions, requiring businesses to navigate complex legal frameworks to ensure compliance. Consumer education and trust also represent hurdles, as some customers remain unfamiliar or wary of digital currencies. User experience must be carefully designed to simplify wallet management, token redemption, and security without compromising the benefits of decentralization (de Boer E., et al, 2025). Interoperability between different blockchain platforms and loyalty ecosystems is another technical challenge that developers are actively addressing through emerging standards and cross-chain solutions. Looking ahead, the future of loyalty programs lies in greater integration with digital currencies and decentralized technologies. Advances in blockchain scalability, privacy-preserving cryptographic methods, and decentralized identity (DID) solutions will enable more secure, personalized, and inclusive loyalty ecosystems. Brands will increasingly leverage artificial intelligence and machine learning to analyze token usage patterns and dynamically tailor rewards that evolve with consumer preferences and lifecycle stages. The rise of metaverse platforms and tokenized virtual goods will open new avenues for loyalty rewards tied to digital experiences, fostering immersive brand engagement in virtual environments (de Boer E., et al, 2025). In conclusion, digital currency-based loyalty programs represent a transformative shift in how businesses reward and engage customers. By leveraging blockchain's transparency, security, and programmability, these programs create more flexible, trustworthy, and personalized reward experiences that benefit both brands and consumers. While challenges remain in regulation, education, and technology integration, ongoing innovation and increasing digital currency adoption are driving the evolution of loyalty into a more dynamic, interoperable, and customer-centric ecosystem. Businesses that embrace this paradigm shift stand to gain a competitive edge by fostering deeper loyalty, enhancing customer lifetime value, and unlocking new growth opportunities in the rapidly digitizing global economy.

New Business Models: Opportunities for Innovation

The rapid evolution and adoption of digital currencies and blockchain technology are catalyzing the emergence of innovative business models that challenge traditional paradigms and unlock new opportunities for value creation, disruption, and competitive advantage. Unlike conventional business frameworks that rely heavily on intermediaries, centralized control, and opaque processes, these emerging models harness the decentralized, transparent, and programmable nature of blockchain and digital currencies to reinvent how goods, services, and value are exchanged, managed, and monetized (de Boer E., et al, 2025). By enabling trustless peer-to-peer interactions, automated contractual agreements, and tokenized ecosystems, digital currencies empower entrepreneurs and established organizations alike to explore novel avenues for innovation that redefine customer engagement, revenue generation, and operational efficiency. One of the most significant transformations driven by digital currencies is the rise of decentralized platforms and marketplaces, which disintermediate traditional middlemen and create more direct connections between producers and consumers. These decentralized marketplaces utilize blockchain's immutable ledger to facilitate transparent transactions, reduce costs, and increase trust. For example, peer-to-peer (P2P) trading platforms for goods, services, and digital assets leverage cryptocurrencies as a medium of exchange, enabling global participation without the friction of traditional payment gateways or currency conversion. Tokenization is another key enabler of new business models, transforming physical and intangible assets into tradable digital tokens on blockchain networks. This process unlocks fractional ownership, liquidity, and programmability for assets that were traditionally illiquid or cumbersome to transfer. Real estate, fine art, intellectual property, and even personal data can be tokenized, creating novel investment opportunities and business ecosystems (Khan & Belk, 2024). For instance, fractional real estate investment platforms allow individuals to purchase and trade shares in properties using digital tokens, lowering entry barriers and diversifying investment portfolios. Tokenized assets also enable innovative revenue models such as pay-per-use, royalty sharing, and subscription-based access, where smart contracts automatically execute payments and distribute income among stakeholders transparently and efficiently. This creates new value chains and revenue streams while enhancing inclusivity and participation. Digital currencies also enable innovative incentive and governance models that empower communities and customers to actively participate in business ecosystems. Decentralized autonomous organizations (DAOs) are a prime example, where governance decisions, funding allocations, and strategic directions are determined collectively by token holders through transparent voting mechanisms encoded on blockchain.

DAOs redefine traditional corporate governance by fostering inclusivity, democratizing decision-making, and aligning incentives among diverse stakeholders. Businesses can leverage DAOs to co-create products, validate ideas, and crowdsource investments, accelerating innovation cycles and reducing market entry risks. Similarly, token-based incentive systems reward users for engagement, content creation, or referrals, transforming consumers into active collaborators and advocates, thereby driving organic growth and deepening brand loyalty (Bouncken R B., et al, 2021). The convergence of digital currencies with emerging technologies such as the Internet of Things (IoT), artificial intelligence (AI), and the metaverse is opening additional frontiers for new business models. For instance, IoT devices integrated with blockchain can autonomously transact using cryptocurrencies, enabling machine-to-machine (M2M) economies where devices pay for services, maintenance, or data sharing without human intervention. This supports innovative models in smart cities, connected vehicles, and energy grids. AI-powered analytics combined with blockchain's transparency enable hyper-personalized services and dynamic pricing models that adapt to consumer behavior and market conditions in real-time. In the metaverse—a virtual, immersive digital universe—digital currencies serve as native means of exchange, supporting virtual real estate markets, digital goods, entertainment, and social interactions (Kanbach D K., et al, 2024). Businesses can monetize experiences, events, and creations within these virtual spaces, crafting entirely new revenue streams and consumer engagement paradigms. Despite these exciting opportunities, the development and scaling of new business models based on digital currencies face several challenges. Regulatory uncertainty remains a major obstacle, as governments worldwide continue to evaluate and implement policies regarding cryptocurrency usage, token offerings, and blockchain-based financial services. Compliance complexity varies across jurisdictions, requiring businesses to adopt flexible legal strategies and proactive stakeholder engagement. Technical challenges related to blockchain scalability, interoperability, and user experience must also be overcome to support mass adoption. Many blockchain platforms are still evolving, and user onboarding can be hindered by the complexity of managing wallets, private keys, and digital assets securely. In summary, new business models enabled by digital currencies and blockchain technology represent a fundamental shift in how value is created, exchanged, and governed across industries (Hubballi R B., et al, 2025). By fostering decentralization, transparency, and automation, these models unlock unprecedented opportunities for innovation in commerce, finance, supply chain, governance, and digital experiences. While challenges remain, ongoing advancements in blockchain scalability, privacy, interoperability, and regulatory frameworks are paving the way for widespread adoption and transformation. Businesses that embrace these novel paradigms can drive competitive advantage by creating more efficient, inclusive, and engaging ecosystems that align with the evolving digital economy and the expectations of a new generation of consumers.

Global Reach: Expanding Market Access

In an increasingly interconnected and digitalized world, businesses are perpetually seeking ways to transcend geographical boundaries and tap into new markets to fuel growth. Digital currencies have emerged as a powerful catalyst for expanding global reach, fundamentally transforming the way companies access, serve, and engage international customers. By leveraging the decentralized, borderless nature of cryptocurrencies and blockchain technology, businesses can overcome many traditional limitations associated with cross-border transactions, payment infrastructure, currency conversions, and regulatory complexities, thereby unlocking unprecedented opportunities for market expansion and global commerce (Maheswaran K., et al, 2025). One of the primary barriers to global trade and market access has historically been the complexity and cost associated with international payments. Traditional cross-border transactions often involve multiple intermediaries such as banks, payment processors, and currency exchanges, each imposing fees, processing delays, and varying compliance requirements. Digital currencies operate on decentralized networks that facilitate peer-to-peer transfers without the need for intermediaries, significantly reducing transaction fees and enabling near-instantaneous payments across borders. This streamlined payment process lowers entry barriers for small and medium-sized enterprises (SMEs) and entrepreneurs, allowing them to competitively offer goods and services to global audiences with minimal overhead. The ability of digital currencies to facilitate seamless currency conversion and eliminate foreign exchange friction further enhances global reach. In traditional trade, currency volatility and exchange costs can erode profit margins or deter customers from purchasing due to confusing pricing (Trischler M F G., et al, 2023). Digital currencies, especially stablecoins pegged to fiat currencies, provide a stable, universally accepted medium of exchange that simplifies pricing and payment settlement. Merchants can price goods in their preferred currency while receiving payment in digital tokens that are easily converted or held without the risk of drastic fluctuations. This stability and predictability encourage international transactions and foster trust among consumers wary of hidden fees or fluctuating exchange rates.

The transparent, immutable nature of blockchain technology undergirding digital currencies enhances cross-border regulatory compliance and reduces risks related to fraud and money laundering. Businesses operating internationally face diverse and often complex regulatory frameworks that can slow down market entry or expose firms to legal risks. Blockchain's ability to provide auditable transaction histories in real time facilitates regulatory reporting, know-your-customer (KYC), and anti-money laundering (AML) compliance across jurisdictions. This transparency builds confidence among regulators, partners, and consumers, smoothing the path to global market access while maintaining legal and ethical standards. The global reach empowered by digital currencies also supports cross-border partnerships and supply chain optimization (Clauss T., et al, 2022). Businesses can coordinate payments with international suppliers and partners efficiently, reducing delays and transactional uncertainties. Smart contracts automate payments and enforce contractual terms, ensuring transparency and accountability in complex international supply chains. This operational efficiency strengthens global networks and enhances competitiveness in global markets, particularly for companies dealing with perishable goods, just-in-time manufacturing, or distributed production. Despite the significant advantages, expanding global reach through digital currencies is not without challenges. Regulatory uncertainty and varying acceptance of cryptocurrencies across countries require businesses to navigate a complex legal landscape. While some nations embrace digital currencies with clear frameworks and incentives, others impose restrictions or bans, complicating international operations. Businesses must adopt flexible compliance strategies and monitor regulatory developments continuously to mitigate risks. Additionally, the technological infrastructure and digital literacy required to support cryptocurrency adoption vary globally, necessitating tailored approaches to education, user experience, and support (Erenstein O., et al, 2022). In conclusion, digital currencies dramatically expand global reach by overcoming traditional barriers related to payments, currency exchange, regulatory compliance, and market access. Their borderless, decentralized nature allows businesses to serve a broader, more diverse customer base, tapping into underserved markets and facilitating innovative, inclusive commerce. While challenges in regulation, infrastructure, and security remain, ongoing advancements in technology and policy are steadily reducing these obstacles, enabling businesses of all sizes to compete and thrive on the global stage. Embracing digital currencies as part of a comprehensive international strategy not only opens new revenue streams but also positions companies at the forefront of the rapidly evolving digital economy, ready to capitalize on emerging opportunities in the global marketplace.

Data-Driven Insights: Informing Marketing Strategies

In the modern era of digital transformation, data-driven insights have become the cornerstone of effective marketing strategies, empowering businesses to understand consumer behavior, optimize campaigns, and drive measurable growth. The explosion of data generated by online interactions, mobile devices, and connected ecosystems presents unprecedented opportunities to capture, analyze, and leverage rich datasets. Digital currencies and blockchain technologies further enrich this data landscape by enabling secure, transparent, and granular tracking of consumer transactions, interactions, and preferences (Aggarwal S., et al, 2024). Together, these advances provide marketers with deep, actionable insights that inform strategy development, personalize customer experiences, and optimize resource allocation across marketing channels, ensuring that every decision is backed by robust evidence rather than intuition alone. At its core, data-driven marketing relies on collecting and analyzing vast amounts of customer data across touchpoints to reveal patterns, preferences, and trends. Traditional sources include website analytics, social media engagement, customer relationship management (CRM) systems, and point-of-sale data. The integration of digital currencies introduces a new dimension of transactional data that is highly accurate and tamper-proof due to blockchain's immutable ledger. Each digital currency transaction is recorded with precise timestamps, amounts, and counterparties, allowing marketers to obtain verified purchase histories and spending behavior that surpass the reliability of conventional data sources. This level of data integrity enhances the quality of insights and minimizes discrepancies caused by data fraud, manipulation, or fragmentation (Rosário & Dias, 2023). Another transformative impact of data-driven insights powered by digital currencies lies in optimizing omnichannel marketing strategies. As consumer journeys become increasingly complex, spanning multiple digital and physical channels, marketers must understand how different touchpoints influence purchasing decisions. Blockchain-based data aggregation provides a unified, real-time view of customer interactions across platforms, enabling attribution modeling that accurately measures the impact of each channel and campaign. This clarity empowers marketing teams to allocate budgets effectively, discontinue underperforming efforts, and double down on high-impact initiatives, ultimately improving overall campaign efficiency and effectiveness.

The granularity and transparency of blockchain data also improve the accuracy and trustworthiness of marketing analytics, addressing long-standing issues related to data fraud, ad fraud, and inflated metrics. With digital currency transaction data and blockchain-verified user actions, marketers can confidently assess the true effectiveness of digital advertising campaigns, eliminating wasted spend on fake clicks or bots. This accountability not only enhances marketing ROI but also strengthens relationships with advertising partners and stakeholders by providing verifiable performance reports. Despite the tremendous potential, deploying data-driven marketing strategies powered by digital currencies presents challenges. Data privacy regulations such as GDPR and CCPA require rigorous compliance frameworks to ensure consumer data is handled ethically and securely (Hossain Q., et al, 2024). Blockchain's immutability, while a strength, also necessitates careful management of data retention policies and the right to be forgotten. Technical integration between legacy marketing systems and blockchain infrastructures can be complex, requiring investment in expertise and tools to bridge data silos and enable real-time analytics. Additionally, ensuring data quality and standardization across disparate blockchain networks is essential to avoid fragmentation and inconsistency in insights. Looking forward, the convergence of artificial intelligence (AI), big data analytics, and blockchain-enabled digital currency data promises to usher in a new era of marketing intelligence. AI algorithms will increasingly automate data interpretation, trend detection, and decision-making, providing marketers with prescriptive insights and optimized campaign recommendations. Real-time data streams combined with blockchain's reliability will enable adaptive marketing strategies that respond instantaneously to shifting consumer behavior, market conditions, and competitive dynamics. The emergence of decentralized identity (DID) systems integrated with digital currencies will further personalize marketing by securely linking consumer identities across platforms while preserving privacy. In conclusion, data-driven insights are revolutionizing marketing strategies by enabling a deep, accurate, and actionable understanding of consumers (Sakthivel M., et al, 2025). The integration of digital currencies and blockchain technology enhances the quality, transparency, and privacy of data, empowering marketers to deliver personalized, efficient, and impactful campaigns that resonate with today's discerning customers. While challenges in regulation, technology, and consumer trust remain, the ongoing maturation of these technologies is setting the stage for a future where marketing decisions are not only informed by robust evidence but also dynamically adapted to maximize value for businesses and consumers alike. Organizations that successfully harness data-driven insights powered by digital currencies will gain a decisive competitive edge, fostering stronger customer relationships and driving sustained growth in the digital economy.

Adoption and Education: Overcoming Barriers

The widespread adoption of digital currencies represents a transformative shift in the global financial system, yet this revolution is still in its formative stages. Despite growing interest and technological advancement, mainstream adoption continues to face significant hurdles, ranging from a lack of awareness and technical complexity to regulatory uncertainty and cultural resistance. Overcoming these barriers requires a deliberate and strategic focus on education, user experience, trust-building, and systemic alignment between stakeholders. Education plays a foundational role in demystifying digital currencies and enabling individuals, businesses, and institutions to confidently engage with this emerging financial ecosystem. Education needs to be tailored to different audiences (Kotte K R., et al, 2025). For consumers, initiatives should focus on how to safely use digital wallets, make transactions, protect private keys, and avoid scams. For businesses, the emphasis should be on integrating cryptocurrency payment options, understanding tax implications, maintaining compliance, and leveraging blockchain for operational efficiencies. Governments and regulators require education on the potential of digital currencies to support financial innovation while maintaining oversight. Tailored, role-specific education helps ensure that all actors in the ecosystem can make informed decisions based on accurate and contextual information (Smith & Sinkford, 2022). A critical challenge for both education and adoption is usability. The technical nature of interacting with cryptocurrencies, such as managing public and private keys, dealing with long hexadecimal addresses, or recovering access to lost wallets, can be daunting, especially for non-technical users. This complexity often leads to fear of irreversible mistakes, discouraging participation. To address this, user education must be paired with product design that emphasizes simplicity, safety, and clarity. Platforms that offer intuitive interfaces, educational tooltips, in-app tutorials, and guided onboarding experiences can significantly reduce the learning curve and increase confidence among new users. Digital literacy also plays a crucial role in adoption. Many users in emerging markets, where cryptocurrencies offer significant benefits such as protection from currency devaluation or access to financial services, may lack the digital skills or internet infrastructure needed to engage with the technology (Menzli L J., et al, 2022). Initiatives that combine digital literacy with cryptocurrency education, such as community workshops, mobile-first apps, and offline tools, can empower underserved populations to participate in the digital economy. This, in turn, supports broader financial inclusion goals, enabling individuals to save, invest, and transact in ways previously inaccessible.

Financial institutions, which have traditionally been cautious in approaching digital currencies, are also undergoing a learning curve. Education for banks, asset managers, and insurance providers is key to enabling them to safely offer crypto-related services, such as custody, trading, and digital asset-backed products. As institutional adoption grows, these organizations can become trusted bridges between the traditional and decentralized finance worlds, provided their staff and leadership have a robust understanding of the technology and its implications. One of the most influential forces in accelerating education and adoption is the role of governments and central banks. Clear, consistent regulatory frameworks provide the stability and legal assurance needed to encourage participation. Education at the governmental level must address not only the technical and economic aspects of digital currencies but also their potential societal impacts, such as the ability to track and combat illicit finance, increase efficiency in aid distribution, and modernize public financial infrastructure (Zhang A., et al, 2022). Public-private partnerships can support this effort by creating regulatory sandboxes, innovation hubs, and collaborative educational platforms. Another promising pathway to broader adoption is through formal education. Integrating blockchain and digital currency topics into school curricula, university programs, and professional certifications will prepare the next generation of workers, entrepreneurs, and policymakers. Educational institutions can offer foundational courses, specialized degrees, and interdisciplinary research opportunities that explore the economic, technical, legal, and ethical dimensions of digital currencies. Academic partnerships with industry can ensure that educational content remains current and aligned with real-world developments. Gamification and experiential learning can also be powerful tools in promoting adoption. Simulated trading environments, blockchain-based games, and earn-to-learn platforms allow users to interact with digital currencies in low-risk or rewarding ways. For example, some platforms offer crypto rewards for completing educational modules, encouraging exploration while building confidence (Grandhi B., et al, 2021). These approaches make learning more engaging and can lead to higher retention and practical application of knowledge. Despite all these efforts, adoption will remain uneven without addressing systemic barriers such as regulatory uncertainty, infrastructure limitations, and economic instability. Governments must establish clear guidelines on how digital currencies can be legally used and taxed. Technology providers must ensure that platforms are accessible across devices and networks, particularly in regions with limited infrastructure. Economic policies must consider how digital currencies interact with inflation, capital controls, and monetary sovereignty. Education alone cannot overcome these structural issues, but it can empower stakeholders to advocate for and contribute to meaningful change.

CONCLUSION

As digital currencies transition from experimental technologies to integral components of the global financial ecosystem, the landscape of marketing is undergoing a profound transformation. Digital currency marketing is no longer just about advertising crypto products or onboarding users to exchanges; it's about leveraging the foundational principles of decentralization, transparency, and programmability to reimagine how brands engage consumers, build loyalty, and create value. The future of digital currency marketing promises to be dynamic, data-rich, and highly personalized, offering exciting opportunities for businesses to innovate while addressing evolving consumer expectations. Emerging trends point toward a future where the integration of blockchain technology, decentralized finance (DeFi), non-fungible tokens (NFTs), and tokenized economies will become central to marketing strategies across virtually every industry. Another future-forward development is the merging of digital currency ecosystems with the creator economy and influencer marketing. As more creators monetize their content via blockchain-based platforms, they are issuing their tokens, forming decentralized autonomous organizations (DAOs), and building communities with financial and governance participation. Brands can collaborate with these creator economies not only through sponsorships but by co-creating digital assets, incentivizing user engagement, and embedding marketing messages within tokenized ecosystems. This shift will redefine brand partnerships, turning consumers into stakeholders and advocates, blurring the lines between advertising, content, and community. Gamification and metaverse integration represent another critical frontier. As virtual environments grow in popularity—whether in gaming, virtual events, or social metaverses—brands will use digital currencies to enable immersive, interactive experiences. Users can earn crypto tokens by completing branded challenges, attending virtual product launches, or purchasing avatar wearables, turning engagement into a monetizable activity. With blockchain underpinning these environments, marketers will benefit from verifiable engagement data, transparent economic flows, and the ability to build long-term digital identities and asset ownership that span multiple platforms. The convergence of VR/AR, NFTs, and digital currency will enable new forms of storytelling, brand expression, and user interaction. Cross-border marketing will be profoundly impacted by the borderless nature of digital currencies. As global commerce becomes more decentralized, digital currencies will facilitate seamless transactions without reliance on intermediaries or traditional banking infrastructure. This opens new opportunities for brands to engage consumers in emerging markets, where traditional financial systems may be limited but mobile adoption is high. Digital currencies reduce transaction friction, lower costs, and enable instant settlement, allowing marketers to deploy campaigns with global reach and localized reward mechanisms. Crypto airdrops, community grants, and blockchain bounties can become localized marketing tools tailored to diverse audiences across continents.

Artificial intelligence (AI) and blockchain integration will further enhance the future of digital currency marketing. AI-driven analytics platforms will process blockchain data to generate real-time consumer insights, optimize campaign performance, and predict trends. Smart contracts will automate marketing transactions—from affiliate payments to user rewards—reducing administrative burden and enhancing trust. For instance, influencers might receive real-time micropayments in digital currency based on engagement metrics verified on-chain. These autonomous, transparent interactions will drive greater efficiency, accountability, and innovation in the marketing supply chain. Regulatory evolution will also shape the trajectory of digital currency marketing. As governments develop frameworks around data protection, token issuance, and digital identity, marketers will need to align with legal standards while advocating for consumer-centric innovation. Regulatory clarity will empower more mainstream brands to enter the digital currency space with confidence, resulting in a surge of experimentation, partnerships, and investment. In response, marketers must prioritize compliance, transparency, and ethical data usage to build sustainable relationships in the Web3 era. Looking ahead, one of the most revolutionary shifts will be toward participatory, community-driven marketing models. In the Web3 paradigm, users will no longer be passive recipients of marketing messages—they will be active contributors, co-creators, and beneficiaries. Community tokens will empower users to influence brand decisions, share in the economic upside, and shape the evolution of marketing strategies. Brands that embrace this shift will transition from centralized ownership to collaborative ecosystems, where value creation is distributed among all participants. DAOs dedicated to brand development, product feedback, or regional marketing efforts will become more common, democratizing the way marketing campaigns are conceived and executed. In conclusion, the future of digital currency marketing is poised to be transformative, characterized by personalization, decentralization, and participatory innovation. As blockchain technology and digital currencies become deeply embedded in the digital fabric of society, marketers will have access to powerful tools to connect with audiences in more meaningful, secure, and transparent ways. From tokenized loyalty systems and metaverse branding to AI-powered insights and creator partnerships, the opportunities are vast, but success will require agility, education, ethical alignment, and a willingness to co-create with communities. Brands that embrace these changes will not only stay ahead of the curve but will help shape the very future of commerce, engagement, and value creation in a digital-first world.

REFERENCES

Agarwal, S., Malik, P., & Gautam, S. (2023). Analysis of Customer Satisfaction and the Customer Experience in Digital Payments: A Meta-Analysis Review. *International Journal of Business Science and Applied Management, 18*(1), 1–17. DOI: 10.69864/ijbsam.18-1.168

Aggarwal, S., Giera, B., Jeong, D., Robinson, J., & Spearot, A. (2024). Market Access, Trade Costs, and Technology Adoption: Evidence from Northern Tanzania. *The Review of Economics and Statistics, 106*(6), 1511–1528. DOI: 10.1162/rest_a_01263

Ajayi, A. J., Joseph, S., Metibemu, O. C., Olutimehin, A. T., Balogun, A. Y., & Olaniyi, O. O. (2025). The Impact of Artificial Intelligence on Cyber Security in Digital Currency Transactions. *Asian Journal of Current Research and Innovation, 25*(2), 45–60. DOI: 10.9734/acri/2025/v25i2300

Alonge, E. O., Eyo-Udo, N. L., Ubanadu, B. C., Daraojimba, A. I., Balogun, E. D., & Ogunsola, K. O. (2021). Enhancing Data Security with Machine Learning: A Study on Fraud Detection Algorithms. *International Journal of Fraud Management and Research, 2*(1), 19–31. DOI: 10.54660/.IJFMR.2021.2.1.19-31

Bouncken, R. B., Kraus, S., & Roig-Tierno, N. (2021). Knowledge- and Innovation-Based Business Models for Future Growth: Digitalized Business Models and Portfolio Considerations. *Review of Managerial Science, 15*(1), 1–14. DOI: 10.1007/s11846-019-00366-z

Brown, D., Batra, G., Zafar, H., & Saeed, K. (2024). Reducing Fraud in Organizations Through Information Security Policy Compliance: An Information Security Controls Perspective. *Computers & Security, 131*, 103958. Advance online publication. DOI: 10.1016/j.cose.2024.103958

Chauhan, S., Akhtar, A., & Gupta, A. (2022). Customer Experience in Digital Banking: A Review and Future Research Directions. *International Journal of Quality and Service Sciences, 14*(2), 311–348. DOI: 10.1108/IJQSS-02-2021-0027

Chen, X., Miraz, M. H., Gazi, M A I., Rahaman, M. A., Habib, M. M., & Hossain, A. I. (2022). Factors Affecting Cryptocurrency Adoption in Digital Business Transactions: The Mediating Role of Customer Satisfaction. *Technology in Society, 70*, 102059. Advance online publication. DOI: 10.1016/j.techsoc.2022.102059

Clauss, T., Breier, M., Kraus, S., Durst, S., & Mahto, R. V. (2022). Temporary Business Model Innovation–SMEs' Innovation Response to the Covid-19 Crisis. *R & D Management, 52*(2), 294–312. DOI: 10.1111/radm.12498

Dadwal, S., Haq, A., Jamal, A., & Nawaz, I. (2021). Value of Data as a Currency and a Marketing Tool. In *Strategy, Leadership, and AI in the Cyber Ecosystem* (pp. 321–336). Elsevier., DOI: 10.1016/B978-0-12-821442-8.00017-3

de Boer, E., & Chin, X. Y. (2025). Designing the Best Currency for Your Loyalty Program. In *Loyalty Programs and the Currency Effect* (pp. 21–47). Springer., DOI: 10.1007/978-3-031-78849-9_2

de Boer, E., & Chin, X. Y. (2025). Measuring Financial Performance and Accounting for Loyalty Program Currencies. In *Loyalty Programs and the Currency Effect* (pp. 133–155). Springer., DOI: 10.1007/978-3-031-78849-9_7

de Boer, E., & Chin, X. Y. (2025). *Loyalty Programs and the Currency Effect: A Comprehensive Guide to Realizing the Power of Points*. Springer., DOI: 10.1007/978-3-031-78849-9

Erenstein, O., Jaleta, M., Sonder, K., Mottaleb, K., & Prasanna, B. M. (2022). Global Maize Production, Consumption and Trade: Trends and R&D Implications. *Food Security*, *14*(5), 1295–1319. DOI: 10.1007/s12571-022-01288-7

Grandhi, B., Patwa, N., & Saleem, K. (2021). Data-Driven Marketing for Growth and Profitability. *EuroMed Journal of Business*, *16*(4), 381–398. DOI: 10.1108/EMJB-09-2018-0054

Hai, T. N., Van, Q. N., & Thi Tuyet, M. N. (2021). Digital Transformation: Opportunities and Challenges for Leaders in the Emerging Countries in Response to COVID-19 Pandemic. *Emerging Science Journal*, *5*(Special Issue), 21–36. DOI: 10.28991/esj-2021-SPER-03

Hossain, Q., Yasmin, F., Biswas, T. R., & Asha, N. B. (2024). Data-Driven Business Strategies: A Comparative Analysis of Data Science Techniques in Decision-Making. *Scholars Journal of Economics, Business and Management*, *11*(9), 257–263. DOI: 10.36347/sjebm.2024.v11i09.002

Hubballi, R. B., Selvakumar, P., Seenivasan, R., Basava Aradhya, S. G., Dinesh, N., & Seelam, P. K. (2025). Overview of Current AI Technologies in Education. In Kyei-Blankson, L., & Ntuli, E. (Eds.), *Transformative AI Practices for Personalized Learning Strategies* (pp. 1–26). IGI Global Scientific Publishing., DOI: 10.4018/979-8-3693-8744-3.ch001

Kanbach, D. K., Heiduk, L., Blueher, G., Schreiter, M., & Lahmann, A. (2024). The GenAI Is Out of the Bottle: Generative Artificial Intelligence from a Business Model Innovation Perspective. *Review of Managerial Science*, *18*(4), 1189–1220. DOI: 10.1007/s11846-023-00696-z

Kehinde, A. O., & Naomi, O. A. (2024). Digital Payment Option Adoption and Customer Experience Management Among SMEs in the Retail Sector. *African Multidisciplinary Journal of Research*, *9*(1), 78–96. DOI: 10.71064/spu.amjr.9.1.300

Khan, J., & Belk, R. (2024). *Digital Currency and Consumption: The Meaning of Money in an Era of Digital Currency*. Routledge., DOI: 10.4324/9781003425021

Kotte, K. R., Attaluri, V., & Selvakumar, P. (2025). Revolutionising Retail Banking with AI and Virtual Technologies: AI-Driven Financial Training. In Nair, A. J., Manohar, S., Limbu, Y. B., & Huhmann, B. A. (Eds.), *Intersecting Natural Language Processing and FinTech Innovations in Service Marketing* (pp. 421–440). IGI Global Scientific Publishing., DOI: 10.4018/979-8-3693-9944-6.ch020

Maheswaran, K., Selvakumar, P., Shaikh, M., Shashank, G., Gupta, S., & Manjunath, T. C. (2025). The Evolution of Smart Urban Mobility With Emerging Technologies. In Ponnusamy, S., Assaf, M., Youssef, B., Jeon, G., & Kalyanaraman, S. (Eds.), *Leveraging VANETs and Blockchain Technology for Urban Mobility* (pp. 1–28). IGI Global Scientific Publishing., DOI: 10.4018/979-8-3373-0265-2.ch001

Menzli, L. J., Smirani, L. K., Boulahia, J. A., & Hadjouni, M. (2022). Investigation of Open Educational Resources Adoption in Higher Education Using Rogers' Diffusion of Innovation Theory. *Heliyon*, *8*(7), e09885. Advance online publication. DOI: 10.1016/j.heliyon.2022.e09885 PMID: 35874064

Nalbant, K. G., & Aydın, S. (2023). Development and Transformation in Digital Marketing and Branding with Artificial Intelligence and Digital Technologies Dynamics in the Metaverse Universe. *Journal of Metaverse*, *3*(1), 9–18. DOI: 10.57019/jmv.1148015

Ogunmokun, A. S., Balogun, E. D., & Ogunsola, K. O. (2022). A Strategic Fraud Risk Mitigation Framework for Corporate Finance Cost Optimization and Loss Prevention. *International Journal of Multidisciplinary Research and Growth Evaluation*, *3*(1), 783–790. DOI: 10.54660/.IJMRGE.2022.3.1.783-790

Osei, L. K., Cherkasova, Y., & Oware, K. M. (2023). Unlocking the Full Potential of Digital Transformation in Banking: A Bibliometric Review and Emerging Trend. *Future Business Journal*, *9*(1), 30. Advance online publication. DOI: 10.1186/s43093-023-00207-2

Rosário, A. T., & Dias, J. C. (2023). How Has Data-Driven Marketing Evolved: Challenges and Opportunities with Emerging Technologies. *International Journal of Information Management Data Insights*, *3*(2), 100203. Advance online publication. DOI: 10.1016/j.jjimei.2023.100203

Sakas, D. P., Giannakopoulos, N. T., Reklitis, D. P., & Dasaklis, T. K. (2021). The Effects of Cryptocurrency Trading Websites on Airlines' Advertisement Campaigns. *Journal of Theoretical and Applied Electronic Commerce Research, 16*(7), 3099–3119. DOI: 10.3390/jtaer16070169

Sakthivel, M., Jaganathan, A., & Mohanraj, M. Mohit, & Manjunath T. (2025). "Brand Loyalty through AI and Personalisation: AI-Generated Content for Financial Marketing." In A. J Nair, S. Manohar, Y. B. Limbu, & B. A. Huhmann (Eds.), Intersecting Natural Language Processing and FinTech Innovations in Service Marketing (pp. 77-96). IGI Global Scientific Publishing. DOI: DOI: 10.4018/979-8-3693-9944-6.ch006

Samal, A., & Misra, L. I. Mohit, Pougajendy S, Chandel P S, Manjunath T C. (2025). "Musealization and Tourism Practices." In S. Oflazoğlu & M. Dora (Eds.), Preserving Cultural Heritage in Post-Disaster Urban Renewal (pp. 115-142). IGI Global Scientific Publishing. DOI: DOI: 10.4018/979-8-3693-9097-9.ch006

Shree, S., Pratap, B., Saroy, R., & Dhal, S. (2021). Digital Payments and Consumer Experience in India: A Survey Based Empirical Study. *Journal of Banking and Financial Technology*, *5*(1), 1–20. DOI: 10.1007/s42786-020-00024-z

Smith, S. G., & Sinkford, J. C. (2022). Gender Equality in the 21st Century: Overcoming Barriers to Women's Leadership in Global Health. *Journal of Dental Education*, *86*(9), 1144–1173. DOI: 10.1002/jdd.13059 PMID: 36165260

Soundariya, A., Behera, N. K., Selvakumar, P., Chandel, P. S., Gupta, S., & Manjunath, T. C. (2025). AI and Environmental Ethics: The Role of AI in Addressing Climate Change—Ethical Considerations in AI-Driven Environmental Solutions. In Chakraborty, S. (Ed.), *Human Values, Ethics, and Dignity in the Age of Artificial Intelligence* (pp. 321–346). IGI Global Scientific Publishing., DOI: 10.4018/979-8-3693-9035-1.ch013

Taherdoost, H. (2021). A Review on Risk Management in Information Systems: Risk Policy, Control and Fraud Detection. *Electronics (Basel), 10*(24), 3065. Advance online publication. DOI: 10.3390/electronics10243065

Thirunagalingam, A., & Selvakumar, P. (2025). Application of Automation and Robotics in the Oil and Gas Industry. In Abdullayev, V., & Khang, A. (Eds.), *Revolutionizing AI and Robotics in the Oil and Gas Industry* (pp. 229–244). IGI Global Scientific Publishing., DOI: 10.4018/979-8-3693-8156-4.ch011

Trischler, M F G., & Li-Ying, J. (2023). Digital Business Model Innovation: Toward Construct Clarity and Future Research Directions. *Review of Managerial Science*, *17*(1), 3–32. DOI: 10.1007/s11846-021-00508-2

Upadhyaya, A. N., Debnath, P., Rani, P., Manjunath, T. C., & Bhattacharya, S. (2025). AI-Based Disaster Prediction and Early Warning Systems. In Mobo, F. (Ed.), *AI and Emerging Technologies for Emergency Response and Smart Cities* (pp. 55–74). IGI Global Scientific Publishing., DOI: 10.4018/979-8-3693-9770-1.ch003

Zhang, A., Alvi, M. F., Gong, Y., & Wang, J. X. (2022). Overcoming Barriers to Supply Chain Decarbonization: Case Studies of First Movers. *Resources, Conservation and Recycling*, *186*, 106536. Advance online publication. DOI: 10.1016/j.resconrec.2022.106536

Compilation of References

Aarts, H., & Dijksterhuis, A. (2000). Habits as knowledge structures : Automaticity in goal-directed behavior. *Journal of Personality and Social Psychology*, *78*(1), 53–63. DOI: 10.1037/0022-3514.78.1.53 PMID: 10653505

Abikari, M., Öhman, P., & Yazdanfar, D. (2023). Negative emotions and consumer behavioral intention to adopt emerging e-banking technology. *Journal of Financial Services Marketing*, *28*(4), 691–704. DOI: 10.1057/s41264-022-00172-x

Afroogh, S., Emsley, I., & Floridi, L. (2024). Trust in AI: Progress, challenges, and future directions. *Humanities & Social Sciences Communications*, *11*(1), 1568. Advance online publication. DOI: 10.1057/s41599-024-04044-8

Agag, G., Brown, A., Hassanein, A., & Shaalan, A. (2020). Decoding travellers' willingness to pay more for green travel products : Closing the intention–behaviour gap. *Journal of Sustainable Tourism*, *28*(10), 1551–1575. DOI: 10.1080/09669582.2020.1745215

Agarwal, J., Sharma, S., & Agarwal, P. (2025). S2E3: Sustainable, Secure, Ethical, e-Branding and e-Advertising System Using Metaverse Techniques and Hybrid Cryptographic Approach. In 2025 International Conference on Intelligent Control, Computing and Communications (IC3), 10–15. DOI: 10.1109/IC363308.2025.10956649

Agarwal, M., & Khatri, M. (2024). From Cash-Centric to Cashless Economy-A Study on User′ s Awareness towards Digital Payment Systems. *Academy of Marketing Studies Journal*, *28*(1).

Agarwal, S., Malik, P., & Gautam, S. (2023). Analysis of customer satisfaction and the customer experience in digital payments: A meta-analysis review. *International Journal of Business Science and Applied Management*, *18*(1), 1–17. DOI: 10.69864/ijbsam.18-1.168

Aggarwal, S., Giera, B., Jeong, D., Robinson, J., & Spearot, A. (2024). Market Access, Trade Costs, and Technology Adoption: Evidence from Northern Tanzania. *The Review of Economics and Statistics*, *106*(6), 1511–1528. DOI: 10.1162/rest_a_01263

Ahmad, M. B., Ali, H. F., Malik, M. S., Humayun, A. A., & Ahmad, S. (2019). Factors affecting impulsive buying behavior with mediating role of positive mood: An empirical study. *European Online Journal of Natural and Social Sciences*, *8*(1), 17–27.

Ahmed, I., Al Masud, A., Hossain, M. A., Hossain, M. A., & Kim, M. (2025). Understanding customer switching behavior: An empirical examination of mobile financial services. *Social Sciences & Humanities Open*, *11*, 101270. DOI: 10.1016/j.ssaho.2024.101270

Aiolfi, S., Bellini, S., & Pellegrini, D. (2021). Data-driven digital advertising: Benefits and risks of online behavioral advertising. *International Journal of Retail & Distribution Management*, *49*(7), 1089–1110.

Aivazpour, Z., Valecha, R., & Chakraborty, R. (2022, November 16). Data breaches: An empirical study of the effect of monitoring services. *The Data Base for Advances in Information Systems*, *53*(4), 65–82. DOI: 10.1145/3571823.3571829

Ajayi, A. J., Joseph, S., Metibemu, O. C., Olutimehin, A. T., Balogun, A. Y., & Olaniyi, O. O. (2025). The Impact of Artificial Intelligence on Cyber Security in Digital Currency Transactions. *Asian Journal of Current Research and Innovation*, *25*(2), 45–60. DOI: 10.9734/acri/2025/v25i2300

Ajzen, I. (1991). The theory of planned behavior. *Organizational Behavior and Human Decision Processes*, *50*(2), 179–211. DOI: 10.1016/0749-5978(91)90020-T

Aker, J. C., & Mbiti, I. M. (2010). Mobile phones and economic development in Africa. *The Journal of Economic Perspectives*, *24*(3), 207–232. DOI: 10.1257/jep.24.3.207

Akinyemi, B. E., & Mushunje, A. (2020). Determinants of mobile money technology adoption in rural areas of Africa. *Cogent Social Sciences*, *6*(1), 1815963. DOI: 10.1080/23311886.2020.1815963

Akpati, C., & Mathew, A. (2023). A multimodal critical discourse study of persuasive strategies and identity construction in admission adverts of select private university websites. In *Transformation of Higher Education Through Institutional Online Spaces*. https://doi.org/DOI: 10.4018/978-1-6684-8122-6.ch013

Alabi, O. F., & David, A. A. (2022, November 23). Framework for detection of fraud at Point of Sale on electronic commerce sites using logistic regression. *EAI Endorsed Transactions on Scalable Information Systems*, *10*(2), e6. DOI: 10.4108/eetsis.v10i2.1596

Alam, N., Gupta, L., & Zameni, A. (2019). *Fintech and Islamic finance. sl.* Springer International Publishing. DOI: 10.1007/978-3-030-24666-2

Alamoudi, H., Glavee-Geo, R., Alharthi, M., Doszhan, R., & Suyunchaliyeva, M. M. (2025). Exploring trust and outcome expectancy in FinTech digital payments: Insights from the stimulus-organism-response model. *International Journal of Bank Marketing*, *43*(4), 897–919. DOI: 10.1108/IJBM-04-2024-0252

Alam, S. S., Ahsan, M. N., Kokash, H. A., & Ahmed, S. (2025). Live streaming characteristics, motivational factors, trust and impulsive hotel booking intention: Mediating role of customer engagement and moderating role of time pressure. *Journal of Hospitality Marketing & Management*, *34*(5), 651–680. DOI: 10.1080/19368623.2025.2472071

Alamsyah, A., Kusuma, G. N. W., & Ramadhani, D. P. (2024, February 26). A review on decentralized finance ecosystems. *Future Internet*, *16*(3), 76. DOI: 10.3390/fi16030076

Alba, J., Lynch, J., Weitz, B., Janiszewski, C., Lutz, R., Sawyer, A., & Wood, S. (1997). Interactive home shopping: Consumer, retailer, and manufacturer incentives to participate in electronic marketplaces. *Journal of Marketing*, *61*(3), 38–53. DOI: 10.1177/002224299706100303

Aldboush, H. H. H.. (2023). Building trust in Fintech: An analysis of ethical and privacy considerations. *Journal of Risk and Financial Management*, *11*(3).

Alexander, S., & Sarrafzadeh, A. (2004). Interfaces that adapt like humans. In *Asia-Pacific Conference on Computer Human Interaction* (pp. 641–645). Berlin, Heidelberg: Springer Berlin Heidelberg.

Allen, R. G. (2006). *Multiple Streams of Internet Income: How Ordinary People Make Extraordinary Money Online*. John Wiley & Sons.

Alnaim, A. F. (2022, July-December). Effects of individual (perceived identity theft, cognitive trust, and attitude) and situational (website quality, perceived reputation, social presence) factors on online purchase intention: Moderating role of cyber security. *International Journal of Cyber Criminology*, *16*(2), 131–148.

Alonge, E. O., Eyo-Udo, N. L., Ubanadu, B. C., Daraojimba, A. I., Balogun, E. D., & Ogunsola, K. O. (2021). Enhancing Data Security with Machine Learning: A Study on Fraud Detection Algorithms. *International Journal of Fraud Management and Research, 2*(1), 19–31. DOI: 10.54660/.IJFMR.2021.2.1.19-31

Alqarni, A. (2024). A blockchain-based solution for transparent intellectual property rights management: Smart contracts as enablers. *Kybernetes*. Advance online publication. DOI: 10.1108/K-04-2024-1074

Al-Qudah, A. A., Al-Okaily, M., Shiyyab, F. S., Taha, A. A., Almajali, D. A., Masa'deh, R. E., & Warrad, L. H. (2024). Determinants of digital payment adoption among Generation Z: An empirical study. *Journal of Risk and Financial Management, 17*(11), 521. DOI: 10.3390/jrfm17110521

Alshater, M. M., Nasrallah, N., Khoury, R., & Joshipura, M. (2024). Deciphering the world of NFTs: A scholarly review of trends, challenges, and opportunities. *Electronic Commerce Research*. Advance online publication. DOI: 10.1007/s10660-024-09900-y

Amudharasan, A. (2023). The Impact of Recommendation Algorithms: Analyzing the Influence of Data on Marketing Strategies in the Media Sector. *Open Journal of Business and Management, 11*(6), 3373–3384. DOI: 10.4236/ojbm.2023.116184

Andaregie, A., Abebe, G. K., Gupta, P., Worku, G., Matsumoto, H., Astatkie, T., & Takagi, I. (2024). Exploring individuals' socioeconomic characteristics and digital infrastructure determinants of digital payment adoption in Ethiopia. *Digital Business, 4*(2), 100092. DOI: 10.1016/j.digbus.2024.100092

An, J., Yuen, C., Dai, L., Di Renzo, M., Debbah, M., & Hanzo, L. (2024, June). Near-Field Communications: Research advances, potential, and challenges. *IEEE Wireless Communications, 31*(3), 100–107. DOI: 10.1109/MWC.004.2300450

Ante, L., & Fiedler, I. (2025). The new digital economy: How decentralized finance (DeFi) and non-fungible tokens (NFTs) are transforming value creation, ownership models, and economic systems. *Digital Business, 5*(1), 100094. DOI: 10.1016/j.digbus.2024.100094

Antonio, B. O., Juan, L. R., Ana, I. D., & Francisco, L. C. (2024). Examining user behavior with machine learning for effective mobile peer-to-peer payment adoption. *Financial Innovation, 10*(1), 94. DOI: 10.1186/s40854-024-00625-3

Arango, C., Huynh, K. P., & Sabetti, L. (2015). Consumer payment choice : Merchant card acceptance versus pricing incentives. *Journal of Banking & Finance, 55*, 130–141. DOI: 10.1016/j.jbankfin.2015.02.005

Ardity, A. (2019, October 24). *Credit card machine, debit card machine*. [Image]. Pixabay. https://pixabay.com/photos/credit-card-machine-4577767 and https://pixabay.com/photos/credit-card-machine-4577768

Arianita, A., Anggarawati, S., & Alfansi, I. (2023). Analysis Factor Affecting The Use Of Digital Payment With The Extended Utaut Model. *Management Review*, *5*(1), 91–108.

Arner, D. W., Auer, R., & Frost, J. (2020). *Stablecoins: Risks, potential and regulation* (BIS Working Papers No. 905). Bank for International Settlements. https://www.bis.org/publ/work905.htm

Assima, C., Herbert, M., & Robert, I. (2023). Use it or lose it?: Exploring the grey area of dormant possessions and the role of rituals in value dynamics of household objects. *Journal of Marketing Management*, *39*(13–14), 1220–1254. DOI: 10.1080/0267257X.2023.2252455

Attanasio, O. P., Guiso, L., & Jappelli, T. (2002). The Demand for Money, Financial Innovation, and the Welfare Cost of Inflation : An Analysis with Household Data. *Journal of Political Economy*, *110*(2), 317–351. DOI: 10.1086/338743

Aurazo, J., & Vega, M. (2021). Why people use digital payments : Evidence from micro data in Peru. *Latin American Journal of Central Banking*, *2*(4), 100044. DOI: 10.1016/j.latcb.2021.100044

Bach, T. A., Khan, A., Hallock, H., Beltrão, G., & Sousa, S. (2023). A systematic literature review of user trust in AI-enabled systems: An HCI perspective. *arXiv*. https://doi.org//arXiv.2304.08795DOI: 10.48550

Badgujar, P. (2023, October 10). Securing customer data and best practices for retail Point-of-Sale systems. *Journal of Technology & Innovation*, *4*(4).

Bagnall, J., Bounie, D., Huynh, K. P., Kosse, A., Schmidt, T., & Schuh, S. (2018). Consumer cash usage : A cross-country comparison with payment diary survey data. *46th issue (December 2016) of the International Journal of Central Banking*. https://www.ijcb.org/journal/ijcb16q4a1.pdf

Bahri, M. I., Dermawan, A. A., & Tambunan, A. A. (2025). Public-Private Collaborations in Smart Cities: Startup-Driven Innovations in E-Government. In *Startup-Driven E-Government: Digital Innovation for Sustainable Ecosystems* (pp. 61–86). IGI Global Scientific Publishing.

Bahri, S., Suroso, A. I., Suhendi, , & Sari, L. K. (2025). Examining Cash Usage Behavior in Metropolitan Greater Jakarta Societies. *Societies (Basel, Switzerland)*, *15*(5), 120. DOI: 10.3390/soc15050120

Baidoun, S. D., & Salem, M. Z. (2024). The moderating role of perceived trust and perceived value on online shopping behavioral intention of Palestinian millennials during COVID-19. *Competitiveness Review*, *34*(1), 125–143. DOI: 10.1108/CR-10-2022-0161

Bajunaied, K., Hussin, N., & Kamarudin, S. (2023). Behavioral intention to adopt FinTech services: An extension of unified theory of acceptance and use of technology. *Journal of Open Innovation*, *9*(1), 100010. DOI: 10.1016/j.joitmc.2023.100010

Bajwa, F. A., Fu, J., Bajwa, I. A., Ahmad, S., & Mahmood, F. (2025). Factors influencing usage and loyalty for payment app customers in Saudi Arabia. *Acta Psychologica*, *255*, 104961. DOI: 10.1016/j.actpsy.2025.104961

Bakhshi, P., Agrawal, R., Mendon, S., Frank, D., Spulbar, C., Birau, R., & Filip, R. D. (2024). Barriers in adoption of FinTech by street vendors and hawkers in India using interpretive structural modeling. *Business: Theory and Practice*, *25*(1), 231–240. DOI: 10.3846/btp.2024.19208

Balasubramaniam, S. (2023). Trust and privacy in AI-enabled financial services: A user-centric perspective. *Journal of Digital Banking*, *8*(2), 112–128.

Bandura, R., & Ramanujam, S. R. (2021). Developing inclusive digital payment systems. *Center for Strategic and International Studies, 21.*

Banerjee, A. K., & Pradhan, H. K. (2024). Influence of demographic profiles in adoption of digital payment system in India : A multigroup invariance analysis. *Technology Analysis and Strategic Management*, *36*(10), 2285–2301. DOI: 10.1080/09537325.2022.2132928

Bank Al-Maghrib. (2023). *Rapport annuel sur les infrastructures des marchés financiers et les moyens de paiement, leur surveillance et l'inclusion financière—Exercice 2023.*

Bank Al-Maghrib. (2024). *Rapport annuel sur les infrastructures des marchés financiers et les moyens de paiement et leur surveillance—Exercice 2024.*

Barkas, P. (2024). Consumer Protection and Financial Innovation: Microeconomic, Policy, and Behavioral Considerations for the Digital Era. In *Finance, Growth and Democracy: Connections and Challenges in Europe and Latin America in the Era of Permacrisis: Democracy, Finance, and Growth* (pp. 263–285). Springer Nature Switzerland.

Barney, N. (2024, May 2). *What is PCI DSS (Payment Card Industry Data Security Standard)?* TechTarget Search Security. https://www.techtarget.com/searchsecurity/definition/PCI-DSS-Payment-Card-Industry-Data-Security-Standard

Bartels, J., & Reinders, M. J. (2011). Consumer innovativeness and its correlates : A propositional inventory for future research. *Journal of Business Research, 64*(6), 601–609. DOI: 10.1016/j.jbusres.2010.05.002

BasuMallick. C. (2022, September 29). *What is NFC (Near Field Communication)? Definition, work, and examples.* Spiceworks. https://www.spiceworks.com/tech/networking/articles/what-is-near-field-communication

Beaird, J., George, J., & Walker, A. (2020). *The principles of beautiful web design* (4th ed.). SitePoint., https://research.ebsco.com/c/i5sifr/ebook-viewer/epub/xh4ffz7xlv

Belch, G. E., & Belch, M. A. (2021). *Advertising and promotion: An integrated marketing communications perspective* (6th ed.). McGraw-Hill.

Belk, R., Humayun, M., & Brouard, M. (2022). Money, possessions, and ownership in the Metaverse: NFTs, cryptocurrencies, Web3 and Wild Markets. *Journal of Business Research, 153*, 198–205. DOI: 10.1016/j.jbusres.2022.08.031

Bell, L., Vogt, J., Willemse, C., Routledge, T., Butler, L. T., & Sakaki, M. (2018). Beyond self-report: A review of physiological and neuroscientific methods to investigate consumer behavior. *Frontiers in Psychology, 9*, 1655. DOI: 10.3389/fpsyg.2018.01655 PMID: 30245657

Ben David, D., Resheff, Y. S., & Tron, T. (2021). Explainable AI and adoption of financial algorithmic advisors: an experimental study. In *Proceedings of the 2021 AAAI/ACM Conference on AI, Ethics, and Society* (pp. 390–400). DOI: 10.1145/3461702.3462565

Berman, B. (2016). Referral marketing: Harnessing the power of your customers. *Business Horizons, 59*(1), 19–28. Advance online publication. DOI: 10.1016/j.bushor.2015.08.001

Bharadwaj, A., El Sawy, O. A., Pavlou, P. A., & Venkatraman, N. V. (2013). Digital business strategy: Toward a next generation of insights. *Management Information Systems Quarterly, 37*(2), 471–482. DOI: 10.25300/MISQ/2013/37:2.3

Bhaskar, N., Bland, M., Levchenko, K., & Schulman, A. (2019, August). Please pay inside: Evaluating Bluetooth-based detection of gas pump skimmers. *Proceedings of the 28th USENIX Security Symposium* (pp. 373-388), Santa Clara, CA. https://www.usenix.org/system/files/sec19-Bhaskar et al., 2019.pdf

Bhavsar, V., & Samanta, P. K. (2022, January). Sustainability of digital payments: Empirical evidence from India. In *2nd International Conference on Sustainability and Equity (ICSE-2021)*, 70-78. DOI: 10.2991/ahsseh.k.220105.009

Bilgihan, A., Leong, A. M. W., Okumus, F., & Bai, J. (2024). Proposing a metaverse engagement model for brand development. *Journal of Retailing and Consumer Services*, *78*, 103781. DOI: 10.1016/j.jretconser.2024.103781

Blanco-Encomienda, F. J., & Rosillo-Díaz, E. (2025). Exploring brand attitude in e-commerce using the SOR model: The role of information quantity, information quality and source credibility. *Journal of Product and Brand Management*, *34*(7), 1041–1055. DOI: 10.1108/JPBM-04-2024-5095

Bleier, A., & Eisenbeiss, M. (2015a). Personalized online advertising effectiveness: The interplay of what, when, and where. *Marketing Science*, *34*(5), 669–688. DOI: 10.1287/mksc.2015.0930

Bleier, A., & Eisenbeiss, M. (2015b). The importance of trust for personalized online advertising. *Journal of Retailing*, *91*(3), 390–409. DOI: 10.1016/j.jretai.2015.04.001

Board of Governors of the Federal Reserve System. (1980, June). *Alice in debitland.* https://files.eric.ed.gov/fulltext/ED235069.pdf

Boateng, H., Kosiba, J. P. B., & Okoe, A. F. (2019). Determinants of consumers' participation in the sharing economy: A social exchange perspective within an emerging economy context. *International Journal of Contemporary Hospitality Management*, *31*(2), 718–733. DOI: 10.1108/IJCHM-11-2017-0731

Boeschoten, W. C. (1998). Cash Management, Payment Patterns and the Demand for Money. *De Economist*, *146*(1), 117–142. DOI: 10.1023/A:1003258026314

Bolonin, A., & Balykin, V. (2021). Payment Systems in Digital Economy. In *Financial Markets Evolution: From the Classical Model to the Ecosystem. Challengers, Risks and New Features* (pp. 97–108). Springer International Publishing. DOI: 10.1007/978-3-030-71337-9_7

Boni, E. K. (2024). *Exploring perceived risk and digital ethics affecting consumer experiences and decision-making in smart retailing* (Doctoral dissertation, Brunel University London).

Bouncken, R. B., Kraus, S., & Roig-Tierno, N. (2021). Knowledge- and Innovation-Based Business Models for Future Growth: Digitalized Business Models and Portfolio Considerations. *Review of Managerial Science*, *15*(1), 1–14. DOI: 10.1007/s11846-019-00366-z

Bounie, D., & Francois, A. (2006). *Cash, Check or Bank Card? The Effects of Transaction Characteristics on the Use of Payment Instruments* (SSRN Scholarly Paper No. 891791). Social Science Research Network. DOI: 10.2139/ssrn.891791

Boutang, J., & De Lara, M. (2016, September 2). Risk marketing. *Journal of CENTRUM Cathedra (JCC) the Business and Economics Research Journal, 9*(1), 27–51. https://doi.org/DOI: 10.1108/JCC-08-2016-0008

Bowles, S., & Gintis, H. (1988). Contested exchange: Political economy and modern economic theory. *The American Economic Review, 78*(2), 145–150. https://www.jstor.org/stable/1818113

Brakus, J. J., Schmitt, B. H., & Zarantonello, L. (2009). Brand experience: What is it? How is it measured? Does it affect loyalty? *Journal of Marketing*, *73*(3), 52–68. DOI: 10.1509/jmkg.73.3.052

Braun, V., & Clarke, V. (2006). Using thematic analysis in psychology. *Qualitative Research in Psychology*, *3*(2), 77–101. DOI: 10.1191/1478088706qp063oa

Brosche, K., & Kumar, A. (2015). Realtime data accelerates online marketing. In *Programmatic Advertising: The Successful Transformation to Automated, Data-Driven Marketing in Real-Time* (pp. 221–232). Springer International Publishing. DOIhttps., DOI: 10.1007/978-3-319-25023-6_18

Brown, D., Batra, G., Zafar, H., & Saeed, K. (2024). Reducing Fraud in Organizations Through Information Security Policy Compliance: An Information Security Controls Perspective. *Computers & Security*, *131*, 103958. Advance online publication. DOI: 10.1016/j.cose.2024.103958

Budak, J., Lugović, Š., Penezić, Z., Rajh, E., Slijepčević, S., & Škrinjarić, B. (2023). *Consumer resilience to privacy violation online* (Budak, J., Ed.). The Institute of Economics., http://www.eizg.hr

Busalim, A., Hollebeek, L. D., & Lynn, T. (2023). The effect of social commerce attributes on customer engagement: An empirical investigation. *Internet Research*, *34*(7), 187–214. Advance online publication. DOI: 10.1108/INTR-03-2022-0165

Cahyani, I. R., & Bahri, M. I. (2022). Change Management and Special Autonomy Paradigm: Opportunities and Challenges of Human Resources in Digital Literacy Implementation in Indonesian Regional Libraries. *Congress of Southeast Asian Librarians (CONSAL) XVIII.*

Cai, J., Liu, Z., & Huang, H. (2023). Live commerce as a new form of online shopping: Review and future research agenda. *Journal of Retailing and Consumer Services*, *70*, 103114. DOI: 10.1016/j.jretconser.2022.103114

Campos, Á. (2024, October 17). *Are POS terminals on the way out?* Valor International. https://valorinternational.globo.com/business/news/2024/10/17/are-pos-terminals-on-their-way-out.ghtml

Candan, B., Ünal, S., & Erciş, A. (2013). Analysing the relationship between consumption values and brand loyalty of young people: A study on personal care products. *Management*, *29*(46), 163–168.

Cao, Z., & Shi, X. (2021). A systematic literature review of entrepreneurial ecosystems in advanced and emerging economies. *Small Business Economics*, *57*(1), 75–110. DOI: 10.1007/s11187-020-00326-y

Capgemini. (2025). *World payments report 2025: Velocity, meet value—Lead in the open and instant future of payments*. Capgemini. https://www.capgemini.com/insights/research-library/world-payments-report-2025/

Carow, K. A., & Staten, M. E. (1999). Debit, credit, or cash : Survey evidence on gasoline purchases. *Journal of Economics and Business*, *51*(5), 409–421. DOI: 10.1016/S0148-6195(99)00016-8

Catalini, C., & Gans, J. S. (2020). Some simple economics of the blockchain. *Communications of the ACM*, *63*(7), 80–90. DOI: 10.1145/3359552

Çelik, A. A. (2025, May 8). Empowering visually impaired consumers in retail: A systematic review of accessible payment solutions for all. *Ege Academic Review*, *25*(2), 437–454. DOI: 10.21121/eab.20250212

Chaiyasoonthorn, W., & Suksa-Ngiam, W. (2019). The diffusion and adoption of electronic payment systems in Bangkok. [IJEBR]. *International Journal of E-Business Research*, *15*(2), 102–115. DOI: 10.4018/IJEBR.2019040106

Chakraborty, P. (n.d.). *The Importance of Ethical Practices in Affiliate Marketing: Building Trust and Credibility*.

Chakraborty, T., Sharada, V. S., & Sai, S. D. D. (2025). Transactions to Relationship: Customers Engagement in the Digital Age. In *Managing Customer-Centric Strategies in the Digital Landscape* (pp. 25–50). IGI Global.

Chand, A., Sharma, S., Agarwal, P., & Mohan, S. (2024). Credit card fraud detection using deep learning for Internet-of-Things enabled smart digital financial system. In Internet of things (pp. 469–485). DOI: 10.1007/978-981-97-0052-3_23

Chandak, D. (2024). Prospects of Unified Payments Interface (UPI) Systems on Business & Digital Payments Across India-A Review Analysis. *Educational Administration: Theory and Practice*, *30*(1), 3197–3206.

Changani, S., & Kumar, R. (2023). How social media marketing enhances brand community engagement: developing an integrated model using the SOR paradigm. In *International Working Conference on Transfer and Diffusion of IT* (pp. 302–318). Cham: Springer Nature Switzerland.

Chauhan, K., Singh, S., & Aggarwal, P. (2025). Reconceptualizing digital payments: The enhanced role of AI in transforming financial transactions ecosystem. *Lloyd Business Review,* 210–219.

Chauhan, K., & Gulati, K. (2025). Global Customer Engagement Through Gamified Mobile Apps and Virtual Experiences. In *Addressing Practical Problems Through the Metaverse and Game-Inspired Mechanics* (pp. 169–196). IGI Global., DOI: 10.4018/979-8-3693-8886-0.ch008

Chauhan, S., Akhtar, A., & Gupta, A. (2022). Customer Experience in Digital Banking: A Review and Future Research Directions. *International Journal of Quality and Service Sciences, 14*(2), 311–348. DOI: 10.1108/IJQSS-02-2021-0027

Chavee, N., & Vongurai, R. (2023). Factors Impacting Trust, Satisfaction, and Purchase Intention Via Social Live Stream Commerce with Thai Influencers. *Journal of Multidisciplinary in Social Sciences, 19*(2).

Chaveesuk, S., Khalid, B., & Chaiyasoonthorn, W. (2021). Digital payment system innovations: A marketing perspective on intention and actual use in the retail sector. *Innovative Marketing, 17*(3), 109–123. DOI: 10.21511/im.17(3).2021.09

Chawla, D., & Joshi, H. (2019). Consumer attitude and intention to adopt mobile wallet in India: An empirical study. *International Journal of Bank Marketing, 37*(7), 1590–1618. DOI: 10.1108/IJBM-09-2018-0256

Chen, Z. (2004). Electronic payment systems: General review and comparative analysis (Doctoral dissertation, Texas Tech University). https://ttu-ir.tdl.org/server/api/core/bitstreams/74ede8b1-e829-44b0-9bea-b8f31b5e30f0/content

Chen, H., Yang, Z., & Lyu, T. (2025). Empirical Investigation of Digital Collectibles Purchase Intention: The Roles of Value, Risks, Identification, and Scarcity. *International Journal of Human-Computer Interaction, 41*(16), 9861–9880. DOI: 10.1080/10447318.2024.2429755

Chen, X., & Lin, Z. (2022). Exploring consumer purchase intention in live streaming commerce: The roles of social presence, perceived value, and entertainment. *Electronic Commerce Research, 22*(3), 647–672. DOI: 10.1007/s10660-021-09489-x

Chen, X., Miraz, M. H., Gazi, M A I., Rahaman, M. A., Habib, M. M., & Hossain, A. I. (2022). Factors Affecting Cryptocurrency Adoption in Digital Business Transactions: The Mediating Role of Customer Satisfaction. *Technology in Society*, *70*, 102059. Advance online publication. DOI: 10.1016/j.techsoc.2022.102059

Chen, Y. (2018). Blockchain tokens and the potential democratization of entrepreneurship and innovation. *Business Horizons*, *61*(4), 567–575. DOI: 10.1016/j.bushor.2018.03.006

Chen, Y., Mislove, A., & Wilson, C. (2016). An empirical analysis of algorithmic pricing on Amazon Marketplace. In *Proceedings of the 25th International Conference on World Wide Web* (pp. 1339–1349). ACM. DOI: 10.1145/2872427.2883089

Cheung, C. M., & Thadani, D. R. (2012). The impact of electronic word-of-mouth communication: A literature analysis and integrative model. *Decision Support Systems*, *54*(1), 461–470. DOI: 10.1016/j.dss.2012.06.008

Choi, Y. (2019). The study of antecedents of consumer engagement and purchase intention in social commerce. *International Journal of Cyber Behavior, Psychology and Learning*, *9*(2), 51–63. Advance online publication. DOI: 10.4018/IJCBPL.2019040104

Chowdhury, A. R., Paul, R., & Rozony, F. Z. (2025). A systematic review of demand forecasting models for retail e-commerce enhancing accuracy in inventory and delivery planning. *International Journal of Scientific Interdisciplinary Research*, *6*(1), 01-27.

Cialdini, R. B. (2007). *Influence: The Psychology of Persuasion* (Rev. ed.). Harper Business.

Clauss, T., Breier, M., Kraus, S., Durst, S., & Mahto, R. V. (2022). Temporary Business Model Innovation–SMEs' Innovation Response to the Covid-19 Crisis. *R & D Management*, *52*(2), 294–312. DOI: 10.1111/radm.12498

Cloarec, J., Meyer-Waarden, L., & Munzel, A. (2024). Transformative privacy calculus: Conceptualizing the personalization–privacy paradox on social media. *Psychology and Marketing*, *41*(7), 1574–1596. Advance online publication. DOI: 10.1002/mar.21998

Coche, E., Kolk, A., & Ocelík, V. (2024). Unravelling cross-country regulatory intricacies of data governance: The relevance of legal insights for digitalization and international business. *Journal of International Business Policy*, *7*(1), 112–127. DOI: 10.1057/s42214-023-00172-1

Cokus, M. S., & Dahlgren, J. W. (2007, June). System evolution in the intelligent enterprise: An historical case study of VISA's Transaction Processing Systems. *INCOSE International Symposium, 17*(1), 776–787. DOI: 10.1002/j.2334-5837.2007.tb02911.x

Cole, S. A., Sampson, T., & Zia, B. (2010). Prices or knowledge? What drives demand for financial services in emerging markets? SSRN Electronic Journal. https://papers.ssrn.com/sol3/papers.cfm?abstract_id=1374078

Connolly, S., & Stavins, J. (2015). Payment Instrument Adoption and Use in the United States, 2009-2013, by Consumers' Demographic Characteristics. *Research Data Reports Paper, 15-6*. https://papers.ssrn.com/sol3/papers.cfm?abstract_id=2709920

Constantinides, E. (2006). The marketing mix revisited: Towards the 21st century marketing. *Journal of Marketing Management, 22*(3–4), 407–438. DOI: 10.1362/026725706776861190

Copeland, D. G., & McKenney, J. L. (1988). Airline reservations systems: Lessons from history. *Management Information Systems Quarterly, 12*(3), 353–370. DOI: 10.2307/249202

Corbet, S., Larkin, C., Lucey, B., Meegan, A., & Yarovaya, L. (2020). Cryptocurrency reaction to FOMC announcements: Evidence of heterogeneity based on blockchain stack position. *Journal of Financial Stability, 46*, 100706. DOI: 10.1016/j.jfs.2019.100706

Cortada, J. W. (2005). The digital hand: Volume II: How computers changed the work of American financial, telecommunications, media, and entertainment industries. Oxford University Press.

Crandall, R. (2010, January). *Cash register history: The incorruptible cashier*. Rick Crandall. https://www.rickcrandall.net/cash-register-history

Crawford, H., Anthony, L., & Orem, T. (2024, July 12). *What is a Point-of-Sale (POS) system and how does it work?* Nerd Wallet. https://www.nerdwallet.com/article/small-business/what-is-a-pos-system

Crede, A. (1995). Electronic commerce and the banking industry: The requirement and opportunities for new payment systems using the Internet. *Journal of Computer-Mediated Communication, 1*(3), JCMC135. DOI: 10.1111/j.1083-6101.1995.tb00171.x

Creswell, J. W., & Plano Clark, V. L. (2017). *Designing and conducting mixed methods research* (3rd ed.). SAGE Publications.

Dabbous, A., Aoun Barakat, K., & Merhej Sayegh, M. (2020). Social Commerce Success: Antecedents of Purchase Intention and the Mediating Role of Trust. *Journal of Internet Commerce*, *19*(3), 262–297. Advance online publication. DOI: 10.1080/15332861.2020.1756190

Dadwal, S., Haq, A., Jamal, A., & Nawaz, I. (2021). Value of Data as a Currency and a Marketing Tool. In *Strategy, Leadership, and AI in the Cyber Ecosystem* (pp. 321–336). Elsevier., DOI: 10.1016/B978-0-12-821442-8.00017-3

Dahlberg, T., Guo, J., & Ondrus, J. (2015). A critical review of mobile payment research. *Electronic Commerce Research and Applications*, *14*(5), 265–284. DOI: 10.1016/j.elerap.2015.07.006

Dahlberg, T., Mallat, N., Ondrus, J., & Zmijewska, A. (2008). Past, present and future of mobile payments research: A literature review. *Electronic Commerce Research and Applications*, *7*(2), 165–181. DOI: 10.1016/j.elerap.2007.02.001

Damasco, L. B.Jr, Lacson, N. R. F., & Intong, R. T. (2023, December 14). Safeguarding transactional data: A comprehensive investigation of VPN integration in Point-of-Sale systems for enhanced security and connectivity. *Eximia*, *12*(1), 570–589. DOI: 10.47577/eximia.v12i1.405

Daniel, J. (2024, November 28). *How to resolve connection issues with your POS.* PayCliq. https://blog.paycliq.africa/how-to-resolve-connection-issues-with-your-pos

Dasawat, S. S., & Sharma, S. (2024). Edge Intelligence Enabled Secured Digital Transformation in Internet of Autonomous Vehicles. In 2024 12th International Conference on Internet of Everything, Microwave, Embedded, Communication and Networks (IEMECON), 1–6. DOI: 10.1109/IEMECON62401.2024.10846320

Davenport, W. B. (1967, Spring). Bank credit cards and the Uniform Commercial Code. *Valparaiso University Law Review. Valparaiso University. School of Law*, *1*(2), 218–254.

Davinson, N., & Sillence, E. (2014). Using the health belief model to explore users' perceptions of 'being safe and secure' in the world of technology mediated financial transactions. *International Journal of Human-Computer Studies*, *72*(2), 154–168. DOI: 10.1016/j.ijhcs.2013.10.003

Davis, F. D. (1989). Perceived usefulness, perceived ease of use, and user acceptance of information technology. *Management Information Systems Quarterly*, *13*(3), 319–340. DOI: 10.2307/249008

Davis, F. D. (1989). Technology acceptance model: TAM. *Al-Suqri, MN, Al-Aufi, AS. Information Seeking Behavior and Technology Adoption*, *205*(219), 5.

de Boer, E., & Chin, X. Y. (2025). Designing the Best Currency for Your Loyalty Program. In *Loyalty Programs and the Currency Effect* (pp. 21–47). Springer., DOI: 10.1007/978-3-031-78849-9_2

de Boer, E., & Chin, X. Y. (2025). *Loyalty Programs and the Currency Effect: A Comprehensive Guide to Realizing the Power of Points*. Springer., DOI: 10.1007/978-3-031-78849-9

De Veirman, M., Cauberghe, V., & Hudders, L. (2017). Marketing through Instagram influencers: Impact of number of followers and product divergence on brand attitude. *International Journal of Advertising, 36*(5), 798–828. DOI: 10.1080/02650487.2017.1348035

Debenedetti, A., Philippe, D., & Dion, D. (2025). Balancing exclusivity and inclusivity through the strategic domestication of the luxury retail experience. *International Journal of Research in Marketing, 42*(1), 32–54. DOI: 10.1016/j.ijresmar.2024.05.001

Delgado, O., Barrera, J., Delgado, O., & Barrera, J. (2023). Determinantes del Uso de Efectivo en América Latina. *Revista Politécnica, 51*(2), 73–86. DOI: 10.33333/rp.vol51n2.07

Demirguc-Kunt, A., & Klapper, L. (2012). Measuring Financial Inclusion: The Global Findex database. In World Bank policy research working paper. DOI: 10.1596/1813-9450-6025

Demirgüç-Kunt, A., Klapper, L., Singer, D., & Ansar, S. (2022). *The Global Findex Database 2021 : Financial Inclusion, Digital Payments, and Resilience in the Age of COVID-19*. World Bank Publications. DOI: 10.1596/978-1-4648-1897-4

Denzin, N., Lincoln, Y., Giardina, M., & Cannella, G. (2023). *The Sage Handbook of Qualitative Research*. Sage Publications.

Department of Justice. (2000, April). *Information Technology and people with disabilities: The current state of federal accessibility. Presented by the Attorney General to the President of the United States*. Civil Rights Division. https://files.eric.ed.gov/fulltext/ED450514.pdf

Deutch, K. (2023, November 30). *What is a POS system and how does it work?* Square: The Bottom Line. https://squareup.com/us/en/the-bottom-line/operating-your-business/what-pos-system

Diep, D. M., & Mai, L. T. T. (2025). How Does Sense of Belonging Bridge the Linkage Between Customer Engagement and Word of Mouth in Online Nonprofit Communities? *Journal of Nonprofit & Public Sector Marketing*, •••, 1–24. DOI: 10.1080/10495142.2025.2488775

Dierksmeier, C., & Seele, P. (2020). Blockchain and business ethics. *Business Ethics (Oxford, England), 29*(2), 348–359. DOI: 10.1111/beer.12259

Dos Santos, P. L., & Harvold Kvangraven, I. (2017). Better than cash, but beware the costs: Electronic payments systems and financial inclusion in developing economies. *Development and Change, 48*(2), 205–227. DOI: 10.1111/dech.12296

Dowling, M. (2022). Is non-fungible token pricing driven by cryptocurrencies? *Finance Research Letters, 44*, 102097. DOI: 10.1016/j.frl.2021.102097

Duan, Y., & Zhu, Q. (2025). Blockchain empowerment: Enhancing consumer trust and outreach through supply chain transparency. *International Journal of Production Research, 63*(14), 5358–5382. DOI: 10.1080/00207543.2024.2434951

Dudczyk, P., Dunston, J. K., & Crosby, G. V. (2024). Blockchain Technology for Global Supply Chain Management: A Survey of Applications, Challenges, Opportunities and Implications. *IEEE Access : Practical Innovations, Open Solutions, 12*, 70065–70088. DOI: 10.1109/ACCESS.2024.3399759

Duplys, P., & Schmitz, R. (2024, January). *TLS cryptography in-depth*. Packt. https://learning.oreilly.com/library/view/tls-cryptography-in-depth/9781804611951

Dwivedi, Y. K., Hughes, L., Wang, Y., Alalwan, A. A., Ahn, S. J., Balakrishnan, J., Barta, S., Belk, R., Buhalis, D., Dutot, V., Felix, R., Filieri, R., Flavián, C., Gustafsson, A., Hinsch, C., Hollensen, S., Jain, V., Kim, J., Krishen, A. S., & Wirtz, J. (2023). Metaverse marketing: How the metaverse will shape the future of consumer research and practice. *Psychology and Marketing, 40*(4), 750–776. DOI: 10.1002/mar.21767

Dzogbenuku, R. K., Amoako, G. K., Kumi, D. K., & Bonsu, G. A. (2022). Digital Payments and Financial Wellbeing of the Rural Poor : The Moderating Role of Age and Gender. *Journal of International Consumer Marketing, 34*(2), 113–136. DOI: 10.1080/08961530.2021.1917468

Egerson, J. I., Williams, M., Aribigbola, A., Okafor, M., & Olaleye, A. (2024, August). Cybersecurity strategies for protecting big data in business intelligence systems: Implication for operational efficiency and profitability. *World Journal of Advanced Research and Reviews, 23*(2), 916–924. DOI: 10.30574/wjarr.2024.23.2.2390

Elangovan, A., & Babu, M. (2024). Decoding users' continuance intentions towards digital financial platforms in the Indian economy. *International Review of Management and Marketing, 14*(6), 213–222. DOI: 10.32479/irmm.17199

Elayat, A. M. A., & Elalfy, R. M. (2025). Using SOR theory to examine the impact of AI chatbot quality on Gen Z's satisfaction and advocacy within the fast-food sector. *Young Consumers, 26*(2), 352–383. DOI: 10.1108/YC-08-2024-2199

Elena-Bucea, A., Cruz-Jesus, F., Oliveira, T., & Coelho, P. S. (2021). Assessing the Role of Age, Education, Gender and Income on the Digital Divide : Evidence for the European Union. *Information Systems Frontiers*, *23*(4), 1007–1021. DOI: 10.1007/s10796-020-10012-9

Emon, M. M. H. (2025). Cybersecurity in the Smart City Era: Overcoming Challenges With Modern Cryptographic Solutions. In *Securing Smart Cities Through Modern Cryptography Technologies* (pp. 43–74). IGI Global Scientific Publishing. DOI: 10.4018/979-8-3373-3326-7.ch002

Emon, M. M. H., Nath, A., & Nipa, M. N. (2025). *AI for a Greener Tomorrow: Harnessing Artificial Intelligence for Environmental Sustainability*. IGI Global Scientific Publishing., DOI: 10.4018/979-8-3373-2387-9.ch010

Enge, E., Spencer, S., & Stricchiola, J. (2023, September). *The art of SEO: Mastering Search Engine Optimization* (4th ed.). O'Reilly Media. https://learning.oreilly.com/library/view/the-art-of/9781098102609

Erenstein, O., Jaleta, M., Sonder, K., Mottaleb, K., & Prasanna, B. M. (2022). Global Maize Production, Consumption and Trade: Trends and R&D Implications. *Food Security*, *14*(5), 1295–1319. DOI: 10.1007/s12571-022-01288-7

Eshet-Alkalai, Y. (2004). Digital literacy: A conceptual framework for survival skills in the digital era. *Journal of Educational Multimedia and Hypermedia*, *13*(1), 93–106.

Eswaran, M. (2019). The role of demographic factors in digital payment adoption in India. *International Journal of Bank Marketing*, *37*(7), 1694–1714.

European Central Bank. (2021). *Study on the payment attitudes of consumers in the euro area (SPACE)*. Publications Office. https://data.europa.eu/doi/10.2866/240571

Evans, D. S., & Schmalensee, R. (2004). *Paying with plastic: the digital revolution in buying and borrowing*. Mit Press. DOI: 10.7551/mitpress/5281.001.0001

Falk, T., Kunz, W. H., Schepers, J. J. L., & Mrozek, A. J. (2016). How mobile payment influences the overall store price image. *Journal of Business Research*, *69*(7), 2417–2423. DOI: 10.1016/j.jbusres.2016.01.011

Farashi, S., & Bashirian, S. (2021, August). A complementary webometric ranking system based on the website quality and traffic measures: A study focusing on top-ranked world universities. *Education for Information*, *37*(3), 227–254. DOI: 10.3233/EFI-200422

Farinloye, T., Omotoye, O., Oginni, A., Moharrak, M., & Mogaji, E. (2024). Bridging the digital divide: Consumer engagement with transportation payment apps in emerging economies. *Journal of Consumer Behaviour, 23*(6), 3011–3029. DOI: 10.1002/cb.2388

Farok, N. A. Z., & Zolkipli, M. F. (2024, June). Incident response planning and procedures. *Borneo International Journal, 7*(2), 69–76. Majmuah Enterprise. http://majmuah.com/journal/index.php/bij/article/view/641/325

Flannigan, S. (2025, May 14). *The wild history of POS systems (and where they're headed next).* The Retail Exec. https://theretailexec.com/payment-processing/pos-system-history

Fornell, C., & Larcker, D. F. (1981). Evaluating structural equation models with unobservable variables and measurement error. *JMR, Journal of Marketing Research, 18*(1), 39–50. DOI: 10.1177/002224378101800104

Fournier, S. (1998). Consumers and their brands: Developing relationship theory in consumer research. *The Journal of Consumer Research, 24*(4), 343–373. DOI: 10.1086/209515

Frączek, B., & Urbanek, A. (2021). Financial inclusion as an important factor influencing digital payments in passenger transport: A case study of EU countries. *Research in Transportation Business & Management, 41*, 100691. DOI: 10.1016/j.rtbm.2021.100691

Franque, F. B., Oliveira, T., & Tam, C. (2023). Continuance Intention of Mobile Payment : TTF Model with Trust in an African Context. *Information Systems Frontiers, 25*(2), 775–793. DOI: 10.1007/s10796-022-10263-8

Frisby, W., Moench, B., Recht, B., & Ristenpart, T. (2012, August). Security analysis of smartphone Point-of-Sale systems. *Proceedings of the 6th USENIX Workshop on Offensive Technologies*, Bellevue, WA.

Frohmann, F. (2023). *Digital Pricing*. Springer International Publishing., DOI: 10.1007/978-3-031-24591-6

Fujiki, H., & Tanaka, M. (2018). How do we choose to pay using evolving retail payment technologies? Evidence from Japan. *Journal of the Japanese and International Economies, 49*, 85–99. DOI: 10.1016/j.jjie.2018.05.001

Gao, L., Waechter, K. A., & Bai, X. (2015). Understanding consumers' continuance intention towards mobile purchase: A theoretical framework and empirical study–A case of China. *Computers in Human Behavior, 53*, 249–262. DOI: 10.1016/j.chb.2015.07.014

García-Monleón, F., Danvila-del-Valle, I., & Lara, F. J. (2021). Intrinsic value in crypto currencies. *Technological Forecasting and Social Change*, *162*, 120393. DOI: 10.1016/j.techfore.2020.120393

Garg, A., Ghanshala, K. K., & Sharma, S. (2024). A hybrid model of integrating sentiment analysis and key market indicators for IPO listing trend prediction. *International Journal of Mathematical Engineering and Management Sciences*, *9*(4), 902–913. DOI: 10.33889/IJMEMS.2024.9.4.047

Gaur, S., & Verma, P. (2023). Understanding trust dynamics in rural mobile payment adoption: Evidence from Haryana. *Journal of Financial Inclusion Research*, *12*(2), 45–62.

Gefen, D. (2000). E-commerce: The role of familiarity and trust. *Omega*, *28*(6), 725–737. DOI: 10.1016/S0305-0483(00)00021-9

Gefen, D., Karahanna, E., & Straub, D. W. (2003). Trust and TAM in online shopping: An integrated model. *Management Information Systems Quarterly*, *27*(1), 51–90. DOI: 10.2307/30036519

George, A. S., George, A. H., & Baskar, T. (2023). The death of analog: Assessing the Impacts of ubiquitous mobile technology. *Partners Universal Innovative Research Publication*, *1*(2), 15–33.

Georgiev, M. (2024, May 9). *The evolution of POS systems: From "cha-ching" cash registers to modern solutions*. NRS Plus. https://nrsplus.com/blog/evolution-of-pos-systems

Gerli, P., Clement, J., Esposito, G., Mora, L., & Crutzen, N. (2022). The hidden power of emotions: How psychological factors influence skill development in smart technology adoption. *Technological Forecasting and Social Change*, *180*, 121721. DOI: 10.1016/j.techfore.2022.121721

Ghouse, S. M., Shekhar, R., & Chaudhary, M. (2025). Driving financial inclusion: Exploring mobile wallet adoption among rural Omani millennials. *Journal of Islamic Marketing*, *16*(4), 1229–1257. DOI: 10.1108/JIMA-02-2024-0073

Giamporcaro, S., & Kuk, G. (2024). Affordances-in-practice: Unravelling blockchain-for-good in sustainable investment. *Information Technology & People*. Advance online publication. DOI: 10.1108/ITP-05-2022-0349

Giangiulio, D., & Cheung, B. (2024, July 4). Major retailers are backtracking on self-checkout. *NBC News*. https://www.nbcnews.com/business/business-news/major-retailers-are-backtracking-self-checkout-rcna160234

Giannakoudi, S. (1999). Internet banking: The digital voyage of banking and money in cyberspace. *Information & Communications Technology Law, 8*(3), 205–243. DOI: 10.1080/13600834.1999.9965811

Gilde, S. (2024, October 16). *A current guide to EMV testing and certification*. Paragon Application Systems. https://www.paragonedge.com/blog/a-current-guide-to-emv-certification-and-testing-requirements

Gioia, D. A., Corley, K. G., & Hamilton, A. L. (2013). Seeking qualitative rigor in inductive research: Notes on the Gioia methodology. *Organizational Research Methods, 16*(1), 15–31. DOI: 10.1177/1094428112452151

Giri, R. R. W., Apriliani, D., & Sofia, A. (2019). Behavioral intention analysis on e-money services in Indonesia : Using the modified UTAUT model. *1st International Conference on Economics, Business, Entrepreneurship, and Finance (ICEBEF 2018)*, 73-76. https://www.atlantis-press.com/proceedings/icebef-18/125908031

Goczek, Ł., & Witkowski, B. (2016). Determinants of card payments. *Applied Economics, 48*(16), 1530–1543. DOI: 10.1080/00036846.2015.1102846

Goldfarb, A., & Tucker, C. (2011). Online display advertising: Targeting and obtrusiveness. *Marketing Science, 30*(3), 389–404. DOI: 10.1287/mksc.1100.0583

Gomber, P., Kauffman, R. J., Parker, C., & Weber, B. W. (2018). On the fintech revolution: Interpreting the forces of innovation, disruption, and transformation in financial services. *Journal of Management Information Systems, 35*(1), 220–265. DOI: 10.1080/07421222.2018.1440766

Grandhi, B., Patwa, N., & Saleem, K. (2021). Data-Driven Marketing for Growth and Profitability. *EuroMed Journal of Business, 16*(4), 381–398. DOI: 10.1108/EMJB-09-2018-0054

Grewal, D., Roggeveen, A. L., & Nordfält, J. (2017). The future of retailing. *Journal of Retailing, 93*(2), 121–146. DOI: 10.1016/j.jretai.2016.12.008

GSMA. (2025). *State of the industry report on mobile money: Mobile money surpasses two billion registered accounts and over half a billion monthly active users globally*. GSMA. https://www.gsma.com

Guo, L., & Liu, X. (2025). Learning of the user behavior structure based on the time granularity analysis model. *PeerJ. Computer Science, 11*, e2573. DOI: 10.7717/peerj-cs.2573

Gupta, S., & Prusty, S. (2024). Does consumer empowerment influence e-payment systems adoption? A digital consumer-centric perspective. *Journal of Financial Services Marketing*, *29*(3), 1–15. DOI: 10.1057/s41264-023-00238-4

Gutfreund, J. (2016). Move over, Millennials: Generation Z is changing the consumer landscape. *Journal of Brand Strategy*, *5*(3), 245–249. DOI: 10.69554/KAHI1951

Guzmán Martínez, F. J., & Orozco Gómez, M. M. (2024). The emotional side of price: Reference price as a decision maker in consumer purchase. *International Journal of Retail & Distribution Management*, *52*(7/8), 771–785. DOI: 10.1108/IJRDM-08-2023-0504

Hair, J. F., Hult, G. T. M., Ringle, C. M., & Sarstedt, M. (2019). *A primer on partial least squares structural equation modeling (PLS-SEM)* (2nd ed.). SAGE Publications. DOI: 10.3926/oss.37

Hai, T. N., Van, Q. N., & Thi Tuyet, M. N. (2021). Digital Transformation: Opportunities and Challenges for Leaders in the Emerging Countries in Response to COVID-19 Pandemic. *Emerging Science Journal*, *5*(Special Issue), 21–36. DOI: 10.28991/esj-2021-SPER-03

Hamari, J., Hanner, N., & Koivisto, J. (2020). Why pay premium in freemium services?" A study on perceived value, continued use and purchase intentions in free-to-play games. *International Journal of Information Management*, *51*, 102040. DOI: 10.1016/j.ijinfomgt.2019.102040

Hamari, J., Koivisto, J., & Sarsa, H. (2014). Does gamification work? A literature review of empirical studies on gamification. In *Proceedings of the 47th Hawaii International Conference on System Sciences* (pp. 3025–3034). IEEE. DOI: 10.1109/HICSS.2014.377

Handoko, A., & Mauritsius, T. (2021, April 19). Factors determining the use of POS (point of sales) systems in small and medium sized enterprises. *Turkish Journal of Computer and Mathematics Education*, *12*(6), 4800–4806.

Han, H., & Ryu, K. (2012). The theory of repurchase decision-making (TRD): Identifying the critical factors in the post-purchase decision-making process. *International Journal of Hospitality Management*, *31*(3), 786–797. DOI: 10.1016/j.ijhm.2011.09.015

Hanna, N. (2018). Urban-rural differences in financial literacy: Evidence from a national survey of high school students. Journal of Rural Studies, 60, 19-29.a.

Hänninen, M. (2020). Review of studies on digital transaction platforms in marketing journals. *International Review of Retail, Distribution and Consumer Research, 30*(2), 164–192. DOI: 10.1080/09593969.2019.1651380

Hassoun, A., Beacock, I., Carmody, T., Kelley, P. G., Goldberg, B., Kumar, D., Murray, L., Park, R. S., Sarmadi, B., & Consolvo, S. (2025). Beyond Digital Literacy: Building Youth Digital Resilience Through Existing "Information Sensibility" Practices. *Social Sciences (Basel, Switzerland), 14*(4), 230. DOI: 10.3390/socsci14040230

Hayashi, F., & Klee, E. (2003). Technology adoption and consumer payments : Evidence from survey data. *Review of Network Economics, 2*(2). Advance online publication. https://scholar.archive.org/work/gvgjurwbtbb4hf3ps3xoh7m3vi/access/wayback/http://time.dufe.edu.cn:80/jingjiwencong/waiwenziliao/31327.pdf. DOI: 10.2202/1446-9022.1025

Hayes, A. (2024, September 24). *What is Point of Sale (POS)?* Investopedia. https://www.investopedia.com/terms/p/point-of-sale.asp

Henseler, J., Ringle, C. M., & Sarstedt, M. (2015). A new criterion for assessing discriminant validity in variance-based structural equation modeling. *Journal of the Academy of Marketing Science, 43*(1), 115–135. DOI: 10.1007/s11747-014-0403-8

Henseler, J., Ringle, C. M., & Sarstedt, M. (2016). Testing measurement invariance of composites using the MICOM procedure. *Journal of Business Research, 69*(9), 3150–3158.

Herhausen, D., Kleinlercher, K., Verhoef, P. C., Emrich, O., & Rudolph, T. (2019). Loyalty formation for different customer journey segments. *Journal of Retailing, 95*(3), 9–29. DOI: 10.1016/j.jretai.2019.05.001

Hewei, T., & Youngsook, L. (2022). Factors affecting continuous purchase intention of fashion products on social E-commerce: SOR model and the mediating effect. *Entertainment Computing, 41*, 100474. DOI: 10.1016/j.entcom.2021.100474

Hobeika, J., Liew, C. Y., & Rajan, M. E. S. (2025). Are cryptocurrencies influenced by stereotypes and values? Evidence of the role of banker stereotypes and human values on cryptocurrency acceptance. *International Journal of Bank Marketing, 43*(8), 1627–1661. Advance online publication. DOI: 10.1108/IJBM-04-2024-0181

Ho, C.-I., Liu, Y., & Chen, M.-C. (2022). Antecedents and consequences of consumers' attitudes toward live streaming shopping: An application of the stimulus–organism–response paradigm. *Cogent Business & Management, 9*(1), 2145673. Advance online publication. DOI: 10.1080/23311975.2022.2145673

Ho, C.-Y., Kim, N., Rong, Y., & Tian, X. (2022, October). Promoting mobile payment with price incentives. *Management Science*, *68*(10), 7614–7630. DOI: 10.1287/mnsc.2021.4233

Hoehle, H., Venkatesh, V., Brown, S., Tepper, B. J., & Kude, T. (2022, February). Impact of customer compensation strategies on outcomes and the mediating role of justice perceptions: A longitudinal study of Target's data breach. *Management Information Systems Quarterly*, *46*(1), 299–340. DOI: 10.25300/MISQ/2022/14740

Hokkanen, H., Hänninen, M., Yrjölä, M., & Saarijärvi, H. (2021). From customer to actor value propositions: An analysis of digital transaction platforms. *International Review of Retail, Distribution and Consumer Research*, *31*(3), 257–279. DOI: 10.1080/09593969.2021.1880463

Hollebeek, L. D., & Macky, K. (2019). Digital content marketing's role in fostering consumer engagement, trust, and value: Framework, fundamental propositions, and implications. *Journal of Interactive Marketing*, *45*(1), 27–41. DOI: 10.1016/j.intmar.2018.07.003

Hopalı, E., Vayvay, Ö., Kalender, Z. T., Turhan, D., & Aysuna, C. (2022). How do mobile wallets improve sustainability in payment services? A comprehensive literature review. *Sustainability (Basel)*, *14*(24), 16541. DOI: 10.3390/su142416541

Hossain, Q., Yasmin, F., Biswas, T. R., & Asha, N. B. (2024). Data-Driven Business Strategies: A Comparative Analysis of Data Science Techniques in Decision-Making. *Scholars Journal of Economics, Business and Management*, *11*(9), 257–263. DOI: 10.36347/sjebm.2024.v11i09.002

Huang, L., Gao, B., & Gao, M. (2023). *Value Realization in the Phygital Reality Market*. Springer Nature Singapore., DOI: 10.1007/978-981-99-4129-2

Hubballi, R. B., Selvakumar, P., Seenivasan, R., Basava Aradhya, S. G., Dinesh, N., & Seelam, P. K. (2025). Overview of Current AI Technologies in Education. In Kyei-Blankson, L., & Ntuli, E. (Eds.), *Transformative AI Practices for Personalized Learning Strategies* (pp. 1–26). IGI Global Scientific Publishing., DOI: 10.4018/979-8-3693-8744-3.ch001

Hudders, L., De Jans, S., & De Veirman, M. (2021). The commercialization of social media stars: A literature review and conceptual framework on the strategic use of social media influencers. *International Journal of Advertising*, *40*(3), 327–375. DOI: 10.1080/02650487.2020.1836925

Hull, C. L. (1943). *Principles Of Behavior : An Introduction to Behavior Theory (New York: Appleton-Century-Crofts;).[Google Scholar].*

Hwang, J., & Kandampully, J. (2012). The role of emotional aspects in younger consumer-brand relationships. *Journal of Product and Brand Management, 21*(2), 98–108. DOI: 10.1108/10610421211215517

Iansiti, M., & Lakhani, K. R. (2017). The truth about blockchain. *Harvard Business Review, 95*(1), 118–127.

Imbayani, I. G. A., & Gama, A. W. S. (2024). The impact of hedonic shopping motivation, discounts, and urge to buy impulsively on impulse buying: SOR perspective. *International Journal of Applied Business and International Management, 9*(3), 349–364. DOI: 10.32535/ijabim.v9i3.3438

Influenciity. (2024, June 25). *Social Responsibility and Ethics in Influencer Marketing: The Role and Impact of Content Creators.*

iResearch. (2022). China's Live Commerce Market Report 2022. iResearch Consulting Group.

Irimia-Diéguez, A., Liébana-Cabanillas, F., Blanco-Oliver, A., & Lara-Rubio, J. (2025). What drives consumers to use P2P payment systems? An analytical approach based on the stimulus–organism–response (S-O-R) model. *European Journal of Management and Business Economics.* https://www.econstor.eu/bitstream/10419/325591/1/1923036564.pdf

Irimia-Diéguez, A., Liébana-Cabanillas, F., Blanco-Oliver, A., & Lara-Rubio, J. (2025). What drives consumers to use P2P payment systems? An analytical approach based on the stimulus–organism–response (S-O-R) model. *European Journal of Management and Business Economics, 34*(2), 149–173. DOI: 10.1108/EJMBE-12-2022-0374

Islam, S. (2024). Impact of online payment systems on customer trust and loyalty in E-commerce analyzing security and convenience. Available at *SSRN* 5064838. DOI: 10.2139/ssrn.5064838

Isler, O., Yilmaz, O., & Dulleck, U. (2023). Scarcity improves economic valuations when cognitively salient. *Journal of Economic Behavior & Organization, 209*, 362–371. DOI: 10.1016/j.jebo.2023.02.019

Ismaeel, B., & Alsariera, W. (2023). The Effect of Marketing Effectiveness, and Consumer Behaviour on Consumer Purchase Preference for Unilever Products in Jordan. *International Journal of Academic Research in Business & Social Sciences, 13*(7), 1014–1032. DOI: 10.6007/IJARBSS/v13-i7/17416

Jack, W., & Suri, T. (2013). Risk Sharing and Transactions Costs: Evidence from Kenya's Mobile Money Revolution. *The American Economic Review, 104*(1), 183–223. DOI: 10.1257/aer.104.1.183

Jannah, N., Bahri, M. I., Kismawadi, E. R., & Handriana, T. (2024). The Effect of Green Brand Image and Green Satisfaction on Green Brand Equity Mediated Green Trust Outpatient's. *Quality - Access to Success, 25*(198). Advance online publication. DOI: 10.47750/QAS/25.198.40

Jaradat, M. I. R. M., & Al-Mashaqba, A. M. (2014). Understanding the adoption and usage of mobile payment services by using TAM3. *International Journal of Business Information Systems, 16*(3), 271–296. DOI: 10.1504/IJBIS.2014.063768

Javalgi, R. G., Martin, C. L., & Todd, P. R. (2004). The export of e-services in the age of technology transformation: Challenges and implications for international service providers. *Journal of Services Marketing, 18*(7), 560–573. DOI: 10.1108/08876040410561884

Jechle, D., Schuetz, S., & Gewald, H. (2025, June). Decoding data breach response strategies: Insights from situational crisis communication theory using topic modeling. In Association for Information Systems, AIS Electronic Library (AISeL), & European Conference on Information Systems (ECIS), *ECIS 2025 Proceedings* [Journal-article]. https://aisel.aisnet.org/ecis2025/security/security/2

Jiang, C., & Phoong, S. W. (2024). Research on tourists' travel intention and usage of VR technologies based on SOR model: Mediating role of personal values. *Journal of Hospitality and Tourism Insights, 7*(2), 949–967. DOI: 10.1108/JHTI-08-2023-0545

Jonnala, A. (2024). Transforming customer experience with digital voice assistants. *Computing in Science & Engineering, 14*(3), 67–74. DOI: 10.5923/j.computer.20241403.03

Joy, A., Zhu, Y., Peña, C., & Brouard, M. (2022). Digital future of luxury brands: Metaverse, digital fashion, and non-fungible tokens. *Strategic Change, 31*(3), 337–343. DOI: 10.1002/jsc.2502

Kabir, M. A., & Han, B. (2016, July). An improved usability evaluation model for Point-of-Sale systems. *International Journal of Smart Home, 10*(7), 269–282. DOI: 10.14257/ijsh.2016.10.7.27

Kadam, V. (2025). *Cyber security breaches in corporate networks: a literature review on recent threats and their impact.* [Student Thesis, LAB University of Applied Sciences]. Theseus. https://www.theseus.fi/handle/10024/887316

Kajdi, L. (2022). *Consumer payment preferences in the euro area*. ECB Working Paper. https://www.econstor.eu/handle/10419/269136

Kalyanaraman, S., & Sundar, S. S. (2006). The psychological appeal of personalized content in web portals: Does customization affect attitudes and behavior? *Journal of Communication*, *56*(1), 110–132. DOI: 10.1111/j.1460-2466.2006.00006.x

Kamble, S., Gunasekaran, A., & Sharma, R. (2020). Modeling the blockchain-enabled traceability in agriculture supply chain. *International Journal of Information Management*, *52*, 101967. DOI: 10.1016/j.ijinfomgt.2019.05.023

Kamboj, S., & Rahman, Z. (2015). Marketing capabilities and firm performance: Literature review and future research agenda. *International Journal of Productivity and Performance Management*, *64*(8), 1041–1067. DOI: 10.1108/IJPPM-08-2014-0117

Kanaparthi, V. (2024). AI-driven personalisation in digital banking: Enhancing user experience and financial decision-making. *International Journal of Financial Studies*, *12*(1), 45–60.

Kanbach, D. K., Heiduk, L., Blueher, G., Schreiter, M., & Lahmann, A. (2024). The GenAI Is Out of the Bottle: Generative Artificial Intelligence from a Business Model Innovation Perspective. *Review of Managerial Science*, *18*(4), 1189–1220. DOI: 10.1007/s11846-023-00696-z

Kaplan, A. M., & Haenlein, M. (2010). Users of the world, unite! The challenges and opportunities of Social Media. *Business Horizons*, *53*(1), 59–68. DOI: 10.1016/j.bushor.2009.09.003

Kapoor, A. P., & Vij, M. (2018). Technology at the dinner table: Ordering food online through mobile apps. *Journal of Retailing and Consumer Services*, *43*, 342–351. DOI: 10.1016/j.jretconser.2018.04.001

Kapron, Z. (2024, April 12). Why cash is still prevalent in Asia. *Forbes*. https://www.forbes.com/sites/zennonkapron/2024/04/12/why-cash-is-still-prevalent-in-asia

Karmaker, S., Oishi, M. E. F., Qasem, A., Sami, S. B. S., & Noor, J. (2025). Exploring influential factors of consumer purchase behavior on the adoption of digital payment apps in Bangladesh. *Computers in Human Behavior Reports*, *17*, 100587. DOI: 10.1016/j.chbr.2025.100587

Kashiramka, S., Gupta, P., Bali, A., & Guha, A. (2025). Gamifying Loyalty: Real World Success Stories. In Evolving Strategies for Organizational Management and Performance Evaluation (pp. 209-236). IGI Global Scientific Publishing.

Kashiramka, S., Gupta, P., Bali, A., & Guha, A. (2025). *Gamifying Loyalty*., DOI: 10.4018/979-8-3373-0149-5.ch011

Kaur, S., & Walia, N. (2020). Exploring tourist adoption behaviour of digital payment system using TAM model. *Solid State Technology*, *63*(6), 4632–4640.

Kehinde, A. O., & Naomi, O. A. (2024). Digital Payment Option Adoption and Customer Experience Management Among SMEs in the Retail Sector. *African Multidisciplinary Journal of Research*, *9*(1), 78–96. DOI: 10.71064/spu.amjr.9.1.300

Keskin, O. F., Caramancion, K. M., Tatar, I., Raza, O., & Tatar, U. (2021, May). Cyber third-party risk management: A comparison of non-intrusive risk scoring reports. *Electronics (Basel)*, *10*(10), 1168. DOI: 10.3390/electronics10101168

Khan, A. N., Kiah, M. M., Khan, S. U., & Madani, S. A. (2013). Towards secure mobile cloud computing: A survey. *Future Generation Computer Systems*, *29*(5), 1278–1299. DOI: 10.1016/j.future.2012.08.003

Khan, A., Ullah, M., & Malik, F. F. (2022). Mediating role of consumer involvement in the relationship between marketing stimuli and consumer purchase behavior. *Journal of Marketing Strategies*, *4*(1), 202–220. DOI: 10.52633/jms.v4i1.185

Khan, B. U. I., Olanrewaju, R. F., Baba, A. M., Langoo, A. A., & Assad, S. (2017). A compendious study of online payment systems: Past developments, present impact, and future considerations. *International Journal of Advanced Computer Science and Applications*, *8*(5).

Khan, J., & Belk, R. (2024). *Digital Currency and Consumption: The Meaning of Money in an Era of Digital Currency*. Routledge., DOI: 10.4324/9781003425021

Khan, M. T. (2014). The concept of'marketing mix'and its elements (a conceptual review paper). *International Journal of Information, Business and Management*, *6*(2), 95.

Khan, W. A., & Abideen, Z. U. (2023). Effects of behavioural intention on usage behaviour of digital wallet: The mediating role of perceived risk and moderating role of perceived service quality and perceived trust. *Future Business Journal*, *9*(1), 73. Advance online publication. DOI: 10.1186/s43093-023-00242-z

Kholod, M., Celani, A., & Ciaramella, G. (2024, December). The analysis of customers' transactions based on POS and RFID data using big data analytics tools in the retail space of the future. *Applied Sciences (Basel, Switzerland)*, *14*(24), 11567. DOI: 10.3390/app142411567

Kiely, T. J. (2025, February 27). *Business reputation management: How to create and maintain a good reputation.* Meltwater. https://www.meltwater.com/en/blog/how-a-business-can-create-and-maintain-a-good-reputation

Kim, C., Mirusmonov, M., & Lee, I. (2010). An empirical examination of factors influencing the intention to use mobile payment. *Computers in Human Behavior*, *26*(3), 310–322. DOI: 10.1016/j.chb.2009.10.013

Kim, J. H., & Park, J. W. (2019). The effect of airport self-service characteristics on passengers' perceived value, satisfaction, and behavioral intention: Based on the SOR model. *Sustainability (Basel)*, *11*(19), 5352. DOI: 10.3390/su11195352

Kim, J., & Lennon, S. J. (2013). Effects of reputation and website quality on online consumers' emotion, perceived risk and purchase intention: Based on the stimulus-organism-response model. *Journal of Research in Interactive Marketing*, *7*(1), 33–56. DOI: 10.1108/17505931311316734

Kim, S. S., & Malhotra, N. K. (2005). A Longitudinal Model of Continued IS Use : An Integrative View of Four Mechanisms Underlying Postadoption Phenomena. *Management Science*, *51*(5), 741–755. DOI: 10.1287/mnsc.1040.0326

Kim, S. S., Malhotra, N. K., & Narasimhan, S. (2005). Research Note—Two Competing Perspectives on Automatic Use : A Theoretical and Empirical Comparison. *Information Systems Research*, *16*(4), 418–432. DOI: 10.1287/isre.1050.0070

King, N., & Horrocks, C. (2018). *Interviews in Qualitative Research*. SAGE Publications.

Kismawadi, E. R., & Irfan, M. (2025). Artificial Intelligence as a Catalyst for Innovation in Islamic Entrepreneurship: Balancing Ethics and Efficiency. In *Improving Entrepreneurial Processes Through Advanced AI* (pp. 1–16). IGI Global.

Kiyotaki, N., & Wright, R. (1989). On money as a medium of exchange. *Journal of Political Economy*, *97*(4), 927–954. https://www.journals.uchicago.edu/doi/abs/10.1086/261634. DOI: 10.1086/261634

Klaus, P. P., & Maklan, S. (2012). EXQ: A multiple item scale for assessing service experience. *Journal of Service Management*, *23*(1), 5–33. DOI: 10.1108/09564231211208952

Koay, K. Y., Tjiptono, F., Teoh, C. W., Memon, M. A., & Connolly, R. (2023). Social Media Influencer Marketing: Commentary on the Special Issue. In *Journal of Internet Commerce* (Vol. 22, Issue S1). https://doi.org/DOI: 10.1080/15332861.2022.2128277

Kosse, A., & Jansen, D.-J. (2013). Choosing how to pay : The influence of foreign backgrounds. *Journal of Banking & Finance*, *37*(3), 989–998. DOI: 10.1016/j.jbankfin.2012.11.005

Kotler, P., & Keller, K. L. (2016). *A framework for marketing management* (6th ed.). Pearson Education Limited.

Kotler, P., & Keller, K. L. (2016). *Marketing management* (15th ed.). Pearson.

Kotte, K. R., Attaluri, V., & Selvakumar, P. (2025). Revolutionising Retail Banking with AI and Virtual Technologies: AI-Driven Financial Training. In Nair, A. J., Manohar, S., Limbu, Y. B., & Huhmann, B. A. (Eds.), *Intersecting Natural Language Processing and FinTech Innovations in Service Marketing* (pp. 421–440). IGI Global Scientific Publishing., DOI: 10.4018/979-8-3693-9944-6.ch020

Kshetri, N. (2017). Can blockchain strengthen the internet of things? *IT Professional*, *19*(4), 68–72. DOI: 10.1109/MITP.2017.3051335

Kumar, K., & Mas, I. (2008). Banking on mobiles: Why, how, for whom? SSRN Electronic Journal, 1–28. https://papers.ssrn.com/sol3/Delivery.cfm/SSRN_ID1655282_code829576.pdf?abstractid=1655282&rulid=42517&mirid=4

Kumari, N., & Biswas, A. (2023). Does M-payment service quality and perceived value co-creation participation magnify M-payment continuance usage intention? Moderation of usefulness and severity. *International Journal of Bank Marketing*, *41*(6), 1330–1359. DOI: 10.1108/IJBM-11-2022-0500

Kumar, J. (2021). Understanding customer brand engagement in brand communities: An application of psychological ownership theory and congruity theory. *European Journal of Marketing*, *55*(4), 969–994. DOI: 10.1108/EJM-04-2018-0290

Kumar, R., Mishra, V., & Saha, S. (2019). Digital financial services in India: An analysis of trends in digital payment. *International Journal of Research and Analytical Reviews*, *6*(2), 2348–1269.

Kumar, S., & Singh, V. P. (2024). Trust and satisfaction with digital payment systems in Gen Z and Gen Y: A comparative study. *Library of Progress–Library Science. Information Technology & Computer*, *44*(3). Advance online publication. DOI: 10.48165/bapas.2024.44.2.1

Kumar, V., Ramachandran, D., & Kumar, B. (2021). Influence of new-age technologies on marketing: A research agenda. *Journal of Business Research*, *125*, 864–877. DOI: 10.1016/j.jbusres.2020.01.007

Kuppelwieser, V. G., & Klaus, P. (2021). Measuring customer experience quality: The EXQ scale revisited. *Journal of Business Research, 126*, 624–633. DOI: 10.1016/j.jbusres.2020.01.042

Lahmeyer, S., & Roemer, E. (2024, March). The link between consumer-facing technologies and customer experience in physical retail environments: A critical literature review. *International Review of Retail, Distribution and Consumer Research, 34*(2), 128–159. DOI: 10.1080/09593969.2024.2332224

Lakshmanan, A., & Krishnan, H. S. (2011). The Aha! Experience : Insight and Discontinuous Learning in Product Usage. *Journal of Marketing, 75*(6), 105–123. DOI: 10.1509/jm.10.0348

Lankton, N. K., Wilson, E. V., & Mao, E. (2010). Antecedents and determinants of information technology habit. *Information & Management, 47*(5), 300–307. DOI: 10.1016/j.im.2010.06.004

Larson, S., Jones, J., & Swauger, J. (2020, October 27). A forensic first look at a POS device: Searching for PCI DSS data storage violations. T*he Journal of Digital Forensics. Security and Law, 15*, 4. DOI: 10.15394/jdfsl.2020.1614

Lavanya, R., & Shrivastava, S. (2024). DEMOGRAPHIC FACTORS INFLUENCING THE ADOPTION OF DIGITAL PAYMENT METHODS : A STATISTICAL ANALYSIS OF USER PREFERENCE. [JOM]. *Journal of Management, 11*(3), 1–15.

Lee, E., & Yoon, Y. (2022). Trusted information project platform based on blockchain for sharing strategy. *Journal of Ambient Intelligence and Humanized Computing, 13*(3), 1575–1585. DOI: 10.1007/s12652-019-01421-z

Lee, H. S. H., Loke, Y. J., & Tan, A. K. G. (2013). The Demand for E-Payments in Malaysia : An Examination of Usage Intensity. *Margin, 7*(4), 371–389. DOI: 10.1177/0973801013500167

Lee, H., Rana, M. R. I., & Xu, Y. (2024). Transforming luxury: Young consumers' motivations towards purchasing virtual luxury non-fungible token wearables. *International Journal of Retail & Distribution Management, 52*(12), 1208–1228. DOI: 10.1108/IJRDM-12-2023-0712

Lee, J. Y. (2019). A decentralized token economy: How blockchain and cryptocurrency can revolutionize business. *Business Horizons, 62*(6), 773–784. DOI: 10.1016/j.bushor.2019.08.003

Lee, S. S., & Johnson, B. K. (2022). Are they being authentic? The effects of self-disclosure and message sidedness on sponsored post effectiveness. *International Journal of Advertising, 41*(1), 30–53. Advance online publication. DOI: 10.1080/02650487.2021.1986257

Lee, S., & Watkins, B. (2022). The rise of live streaming commerce: How platforms and influencers shape consumer engagement. *Journal of Interactive Marketing, 58*, 68–84. DOI: 10.1016/j.intmar.2022.03.002

Lemon, K. N., & Verhoef, P. C. (2016). Understanding customer experience throughout the customer journey. *Journal of Marketing, 80*(6), 69–96. DOI: 10.1509/jm.15.0420

Lestariningati, S. I. (2018). Mobile point of sale design and implementation. *IOP Conference Series. Materials Science and Engineering, 407*, 012094. DOI: 10.1088/1757-899X/407/1/012094

Lestari, S., Adawiyah, W. R., Alhamidi, A. L., Prayogi, J., & Haryanto, R. (2024). Navigating perilous seas: Unmasking online banking frauds, perceived usefulness, fear of cybercrime and distrust in online banking. *Safer Communities, 23*(4), 444–464. DOI: 10.1108/SC-04-2024-0018

Li, D., Wong, W. E., Chau, M., Pan, S., & Koh, L. S. (2020). A survey OFNFC mobile payment: Challenges and solutions using blockchain and cryptocurrencies. In *2020 7th International conference on dependable systems and their applications (DSA)*, 69-77. IEEE.

Liao, S. H., Chu, P. H., Chen, Y. J., & Chang, C. C. (2012). Mining customer knowledge for exploring online group buying behavior. *Expert Systems with Applications, 39*(3), 3708–3716. DOI: 10.1016/j.eswa.2011.09.066

Liébana-Cabanillas, F., Ramos de Luna, I., & Montoro-Ríos, F. (2017). Intention to use new mobile payment systems: A comparative analysis of SMS and NFC payments. *Economic research-. Ekonomska Istrazivanja, 30*(1), 892–910. DOI: 10.1080/1331677X.2017.1305784

Li, H., Zhang, X., & Li, Y. (2021). Live streaming commerce and consumer purchase behavior: Evidence from China. *Electronic Commerce Research and Applications, 49*, 101086. DOI: 10.1016/j.elerap.2021.101086

Li, K., Zhang, X., & Wen, Y. (2025). Does peer pressure increase pro-environmental engagement? An extended stimulus-organism-response model. *Journal of Environmental Management, 392*, 126719. DOI: 10.1016/j.jenvman.2025.126719

Limayem, M., Hirt, S. G., & Cheung, C. M. (2007). How habit limits the predictive power of intention : The case of information systems continuance. *Management Information Systems Quarterly, 31*(4), 705–737. DOI: 10.2307/25148817

Ling, J., Xue, Y., Yang, C., & Zhang, Y. (2022). Effect of farmers' awareness of climate change on their willingness to adopt low-carbon production: Based on the TAM-SOR model. *International Journal of Environmental Research and Public Health, 20*(1), 619. DOI: 10.3390/ijerph20010619

Linh, T. T., & Huyen, N. T. T. (2025). An extension of Trust and TAM model with TPB in the adoption of digital payment: An empirical study in Vietnam. *F1000 Research, 14*, 127. DOI: 10.12688/f1000research.157763.2

Lin, K.-Y., Wang, Y.-T., & Huang, T. K. (2020). Exploring the antecedents of mobile payment service usage. *Online Information Review, 44*(1), 299–318. DOI: 10.1108/OIR-05-2018-0175

Liu, Y., Li, H., & Hu, F. (2021). Enhancing cross-border e-commerce with digital payment systems: Evidence from China. *Electronic Commerce Research and Applications, 48*, 101058. DOI: 10.1016/j.elerap.2021.101058

Lou, C., & Yuan, S. (2019). Influencer marketing: How message value and credibility affect consumer trust and purchase intention. *Journal of Interactive Advertising, 19*(1), 58–73. DOI: 10.1080/15252019.2018.1533501

Loutfi, S. A., Jose, S., Jabeen, F., & Piccardi, P. (2025). A stimulus-organism-response (SOR) perspective on healthcare telemedicine usage towards sustainable development. *Management of Environmental Quality, 36*(7), 1848–1870. DOI: 10.1108/MEQ-01-2025-0015

Lowden, G. A. (1957, May). National Cash Register. *The Analysts Journal; Proceedings of the Tenth Annual Convention, National Federation of Financial Analysts Societies, 13*(3). 145-148. https://www.jstor.org/stable/4468825

Luangrath, A. W., Peck, J., Hedgcock, W., & Xu, Y. (2022). Observing Product Touch: The Vicarious Haptic Effect in Digital Marketing and Virtual Reality. *JMR, Journal of Marketing Research, 59*(2), 306–326. DOI: 10.1177/00222437211059540

Lusardi, A., & Mitchell, O. S. (2014). The Economic Importance of Financial Literacy: Theory and evidence. *Journal of Economic Literature, 52*(1), 5–44. DOI: 10.1257/jel.52.1.5 PMID: 28579637

Mäder, A., Jüttner, M., & Gatica-Perez, D. (2024). *You Are How You Pay : Understanding and Identifying the Payment Behavior of Sociodemographic Groups* (SSRN Scholarly Paper No. 5064475). Social Science Research Network. DOI: 10.2139/ssrn.5064475

Madhani, P. M. (2022). Blockchain Deployment in Marketing: Developing Conceptual Frameworks and Research Propositions. *IUP Journal of Business Strategy, 19*(3).

Madine, M., Salah, K., Jayaraman, R., Battah, A., Hasan, H., & Yaqoob, I. (2022). Blockchain and NFTs for Time-Bound Access and Monetization of Private Data. *IEEE Access : Practical Innovations, Open Solutions, 10*, 94186–94202. DOI: 10.1109/ACCESS.2022.3204274

Maheswaran, K., Selvakumar, P., Shaikh, M., Shashank, G., Gupta, S., & Manjunath, T. C. (2025). The Evolution of Smart Urban Mobility With Emerging Technologies. In Ponnusamy, S., Assaf, M., Youssef, B., Jeon, G., & Kalyanaraman, S. (Eds.), *Leveraging VANETs and Blockchain Technology for Urban Mobility* (pp. 1–28). IGI Global Scientific Publishing., DOI: 10.4018/979-8-3373-0265-2.ch001

Mahida, A., Mandala, V., Bauskar, S., Konkimalla, S., & Reddy, M. S. (2024). Real-time fraud mitigation in digital payments: Big data and AI-driven biometric authentication. *Nanotechnology Perceptions, 20*, 1176–1193.

Maji, G., Dutta, L., & Sen, S. (2019). Targeted marketing and market share analysis on pos payment data using dw and olap. In A. Abraham, P. Dutta, J. K. Mandal, A. Bhattacharya, & S. Dutta (Eds.), *Emerging Technologies in Data Mining and Information Security: Proceedings of IEMIS 2018,* Volume 2 (pp. 189-199). Springer Singapore. DOI: 10.1007/978-981-13-1498-8_17

Malaquias, R. F., & Silva, A. F. (2020). Understanding the use of mobile banking in rural areas of Brazil. *Technology in Society, 62*, 101260. DOI: 10.1016/j.techsoc.2020.101260

Malempati, M. (2022). AI Neural Network Architectures For Personalized Payment Systems: Exploring Machine Learning's Role In Real-Time Consumer Insights. *Migration Letters : An International Journal of Migration Studies, 19*(S8), 1934–1948.

Malhotra, N. K., Kim, S. S., & Agarwal, J. (2004). Internet users' information privacy concerns (IUIPC): The construct, the scale, and a causal model. *Information Systems Research, 15*(4), 336–355. DOI: 10.1287/isre.1040.0032

Mallat, N. (2007). Exploring consumer adoption of mobile payments: A qualitative study. *The Journal of Strategic Information Systems, 16*(4), 413–432. DOI: 10.1016/j.jsis.2007.08.001

Manser Payne, E. H., Peltier, J., & Barger, V. A. (2021). Enhancing the value co-creation process: Artificial intelligence and mobile banking service platforms. *Journal of Research in Interactive Marketing*, *15*(1), 68–85. DOI: 10.1108/JRIM-10-2020-0214

Manta, O., Vasile, V., & Rusu, E. (2025). Banking transformation through FinTech and the integration of artificial intelligence in payments. *FinTech*, *4*(2), 13. DOI: 10.3390/fintech4020013

Mantel, B. (2001). *Why Do Consumers Pay Bills Electronically? An Empirical Analysis* (SSRN Scholarly Paper No. 248409). Social Science Research Network. DOI: 10.2139/ssrn.248409

Mao, W. (2016). Sometimes "fee" is better than "free": Token promotional pricing and consumer reactions to price promotion offering product upgrades. *Journal of Retailing*, *92*(2), 173–184. DOI: 10.1016/j.jretai.2015.09.001

Mariana, C. D., Husodo, Z. A., Ekaputra, I. A., & Fahlevi, M. (2025). The advancement of digital payment ecosystem in metaverse: A literature review. *Computers in Human Behavior Reports*, *17*, 100570. DOI: 10.1016/j.chbr.2024.100570

Marikyan, D., & Papagiannidis, S. (2023, September 23). Unified Theory of Acceptance and Use of Technology: A review. In Papagiannidis, S. (Ed.), *TheoryHub Book* (pp. 263–280)., https://open.ncl.ac.uk/theoryhub-book

Markus, M. L. (1987). Toward a "critical mass" theory of interactive media: Universal access, interdependence and diffusion. *Communication Research*, *14*(5), 491–511. DOI: 10.1177/009365087014005003

Martin, K. D., & Murphy, P. E. (2017). The role of data privacy in marketing. *Journal of the Academy of Marketing Science*, *45*(2), 135–155. DOI: 10.1007/s11747-016-0495-4

Masihuddin, M., Sharma, A., & Gupta, R. (2017). Socioeconomic factors influencing mobile banking adoption: An empirical study. *International Journal of Business and Management*, *12*(8), 55–67.

Mathur, M. (2018, March). Where is the security blanket? Developing social media marketing capability as a shield from perceived cybersecurity risk. *Journal of Promotion Management*, *25*(2), 200–224. DOI: 10.1080/10496491.2018.1443310

Mattila, M., Yrjölä, M., & Hautamäki, P. (2021). Digital transformation of business-to-business sales: What needs to be unlearned? *Journal of Personal Selling & Sales Management*, *41*(2), 113–129. DOI: 10.1080/08853134.2021.1916396

Mavlutova, I., Spilbergs, A., Verdenhofs, A., Natrins, A., Arefjevs, I., & Volkova, T. (2022). Digital transformation as a driver of the financial sector sustainable development: An impact on financial inclusion and operational efficiency. *Sustainability (Basel), 15*(1), 207. DOI: 10.3390/su15010207

Mbiti, I., & Weil, D. (2011). Mobile Banking: The impact of M-PESA in Kenya. DOI: 10.3386/w17129

McCarthy, E. J. (1960). *Basic marketing: A managerial approach*. Irwin.

McChesney, R. W. (2016). *Rich media, poor democracy: Communication politics in dubious times*. New Press.

McKinney, R. E., Jr., Shao, L. P., Shao, D. H., & Rosenlieb, D. C., Jr. (2015). The evolution of financial instruments and the legal protection against counterfeiting: a look at coin, paper, and virtual currencies. U. Ill. JL Tech. & Pol'y, 273. https://heinonline.org/HOL/LandingPage?handle=hein.journals/jltp2015&div=13&id=&page=

Mehrabian, A., & Russell, J. A. (1974). *An approach to environmental psychology*. MIT Press.

Melnyk, V., Carrillat, F. A., & Melnyk, V. (2022). The Influence of Social Norms on Consumer Behavior: A Meta-Analysis. *Journal of Marketing, 86*(3), 98–120. DOI: 10.1177/00222429211029199

Menzli, L. J., Smirani, L. K., Boulahia, J. A., & Hadjouni, M. (2022). Investigation of Open Educational Resources Adoption in Higher Education Using Rogers' Diffusion of Innovation Theory. *Heliyon, 8*(7), e09885. Advance online publication. DOI: 10.1016/j.heliyon.2022.e09885 PMID: 35874064

Meyer, D. (2019). The marketing funnel versus the flywheel: Generating consistent leads through a new model of engagement. *Journal of Digital and Social Media Marketing, 7*(2), 106. Advance online publication. DOI: 10.69554/TLAB6890

Migkos, S. P., Giannakopoulos, N. T., & Sakas, D. P. (2025). Impact of Influencer Marketing on Consumer Behavior and Online Shopping Preferences. *Journal of Theoretical and Applied Electronic Commerce Research, 20*(2), 111. DOI: 10.3390/jtaer20020111

Mikhaylov, A. (2020). Cryptocurrency market analysis from the open innovation perspective. *Journal of Open Innovation, 6*(4), 197. DOI: 10.3390/joitmc6040197

Milfeld, T., & Flint, D. J. (2021). When brands take a stand: The nature of consumers' polarized reactions to social narrative videos. *Journal of Product and Brand Management, 30*(4), 532–548. DOI: 10.1108/JPBM-10-2019-2606

Molitor, D., Saharia, A., Raghupathi, V., & Raghupathi, W. (2024, April). Characteristics of data breaches: A descriptive analytic study. *Journal of Information Security, 15*(2), 168–195. DOI: 10.4236/jis.2024.152011

Mondego, D., & Gide, E. (2024). The impact of security, service quality, perceived usefulness, perceived ease of use, trust, and price value on users' satisfaction in Cloud-Based payment systems in Australia: A PLS-SEM analysis. *Journal of Information Systems and Technology Management, 21*. Advance online publication. DOI: 10.4301/S1807-1775202421004

Montag, C., Lachmann, B., Herrlich, M., & Zweig, K. A. (2021). Addictive features of social media/messenger platforms and freemium games against the background of psychological and economic theories. *International Journal of Environmental Research and Public Health, 18*(14), 7723. DOI: 10.3390/ijerph18147723 PMID: 31340426

Morabito, V. (2017). *Business innovation through blockchain*. Springer International Publishing.

Moro-Visconti, R., & Cesaretti, A. (2023). FinTech and Digital Payment Systems Valuation. In *Digital Token Valuation* (pp. 411–458). Springer Nature Switzerland., DOI: 10.1007/978-3-031-42971-2_13

Murari, K., Bhatt, V., & Kumar, P. (2020). Does socio-demography affect preferences and concerns towards cashless payment? Evidence from the north-eastern region of India. *Journal of Payments Strategy & Systems, 14*(3), 207–222. DOI: 10.69554/LSVO4282

Murphy, N. B. (1991). Determinants of household check writing : The impacts of the use of electronic banking services and alternative pricing of services. *Financial Services Review, 1*(1), 35–44. DOI: 10.1016/1057-0810(91)90006-K

Naeem, M., Hameed, M., & Taha, M. S. (2020, February). A study of electronic payment system. [). IOP Publishing.]. *IOP Conference Series. Materials Science and Engineering, 767*(1), 012008. DOI: 10.1088/1757-899X/767/1/012008

Nalbant, K. G., & Aydın, S. (2023). Development and Transformation in Digital Marketing and Branding with Artificial Intelligence and Digital Technologies Dynamics in the Metaverse Universe. *Journal of Metaverse, 3*(1), 9–18. DOI: 10.57019/jmv.1148015

Nalini, R., Kala, G. S., Prakash, S., Joseph, A. V., & Alamelu, R. (2018). Digital Payments-Embellishing Customer Experience. In *2018 International Conference on Computation of Power, Energy, Information and Communication (ICCPEIC)*, 095-097. IEEE. DOI: 10.1109/ICCPEIC.2018.8525226

Nanda, P., & Kumar, V. (2024). New generation technologies: Development vs. ethical challenges. *International Journal of Business Information Systems*, *46*(3), 411–434. DOI: 10.1504/IJBIS.2024.140437

Nandru, P., SA, S. K., & Chendragiri, M. (2024). Adoption intention of mobile QR code payment system among marginalized street vendors: An empirical investigation from an emerging economy. *Journal of Science and Technology Policy Management*, *15*(6), 1709–1733. DOI: 10.1108/JSTPM-03-2023-0035

Näränen, M. (2013, November). *"What can I get you?" A study on POS marketing effectiveness in a restaurant environment.* [Student Thesis, Helsinki Metropolia University of Applied Sciences]. Theseus. https://www.theseus.fi/bitstream/handle/10024/67963/Naranen_Mikko.pdf

Narayanan, A., Bonneau, J., Felten, E., Miller, A., & Goldfeder, S. (2016). *Bitcoin and cryptocurrency technologies: a comprehensive introduction*. Princeton University Press.

Naslednikov, M. (2024). The Impact of Artificial Intelligence on Customer Relationship Management (CRM). *Strategies (La Jolla, Calif.)*.

Nästesjö, J. (2023). Managing the rules of recognition: How early career academics negotiate career scripts through identity work. *Studies in Higher Education*, *48*(4), 657–669. DOI: 10.1080/03075079.2022.2160974

Nayak, S. (2025). Enhancing customer experience in FinTech through AI, machine learning, and data-driven insights. *The Es Economics and Entrepreneurship*, *3*(03), 437–462.

Newman, N. (2014). Search, antitrust, and the economics of the control of user data. *Yale Journal on Regulation*, *31*, 401.

Nguyen, M.-H., Duong, M.-P. T., Tran, T. T., Sari, N. P. W. P., Tedjasuksmana, H., La, V.-P., & Vuong, Q.-H. (2025). Beyond trust: How usefulness and immersiveness drive space tourism intentions in high-risk contexts. *Current Issues in Tourism*, •••, 1–23. DOI: 10.1080/13683500.2025.2487634

Nikkhah, H. R.Jr, & Grover, V. (2022, December 1). An empirical investigation of company response to data breaches. *Management Information Systems Quarterly*, *46*(4), 2163–2196. DOI: 10.25300/MISQ/2022/16609

Nusrat, F., & Huang, Y. (2024, January). Feeling rewarded and entitled to be served: Understanding the influence of self- versus regular checkout on customer loyalty. *Journal of Business Research, 170*, 114293. DOI: 10.1016/j.jbusres.2023.114293

O'Brien, D., & Kunz, R. E. (2024). Will they stay or will they go? application of machine learning classifiers of purchase and churn prediction to a digital journalistic platform. *Journal of Media Business Studies*, •••, 1–29.

Obi, E. N. (2023, June 30). Impact of criminal acts on point of sales (POS) business operations in Karu Local government area of Nasarawa State. *NIU Journal of Social Sciences*, *9*(2), 119–127. DOI: 10.58709/niujss.v9i2.1632

Ockwell, D., Atela, J., Mbeva, K., Chengo, V., Byrne, R., Durrant, R., Kasprowicz, V., & Ely, A. (2019). Can pay-as-you-go, digitally enabled business models support sustainability transformations in developing countries? Outstanding questions and a theoretical basis for future research. *Sustainability (Basel)*, *11*(7), 2105. DOI: 10.3390/su11072105

Ofori, K. S., Anyigba, H., Adeola, O., Junwu, C., Osakwe, C. N., & David-West, O. (2022). Understanding post-adoption behaviour in the context of ride-hailing apps: The role of customer perceived value. *Information Technology & People*, *35*(5), 1540–1562. DOI: 10.1108/ITP-06-2019-0285

Ogunmokun, A. S., Balogun, E. D., & Ogunsola, K. O. (2022). A Strategic Fraud Risk Mitigation Framework for Corporate Finance Cost Optimization and Loss Prevention. *International Journal of Multidisciplinary Research and Growth Evaluation*, *3*(1), 783–790. DOI: 10.54660/.IJMRGE.2022.3.1.783-790

Okeke, N. I., Alabi, O. A., Igwe, A. N., Ofodile, O. C., & Ewim, C. P. M.Nnenna Ijeoma OkekeOlufunke Anne AlabiAbbey Ngochindo IgweOnyeka Chrisanctus OfodileChikezie Paul-Mikki Ewim. (2024). Customer journey mapping framework for SMEs: Enhancing customer satisfaction and business growth. *World Journal of Advanced Research and Reviews*, *24*(1), 2004–2018. DOI: 10.30574/wjarr.2024.24.1.3206

Olayinka, O. H.Olalekan Hamed Olayinka. (2021). Data driven customer segmentation and personalization strategies in modern business intelligence frameworks. *World Journal of Advanced Research and Reviews*, *12*(3), 711–726. DOI: 10.30574/wjarr.2021.12.3.0658

Olipane, H. G. E., & Inocencio, M. D. (2023), DIGITAL PAYMENT AND ITS INFLUENCE TO CUSTOMER CONVENIENCE. https://tinyurl.com/mr42357p

Olipas, C. N. P., & Esperon, R. M. (2020, March). The design and development of a cashless payment system with an Automatic Identification and Data Collection (AIDC) technology. *International Journal of Scientific & Technology Research*, *9*(3), 6718–6723.

Oliveira, T., Thomas, M., Baptista, G., & Campos, F. (2016). Mobile payment: Understanding the determinants of customer adoption and intention to recommend the technology. *Computers in Human Behavior*, *61*, 404–414. DOI: 10.1016/j.chb.2016.03.030

Oloveze, A. O., Oteh, O. U., Nwosu, H. E., & Obasi, R. O. (2022). How user behaviour is moderated by affective commitment on point of sale terminal. *Rajagiri Management Journal*, *16*(1), 2–20. DOI: 10.1108/RAMJ-05-2020-0019

Omarini, A. E. (2018). Fintech and the future of the payment landscape: The mobile wallet ecosystem. A challenge for retail banks? *International Journal of Financial Research*, *9*(4), 97–116. DOI: 10.5430/ijfr.v9n4p97

Omran, W. (2025). Attributions of QR code-driven perceived flow on customer satisfaction. *International Journal of Markets and Business Systems*, *6*(1), 46–65. DOI: 10.1504/IJMABS.2025.146953

Onat, I., & Ciaccio, J. (2025, April 25). *An analysis of ATM and Point-of-Sale skimming*. Orion Policy Institute. https://orionpolicy.org/an-analysis-of-atm-and-point-of-sale-skimming

Onumadu, P., & Abroshan, H. (2024, November). Near-Field Communication (NFC) Cyber threats and mitigation solutions in payment transactions: A review. *Sensors (Basel)*, *24*(23), 7423. DOI: 10.3390/s24237423

Ooi, K. B., Lee, V. H., Hew, J. J., Leong, L. Y., Tan, G. W. H., & Lim, A. F. (2023). Social media influencers: An effective marketing approach? *Journal of Business Research*, *160*, 113773. Advance online publication. DOI: 10.1016/j.jbusres.2023.113773

Osei, L. K., Cherkasova, Y., & Oware, K. M. (2023). Unlocking the Full Potential of Digital Transformation in Banking: A Bibliometric Review and Emerging Trend. *Future Business Journal*, *9*(1), 30. Advance online publication. DOI: 10.1186/s43093-023-00207-2

Oumaima, J., & Lamari, S. (2024). Customer Experience in the Digital Transformation Era: Insights on Personalization, Digital Marketing, and Customer Relationship Management. *International Journal of Economics* [IJEMF]. *Managerial Finance*, *3*(2), 52–69.

Pahnila, S., Siponen, M., & Zheng, X. (2011). Integrating Habit into UTAUT : The Chinese eBay Case. *Pacific Asia Journal of the Association for Information Systems*, *3*(2), 1–30. Advance online publication. DOI: 10.17705/1pais.03201

Pal, A., Herath, T., De', R., & Rao, H. R. (2020). Contextual facilitators and barriers influencing the continued use of mobile payment services in a developing country: Insights from adopters in India. *Information Technology for Development*, *26*(2), 394–420. DOI: 10.1080/02681102.2019.1701969

Pandey, A. K., Sawhney, A., Sharma, G., & Singh, D. (2025). Web3 and Metaverse in the Online Marketing for Consumer Support and Prediction. In *Practical Strategies and Case Studies for Online Marketing 6.0* (pp. 369–388). IGI Global Scientific Publishing., DOI: 10.4018/979-8-3373-2058-8.ch016

Pangarkar, A., & Shukla, P. (2023). Conspicuous and inconspicuous consumption of luxury goods in a digital world: Insights, implications, and future research directions. *International Journal of Advertising*, *42*(7), 1226–1238. DOI: 10.1080/02650487.2023.2246260

Para, R., Bhatia, R., & Sandiri, S. (2025). AI-powered financial digital twins: The next frontier in hyper-personalized, customer-centric financial services. *Journal of Next-Generation Research 5.0.*

Paraman, P., Annamalah, S., Ahmed, S., & Baharein, K. (2024). Unveiling the Nexus: Disentangling the Impact of Touch Screen Frequency in Electronic Media on Amplifying Control Perception, Elevating Psychological Ownership, and Igniting Intentions to Reuse Payment Services. *International Journal of Human-Computer Interaction*, *40*(22), 7145–7161. DOI: 10.1080/10447318.2023.2261733

Paranavithana, I., Rupasinghe, T. D., & Prior, D. D. (2022, September). A machine-learnt approach to market segmentation and purchase prediction using Point-of-Sale (POS) data. In S. I. Ao, & L. Gelman (Eds.), *Transactions on Engineering Technologies, Proceedings of World Congress on Engineering,* 919 (pp. 49-61). http://dx.doi.org/DOI: 10.1007/978-981-19-3579-4_4

Parise, S., Guinan, P. J., & Kafka, R. (2016). Solving the crisis of immediacy: How digital technology can transform the customer experience. *Business Horizons*, *59*(4), 411–420. DOI: 10.1016/j.bushor.2016.03.004

Pasirayi, S., & Fennell, P. B. (2023). The Effect of Mobile Payments on Retailer Firm Value: The Moderating Role of Promotions, Customer Segment, and Rollout Strategy. *Journal of Interactive Marketing*, *58*(1), 90–107. DOI: 10.1177/10949968221134805

Pavlou, P. A. (2003). Consumer acceptance of electronic commerce: Integrating trust and risk with the technology acceptance model. *International Journal of Electronic Commerce, 7*(3), 101–134. DOI: 10.1080/10864415.2003.11044275

Pazarbasioglu, C., Mora, A. G., Uttamchandani, M., Natarajan, H., Feyen, E., & Saal, M. (2020). Digital financial services. *World Bank, 54*(1), 1-54.

PCI Security Standards Council, LLC. (2025). PCI PIN Transaction Security Device Testing and Approval Program Guide. In *PCI PIN Transaction Security Device Testing and Approval Program Guide* [Report].

PCI Security Standards Council. (2024, June). *Payment Card Industry Data Security Standard, version 4.0.1.* https://docs-prv.pcisecuritystandards.org/PCI%20DSS/Standard/PCI-DSS-v4_0_1.pdf

Peneder, M. (2022). Digitization and the evolution of money as a social technology of account. Journal of Evolutionary Economics, 32(1), 175-203. DOIhttps://DOI: 10.1007/s00191-021-00729-4

Perkins-Southam, T., & Lupini, C. (2025, April 10). *What is a contactless credit card?* Forbes Advisor. https://www.forbes.com/advisor/credit-cards/contactless-credit-cards

Pintér, Z., Nagy, M. Z., Tóth, K., & Varga, J. (2022). The Struggle between Cash and Electronic Payments. *Economies, 10*(12), 304. DOI: 10.3390/economies10120304

Pires, P. B., Prisco, M., Delgado, C., & Santos, J. D. (2024). A conceptual approach to understanding the customer experience in e-commerce: An empirical study. *Journal of Theoretical and Applied Electronic Commerce Research, 19*(3), 1943–1983. DOI: 10.3390/jtaer19030096

PMGDISHA. (2023). *Annual Digital Literacy Report 2023: Haryana State Edition.* Government of India.

Podsakoff, P. M., MacKenzie, S. B., Lee, J.-Y., & Podsakoff, N. P. (2003). Common method biases in behavioral research: A critical review. *The Journal of Applied Psychology, 88*(5), 879–903. DOI: 10.1037/0021-9010.88.5.879

Popescu, A. D. (2022). Understanding FinTech and Decentralized Finance (DeFi) for Financial Inclusion. In Popescu, A. (Ed.), *FinTech Development for Financial Inclusiveness* (pp. 1–13). IGI Global., URL https://www.igi-global.com/book/fintech-development-financial-inclusiveness/285349 DOI: 10.4018/978-1-7998-8447-7.ch001

Prelec, D., & Simester, D. (2001). Always leave home without it: A further investigation of the credit-card effect on willingness to pay. *Marketing Letters*, *12*(1), 5–12. DOI: 10.1023/A:1008196717017

Putrevu, J., & Mertzanis, C. (2024). The adoption of digital payments in emerging economies: Challenges and policy responses. Digital Policy. *Regulation & Governance*, *26*(5), 476–500. DOI: 10.1108/DPRG-06-2023-0077

Rahardja, U., Sigalingging, C. T., Putra, P. O. H., Nizar Hidayanto, A., & Phusavat, K. (2023). The impact of mobile payment application design and performance attributes on consumer emotions and continuance intention. *SAGE Open*, *13*(1), 21582440231151919. DOI: 10.1177/21582440231151919

Rahayu, S. (2024). Digital Marketing Strategies to Build Customer Loyalty: A Systematic Review of Sustainable Financial Benefits. *Atestasi: Jurnal Ilmiah Akuntansi*, *7*(1), 792–806. DOI: 10.57178/atestasi.v7i1.951

Ramachandran, K. K. (2023). Evaluating ROI in digital marketing campaigns: Metrics, measurement, and insights. [IJM]. *International Journal of Management*, *14*(7).

Ramrakhyani, A., & Shrivastava, N. K. (2024). Artificial intelligence: Revolutionizing the future of Fintech. *Commerce Research Review*, *1*(2), 10–22. DOI: 10.21844/crr.v1i02.1109

Ranpariya, T., Joshi, A., & Rajdev, A. (2021). *Factors driving the adoption of UPI services based on UTAUT model*. http://library.atmiya.net:8080/dspace/handle/atmiyauni/2113

Rathod, V. H. (2023). Marketing intelligence redefined: Leveraging the power of AI for smarter business growth. *Towards Excellence*, *15*(2), 378–386. DOI: 10.37867/TE150239

Ratica, A. (2024, December 2). *Skimmer detection on payment terminals*. Technical Disclosure Commons. https://www.tdcommons.org/dpubs_series/7608

Rawat, R., Goyal, H. R., & Sharma, S. (2024c). Hofstede's Cultural Dimension Driven Artificial Narrow Intelligence IDFIS for Industry 5.0 Empowered Digital Society. *Recent Patents on Engineering*, *19*. Advance online publication. DOI: 10.2174/0118722121334543241001050846

Rawat, R., Goyal, H. R., Sharma, S., & Kotiyal, B. (2025c). An Intelligent Self-Aware Financial Inclusion System for Digital Society. In *2025 International Conference on Automation and Computation (AUTOCOM)*, 1164–1168. DOI: 10.1109/AUTOCOM64127.2025.10956854

Rawat, R., Goyal, H. R., Sharma, S., & Kotiyal, B. (2025c). SQRCD: Building Sustainable and Customer Centric DFIS for the Industry 5.0 ERA. *International Journal of Advanced Computer Science and Applications, 16*(2). Advance online publication. DOI: 10.14569/IJACSA.2025.0160293

Raya, J. M., & Vargas, C. (2022). How to become a cashless economy and what are the determinants of eliminating cash. *Journal of Applied Econometrics, 25*(1), 543–562. DOI: 10.1080/15140326.2022.2052000

Regulation, P. (2018). General data protection regulation. *InTouch (Melville, N.Y.), 25*, 1–5.

Reyneke, M., Robertson, J., Ferreira, C., & Näppä, A. (2023). Whose Role Is It Anyway? A Consideration of Responsibility in Digital Addiction: An Abstract. In *Developments in Marketing Science: Proceedings of the Academy of Marketing Science*. https://doi.org/DOI: 10.1007/978-3-031-24687-6_47

RHAFEL, A., Mustapha, Z. I. K. Y., & LIMAM, A. (2025). Analyse du comportement de paiement au Maroc. *Dossiers de Recherches en Économie et Gestion, 13*(02), 213–232.

Riefel, M. T. (2023). *A quantitative study on the impact of data breaches on customer satisfaction.* (Publication No. 95363) [Student Thesis, University of Twente]. University of Twente, Netherlands. https://essay.utwente.nl/95363/1/Riefel_BA_BMS.pdf

Riegger, A. S., Merfeld, K., Klein, J. F., & Henkel, S. (2022). Technology-enabled personalization: Impact of smart technology choice on consumer shopping behavior. *Technological Forecasting and Social Change, 181*, 121752. DOI: 10.1016/j.techfore.2022.121752

Rosário, A. T., & Dias, J. C. (2023). How has data-driven marketing evolved: Challenges and opportunities with emerging technologies. *International Journal of Information Management Data Insights, 3*(2), 100203. DOI: 10.1016/j.jjimei.2023.100203

Rose, D., Bodor, J. N., Hutchinson, P. L., & Swalm, C. M. (2010). The importance of a multi-dimensional approach for studying the links between food access and consumption. *The Journal of Nutrition, 140*(6), 1170–1174. DOI: 10.3945/jn.109.113159 PMID: 20410084

Roy, S. (2014). Understanding mobile payment adoption: An application of TAM. *International Journal of Emerging Markets, 9*(1), 15–33.

Rui, Z., Yang, C., Zhenying, G., & Bhaumik, A. (2023). A look at the rising popularity of cashless economies around the world. [IJRTBT]. *International Journal on Recent Trends in Business and Tourism, 7*(3), 16–32. DOI: 10.31674/ijrtbt.2021.v07i03.002

Saberi, S., Kouhizadeh, M., Sarkis, J., & Shen, L. (2019). Blockchain technology and its relationships to sustainable supply chain management. *International Journal of Production Research, 57*(7), 2117–2135. DOI: 10.1080/00207543.2018.1533261

Sadeghpour, S., & Vlajic, N. (2021). Ads and fraud: A comprehensive survey of fraud in online advertising. *Journal of Cybersecurity and Privacy, 1*(4), 804–832. DOI: 10.3390/jcp1040039

Sadok, H., & Elouaourti, Z. (2025). *The cash vs. digital Paradox : Why financial literacy matters in uncertain times*. Borsa Istanbul Review., DOI: 10.1016/j.bir.2025.08.001

Saeed, S., Shaikh, A., Memon, M. A., Nizamani, M. A., Abbasi, F. A., & Naqvi, S. M. R. (2019, April 30). Evaluating the quality of Point of Sale (POS) software. *University of Sindh Journal of Information and Communication Technology, 3*(2), 69-75. https://sujo2.usindh.edu.pk/index.php/USJICT

Said, R., Najdawi, A., & Chabani, Z. (2021). Analyzing the Adoption of E-payment Services in Smart Cities using Demographic Analytics : The Case of Dubai. *Advances in Science, Technology and Engineering Systems Journal, 6*(2), 113–121. DOI: 10.25046/aj060214

Sajan, V., & PS, R. (2024). Smart moves: Flipkart's name-personalized nudging techniques to combat cart abandonment. Journal of Information Technology Teaching Cases, 20438869241301078.

Sakas, D. P., Giannakopoulos, N. T., Reklitis, D. P., & Dasaklis, T. K. (2021). The Effects of Cryptocurrency Trading Websites on Airlines' Advertisement Campaigns. *Journal of Theoretical and Applied Electronic Commerce Research, 16*(7), 3099–3119. DOI: 10.3390/jtaer16070169

Sakthivel, M., Jaganathan, A., & Mohanraj, M. Mohit, & Manjunath T. (2025). "Brand Loyalty through AI and Personalisation: AI-Generated Content for Financial Marketing." In A. J Nair, S. Manohar, Y. B. Limbu, & B. A. Huhmann (Eds.), Intersecting Natural Language Processing and FinTech Innovations in Service Marketing (pp. 77-96). IGI Global Scientific Publishing. DOI: DOI: 10.4018/979-8-3693-9944-6.ch006

Samal, A., & Misra, L. I. Mohit, Pougajendy S, Chandel P S, Manjunath T C. (2025). "Musealization and Tourism Practices." In S. Oflazoğlu & M. Dora (Eds.), Preserving Cultural Heritage in Post-Disaster Urban Renewal (pp. 115-142). IGI Global Scientific Publishing. DOI: DOI: 10.4018/979-8-3693-9097-9.ch006

Sanchez, J. (2025, May 2). *The Target breach: A historic cyberattack with lasting consequences.* Framework Security. https://frameworksecurity.com/post/the-target-breach-a-historic-cyberattack-with-lasting-consequences

Santosa, A. D., Taufik, N., Prabowo, F. H. E., & Rahmawati, M. (2021). Continuance intention of baby boomer and X generation as new users of digital payment during COVID-19 pandemic using UTAUT2. *Journal of Financial Services Marketing, 26*(4), 259–273. DOI: 10.1057/s41264-021-00104-1

Santos, V., & Mendes, L. (2023, July). Digital transformation of the retail Point of Sale in the Artificial Intelligence era. In Santos, J. D., Pereira, I. V., & Pires, P. B. (Eds.), *Management and Marketing for Improved Retail Competitiveness and Performance* (pp. 200–216)., DOI: 10.4018/978-1-6684-8574-3.ch010

Saunders, M., Lewis, P., & Thornhill, A. (2019). *Research methods for business students* (8th ed.). Pearson Education.

Saura, J. R., Škare, V., & Ozretic Dosen, D. (2024). Is AI-based digital marketing ethical? Assessing a new data privacy paradox. *Journal of Innovation & Knowledge, 9*(4), 100597. DOI: 10.1016/j.jik.2024.100597

Scaife, N., Peeters, C., & Traynor, P. (2018, August). Fear the reaper: Characterization and fast detection of card skimmers. *Proceedings of the 27th USENIX Security Symposium* (pp. 1-14), Baltimore, MD. https://www.usenix.org/system/files/conference/usenixsecurity18/sec18-Scaife et al., 2018.pdf

Schiffman, G. L., & Kanuak, L. L. (2009). *Consumer Decision Making and Beyond in Consumer Behaviour* (9th ed.). Prentice Hall.

Schoenbachler, D. D., & Gordon, G. L. (2002). Multi-channel shopping: Understanding what drives channel choice. *Journal of Consumer Marketing, 19*(1), 42–53. DOI: 10.1108/07363760210414943

Schwartz, B. (2001). *The costs of living: How market freedom erodes the best things in life*. Xlibris Corporation.

Scott, D. M. (2011). *Real-time marketing and PR: How to instantly engage your market, connect with customers, and create products that grow your business now*. John Wiley & Sons.

Sears, C. R., & Cunningham, D. R. (2024, August). Individual differences in psychological stress associated with data breach experiences. *Journal of Cybersecurity and Privacy, 4*(3), 594–614. DOI: 10.3390/jcp4030028

Seema, R., & Sheetal, K. (2023). Challenges in mobile payment adoption in rural Haryana: Empirical evidence. *Journal of Rural Development and Technology, 8*(1), 105–124.

Shafinah, K., Sahari, N., Sulaiman, R., Yusoff, M. S. M., & Ikram, M. M. (2013). Determinants of User Behavior Intention (BI) on Mobile Services : A Preliminary View. *Procedia Technology, 11*, 127–133. DOI: 10.1016/j.protcy.2013.12.171

Shah, D., Kumar, V., & Kim, K. H. (2014). Managing Customer Profits : The Power of Habits. *JMR, Journal of Marketing Research, 51*(6), 726–741. DOI: 10.1509/jmr.13.0423

Shahid, Z. A., Tariq, M. I., Paul, J., Naqvi, S. A. N., & Hallo, L. (2024). Signaling theory and its relevance in international marketing: A systematic review and future research agenda. *International Marketing Review, 41*(2), 514–561. DOI: 10.1108/IMR-04-2022-0092

Shahzad, M. F., Xu, S., An, X., & Javed, I. (2024). Assessing the impact of AI-chatbot service quality on user e-brand loyalty through chatbot user trust, experience and electronic word of mouth. *Journal of Retailing and Consumer Services, 79*, 103867. DOI: 10.1016/j.jretconser.2024.103867

Shamim, A., Abid, M. F., & Ahmad, F. (2024). S–O–R based experiential framework for measuring in-store customer satisfaction in non-fuel retailing. *Journal of Retailing and Consumer Services, 77*, 103672. DOI: 10.1016/j.jretconser.2023.103672

Shankar, A., & Datta, B. (2018). Factors affecting mobile payment adoption intention: An Indian perspective. *Global Business Review, 19*(3, suppl), S72–S89. DOI: 10.1177/0972150918757870

Shankar, A., & Datta, S. (2018). Technology acceptance model in rural financial inclusion: A study of mobile banking adoption in India. *Asian Journal of Technology Innovation, 26*(1), 90–112.

Sharma, S., & Tyagi, R. (2023). Digitalization of Farming Knowledge Using Artificial Intelligence and Vedic Scripture. In 2023 IEEE International Conference on ICT in Business Industry & Government (ICTBIG), 1–6. DOI: 10.1109/ICTBIG59752.2023.10456219

Sharma, S., Gautam, A., & Tyagi, R. (2024). Artificial Intelligence and Cyber Security Driven Social Innovations in Digital Adventure Tourism. In 2024 First International Conference on Pioneering Developments in Computer Science & Digital Technologies (IC2SDT), 1–6. DOI: 10.1109/IC2SDT62152.2024.10696759

Sharma, S., Tyagi, R., & Mohan, S. (2024a). Artificial intelligence and green 6G network-enabled architectures, scenarios, and applications for autonomous connected vehicles. In River Publishers eBooks (pp. 187–212). DOI: 10.1201/9781003515920-9

Sharma, N., Oriaku, E. A., & Oriaku, N. (2020, January 21). Cost and effects of data breaches, precautions, and disclosure laws. *International Journal of Emerging Trends in Social Sciences, 8*(1), 33–41. DOI: 10.20448/2001.81.33.41

Sharma, S., Chauhan, Y., & Tyagi, R. (2023). Machine Vision Techniques for Digital Mesothelioma Diagnostic System. In *2023 Annual International Conference on Emerging Research Areas: International Conference on Intelligent Systems (AICERA/ICIS)*, 1–6. DOI: 10.1109/AICERA/ICIS59538.2023.10420023

Sharma, S., Tomar, M., & Tyagi, R. (2024). Artificial Intelligence and Vedic Scripture in Digital Agriculture for Global Economy. In *2024 International Conference on Intelligent and Innovative Technologies in Computing, Electrical and Electronics (IITCEE)*, 1–6. DOI: 10.1109/IITCEE59897.2024.10467916

Shih, I. T., Silalahi, A. D. K., Baljir, K., & Jargalsaikhan, S. (2024). Exploring the impact of perceived sustainability on customer satisfaction and the mediating role of perceived value. *Cogent Business & Management, 11*(1), 2431647. DOI: 10.1080/23311975.2024.2431647

Shree, S., Pratap, B., Saroy, R., & Dhal, S. (2021). Digital payments and consumer experience in India: A survey based empirical study. *Journal of Banking and Financial Technology, 5*(1), 1–20. DOI: 10.1007/s42786-020-00024-z

Singh, N., Misra, R., Quan, W., Radic, A., Lee, S., & Han, H. (2024). An analysis of consumer's trusting beliefs towards the use of e-commerce platforms. *Humanities & Social Sciences Communications, 11*(1), 899. Advance online publication. DOI: 10.1057/s41599-024-03395-6

Siqueira, J. R.Jr, Peña, N. G., ter Horst, E., & Molina, G. (2019). Spreading the word: How customer experience in a traditional retail setting influences consumer traditional and electronic word-of-mouth intention. *Electronic Commerce Research and Applications, 37*, 100870. DOI: 10.1016/j.elerap.2019.100870

Slade, E. L., Williams, M. D., Dwivedi, Y. K., & Piercy, N. C. (2015). Exploring consumer adoption of proximity mobile payments. *Journal of Strategic Marketing, 23*(3), 209–223. DOI: 10.1080/0965254X.2014.914075

Smith, H. J., Dinev, T., & Xu, H. (2011). Information privacy research: An interdisciplinary review. *Management Information Systems Quarterly, 35*(4), 989–1016. DOI: 10.2307/41409970

Smith, S. G., & Sinkford, J. C. (2022). Gender Equality in the 21st Century: Overcoming Barriers to Women's Leadership in Global Health. *Journal of Dental Education, 86*(9), 1144–1173. DOI: 10.1002/jdd.13059 PMID: 36165260

Soetan, T. O., & Mogaji, E. (2024). Financial services consumers. In *Financial Services in Nigeria: The Path Towards Financial Inclusion, Economic Development and Sustainable Growth* (pp. 163–188). Springer Nature Switzerland. DOI: 10.1007/978-3-031-62340-0_6

Sohanda, M. K., Gupta, M., & Kalpana, A. V. (2024). Interactive e-commerce: Addressing cart abandonment through gamified user engagement. In *Proceedings of the International Conference on Communication, Computing and Internet of Things* (pp. 1–5). IEEE. DOI: 10.1109/IC3IoT60841.2024.10550373

Song, L., & Chang, T.-Y. (2012). Do resources of network members help in help seeking? Social capital and health information search. *Social Networks, 34*(4), 658–669. DOI: 10.1016/j.socnet.2012.08.002

Soni, S., & Mitchell, A. M. (2022). Women and digital government in rural India: Extending TAM with user assisted service. *Information Polity, 27*(1), 59–78. DOI: 10.3233/IP-200302

Sorescu, A. (2017). Data-driven business model innovation. *Journal of Product Innovation Management, 34*(5), 691–696. DOI: 10.1111/jpim.12398

Soundariya, A., Behera, N. K., Selvakumar, P., Chandel, P. S., Gupta, S., & Manjunath, T. C. (2025). AI and Environmental Ethics: The Role of AI in Addressing Climate Change—Ethical Considerations in AI-Driven Environmental Solutions. In Chakraborty, S. (Ed.), *Human Values, Ethics, and Dignity in the Age of Artificial Intelligence* (pp. 321–346). IGI Global Scientific Publishing., DOI: 10.4018/979-8-3693-9035-1.ch013

Stagg, K., Douglas, J., & Iacono, T. (2023). Living with stroke during the first year after onset: An instrumental case study exploring the processes that influence adjustment. *Disability and Rehabilitation, 45*(21), 3610–3619. DOI: 10.1080/09638288.2022.2131005 PMID: 36222354

Statista (2025). Digital Payments – Worldwide. Available at: https://www.statista.com/outlook/fmo/payments/digital-payments/worldwide?currency=USD

Stavins, J. (2001). Effect of Consumer CharacteristicsontheUse of Payment Instruments. *New England Economic Review(Issue Number 3)*, 19-31.

Stix, H. (2004). The impact of ATM transactions and cashless payments on cash demand in Austria. *Monetary Policy & the Economy, 1*(04), 90–105.

Stylios, I., Kokolakis, S., Thanou, O., & Chatzis, S. (2022). Key factors driving the adoption of behavioral biometrics and continuous authentication technology: An empirical research. *Information and Computer Security, 30*(4), 562–582. DOI: 10.1108/ICS-08-2021-0124

Sultan, P., Wong, H. Y., & Azam, M. S. (2021). How perceived communication source and food value stimulate purchase intention of organic food: An examination of the stimulus-organism-response (SOR) model. *Journal of Cleaner Production, 312*, 127807. DOI: 10.1016/j.jclepro.2021.127807

Sumanjeet. (2009, November). Emergence of payment systems in the age of electronic commerce: The state of art. In 2009 First Asian Himalayas International Conference on Internet (pp. 1-18). IEEE.

Sun, Y., Shao, X., Li, X., Guo, R., & Nie, F. (2019). How live streaming influences consumers' purchase intentions in social commerce: An IT affordance perspective. *Electronic Commerce Research and Applications, 37*, 100886. DOI: 10.1016/j.elerap.2019.100886

Sun, Y., Xue, W., Bandyopadhyay, S., Peng, Z., & Su, Y. (2021). WeChat mobile-payment-based smart retail customer experience: An integrated framework. *Information Technology and Management, 23*(2), 77–94. DOI: 10.1007/s10799-021-00346-4

Susilawati, A. D., Wahyudi, F., Putra, W. P., Supriyanto, W., & Limpo, L. (2024). The impact of digital user experience on brand perception and consumer loyalty in the e-commerce industry in Indonesia. *The Eastasouth Journal of Information System and Computer Science, 1*(03), 109–122. DOI: 10.58812/esiscs.v1i03.244

Taddesse, W., & Kidan, T. G. (2005). e-Payment: Challenges and opportunities in Ethiopia. United Nations Economic Commission for Africa. Tandfonline. (2024). Decentralized Finance (DeFi): A paradigm shift in the fintech.

Tagat, A., Özmen, M., & Trivedi, P. L. (2019). Consumer payments survey of India : A closer look at household finances and payment instruments. *NCAER National Data Innovation Centre Report*. https://www.researchgate.net/profile/Anirudh-Tagat/publication/342956792_Consumer_payments_survey_of_India_A_closer_look_at_household_finances_and_payment_instruments/links/5f16926a4585151299ad5b6c/Consumer-payments-survey-of-India-A-closer-look-at-household-finances-and-payment-instruments.pdf

Taherdoost, H. (2021). A Review on Risk Management in Information Systems: Risk Policy, Control and Fraud Detection. *Electronics (Basel), 10*(24), 3065. Advance online publication. DOI: 10.3390/electronics10243065

Tandon, A., Dhir, A., Almugren, I., AlNemer, G. N., & Mäntymäki, M. (2021). Fear of missing out (FoMO) among social media users: A systematic literature review, synthesis and framework for future research. *Internet Research*, *31*(3), 782–821. DOI: 10.1108/INTR-11-2019-0455

Tan, G. K. S.TAN. (2025). Assetizing the video game: Play-to-earn (P2E) games and blockchain rentiership. *Progress in Economic Geography*, *3*(1), 100036. DOI: 10.1016/j.peg.2025.100036

Tapscott, D., & Tapscott, A. (2017). *Blockchain revolution: How the technology behind bitcoin is changing money, business, and the world*. Penguin.

Taylor, B. (2016). Starbucks' mobile payment and rewards strategy: Creating customer loyalty through innovation. *International Journal of Business and Management*, *11*(4), 10–20. DOI: 10.5539/ijbm.v11n4p10

Taylor, E. (2016). Mobile payment technologies in retail: A review of potential benefits and risks. *International Journal of Retail & Distribution Management*, *44*(2), 159–177. DOI: 10.1108/IJRDM-05-2015-0065

Teh, G. M., Kalidas, V., & Zeeshan, M. (2014). Personality as a moderator of SOR model. *Review of Integrative Business and Economics Research*, *3*(2), 67.

Tejeda, V. (2025, June 23). *What is a customer-facing display & do I need one? (2025)*. Shopify. https://www.shopify.com/retail/customer-facing-display

Thaichon, P., Liyanaarachchi, G., Quach, S., Weaven, S., & Bu, Y. (2020). Online relationship marketing: Evolution and theoretical insights into online relationship marketing. *Marketing Intelligence & Planning*, *38*(6), 676–698. DOI: 10.1108/MIP-04-2019-0232

Thaler, R. (1985). Mental accounting and consumer choice. *Marketing Science*, *4*(3), 199–214. DOI: 10.1287/mksc.4.3.199

Thirunagalingam, A., & Selvakumar, P. (2025). Application of Automation and Robotics in the Oil and Gas Industry. In Abdullayev, V., & Khang, A. (Eds.), *Revolutionizing AI and Robotics in the Oil and Gas Industry* (pp. 229–244). IGI Global Scientific Publishing., DOI: 10.4018/979-8-3693-8156-4.ch011

Tiwari, T., Srivastava, A., & Kumar, S. (2019). Adoption of digital payment methods in India. *International Journal of Electronic Finance*, *9*(3), 217. DOI: 10.1504/IJEF.2019.099058

Tordoya-Espinoza, R., Alvarez-Risco, A., & Guevara-Zavaleta, V. (2025). Quality, loyalty and repurchase behavior in food convenience stores: Use of SOR model. *British Food Journal*, *127*(12), 1–19. DOI: 10.1108/BFJ-04-2025-0401

Treiblmaier, H., & Petrozhitskaya, E. (2023). Is it time for marketing to reappraise B2C relationship management? The emergence of a new loyalty paradigm through blockchain technology. *Journal of Business Research*, *159*, 113725. DOI: 10.1016/j.jbusres.2023.113725

Triandis, H. C. (1979). Values, attitudes, and interpersonal behavior. *Nebraska Symposium on Motivation*, *27*, 195–259. PMID: 7242748

Trischler, M F G., & Li-Ying, J. (2023). Digital Business Model Innovation: Toward Construct Clarity and Future Research Directions. *Review of Managerial Science*, *17*(1), 3–32. DOI: 10.1007/s11846-021-00508-2

Truman, G. E., Sandoe, K., & Rifkin, T. (2003). An empirical study of smart card technology. *Information & Management*, *40*(6), 591–606. DOI: 10.1016/S0378-7206(02)00046-0

Turley, L. W., & Milliman, R. E. (2000). Atmospheric effects on shopping behavior: A review of the experimental evidence. *Journal of Business Research*, *49*(2), 193–211. DOI: 10.1016/S0148-2963(99)00010-7

Uche, D. B., Osuagwu, O. B., Nwosu, S. N., & Otika, U. S. (2021). Integrating trust into technology acceptance model (TAM), the conceptual framework for e-payment platform acceptance. *British Journal of Management and Marketing Studies*, *4*(4), 34–56.

Unal, M., & Park, Y. H. (2023). Fewer clicks, more purchases. *Management Science*, *69*(12), 7317–7334. DOI: 10.1287/mnsc.2023.4716

Ünalmış, E., Dirsehan, T., & Erdoğmuş, İ. E. (2024). Consequences of influencer-created content on influencers' authenticity in the beauty and personal care industry. *Journal of Marketing Communications*, 1–27. Advance online publication. DOI: 10.1080/13527266.2024.2371833

Upadhyaya, A. N., Debnath, P., Rani, P., Manjunath, T. C., & Bhattacharya, S. (2025). AI-Based Disaster Prediction and Early Warning Systems. In Mobo, F. (Ed.), *AI and Emerging Technologies for Emergency Response and Smart Cities* (pp. 55–74). IGI Global Scientific Publishing., DOI: 10.4018/979-8-3693-9770-1.ch003

Upadhyay, U., Kumar, A., Sharma, G., Sharma, S., Arya, V., Panigrahi, P. K., & Gupta, B. B. (2024). A systematic data-driven approach for targeted marketing in enterprise information system. *Enterprise Information Systems, 18*(8), 2356770. DOI: 10.1080/17517575.2024.2356770

Valor, C., Antonetti, P., & Crisafulli, B. (2022). Emotions and consumers' adoption of innovations: An integrative review and research agenda. *Technological Forecasting and Social Change, 179*, 121609. DOI: 10.1016/j.techfore.2022.121609

van der Cruijsen, C., Hernandez, L., & Jonker, N. (2017). In love with the debit card but still married to cash. *Applied Economics, 49*(30), 2989–3004. DOI: 10.1080/00036846.2016.1251568

Van Der Cruijsen, C., & Van Der Horst, F. (2019). Cash or Card? Unravelling the Role of Socio-Psychological Factors. *De Economist, 167*(2), 145–175. DOI: 10.1007/s10645-019-09340-2

Van der Horst, F., & Matthijsen, E. (2013). The irrationality of payment behaviour. *DNB Occasional Studies, 11*(4). https://www.researchgate.net/profile/Jakub-Gorka-2/publication/305073403_Merchant_Indifference_Test_Application_-_A_Case_For_Revising_Interchange_Fee_Level_in_Poland/links/5780e4a608ae69ab88262a1d/Merchant-Indifference-Test-Application-A-Case-For-Revising-Interchange-Fee-Level-in-Poland.pdf#page=133

Végso, T., Belházyné Illés, Á., & Bódi-Schubert, A. (2018). Cash or card?–An explorative analysis of consumers' payment behaviour in Hungary. *Public Finance Quarterly= Pénzügyi Szemle, 63*(4), 448-472.

Venkatesh, V., & Bala, H. (2008). Technology Acceptance Model 3 and a Research Agenda on Interventions. *Decision Sciences, 39*(2), 273–315. DOI: 10.1111/j.1540-5915.2008.00192.x

Venkatesh, V., Brown, S. A., & Bala, H. (2013). Bridging the qualitative–quantitative divide: Guidelines for conducting mixed methods research in information systems. *Management Information Systems Quarterly, 37*(1), 21–54. DOI: 10.25300/MISQ/2013/37.1.02

Venkatesh, V., & Davis, F. D. (2000). A theoretical extension of the technology acceptance model: Four longitudinal field studies. *Management Science, 46*(2), 186–204. DOI: 10.1287/mnsc.46.2.186.11926

Venkatesh, V., Morris, M. G., Davis, G. B., & Davis, F. D. (2003). User Acceptance of Information Technology : Toward a Unified View. *Management Information Systems Quarterly, 27*(3), 425–478. DOI: 10.2307/30036540

Venkatesh, V., Thong, J. Y. L., & Xu, X. (2012). Consumer Acceptance and Use of Information Technology : Extending the Unified Theory of Acceptance and Use of Technology. *Management Information Systems Quarterly*, *36*(1), 157–178. DOI: 10.2307/41410412

Venkatraman, V., Dimoka, A., Pavlou, P. A., Vo, K., Hampton, W., Bollinger, B., Hershfield, H. E., Ishihara, M., & Winer, R. S. (2015). Predicting advertising success beyond traditional measures: New insights from neurophysiological methods and market response modeling. *JMR, Journal of Marketing Research*, *52*(4), 436–452. DOI: 10.1509/jmr.13.0593

Verhoef, P. C., Kannan, P. K., & Inman, J. J. (2015). From multi-channel retailing to omni-channel retailing. *Journal of Retailing*, *91*(2), 174–181. DOI: 10.1016/j.jretai.2015.02.005

Verma, R. K., & Kumari, N. (2023). Generative AI as a Tool for Enhancing Customer Relationship Management Automation and Personalization Techniques. *International Journal of Responsible Artificial Intelligence*, *13*(9).

Verma, R., Khound, K., & Mishra, V. (2025). Subscription Economy and the Transformation of Ownership: A Review of Value Perceptions and Retention Strategies. *Advances in Consumer Research. Association for Consumer Research (U. S.)*, *2*, 935–951. DOI: 10.61336/acr-25-05-08

Verma, S., & Mehta, R. (2023). The moderating role of demographic variables on digital payment adoption in India. *International Journal of Digital Economy*, *15*(3), 128–145.

Verplanken, B., & Aarts, H. (1999). Habit, Attitude, and Planned Behaviour : Is Habit an Empty Construct or an Interesting Case of Goal-directed Automaticity? *European Review of Social Psychology*, *10*(1), 101–134. DOI: 10.1080/14792779943000035

Verplanken, B., Aarts, H., & Van Knippenberg, A. (1997). Habit, information acquisition, and the process of making travel mode choices. *European Journal of Social Psychology*, *27*(5), 539–560. DOI: 10.1002/(SICI)1099-0992(199709/10)27:5<539::AID-EJSP831>3.0.CO;2-A

Verplanken, B., & Orbell, S. (2003). Reflections on Past Behavior : A Self-Report Index of Habit Strength[1]. *Journal of Applied Social Psychology*, *33*(6), 1313–1330. DOI: 10.1111/j.1559-1816.2003.tb01951.x

Vinitha, K., & Vasantha, S. (2017). Influence of demographic variables on usage of e-payment system. [IJMET]. *Int J Mech Eng Technol*, *8*, 265–276.

Visconti-Caparrós, J. M., & Campos-Blázquez, J. R. (2022). The development of alternate payment methods and their impact on customer behavior: The Bizum case in Spain. *Technological Forecasting and Social Change*, *175*, 121330. DOI: 10.1016/j.techfore.2021.121330

Walker-Munro, B., Ferrill, J., & Tiwari, M. (2025). Sensitive research & anti-money laundering: A possible marriage of convenience? *Current Issues in Criminal Justice*, *37*(2), 308–333. DOI: 10.1080/10345329.2025.2467483

Wang, H., Wang, L., & Zhou, W. (2022). The influence of live streaming commerce on consumers' purchase intention: The mediating role of trust and perceived value. *Computers in Human Behavior*, *135*, 107388. DOI: 10.1016/j.chb.2022.107388

Wang, Q., Niu, G., Zhou, Y., & Gan, X. (2024). Education and FinTech adoption : Evidence from China. *China Finance Review International*, *15*(1), 140–165. DOI: 10.1108/CFRI-06-2023-0141

Wang, R. J. H., Krishnamurthi, L., & Malthouse, E. C. (2018). When reward convenience meets a mobile app: Increasing customer participation in a coalition loyalty program. *Journal of the Association for Consumer Research*, *3*(3), 314–329. DOI: 10.1086/698331

Wang, X., Sung, B., & Phau, I. (2022). Examining the influences of perceived exclusivity and perceived rarity on consumers' perception of luxury. *Journal of Fashion Marketing and Management*, *26*(2), 365–382. DOI: 10.1108/JFMM-12-2020-0254

Wedel, M., & Kannan, P. K. (2016). Marketing analytics for data-rich environments. *Journal of Marketing*, *80*(6), 97–121. DOI: 10.1509/jm.15.0413

Weichert, M. (2017). The future of payments: How FinTech players are accelerating customer-driven innovation in financial services. *Journal of Payments Strategy & Systems*, *11*(1), 23–33. https://www.ingentaconnect.com/content/hsp/jpss/2017/00000011/00000001/art00004. DOI: 10.69554/JHTK5372

Weng, S. S., & Liu, M. J. (2004). Feature-based recommendations for one-to-one marketing. *Expert Systems with Applications*, *26*(4), 493–508. DOI: 10.1016/j.eswa.2003.10.008

Wongyai, P. H., Suwannawong, K., Wannakul, P., Thepchalerm, T., & Arreeras, T. (2024, October 15). The adoption of self-service check-in kiosks among commercial airline passengers. *Heliyon*, *10*(19), e38676. DOI: 10.1016/j.heliyon.2024.e38676

Wood, W., & Neal, D. T. (2009). The habitual consumer. *Journal of Consumer Psychology*, *19*(4), 579–592. DOI: 10.1016/j.jcps.2009.08.003

Wood, W., Quinn, J. M., & Kashy, D. A. (2002). Habits in everyday life : Thought, emotion, and action. *Journal of Personality and Social Psychology*, *83*(6), 1281–1297. DOI: 10.1037/0022-3514.83.6.1281 PMID: 12500811

World Bank. (2017). *The Global Findex Database 2017: Measuring Financial Inclusion and the Fintech Revolution*. World Bank.

World Bank. (2022). Digital Financial Inclusion. Available at: https://www.worldbank.org/en/topic/financialinclusion/publication/digital-financial-inclusion

Wu, Y. L., & Li, E. Y. (2018). Marketing mix, customer value, and customer loyalty in social commerce: A stimulus-organism-response perspective. *Internet Research*, *28*(1), 74–104. DOI: 10.1108/IntR-08-2016-0250

Xia, L., Monroe, K. B., & Cox, J. L. (2004). The price is unfair! A conceptual framework of consumers' perceptions of price unfairness. *Journal of Marketing*, *68*(4), 1–15. DOI: 10.1509/jmkg.68.4.1.42733

Xiao, G., Lee, M., Lee, C.-K., & Kim, M. (2025). Tourism live streamers as brand ambassadors: Investigating their impact on destination brand value through social exchange theory. *Journal of Hospitality and Tourism Technology*, *16*(1), 174–193. DOI: 10.1108/JHTT-11-2023-0371

Yacek, H. (2023, October). Points of Surprise! *Shooting Industry*, *68*(10), 34–37.

Yadav, M., Pandey, D., Kakkar, A., & Farhan, M. (2025). Linking brand trust and attitude with purchase intention of Indian organic food customers: An S–OR model based analysis. *Organic Agriculture*, *15*(1), 11–27. DOI: 10.1007/s13165-024-00480-7

Yanagawa, E. (2018). Digital transformation in Japan's banking industry. *Journal of Payments Strategy & Systems*, *12*(4), 351–364. DOI: 10.69554/NHNO1734

Yaqub, R. M. S., Atif, M. A., Waseem, F., & Ehtisham, . (2023). Unveiling the dynamics: Exploring the Stimulus-Organism-Response (SOR) model in the context of social media influencer marketing, electronic word of mouth, and purchase decisions, with a focus on the mediating role of brand awareness. [BBE]. *Bulletin of Business and Economics*, *12*(2), 435–446. DOI: 10.61506/01.00048

Yeo, W. J., Van Der Heever, W., Mao, R., Cambria, E., Satapathy, R., & Mengaldo, G. (2025). A comprehensive review on financial explainable AI. *Artificial Intelligence Review*, *58*(6), 1–49. DOI: 10.1007/s10462-024-11077-7

Yermack, D. (2016). Corporate governance and blockchains. *Review of Finance*, •••, rfw074. Advance online publication. DOI: 10.1093/rof/rfw074

Yokeshwaran, M., & Murugachandravel, J. (2022, April 10). Enhancement of business using e-commerce with Point of Sale. *International Journal of Health Sciences, 6*(S2), 3250–3260. DOI: 10.53730/ijhs.v6nS2.5811

Yum, K., & Kim, J. (2024). The influence of perceived value, customer satisfaction, and trust on loyalty in entertainment platforms. *Applied Sciences (Basel, Switzerland), 14*(13), 5763. DOI: 10.3390/app14135763

Zamora-Pérez, A., Marini, A., & Honkkila, J. (2024). Is there a digital divide in payments? Understanding why cash remains important for so many. *Economic Bulletin Articles, 2*. https://ideas.repec.org//a/ecb/ecbart/202400022.html

Zamoroka, V. (2023, September 21). *Modern approaches to website development: Techniques and tools.* AlterEgo. https://alterego.digital/blog/modern-approaches-web-development-techniques-and-tools

Zarifis, A., & Cheng, X. (2022). The business models of NFTs and fan tokens and how they build trust. *Journal of Electronic Business & Digital Economics, 1*(1/2), 138–151. DOI: 10.1108/JEBDE-07-2022-0021

Zehra, F., Khan, F., Mazhar, S., Akhlaue, N., Haque, E., & Singh, A. (2024). Exploring consumer preferences and behaviour toward digital payment gateways in India. *International Academic Publishing House, 41*(Spl Vol), 158–167. DOI: 10.52756/ijerr.2024.v41spl.013

Zeithaml, V. A., Berry, L. L., & Parasuraman, A. (1996). The behavioral consequences of service quality. *Journal of Marketing, 60*(2), 31–46. DOI: 10.1177/002224299606000203

Zhang, A., Alvi, M. F., Gong, Y., & Wang, J. X. (2022). Overcoming Barriers to Supply Chain Decarbonization: Case Studies of First Movers. *Resources, Conservation and Recycling, 186*, 106536. Advance online publication. DOI: 10.1016/j.resconrec.2022.106536

Zhang, J. Z., Chang, C. W., & Neslin, S. A. (2022). How physical stores enhance customer value: The importance of product inspection depth. *Journal of Marketing, 86*(2), 166–185. DOI: 10.1177/00222429211012106

Zhang, Q., Ariffin, S. K., Richardson, C., & Wang, Y. (2023). Influencing factors of customer loyalty in mobile payment: A consumption value perspective and the role of alternative attractiveness. *Journal of Retailing and Consumer Services, 73*, 103302. DOI: 10.1016/j.jretconser.2023.103302

Zhang, Y., Lin, N., & Li, T. (2012). Markets or networks : Households' choice of financial intermediary in Western China. *Social Networks*, *34*(4), 670–681. DOI: 10.1016/j.socnet.2012.08.003

Zhao, M., & Abeysekera, I. (2024). The behaviour of FinTech users in the Alipay Ant Forest platform towards environmental protection. *Journal of Open Innovation*, *10*(1), 100201. DOI: 10.1016/j.joitmc.2023.100201

Zhou, T. (2014). Understanding the determinants of mobile payment continuance usage. *Industrial Management & Data Systems*, *114*(6), 936–948. DOI: 10.1108/IMDS-02-2014-0068

Zimmel, S. (2025, April 2). *The impact of cybersecurity on business and brand risk reduction*. Recorded Future. https://www.recordedfuture.com/blog/impact-cybersecurity-business-brand-risk-reduction

Zyskind, G., & Nathan, O. (2015, May). *Decentralizing privacy: Using blockchain to protect personal data. In 2015 IEEE security and privacy workshops*. IEEE.

About the Contributors

Fatih Sahin, PhD Associate Professor of Marketing Bandirma Onyedi Eylul University, Gonen Vocational School Fatih Sahin is an Associate Professor of Marketing at Bandirma Onyedi Eylul University, where he specializes in consumer behavior, digital marketing, branding, and relationship marketing. His research delves into the evolving dynamics of consumer decision-making in the digital era, with a particular focus on strategic marketing and competitive advantage. With a Doctorate in Business Administration from Dumlupinar University, an MSc in Business Administration from Bahcesehir University, and a Bachelor's in Statistics from Yildiz Technical University, Dr. Şahin brings a strong analytical and strategic perspective to marketing research. Dr. Sahin has authored numerous academic publications, including books, book chapters, and peer-reviewed articles, contributing significantly to the fields of marketing strategy, brand management, and digital transformation. His recent works explore the intersection of technology and consumer behavior, investigating topics such as AI-driven marketing, social media influence, and brand transgressions. As a prolific researcher with over 40 publications and extensive academic collaborations, Dr. Şahin focuses on consumer behavior, besides bridging academic theory with practical industry applications. His latest books, Globalized Consumer Insights in the Digital Era and Navigating the Shifting Landscape of Consumer Behavior.

***

Roni Adi A seasoned scholar and sociopreneur, Roni Adi is a lecturer at the Batam Institute of Technology (ITEBA). He has a Master's degree in Management (Finance) from UPI YAI, Jakarta, and is currently enrolled in Universitas Negeri Jakarta's Doctoral Programme in Management Science. He has experience in accounting and finance. Roni has worked for PT Cmira Mandiri Solusinko and PT Garuda Indonesia, giving him extensive professional expertise in senior finance management. His recent articles have concentrated on management accounting systems and the use of accounting in small enterprises, particularly coffee shops, demonstrating his commitment to furthering research in a variety of fields. Creating circular, blue, and green economies as well as digital transformation for micro and small businesses, farmers, fishers, and the general public are his passions.

M. Irvanni Bahri is an academic and researcher from the Batam Institute of Technology who specialises in management science, human resource development, public policy, and regional planning. He obtained his bachelor's degree in management from Malikussaleh University in 2017 and his master's degree in regional and rural development planning from the University of North Sumatra in 2020. Currently, he is pursuing a doctoral programme in management science at Airlangga University.

Sanchita Banga is a PhD scholar with over 10 years of teaching experience as a PGT Commerce. Her research focuses on digital literacy, mobile payment adoption, and rural financial inclusion in India. ORCID: [https://orcid.org/0009-0009-9572-5285]

Lana Beikverdi (D.Sc.) is a dedicated senior lecturer and researcher exploring the complexities of digital marketing, consumer experience and sustainable consumption. Additionally, she investigates the role of emerging technologies such as service robots and AI in shaping consumer behaviours, particularly their potential to promote or deter sustainable practices, meat consumption, and green impulse purchases. She seeks to provide actionable insights into fostering sustainable consumer habits through this research.

Most. Sharmin Ara Chowdhury is a research practitioner with a strong focus on key areas such as Artificial Intelligence, strategic innovation, management information systems, and sustainability. Her research explores the intersection of emerging technologies and organizational strategy. By aligning innovation with practical needs, she addresses real-world challenges through applied and impactful research. Sharmin's expertise also extends to education technology, human-computer interaction, the Internet of Things (IoT), cognitive behavior, and intelligent computing. Her work bridges academic theory and professional practice, making significant contributions to the evolving landscape of digital transformation and sustainable development.

Trupti Dandekar-Humnekar is an accomplished academician and researcher with over 15 years of professional experience spanning academia and industry. She currently serves as an Associate Professor at CMS Business School, Jain (Deemed-to-be University), Bangalore. An alumna of the Indian Institute of Management (IIM) Indore, Dr. Dandekar-Humnekar holds a Ph.D. in Service Marketing from Savitribai Phule Pune University, along with an MBA in Marketing and HR, and a B.Sc. in Statistics. She is presently pursuing an M.A. in Psychology from IGNOU. Her research primarily focuses on service quality measurement, higher education marketing, digital transformation, consumer behavior, and the integration of emerging technologies such as AI, Blockchain, and IoT into management practices. She has authored more than 30 national and international publications and has contributed extensively to curriculum development, corporate training, and academic innovation. A recognized academic innovator, she developed a 12-week, 4-credit MOOC titled "Marketing to the Bottom of the Pyramid" on the SWAYAM-IIMBx platform. She is also a co-inventor of 12 granted design patents and multiple filed patents spanning domains such as smart education, renewable energy, and wearable technology. In addition to serving as a reviewer for leading Emerald journals and the Spanish Journal of Marketing, she has contributed as an editor for the Vishwakarma Business Review (VBR) and actively participates as a resource person in FDPs under UGC and AICTE-ATAL. Her multidisciplinary expertise bridges marketing, technology, and innovation, making her a key contributor to the evolving landscape of management education and research in India and beyond.

Md Mehedi Hasan Emon is a research practitioner specializing in business and management, supply chain management, and digital transformation. His work has significantly contributed to the fields of marketing, artificial intelligence, and supply chain analytics, with a strong focus on sustainable business practices. Through his research, he explores the intersection of technology and business, aiming to enhance operational efficiency, digital marketing strategies, and sustainable supply chain practices. As an active researcher, he has authored numerous publications in Scopus and Web of Science-indexed journals. His research covers areas such as artificial intelligence adoption, digital marketing strategies, sustainable supply chain management, and the economic implications of technological advancements. His work has gained wide recognition, reflecting his growing influence within the academic community. He has been involved in multiple research projects, including those funded by the University Grants Commission of Bangladesh and European Union-funded initiatives. His research aligns with his vision of leveraging technology to drive sustainability, efficiency, and innovation in business ecosystems. Additionally, he has contributed to academia as a peer reviewer for high-impact journals, ensuring rigorous academic standards and fostering knowledge dissemination in his field.

Tamara Phillips Fudge is a full-time professor in the graduate technology programs at Purdue University Global. She has won fellowships and awards for innovation and teaching, and has taught a wide variety of topics, including web development, human-computer interaction, systems analysis and design, and those featuring documentation, diagramming, problem-solving, and presentation. Her career started with music degrees from Indiana University and Florida State University. She sang opera, oratorio, and in recital; her compositions have been heard on Public Radio, at various universities, and a state choral convention. She taught vocal and choral music, piano, pedagogy, foreign language diction, opera production, music theory, composition, and related courses in the traditional college classroom for 20-odd years. For 7 years, she was a weekend correspondent for the Quad-City Times (Davenport, IA). Following a brief stint as an agent/registered representative selling life and health insurance and variable products, she returned to school with a keen interest in technology and has since distinguished herself in online teaching, coordination of large projects, and curriculum development.

Amey Gholpe is a student at the Datta Meghe Institute of Management Studies, Nagpur, where he is pursuing his academic training in management studies. He has developed a strong academic interest in the domains of Marketing and Operations Management, which constitute his core areas of specialization. As a management student, Mr. Gholpe is actively engaged in building conceptual and practical understanding of contemporary business practices, with a particular focus on market analysis, consumer behavior, operational efficiency, and process optimization. His academic pursuits reflect an eagerness to integrate theoretical knowledge with real-world business applications. Through coursework, academic projects, and institutional learning activities, he continues to enhance his analytical, managerial, and problem-solving skills. His interest in Marketing and Operations Management positions him to explore interdisciplinary perspectives within the management field and contributes to his overall professional development. Mr. Gholpe aspires to further his academic and professional journey by applying management principles to organizational challenges and contributing meaningfully to the business and management discipline.

Suhail Mohammad Ghouse is an Associate Professor of Marketing and Entrepreneurship at Dhofar University, Oman. He holds dual PhDs in Entrepreneurship (UK) and Marketing (India), along with dual Master's degrees in International Business Management. A Fellow of the Higher Education Academy (FHEA, UK) and the Chartered Institute of Marketing (FCIM, UK), Dr. Ghouse's research focuses on entrepreneurship, consumer behavior, sustainability, and digital transformation, particularly within emerging and rural contexts. He has published in leading journals such as The International Journal of Management Education, International Journal of Entrepreneurial Behaviour & Research, Young Consumers, and The Bottom Line, and serves as an active reviewer for several international journals. His current projects explore sustainable consumption, fintech adoption, entrepreneurial ecosystems, and the intersection of artificial intelligence with marketing and entrepreneurship.

Linnea Hall is an evaluator at Western Governors University. She earned a doctorate in law from California Western School of Law in San Diego, CA and a doctorate in computer science from Colorado Technical University in Colorado Springs, CO. Her research interests include artificial intelligence, cybersecurity, and ethics.

Mohammad Shariful Islam is an academic enthusiast based in Australia, with a strong focus on core areas of business and management, and a keen interest in the evolving digital entrepreneurship ecosystem. He holds a double master's degree from the University of Wollongong, specializing in Supply Chain Management and Project Management. His academic interests encompass strategic management, operations, digital transformation, and organizational behavior fields essential to navigating today's dynamic and technology-driven business landscape. During his postgraduate studies, Shariful engaged in a broad range of critical subjects, including Strategic Procurement, Advanced Supply Chain Management, Project Portfolio Management, Logistics Systems, and Total Quality Management. These experiences strengthened his ability to analyze and manage complex business systems, fostering expertise in evidence-based decision-making and operational efficiency. Shariful's research centers on the transformative role of digital technologies in business and management. He is particularly focused on how automation, artificial intelligence, analytics, and integrated platforms reshape supply chains, enhance organizational agility, and drive sustainable practices. His forthcoming doctoral research aims to further explore the intersection of digital transformation, strategic operations, and organizational performance. Australia's inclusive and innovation-driven academic environment has significantly shaped Shariful's global perspective, critical thinking, and applied research capabilities. He is committed to developing a research portfolio that advances scholarship while delivering practical value to organizations adapting to digital and managerial change. His foundational academic journey began at the American International University-Bangladesh (AIUB), where he earned a Bachelor of Business Administration in Operations Management and was honored with the prestigious Chairman's (Leadership) Award (Gold Medal) for outstanding academic performance and leadership.

Sujay Mugaloremutt Jayadeva is an Associate Professor in the Department of Health System Management Studies at JSS Academy of Higher Education & Research, Mysuru. With a strong academic foundation and a passion for advancing healthcare management education, Dr. Sujay has been actively involved in teaching, research, and academic leadership. His areas of expertise include health systems management, public health policy, and healthcare quality improvement. Committed to bridging the gap between academic theory and real-world healthcare challenges, he has contributed to numerous research projects and scholarly publications. At JSS Academy, he plays a pivotal role in nurturing future healthcare leaders and promoting excellence in health management education.

Mahesh Kumar K.R. is a distinguished educationist with over 32 + years of experience in teaching, administration, and academic leadership. Currently serving as Director at IZee Business school, Bangalore, he is a prolific author, researcher, and Ph.D. guide, with numerous awards recognizing his contributions to higher education and management.

Kalavathy K.S, an Associate Professor at CMS Business School, Jain (Deemed-to-be University), brings over 29 years of rich teaching and research experience to her role. She excels in guiding postgraduate students through interactive discussions, case studies, and role plays, creating a dynamic learning environment. Dr. Kalavathy finds fulfillment in mentoring inquisitive minds and actively engages in academic pursuits. Her scholarly work includes published articles in prestigious national and international journals, focusing on Marketing Management, Consumer Behavior, Services Marketing, and Strategic Marketing .She is also a Reviewer for prestigious journals.

Shivangi Kashyap is a passionate academician and researcher with a strong foundation in commerce and finance. She holds a Bachelor of Commerce (Honours) in Accounts and an MBA in Financial Management. Currently, she is pursuing her doctoral research as a Research Scholar at the University of Hyderabad, where she continues to explore emerging dimensions of finance, sustainability, and corporate responsibility. Before joining her Ph.D. program, Shivangi worked as an Assistant Professor at Brainware University, Barasat, Kolkata, where she taught courses in Financial Management, Accounting, and Cost and Management Accounting. Her teaching approach emphasizes conceptual clarity, analytical thinking, and the integration of real-world business applications. Shivangi's research interests lie in sustainable finance, ESG investing, financial behavior, and the intersection of technology and management education. She has authored multiple research papers and book chapters published by reputed journals and international publishers.

An advocate of lifelong learning, she actively leverages digital tools and online platforms to enhance both teaching and research outcomes. Her long-term vision is to contribute to academia through impactful research, interdisciplinary collaboration, and mentoring future scholars in the evolving landscape of finance and sustainability.

Pravin Kumar is an experienced academic with over a decade of teaching Marketing Management and related subjects at reputed institutions across India. He currently serves as Assistant Professor at Brainware University, Kolkata, and has previously taught at different Management Institutions. He holds a Ph.D. in Management from T.M. Bhagalpur University, focusing on visual merchandising on consumer behavior in branded apparel. He also holds an MBA in Marketing and an M.A. in English, reflecting his interdisciplinary background. Dr. Kumar has published widely in Scopus, UGC-CARE, and peer-reviewed journals on topics including consumer behavior, brand personality, and AI in sales. He has presented at numerous national and international seminars and FDPs on subjects like cross-cultural marketing and leadership. His industry experience in the automobile and pharmaceutical sectors adds practical value to his teaching. Known for his engaging teaching style and commitment to Universal Design for Learning.

Pritam Lanjewar is an Assistant Professor at the Datta Meghe Institute of Management Studies, Nagpur, with over 14 years of experience in academics. Over the course of his career, he has been actively involved in teaching, mentoring, and contributing to the academic development of students in the field of management education. He has consistently demonstrated a strong commitment to higher education through his engagement in academic instruction, curriculum delivery, and scholarly activities. His experience spans classroom teaching as well as participation in institutional academic initiatives, reflecting a sustained dedication to excellence in management education. Mr. Lanjewar is currently in the advanced stage of his academic journey, having submitted his Ph.D. thesis to the University, marking a significant milestone in his pursuit of research and scholarly advancement. His doctoral work underscores his interest in contributing original research and expanding knowledge within his academic discipline. As an educator and researcher, he continues to strive toward academic growth, professional development, and meaningful contributions to the field of management studies.

Anas Limam is a PhD in economics at the INREDD Research Laboratory for Innovation Responsibility and Sustainable Development, Cadi Ayyad University, Marrakesh, Morocco, with research interests focused on finance and insurance.

Raj Kumar Mishra is an Assistant Professor at the Institute of Management Sciences, University of Lucknow, with 13 years of teaching experience. He has also gained valuable experience working at prestigious institutions like PSIT Kanpur and BBD University. He has authored one book, holds one patent, and has published 12 research papers in esteemed national and international journals. With a Ph.D. in Management specializing in Marketing, his current research focuses on the influence of emerging technologies such as AI and Big Data on consumer behaviour and brand engagement. He also explores the effectiveness of personalized and healthcare marketing, as well as the role of ethical advertising in fostering consumer trust.

Avishek Nath is an enthusiastic scholar specializing in Business and Management as well as Business Administration. His scholarly interests focus on key contemporary areas including Digital Transformation, Entrepreneurial Growth, Innovation Strategies, and Strategic Management. Avishek's expertise extends to Supply Chain Innovation, Digital Entrepreneurship, and Organizational Behavior, with particular emphasis on AI adoption and technology-driven business models. His research also encompasses critical themes such as Operational Efficiency, Strategic Procurement, Business Sustainability, Automation, Data Analytics, and the development of Agile Organizations. With a strong foundation in Project Management and emerging technologies, Avishek investigates how these elements collectively drive business innovation and competitive advantage in rapidly evolving markets. He has more than seven years of industry and market research experience and has numerous research publications. He has also actively participated in international conferences, contributing to global academic discourse in his fields of expertise. Committed to advancing academic knowledge and practical solutions, Avishek's work aims to support sustainable growth and transformative change in today's technology-infused business environment.

P. Selvakumar is currently an Associate Professor in the Department of Science and Humanities at Nehru Institute of Technology, Coimbatore, Tamil Nadu, India. Prior to joining Nehru Institute of Technology, he held positions as an Assistant Professor in the Department of Science and Humanities various engineering colleges in Coimbatore, Tamil Nadu, India, spanning from 2011 to 2023. Dr.P. Selvakumar's academic and research interests lie primarily in the field of phytochemistry. He has made significant contributions to his field, with an impressive record that includes 30 journal publications, 12 patents, and 20 conference papers. Additionally, he has organized 5 seminars and conferences, further demonstrating his commitment to academic and scientific exchange. Throughout his career, Dr. P.Selvakumar has been recognized for his dedication and excellence in teaching, having been honored with several awards, including the Best Teacher Awards. He holds a prominent position in the academic community, contributing actively to research.

Abderrahim Rhafel is a PhD student in economics at the INREDD Research Laboratory for Innovation, Responsibility and Sustainable Development, Cadi Ayyad University, Marrakesh, Morocco, with research interests focused on monetary economics and payment behavior.

Sachin Sharma currently working as a Professor at Amity School of Engineering and Technology, Amity University, Punjab, Mohali, India since January 2025. He also worked as Associate Dean of International Affairs and Professor, Department of Computer Science and Engineering at Graphic Era Deemed to be University, Dehradun, Uttarakhand, India from February, 2018 to January 2025. He has experience of 15+ years in industry and academia. He is also Co-founder and Chief Technology officer (CTO) of IntelliNexus LLC, Arkansas, USA based company. He also worked as a Senior Systems Engineer at Belkin International, Inc., Irvine, California, USA. He received his Philosophy of Doctorate (Ph.D.) degree in Engineering Science and Systems specialization in Systems Engineering from University of Arkansas at Little Rock, USA with 4.0 out 4.0 GPA and M.S. degree in Systems engineering from University of Arkansas at Little Rock with 4.0 out 4.0 GPA. Dr. Sharma has authored/coauthored over 270+ publications in the form of books, patents, and papers in refereed journals and conference proceedings. He holds 70+ US, Indian and international patents in the area of IoT, wireless communication.

Partap Singh is a professor of management whose areas of expertise include consumer behavior, digital finance, and marketing. Besides widely publishing in renowned books and journals, he has guided several PhD scholars.

Bandana Thapa is a doctoral researcher specialising in the intersection of financial technology (FinTech) and human-computer interaction. Her work focuses on the evolution of digital payment ecosystems, specifically exploring how AI-driven user experiences influence trust and adoption among Gen Z consumers in the Nordic region. Her research seeks to bridge the gap between algorithmic transparency and consumer behaviour, providing strategic insights for the digital transformation of financial services.

Mustapha Ziky, PhD, is Professor of Economics and Director of the INREDD Research Laboratory for Innovation Responsibility and Sustainable Development at Cadi Ayyad University in Marrakesh, Morocco. His research focuses on macroeconomics, finance, and human development.

Index

A

B

C

D

R

S

T

U

V

X

www.ingramcontent.com/pod-product-compliance
Lightning Source LLC
LaVergne TN
LVHW061527290926
849162LV00004B/696